Advanced DC-DC Power Converters and Switching Converters II

Advanced DC-DC Power Converters and Switching Converters II

Editor

Salvatore Musumeci

MDPI

Basel • Beijing • Wuhan • Barcelona • Belgrade • Novi Sad • Cluj • Manchester

Editor
Salvatore Musumeci
Politecnico di Torino
Torino
Italy

Editorial Office
MDPI
St. Alban-Anlage 66
4052 Basel, Switzerland

This is a reprint of articles from the Special Issue published online in the open access journal *Energies* (ISSN 1996-1073) (available at: https://www.mdpi.com/journal/energies/special_issues/9B4QCLQ7D1).

For citation purposes, cite each article independently as indicated on the article page online and as indicated below:

Lastname, A.A.; Lastname, B.B. Article Title. *Journal Name* **Year**, *Volume Number*, Page Range.

ISBN 978-3-7258-1353-7 (Hbk)
ISBN 978-3-7258-1354-4 (PDF)
doi.org/10.3390/books978-3-7258-1354-4

Cover image courtesy of Salvatore Musumeci

Contents

About the Editor

Salvatore Musumeci

Salvatore Musumeci received his MS Laurea degree and PhD in Electrical Engineering from the University of Catania in 1991 and 1995, respectively. From 1996 to 2001, he worked at the Research and Development Department of STMicroelectronics of Catania with his industrial research activity focused on the applications of silicon power devices. From 2001 to 2017, he participated in various activities and research collaborations with the Department of Electrical, Electronics, and Systems Engineering at the University of Catania. From 2018 to 2022, he worked as a researcher at the Energy Department of the Politecnico di Torino. He is currently an Associate Professor in the same department, focusing on the topics of electrical converters, machines, and drives. He is also involved with the Power Electronics Innovation Center (PEIC), an interdepartmental competence center of the Politecnico di Torino focused on power electronics applications.

His research interests include advanced power devices, switching power converter applications, high-efficiency industrial motors, magnetic materials and their applications, battery management systems, and automotive power electronics applications.

energies

MDPI

Editorial

Energy Conversion Using Electronic Power Converters: Technologies and Applications

Salvatore Musumeci

Department of Energy "Galileo Ferraris", Politecnico di Torino, 10129 Torino, Italy; salvatore.musumeci@polito.it; Tel.: +39-011-090-7127

1. Introduction

Nowadays, energy conversion plays a crucial role in sustainable growth and development. In the past, energy conversion was achieved using electromechanical converters based primarily on rotating machines. In recent years, conversely, the energy conversion process has been performed by several power electronic circuits [1].

Power electronic converters are switching circuital structures used to achieve efficient energy conversion systems for various applications such as renewable energy conversion, smart grid arrangement, energy storage management, and sustainable transport. Power electronic converter systems are composed of several switching topologies, each related to the specific application. Power electronic circuit solutions are continuously under study to improve existing converter topologies or create new ones. Furthermore, the improvement in technologies of power electronic devices and passive components is leading to the unceasing development of the qualities of converters such as high efficiency, high gain, high power density, and fast transient response. To use a corporeal analogy, muscles are represented by the topological structure, whereas the brain function of the power converters is performed by increasingly performing control techniques. Advanced topologies and control methods are necessary to respond to the progressively more challenging demands of modern applications. Thus, the research of advanced design criteria, the use of innovative technologies, and improved regulating techniques are required to reach the target of obtaining more efficient, compact, cost-effective, and sustainable energy conversion systems [2].

In the field of power converter application to energy conversion, several articles have contributed to the growth of knowledge within the portion of the scientific community that is involved in the publications and use *Energies* to exchange and build knowledge and skills in this strategic technological area of development. In this Editorial, a variety of articles have been selected to disseminate the technical-scientific contributions most read and cited by the scientific community, whether belonging to the journal *Energies* or another publication. The time frame considered in the choice of the selection of the significant articles ranges from 2020 to 2022.

The next section provides a classification of the considered paper contributions according to the main topics. Furthermore, the specific focus and the worth of each article are summarized.

Citation: Musumeci, S. Energy Conversion Using Electronic Power Converters: Technologies and Applications. *Energies* **2023**, *16*, 3590. https://doi.org/10.3390/en16083590

Received: 2 November 2022
Accepted: 5 February 2023
Published: 21 April 2023

2. Overview of the Contributions

The articles included in this editorial on energy conversion using power converters review some of the significant contemporary trends in the strategic power engineering field.

For the sake of clarity and the effectiveness of this Editorial, the paper's contributions have been selected and thus inserted into four, more general categories. For each energy conversion category considered, the contribution of each article to evolving the state of the art in the related topics is indicated.

2.1. Renewable Energy Conversion System

Solar photovoltaic (PV) systems are much more attractive in the current energy conversion scenario due to their continuous technological development. In renewable energy conversion applications from PV sources, DC–DC power converters are used in several non-isolated and isolated topologies [3]. These electronic power converters are necessary to regulate the voltage of the source to bring it to the values required by the DC or AC bus onto which the photovoltaic systems are inserted. The basic boost power converter and its circuit evolution used to adjust the PV output voltage [4] is one of the stimulating subjects of the current research objectives. In PV source systems, the maximum power point tracking (MPPT) methods are implemented to extract maximum power. The boost converter, regulated by the MPPT control technique, is an optimum workbench to investigate the maximum power extraction in a PV system. In [5], C.H. Hussaian Basha and C. Rani investigate several MPPT techniques suitable for boost converter topology. Their paper carried out a comprehensive comparative analysis of the recent MPPT techniques developed. Furthermore, the MPPT methods in static and dynamic irradiation conditions were evaluated to find the optimum duty cycle control.

The limit of the basic boost converter topology is the step-up gain achievable by the actual converter operation compared with the ideal one. Generally, PV systems need step-up voltage management to grid connection. To further increase the boost converter's gain, in [6] Amir Farakhor et al. used three winding-coupled inductors without increasing the number of switches. The topology solution, at the expense of greater passive components circuit complexity, increased the step-up voltage attainable, reducing the voltage stress across the driven single switch [7].

An improvement of the boost converter efficiency may be achieved by the reduction of switching power losses. The soft switching techniques address the improvement of the energy conversion efficiency of the power converter interfacing with the PV system [8]. Following this design approach, Khairy Sayed et al. in [9] present an advanced soft-switched boost-type converter using an additional resonant cell. The circuital topology achieved is denominated edge resonance switched-capacitor (ER-SC) boost converter. The boost circuit is composed of coupled resonant inductors, an auxiliary switch, some auxiliary diodes, and a resonant capacitor. The coupled inductors increase the gain of the DC–DC converter according to the winding turns ratio [10]. The boost operation conditions have been analyzed and discussed in continuous and discontinuous modes and the PWM modulation technique. Furthermore, the effectiveness of the proposed ER-SC boost converter has been experimentally tested. Finally, a comprehensive comparison with the state-of-the-art soft-switching topologies is presented and analyzed, highlighting both advantages and drawbacks.

From the above considerations, meeting high step-up requirements and attaining satisfactory efficiency are significant challenges in these power converter designs. An attractive suggestion is given by M. Karthikeyan et al. in [11] to achieve these design targets. The authors investigate a circuit integration between a Cuk with a boost converter, maintaining a single-switch circuit solution. This way, the high gain requirement is met with substantial efficiency. Furthermore, the converter topology obtained complements the benefits of boost and Cuk converters. The experimental results, obtained on the integrated power converter circuit, feature continuous current mode operation with low voltage stress on the controlled switch and diodes.

Usually, buck-boost DC–DC converters which are used to regulate the output voltage in PV systems, do not provide the isolation between input and output sides [12]. To achieve AC load utilization, the conventional converters produce DC voltages from the PV panels and then invert them into AC through open-loop inverters. A circuital isolation settlement is implemented by Tehzeeb-ul Hassan et al. The authors in [13] propose a push–pull boost converter topology with a computationally fast, rugged, and efficient MPPT algorithm using a fuzzy logic controller.

The MPPT method tracks the maximum power point (MPP) in real-time, while the push–pull boost converter allows for a high gain by adjusting the turn ratio of the high-frequency transformer [14].

2.2. Smart Grid Components and Applications

Nowadays, high-performance power conversion requirements are of increasing interest in microgrid and smart grid applications [15].

Issues like the analysis of power converter topologies and components, protection and monitoring, power quality, and optimization, when applied to smart distribution systems, are continuously under study by academic and industrial researchers. The research effort is oriented to keeping power converter solutions updated, with the latest technological developments and new challenges in energy conversion applied to power systems [16].

DC microgrids are a topic of great attention in the generation–storage and distribution of electrical energy. The DC–DC power converters play a crucial role in regulating the output voltage of various distributed generation (DG) units.

In [17], Javed Ahmad et al. present an improved DC–DC converter with quadratic voltage gain and reduced voltage stress across switching devices. The described converter topology is composed of a merger of a conventional boost converter and a quadratic boost converter (QBC) [18,19]. The high gain step-up converter is implemented to regulate the voltage of DG sources connected to DC microgrids [20]. Furthermore, in the paper, an interesting comparison with several step-up DC–DC converters applied in DG sources to interface the DC grid is made.

In [21], Sara J. Ríos et al. deal with a dual-active bridge (DAB) DC–DC converter applied to a DC microgrid, with a comprehensive point of view on the converter application. The DAB converter interfaces with the renewable energy source (RES) systems and the battery energy storage BES systems in a DC microgrid in a bidirectional way [22]. Thus, the power flow control investigation is carried out. Furthermore, the authors introduce a power management algorithm to select the proper operation of the RES system and BES system, based on load/generation power and state-of-charge of the battery conditions.

In energy conversion systems for AC grids, the quality of the waveform with reduced harmonic distortion and a convenient trade-off between cost and performance are fundamental targets to reach [23]. A three-level neutral-point-clamped (NPC) converter is widely used to achieve these design constraints [24].

In [25], Hao Lin et al. propose an integral sliding-mode control (ISMC)-based direct power control (DPC) strategy for application NPC converters to improve the system disturbance rejection ability. Experimental results show the effectiveness and the advantage of the control strategy developed for the NPC power converter, compared with a conventional PI-based DPC strategy. In AC grids, the integration of the high-power static synchronous compensator (STATCOM) and energy storage in the same system arouses particular interest.

Modular multilevel converter (MMC) topologies represent some attractive power converter topologies for energy storage and static synchronous compensator (ES-STATCOM) arrangement, providing a modular and scalable solution with high efficiency that manages high-power and high-voltage ratings in grid applications [26]. In [27], Sanjay K. Chaudhary et al. provide a dissertation on the MMC converter family application in ES-STATCOM applications. The paper describes and compares several MMC-based ES-STATCOM topologies for both centralized and distributed energy storage needs.

In power system applications, passive transformers are generally used to adjust the voltage request. They exhibit some drawbacks such as a possible DC offset and difficulty in controllability [28]. Noticeable volumes and costs are further limits which must be overcome in modern smart grids.

Solid-state transformers (SSTs) can be identified as promising power electronic structures to substitute for industrial transformers. Furthermore, SSTs improve and harmonize

AC and DC electrical grid interconnection. SST is composed of power converter advanced circuits and high-frequency transformers [29].

Currently, SST demands are growing significantly, as well as those of circuital structures.

In [30], Mohammed Azharuddin Shamshuddin et al. provide an overview of the concepts, topologies, control strategy, materials, and classification of SSTs. The paper's aim is to recognize and describe the several power electronic transformer structures available in the energy conversion area. Furthermore, the paper proposes a simplified terminology to identify and standardize the wide number of definitions and SST structures introduced in the current literature.

In electrical networks, the brief voltage reduction appearance (the so-called voltage sags) is a serious issue that lowers the power quality [31].

The short-time voltage reduction can be solved by a dynamic voltage recovery (DVR) system [32]. The DVR arrangement is a controlled power converter that can inject 3-phase voltage in series and synchronism with the distribution of supplied voltages to correct short voltage sags. In [33], Ali Moghassemi and Sanjeevikumar Padmanaban propose a comprehensive review of the DVR system. The topologies of the power converters involved and the control methods used are explored.

2.3. *Energy Store Management and e-Mobility*

Energy storage management (ESM) is a critical issue in environmental sustainability [34]. Battery energy storage systems (BESSs) are currently fundamental fundamental building blocks of ESM. BESSs are standalone system devices that permit energy from renewables (like solar, wind, etc) to be accumulated and then returned when customers need a surplus of power. It is composed of battery systems, bidirectional power converters (BPC), and a control system [35].

The prevalent storage technology is lithium-ion batteries, used in mobile phones, electronic equipment, and electric cars and in large-scale plants to help electricity grids guarantee a safe and reliable supply of energy [36].

In [37], Andrés Peña Asensio et al. furnish an overall analysis of the effect of the converter synchronizing methods on the contribution that BESSs provide for the support of the inertial response of a power system. System solutions, based on phase-locked loop (PLL) synchronization, virtual synchronous machine (VSM) [38] and synchronization without PLL, are described and then compared. For this, time-domain simulations are used for an isolated microgrid (MG) effective example.

The increase in demand, both for energy storage technologies and electric vehicles, requires high-power DC–DC converters. In EVs, the power can come from fuel cell stacks, supercapacitors, and battery systems [39].

The described energy sources need to step up the voltages. The DC–DC converter for high-power conversion applications (i.e., resonant, full-bridge, or dual-active bridge) requires a high-frequency magnetic transformer for separating and coupling with the requested turn ratio of the input and output converter voltage [40]. The high-frequency transformer is a critical design matter. In [41] Muhammad Zeeshan Malik et al. propose a transformer-less high step-up boost converter. The single command switch topology presented features including a specific charge pump capacitor circuit and a capacitor–inductor–diode (CLD) cell, delivering high gain and quite a satisfactory efficiency. The modified boost converter has been investigated by several simulation results. Furthermore, a small-scale laboratory prototype has been developed for experimental validation.

Battery charging in electrical vehicles (EVs) is a meaningful topic to consider in the e-mobility field. Several converter solutions are presented in the literature [42]. In [43], a multi-leg interleaved DC–DC buck converter with digital control is introduced. Stefania Cuoghi et al. describe the design/tuning procedure for the control of an interleaved buck converter in the EV charger system. In the paper, the power converter's continuous and discrete-time model is carried out. Furthermore, reduced power converter operations, are explored using an experimental set-up of a three-switching legs interleaved buck converter.

From the experimental results arise an acceptable level of robustness under load variations of the control approach defined.

In [44] (by Khairy Sayed et al.), a full bridge phase-shift PWM DC–DC converter with a high-frequency transformer and current doubler rectifier in the second stage, designed for an onboard charger, is investigated.

The converter circuit operation is analyzed in detail and the design guidelines are provided. The effectiveness of the proposed converter is validated by several experimental results. In the paper, the current doubler rectifier solution presented achieves an improvement in the converter efficiency.

The battery source EVs are in full development. Battery-powered EVs have a high weight factor for available energy [45]. Currently, the range at full charge is limited to around 500 km. Fuel cells are an energy source substitute for battery-powered systems.

Fuel cells are particularly interesting for long-distance driving [46]. Furthermore, in EV applications, fuel cell stacks with hybrid energy storage systems, composed of batteries and supercapacitors, can be merged to fit the dynamic power demand required by the electric motor and auxiliary systems [47]. In [48], Ioan-Sorin Sorlei et al. explored fuel cell-based EV topologies and energy management strategies. In the paper, the main DC–DC converters' topologies interfacing with the fuel cells were investigated and discussed. Furthermore, the advantages and disadvantages of three types of strategies (rule-based strategies, optimization-based strategies, and learning-based strategies) are described and evaluated.

Considering the power switches used in converter topologies, it is important to evaluate the impact of the new wide bandgap (WBG) semiconductor technologies. WBG power electronic devices such as silicon carbide (SiC) MOSFET and gallium nitride (GaN) FET represents the technological evolution of pure silicon devices [49]. SiC MOSFET and GaN-based devices display superior thermal performances compared to silicon MOSFETs [50,51].

WBG devices can be used to obtain high power density and efficient power converters. GaN technology devices are currently undergoing a significant evolution. In recent years, the need to increase the voltage limits of the GaN FET devices has represented a challenge to semiconductor manufacturers, who require these devices for their use in high-voltage power converters applications. Currently, technology with a maximum voltage of up to 650 V is available on the market, and a higher voltage can be obtained with cascode solution [52,53]. At low voltage (up to 200 V) the GaN FETs are much attractive not only in DC–DC converters for their high switching frequency that can be reached (up to tens of MHz) but also in DC–AC converters for AC motor control as described by the authors in [54]. In the paper, the efficiency and size of the electric motor control system are improved compared with Silicon (Si) MOSFET devices for small battery-powered electric vehicle systems (so-called micro-mobility or light mobility) such as e-kick scooters, e-bikes, electronic skateboards, hover boards, etc. This strengthening of qualities stems from the use of GaN FET switches.

Wireless Power Transfer Applications

A wireless power transfer (WPT) system transmits electrical power based on technologies using time-varying electric, magnetic, or electromagnetic fields. There are different applications where WPT technology are applied such as electronic equipment (mobile phone, home appliances), Unmanned Aerial Vehicles (UAV), and charging system [55]. Nowadays, EV using WPT technologies to charge the battery is quite widespread. Generally, WPT systems are composed of a primary power converter that transmits the power, a coupling system, and a secondary power converter that releases the power to a load. In some specific applications, the power flow can be unidirectional or bidirectional [56]. The main WPT solutions on the market are achieved by the use of the inductive coupling method, the so-called inductive power transfer (IPT). However, capacitive power transfer (CPT) at the present time is growing in use, especially for lower power [57]. Furthermore, the WPT methods in EV battery chargers are classified into the static charging technique and dynamic charging technique. The static charging technique is achieved by IPT and CPT solutions, while the dynamic wireless power transfer (DWPT) technique is obtained by the IPT system.

Nowadays, DWPT systems are becoming increasingly pivotal for moving electric vehicle (EV) charging solutions. This attractive charging technique features an issue with the misalignment tolerance between the coupling inductors. In [58], Eiman ElGhanam et al. propose a DWPT charging system for EVs with improved misalignment tolerance based on inductor–capacitor–capacitor (LCC) compensation. A class D inverter is used on the primary side of the coupling inductors, while a boost rectifier circuit is selected on the secondary side to obtain the AC–DC conversion stage. The detailed DWPT charging system design is discussed. The design constraints and solutions include EV specifications, inductive links, compensation network constraints, and power electronic converter circuits. CPT is a valid charging technique in several wireless power transfer applications. In perspective, CPT can result in lower cost and higher reliability than IPT because it does not need coupling coils and related shields. Furthermore, CPT features the capability to transfer power through metal barriers thanks to the coupling capacitive effect. In low-power transfer solutions, a single-device converter structure may be used. In [59], F. Corti et al. present a complete, detailed design of an E-Inverter for CPT charging systems which is suitable for electronic equipment or small drones. The design procedure guarantees both zero voltage switching (ZVS) and zero derivative switching (ZDS) conditions for the switching device (a silicon MOSFET) at an optimum coupling coefficient, thus enabling high transmission and conversion efficiency. An experimental prototype at 24 V of input voltage with 100 kHz of switching frequency is used for the design validation. The tested prototype can transfer 83.5 Wat at optimal capacitive coupling with a maximum efficiency of 92.5%.

2.4. *Power Converters Prototyping Methodology*

In power converters for energy conversion, the design methodology will need a noticeable effort to achieve the targets of satisfactory static and dynamic performances, high efficiency, thermal management, and reduced volumes involved. The hardware in the loop (HIL) is a cost-effective technique that allows for the simulation of the system under design, avoiding setting up expensive and cumbersome actual converter structures to validate the effectiveness and accuracy of the control methods [60]. HIL utilizes a hardware platform to build, e.g., a microprocessor or FPGA, and the required number of channels and I/O types to test the implemented system. The hardware platform and the real-time model of the power converter topology are interfaced to simulate the whole system under investigation with a quite satisfactory approximation with respect to the real one system. In [61] Leonel Estrada et al. use a well-known and flexible platform such as LabVIEW application software for HIL simulation. The topology of the considered power converter is obtained by means of differential equations that define each state of the converter circuit. The developed method takes place in 5 steps:

- Converter under test design
- Numerical modeling
- off-line simulation of the numerical model using fixed-point representation
- implementation in a Field-Programmable Gate Array (FPGA).

In the paper, two examples of the HIL application technique are detailed: a buck converter and a three-phase Voltage Source Inverter (VSI) system. The results obtained are compared with the simulation of commercial software and actual power converter tests, showing the effectiveness of the HIL technique proposed. The methodology presented is suitable for people with not quite enough experience in the use of software languages and tools such as very high-speed integrated circuit hardware description language (VHDL), real-time simulation (RTS), and HIL simulation tools.

3. Conclusions

The collection of articles in this Editorial provides some indications of the directions of the current research development for the power converters topologies, control methods and technologies applied to the energy conversion field. For homogeneity of investigations presented, three of the most significant sectors have been identified in the area of energy conversion, and the Special Issue is arranged along these lines.

1. Renewable Energy Conversion System
2. Smart Grid Components and Applications
3. Energy Store Management and e-Mobility

The last section is dedicated to the power converter's prototyping methodology. It will be helpful for giving useful suggestions to the designers of power converters in this strategic research field.

The articles considered in the different sections show the vitality of the research on this trending topic and will contribute noticeable development opportunities, ideas, and valuable skills to the international research community.

Funding: the authors have received no funding for preparing this Editorial.

Conflicts of Interest: The authors declare no conflict of interest.

References

1. Jørgensen, K.L.; Zhang, Z.; Andersen, M.A.E. Next generation of power electronic-converter application for energy-conversion and storage units and systems. *Clean Energy* **2019**, *3*, 307–315. [CrossRef]
2. Luo, B.; Ye, D.; Wang, L. Recent Progress on Integrated Energy Conversion and Storage Systems. *Adv. Sci.* **2017**, *4*, 1700104. [CrossRef] [PubMed]
3. Sutikno, T.; Purnama, H.S.; Widodo, N.S.; Padmanaban, S.; Sahid, M.R. A review on non-isolated low-power DC–DC converter topologies with high output gain for solar photovoltaic system applications. *Clean Energy* **2022**, *6*, 557–572. [CrossRef]
4. Hashim, N.; Salam, Z.; Johari, D.; Ismail, N.F.N. DC-DC Boost Converter Design for Fast and Accurate MPPT Algorithms in Stand-Alone Photovoltaic System. *Int. J. Power Electron. Drive Syst. (IJPEDS)* **2018**, *9*, 1038–1050. [CrossRef]
5. Basha, C.H.; Rani, C. Different Conventional and Soft Computing MPPT Techniques for Solar PV Systems with High Step-Up Boost Converters: A Comprehensive Analysis. *Energies* **2020**, *13*, 371. [CrossRef]
6. Farakhor, A.; Abapour, M.; Sabahi, M.; Farkoush, S.G.; Oh, S.-R.; Rhee, S.-B. A Study on an Improved Three-Winding Coupled Inductor Based DC/DC Boost Converter with Continuous Input Current. *Energies* **2020**, *13*, 1780. [CrossRef]
7. Danyali, S.; Aazami, R.; Moradkhani, A.; Haghi, M. A new dual-input three-winding coupled-inductor based DC-DC boost converter for renewable energy applications. *Int. Trans. Electr. Energy Syst.* **2020**, *31*, e12686. [CrossRef]
8. Umadevi, K.; Nagarajan, C. Design and implementation of novel soft switching method based DC-DC converter with non-isolated coupled inductor in solar system using FPGA. *Microprocess. Microsystems* **2019**, *73*, 102952. [CrossRef]
9. Sayed, K.; Gronfula, M.G.; Ziedan, H.A. Novel Soft-Switching Integrated Boost DC-DC Converter for PV Power System. *Energies* **2020**, *13*, 749. [CrossRef]
10. Chen, Y.-T.; Li, Z.-M.; Liang, R.-H. A Novel Soft-Switching Interleaved Coupled-Inductor Boost Converter With Only Single Auxiliary Circuit. *IEEE Trans. Power Electron.* **2017**, *33*, 2267–2281. [CrossRef]
11. Karthikeyan, M.; Elavarasu, R.; Ramesh, P.; Bharatiraja, C.; Sanjeevikumar, P.; Mihet-Popa, L.; Mitolo, M. A Hybridization of Cuk and Boost Converter Using Single Switch with Higher Voltage Gain Compatibility. *Energies* **2020**, *13*, 2312. [CrossRef]
12. Yari, K.; Mojallali, H.; Shahalami, S.H. A New Coupled-Inductor-Based Buck–Boost DC–DC Converter for PV Applications. *IEEE Trans. Power Electron.* **2021**, *37*, 687–699. [CrossRef]
13. Hassan, T.-U.; Abbassi, R.; Jerbi, H.; Mehmood, K.; Tahir, M.; Cheema, K.; Elavarasan, R.; Ali, F.; Khan, I. A Novel Algorithm for MPPT of an Isolated PV System Using Push Pull Converter with Fuzzy Logic Controller. *Energies* **2020**, *13*, 4007. [CrossRef]
14. Lim, J.-W.; Hassan, J.; Kim, M. Bidirectional Soft Switching Push–Pull Resonant Converter Over Wide Range of Battery Voltages. *IEEE Trans. Power Electron.* **2021**, *36*, 12251–12267. [CrossRef]
15. Alam, M.S.; Al-Ismail, F.S.; Salem, A.; Abido, M.A. High-Level Penetration of Renewable Energy Sources Into Grid Utility: Challenges and Solutions. *IEEE Access* **2020**, *8*, 190277–190299. [CrossRef]
16. Kanchev, H.; Lu, D.; Colas, F.; Lazarov, V.; Francois, B. Energy Management and Operational Planning of a Microgrid With a PV-Based Active Generator for Smart Grid Applications. *IEEE Trans. Ind. Electron.* **2011**, *58*, 4583–4592. [CrossRef]
17. Ahmad, J.; Zaid, M.; Sarwar, A.; Lin, C.-H.; Asim, M.; Yadav, R.; Tariq, M.; Satpathi, K.; Alamri, B. A New High-Gain DC-DC Converter with Continuous Input Current for DC Microgrid Applications. *Energies* **2021**, *14*, 2629. [CrossRef]
18. Kumar, M.A.B.; Krishnasamy, V. Quadratic Boost Converter with Less Input Current Ripple and Rear-End Capacitor Voltage Stress for Renewable Energy Applications. *IEEE J. Emerg. Sel. Top. Power Electron.* **2021**, *10*, 2265–2275. [CrossRef]

19. Awaad, M.I.; Afifi, Z.E. Design, Simulation and Implementation of a DC Microgrid based on Quadrupler DC Converter. *Comput. Electr. Eng.* **2020**, *89*, 106948. [CrossRef]
20. Khan, S.; Zaid, M.; Mahmood, A.; Nooruddin, A.S.; Ahmad, J.; Alghaythi, M.L.; Alamri, B.; Tariq, M.; Sarwar, A.; Lin, C.-H. A New Transformerless Ultra High Gain DC–DC Converter for DC Microgrid Application. *IEEE Access* **2021**, *9*, 124560–124582. [CrossRef]
21. Ríos, S.; Pagano, D.; Lucas, K. Bidirectional Power Sharing for DC Microgrid Enabled by Dual Active Bridge DC-DC Converter. *Energies* **2021**, *14*, 404. [CrossRef]
22. Kurm, S.; Agarwal, V. Interfacing Standalone Loads With Renewable Energy Source and Hybrid Energy Storage System Using a Dual Active Bridge Based Multi-Port Converter. *IEEE J. Emerg. Sel. Top. Power Electron.* **2021**, *10*, 4738–4748. [CrossRef]
23. Khajeh, K.G.; Solatialkaran, D.; Zare, F.; Mithulananthan, N. Harmonic analysis of grid-connected inverters considering external distortions: Addressing harmonic emissions up to 9 kHz. *IET Power Electron.* **2020**, *13*, 1934–1945. [CrossRef]
24. Zhang, B.; Du, X.; Zhao, J.; Zhou, J.; Zou, X. Impedance modeling and stability analysis of three-phase three-level NPC inverter connected to grid. *CSEE J. Power Energy Syst.* **2020**, *6*, 270–278. [CrossRef]
25. Lin, H.; Leon, J.I.; Luo, W.; Marquez, A.; Liu, J.; Vazquez, S.; Franquelo, L.G. Integral Sliding-Mode Control-Based Direct Power Control for Three-Level NPC Converters. *Energies* **2020**, *13*, 227. [CrossRef]
26. Perez, M.A.; Ceballos, S.; Konstantinou, G.; Pou, J.; Aguilera, R.P. Modular Multilevel Converters: Recent Achievements and Challenges. *IEEE Open J. Ind. Electron. Soc.* **2021**, *2*, 224–239. [CrossRef]
27. Chaudhary, S.K.; Cupertino, A.F.; Teodorescu, R.; Svensson, J.R. Benchmarking of Modular Multilevel Converter Topologies for ES-STATCOM Realization. *Energies* **2020**, *13*, 3384. [CrossRef]
28. Elghareeb, A.O.; Elrefaey, A.M.; Moussa, M.F.; Dessouky, Y.G. Review of DC Offset Compensation Techniques for Grid Connected Inverters. *Int. J. Power Electron. Drive Syst. (IJPEDS)* **2018**, *9*, 478–494. [CrossRef]
29. Hannan, M.A.; Ker, P.J.; Lipu, M.S.H.; Choi, Z.H.; Rahman, M.S.A.; Muttaqi, K.M.; Blaabjerg, F. State of the Art of Solid-State Transformers: Advanced Topologies, Implementation Issues, Recent Progress and Improvements. *IEEE Access* **2020**, *8*, 19113–19132. [CrossRef]
30. Shamshuddin, M.A.; Rojas, F.; Cardenas, R.; Pereda, J.; Diaz, M.; Kennel, R. Solid State Transformers: Concepts, Classification, and Control. *Energies* **2020**, *13*, 2319. [CrossRef]
31. Han, Y.; Feng, Y.; Yang, P.; Xu, L.; Xu, Y.; Blaabjerg, F. Cause, Classification of Voltage Sag, and Voltage Sag Emulators and Applications: A Comprehensive Overview. *IEEE Access* **2019**, *8*, 1922–1934. [CrossRef]
32. Jiang, F.; Tu, C.; Shuai, Z.; Cheng, M.; Lan, Z.; Xiao, F. Multilevel Cascaded-Type Dynamic Voltage Restorer With Fault Current-Limiting Function. *IEEE Trans. Power Deliv.* **2015**, *31*, 1261–1269. [CrossRef]
33. Moghassemi, A.; Padmanaban, S. Dynamic Voltage Restorer (DVR): A Comprehensive Review of Topologies, Power Converters, Control Methods, and Modified Configurations. *Energies* **2020**, *13*, 4152. [CrossRef]
34. Byrne, R.H.; Nguyen, T.A.; Copp, D.A.; Chalamala, B.R.; Gyuk, I. Energy Management and Optimization Methods for Grid Energy Storage Systems. *IEEE Access* **2018**, *6*, 13231–13260. [CrossRef]
35. Lawder, M.T.; Suthar, B.; Northrop, P.W.C.; DE, S.; Hoff, C.M.; Leitermann, O.; Crow, M.L.; Santhanagopalan, S.; Subramanian, V.R. Battery Energy Storage System (BESS) and Battery Management System (BMS) for Grid-Scale Applications. *Proc. IEEE* **2014**, *102*, 1014–1030. [CrossRef]
36. Sun, Q.; Lv, H.; Wang, S.; Gao, S.; Wei, K. Optimized State of Charge Estimation of Lithium-Ion Battery in SMES/Battery Hybrid Energy Storage System for Electric Vehicles. *IEEE Trans. Appl. Supercond.* **2021**, *31*, 5700606. [CrossRef]
37. Asensio, A.P.; Gonzalez-Longatt, F.; Arnaltes, S.; Rodríguez-Amenedo, J.L. Analysis of the Converter Synchronizing Method for the Contribution of Battery Energy Storage Systems to Inertia Emulation. *Energies* **2020**, *13*, 1478. [CrossRef]
38. Mandrile, F.; Musumeci, S.; Carpaneto, E.; Bojoi, R.; Dragicevic, T.; Blaabjerg, F. State-Space Modeling Techniques of Emerging Grid-Connected Converters. *Energies* **2020**, *13*, 4824. [CrossRef]
39. Marchesoni, M.; Vacca, C. New DC–DC Converter for Energy Storage System Interfacing in Fuel Cell Hybrid Electric Vehicles. *IEEE Trans. Power Electron.* **2007**, *22*, 301–308. [CrossRef]
40. Abu-Siada, A.; Mosaad, M.I.; Kim, D.W.; El-Naggar, M.F. Estimating Power Transformer High Frequency Model Parameters Using Frequency Response Analysis. *IEEE Trans. Power Deliv.* **2019**, *35*, 1267–1277. [CrossRef]
41. Malik, M.Z.; Chen, H.; Nazir, M.S.; Khan, I.A.; Abdalla, A.N.; Ali, A.; Chen, W. A New Efficient Step-Up Boost Converter with CLD Cell for Electric Vehicle and New Energy Systems. *Energies* **2020**, *13*, 1791. [CrossRef]
42. Habib, S.; Khan, M.M.; Abbas, F.; Ali, A.; Faiz, M.T.; Ehsan, F.; Tang, H. Contemporary trends in power electronics converters for charging solutions of electric vehicles. *CSEE J. Power Energy Syst.* **2020**, *6*, 911–929. [CrossRef]
43. Cuoghi, S.; Mandrioli, R.; Ntogramatzidis, L.; Gabriele, G. Multileg Interleaved Buck Converter for EV Charging: Discrete-Time Model and Direct Control Design. *Energies* **2020**, *13*, 466. [CrossRef]
44. Sayed, K.; Ali, Z.M.; Aldhaifallah, M. Phase-Shift PWM-Controlled DC–DC Converter with Secondary-Side Current Doubler Rectifier for On-Board Charger Application. *Energies* **2020**, *13*, 2298. [CrossRef]
45. Deng, J.; Bae, C.; Denlinger, A.; Miller, T. Electric Vehicles Batteries: Requirements and Challenges. *Joule* **2020**, *4*, 511–515. [CrossRef]
46. Müller, H.; Bernt, A.-O.; Salman, P.; Trattner, A. Fuel cell range extended electric vehicle fcreev long driving ranges without emissions. *ATZ Worldw.* **2017**, *119*, 56–60. [CrossRef]

47. Lencwe, M.J.; Chowdhury, S.P.D.; Olwal, T.O. Hybrid energy storage system topology approaches for use in transport vehicles: A review. *Energy Sci. Eng.* **2022**, *10*, 1449–1477. [CrossRef]
48. Sorlei, I.-S.; Bizon, N.; Thounthong, P.; Varlam, M.; Carcadea, E.; Culcer, M.; Iliescu, M.; Raceanu, M. Fuel Cell Electric Vehicles—A Brief Review of Current Topologies and Energy Management Strategies. *Energies* **2021**, *14*, 252. [CrossRef]
49. Zhang, L.; Zheng, Z.; Lou, X. A review of WBG and Si devices hybrid applications. *Chin. J. Electr. Eng.* **2021**, *7*, 1–20. [CrossRef]
50. Shenai, K. Future Prospects of Widebandgap (WBG) Semiconductor Power Switching Devices. *IEEE Trans. Electron Devices* **2014**, *62*, 248–257. [CrossRef]
51. Raciti, A.; Musumeci, S.; Chimento, F.; Privitera, G. A new thermal model for power MOSFET devices accounting for the behavior in unclamped inductive switching. *Microelectron. Reliab.* **2016**, *58*, 3–11. [CrossRef]
52. Buonomo, S.; Musumeci, S.; Pagano, R.; Porto, C.; Raciti, A.; Scollo, R. Driving a New Monolithic Cascode Device in a DC–DC Converter Application. *IEEE Trans. Ind. Electron.* **2008**, *55*, 2439–2449. [CrossRef]
53. Alharbi, S.S.; Matin, M. Experimental evaluation of medium-voltage cascode gallium nitride (GaN) devices for bidirectional DC-DC converters. *CES Trans. Electr. Mach. Syst.* **2021**, *5*, 232–248. [CrossRef]
54. Musumeci, S.; Mandrile, F.; Barba, V.; Palma, M. Low-Voltage GaN FETs in Motor Control Application; Issues and Advantages: A Review. *Energies* **2021**, *14*, 6378. [CrossRef]
55. Faraci, G.; Raciti, A.; Rizzo, S.A.; Schembra, G. Green wireless power transfer system for a drone fleet managed by reinforcement learning in smart industry. *Appl. Energy* **2019**, *259*, 114204. [CrossRef]
56. Huang, M.; Lu, Y.; Martins, R.P. A Reconfigurable Bidirectional Wireless Power Transceiver for Battery-to-Battery Wireless Charging. *IEEE Trans. Power Electron.* **2018**, *34*, 7745–7753. [CrossRef]
57. Dai, J.; Ludois, D.C. A Survey of Wireless Power Transfer and a Critical Comparison of Inductive and Capacitive Coupling for Small Gap Applications. *IEEE Trans. Power Electron.* **2015**, *30*, 6017–6029. [CrossRef]
58. ElGhanam, E.; Hassan, M.; Osman, A. Design of a High Power, LCC-Compensated, Dynamic, Wireless Electric Vehicle Charging System with Improved Misalignment Tolerance. *Energies* **2021**, *14*, 885. [CrossRef]
59. Corti, F.; Reatti, A.; Wu, Y.-H.; Czarkowski, D.; Musumeci, S. Zero Voltage Switching Condition in Class-E Inverter for Capacitive Wireless Power Transfer Applications. *Energies* **2021**, *14*, 911. [CrossRef]
60. Li, G.; Zhang, D.; Xin, Y.; Jiang, S.; Wang, W.; Du, J. Design of MMC Hardware-in-the-Loop Platform and Controller Test Scheme. *CPSS Trans. Power Electron. Appl.* **2019**, *4*, 143–151. [CrossRef]
61. Estrada, L.; Vázquez, N.; Vaquero, J.; de Castro, Á.; Arau, J. Real-Time Hardware in the Loop Simulation Methodology for Power Converters Using LabVIEW FPGA. *Energies* **2020**, *13*, 373. [CrossRef]

energies MDPI

Article

Temperature Characteristic Analysis of the Output Intrinsically Safe Buck Converter and Its Design Consideration

Yan Li [1,2,*], Shulin Liu [1], Zhenqin Liang [1], Yuqing Zhang [1] and Yijun Shen [1]

1 School of Electrical and Control Engineering, Xi'an University of Science & Technology, Xi'an 710054, China
2 School of Energy Engineering, Yulin University, Yulin 719000, China
* Correspondence: lisalee@yulinu.edu.cn

Abstract: Aiming at the unreliability resulting from ignoring the temperature effect and randomness of switching frequency in the traditional design method of an intrinsically safe Buck converter, a reliable design method based on the minimum frequency and considering the temperature characteristic is proposed. The theoretical design range of capacitance is deduced according to the maximum output ripple voltage and output intrinsic safety performance requirements. Considering the temperature characteristics of the capacitor, the actual maximum and minimum capacitances are obtained corresponding to the theoretical design capacitance within a given temperature range. It is pointed out that the actual minimum capacitance increases with the decrease of switching frequency, while the actual maximum capacitance is independent of frequency. Therefore, it can be deduced that there exists a minimum frequency which can meet the requirements of both output ripple voltage and intrinsically safe performance. When the actual maximum capacitance equals the actual minimum capacitance, the analytic expression of the minimum frequency is obtained. Assuming a capacitance adjustment, the actual working frequency of the converter corresponding to the minimum frequency is deduced. The design flow of an intrinsically safe Buck converter based on the minimum switching frequency considering the temperature characteristic is presented. The correctness of the theoretical analysis and the feasibility of the proposed design method are verified by experimental results. This design method can also be applied to other types of intrinsically safe converters.

Keywords: buck converter; output intrinsically safe; the minimum frequency; temperature characteristic

Citation: Li, Y.; Liu, S.; Liang, Z.; Zhang, Y.; Shen, Y. Temperature Characteristic Analysis of the Output Intrinsically Safe Buck Converter and Its Design Consideration. *Energies* **2022**, *15*, 7308. https://doi.org/10.3390/en15197308

Academic Editor: Salvatore Musumeci

Received: 9 September 2022
Accepted: 1 October 2022
Published: 5 October 2022

1. Introduction

The switching power supply used in coal mines, petrochemical plants and other hazardous environments must meet anti-explosive requirements [1,2]. An intrinsically safe anti-explosive power supply guarantees compliance with the anti-explosive requirements because of the electrical parameters of the circuit [3]. Because there is no bulky explosion-proof shell, the safe power supply has the remarkable advantages of small size, light weight, high safety performance, low cost, simple manufacturing process and easy maintenance. Because of this, the anti-explosive power supply should be designed as intrinsically safe as possible [4]. The Buck converter is one of the most commonly used topologies of the switching power supply, and it contains inductor and capacitor. When the inductor is broken or the capacitor is short-circuited, the fault energy of the converter may ignite the specified explosive gas resulting in fire and explosion [5]. Therefore, it is very important to study the intrinsic safety performance of the converter. Generally, large capacitors are selected to meet the output ripple voltage index. However, there is always the possibility that the converter will ignite flammable and explosive gas due to an output short-circuit. Therefore, the intrinsic output safety characteristic corresponding to the output short-circuit discharge in a Buck converter must be considered.

When designing the converter, it is contradictory to choose both the energy storage element to satisfy the output ripple voltage index and the intrinsic safety requirement. That is, the inductance and capacitance required to meet the intrinsic safety performance should be as small as possible. In order to meet the output ripple voltage index, the inductance and capacitance should be as large as possible. This contradiction makes the parameter design of an intrinsically safe converter more complicated [4–6].

At present, according to the output ripple voltage index and the output intrinsic safety requirements, the parameter design range of the converter is obtained under ideal conditions [7,8]. Because the converter is a multi-objective and multi-parameter nonlinear system, some scholars use a genetic algorithm and particle swarm optimization algorithm to optimize the parameters of the converter [9–12]. In addition, in the design of an intrinsically safe switching converter, the maximum intrinsically safe output power under the given parameters is specified to avoid the problem of blindly proposing the intrinsically safe output power [13,14]. The design method of the intrinsically safe converter based on the maximum intrinsically safe output power is proposed [15]. References [16–19] analyzed the performance and reliability of the converter by capacitor and illustrated the effect of temperature on capacitor life. However, the influence of temperature on capacitance, electrical performance and intrinsic safety requirements of the converter was not analyzed.

Although the design methods mentioned above solve some practical problems, the existing design methods usually select the operating frequency randomly according to experience. The design range of inductor and capacitor is obtained according to the electrical performance index and intrinsic safety requirements of the converter. Due to the arbitrary choice of switching frequency, the value range of inductor and capacitor parameters may not exist because the frequency is too small or the switching loss of the circuit may too large due to the high frequency, which reduces the efficiency of the converter. In addition, the capacitance at the output of the converter will change with the temperature. Even if the switching frequency is constant, with the change of temperature, the actual capacitance may be larger than the nominal value, leading to the converter not meeting the requirements of output intrinsic safety. Likewise, the actual capacitance may be smaller than the nominal value, which makes the converter not meet the requirements of output ripple voltage index. However, the existing design method does not consider the influence of temperature. As a result, the range of the designed capacitance is unreliable.

In order to solve this problem, a design method based on the minimum frequency and considering the temperature characteristic is proposed. The main contributions of this paper are as follows.

The reliable actual capacitance value range, considering the temperature characteristics corresponding to the theoretical maximum and minimum capacitance is deduced based on the requirements of the output intrinsic safety and maximum output ripple voltage index This can solve the unreliability issue. The minimum switching frequency is defined, and its expression is derived, which can avoid the blind selection problem of operating frequency in parameter design. The capacitance adjustment is introduced, which can avoid the unreasonable problem of the capacitance design range.

2. Theoretical Parameter Design of the Capacitor

The composition circuit diagram of a Buck converter is shown in Figure 1.

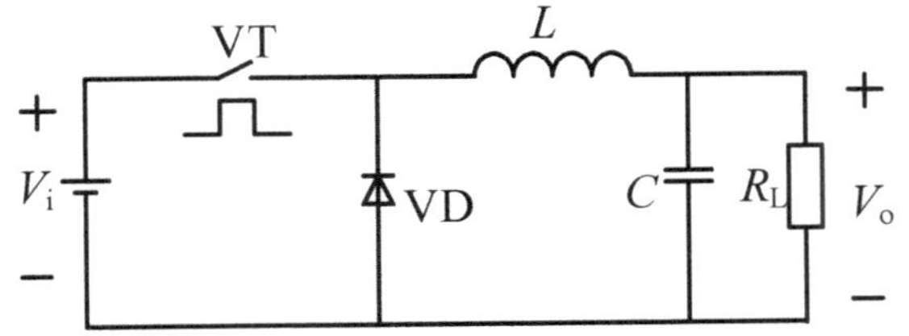

Figure 1. Schematic diagram of a Buck converter.

From Figure 1, V_i and V_o are the input and output voltages, respectively. VT and VD are the switching tube and current-continuing diode of the converter, respectively. L and C are the inductor and capacitor of the converter, respectively. R_L is the load resistance.

For the output intrinsically safe switching converters, it is necessary to meet both the specified electrical performance index and the output intrinsic safety requirements [20–22]. Through analyzing the electrical characteristics of the converter, the minimum theoretical capacitance C_{min} can be obtained, and the maximum theoretical capacitance C_{max} can be generated by analyzing the output intrinsic safety requirements.

2.1. Minimum Theoretical Capacitance

The output ripple voltage of the switching converter is not only relevant to the inductor and capacitor, but also to the operating mode of the converter. In the dynamic operating range, when the designed converter works in CCM, the output ripple voltage V_{pp} can be expressed as shown in Equation (1) [8,23].

$$V_{PP} = \frac{V_o(V_i - V_o)}{8LCf^2V_i} \tag{1}$$

where f is the switching frequency.

In Equation (1), the maximum output ripple voltage is obtained when the inductor L is the minimum value. If the Buck converter works in CCM, the inductor L needs to meet [23]

$$L \geq L_C = \frac{R_L(V_i - V_o)}{2fV_i} \tag{2}$$

where L_C is the critical inductance of the Buck converter working in CCM and DCM. When $L = L_C$, the maximum output ripple voltage $V_{pp,max}$ is

$$V_{PP,max} = \frac{V_o}{4fCR_{L,min}} \tag{3}$$

It can be seen from Equation (3) that the maximum output ripple voltage of the Buck converter increases with decreasing the capacitance C. It is noted that there must be a minimum capacitance to satisfy the maximum output ripple voltage index. According to Equation (3), the minimum theoretical capacitance C_{min} can be obtained as

$$C_{min} = \frac{1}{4mfR_{L,min}} \tag{4}$$

where m is $V_{pp,max}/V_o$.

2.2. Maximum Theoretical Capacitance

Since the output short-circuit discharge energy of the converter is related to inductors [6,23–26], capacitors, the switching states and the operating mode of the converter, the intrinsic safety criterion of simple capacitor circuit cannot be directly used to judge the intrinsic safety performance of the converter. However, by analyzing the maximum output short-circuit discharge energy of the converter and using the energy equivalence principle, the Buck converter can be equivalent to a simple capacitor circuit, and then the output intrinsic safety criterion of the converter is obtained according to the critical ignition voltage curve of the simple capacitor circuit.

The output short test circuit of the Buck converter is shown in Figure 2.

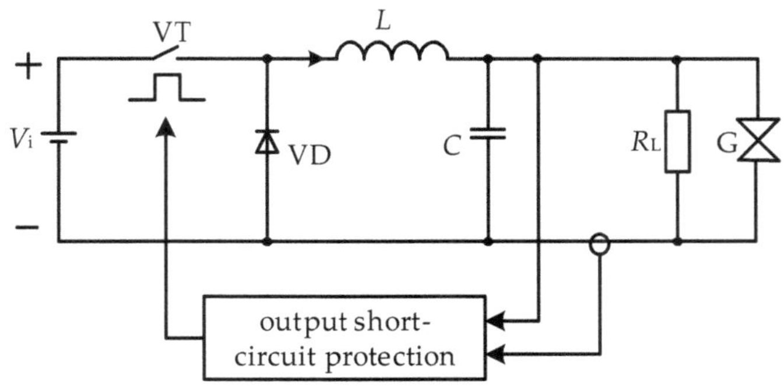

Figure 2. Output short test circuit of the Buck converter.

From Figure 2, G is the safety spark test device. When the two electrodes of G contact, the output is short-circuited.

Assuming that the load changes from $R_{L,min}$ to $R_{L,max}$, the input voltage changes from $V_{i,min}$ to $V_{i,max}$, the maximum output short-circuit discharge energy [8] of the Buck converter is

$$W_{max} = \frac{V_{i,max}^2 V_o^2 \Delta t^2 (V_{i,max} - V_o)}{2L\left[V_{i,max} f V_H^2 T_C - V_o^2 (V_{i,max} - V_o)\right]} + \frac{V_{i,max}^2 \Delta t^2}{L} + \frac{CV_o^2}{2} = W'_{max} + \frac{CV_o^2}{2} \quad (5)$$

where Δt is the response time of the output short-circuit protection circuit. T_C is the output short-circuit spark discharge time. V_H is the average value of the spark discharge voltage. W'_{max} is the equivalent energy in addition to the energy released by the capacitor. So, W'_{max} is

$$W'_{max} = \frac{V_{i,max}^2 V_o^2 \Delta t^2 (V_{i,max} - V_o)}{2L\left[V_{i,max} f V_H^2 T_C - V_o^2 (V_{i,max} - V_o)\right]} + \frac{V_{i,max}^2 \Delta t^2}{L} \quad (6)$$

According to the energy equivalence principle, the Buck converter is equivalent to a simple capacitor circuit (as shown in Figure 3), the equivalent capacitance C_e can be obtained as

$$C_e = \frac{2W_{max}}{V_o^2} = C + \frac{2W'_{max}}{V_o^2} = C + C'_e \quad (7)$$

where C'_e is the equivalent capacitance corresponding to the effective energy transferred from the input power supply and inductor to the output short-circuit point, that is $C'_e = \frac{2W'_{max}}{V_o^2}$.

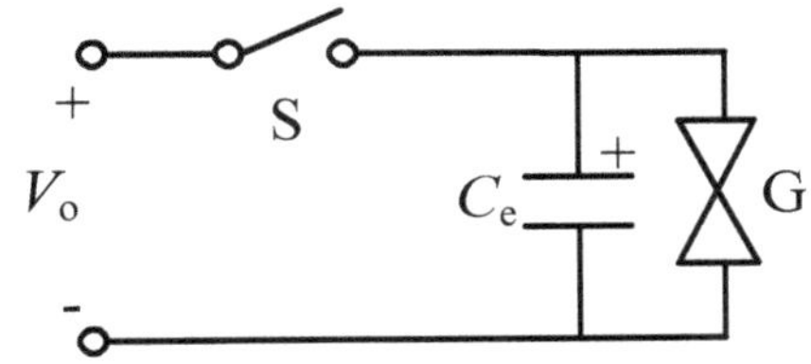

Figure 3. Equivalent simple capacitor circuit.

From Figure 3, S is the equivalent switch of a simple capacitor circuit. When the two electrodes of G are disconnected and S is closed, C_e is charged. When the two electrodes of G are closed and S is disconnected, C_e is discharged.

The non-explosive intrinsic safety criterion to meet the output intrinsic safety requirements is

$$C_e < C_B \quad (8)$$

where, C_B is the critical ignition capacitance corresponding to the output voltage V_o. It can be found from the critical ignition voltage curve of a simple capacitor circuit. When the equivalent capacitance C_e is less than the critical ignition capacitance C_B, the Buck converter satisfies the requirement of output intrinsic safety. Therefore, the maximum equivalent capacitance $C_{e,max}$ of the Buck Converter is C_B. Substituting Equations (5) and (6) into Equation (7), the maximum theoretical capacitance C_{max} of the output intrinsically safe Buck converter is

$$C_{max} = C_B - C'_e = C_B - \frac{2V_{i,max}^2 \Delta t^2}{LV_o^2} - \frac{V_{i,max}^2 \Delta t^2 (V_{i,max} - V_o)}{L\left[V_{i,max} f V_H^2 T_C - V_o^2 (V_{i,max} - V_o)\right]} \tag{9}$$

At present, the output power of the intrinsically safe switching converter is not too high; V_i and V_o are relatively low. It makes C_B corresponding to V_o relatively large (for example, assuming that V_o is 18 V, the corresponding C_B is 50 μF), and in the practical application [4], Δt is 1–2 μs. Here, f is from a few hundred kilohertz to several megahertz; L is tens of microseconds; T_C is tens of microseconds; V_H is 8–9 V. According to the above parameters, C'_e is at least one order of magnitude smaller than C_B. Therefore. C_{max} can be expressed as

$$C_{max} \approx C_B \tag{10}$$

From Equation (10), it can be seen that the maximum theoretical capacitance C_{max} of the output intrinsically safe Buck converter is only determined by the output intrinsic safety requirements C_B, which is independent on the switching frequency f.

Combined with Equation (4) and Equation (10), it can be concluded that the theoretical range of capacitance C satisfying both output ripple voltage and intrinsic safety requirements is

$$\frac{1}{4mfR_{L,min}} \leq C \leq C_B \tag{11}$$

According to Equation (11), the maximum and minimum capacitance can be determined. In addition, the maximum capacitance is independent of switching frequency, while the minimum capacitance is inversely proportional to switching frequency.

2.3. Actual Value Range of Capacitance Considering Temperature Characteristics

With the change of ambient temperature, the capacitance of the switching converter also changes. If the capacitance exceeds the upper limit C_{max}, the converter cannot meet the requirements of output intrinsic safety. Similarly, if the capacitance is less than the lower limit C_{min}, the converter cannot meet the requirements of output ripple voltage. Therefore, the range of theoretical capacitance is not reliable. The temperature characteristics of capacitance must be considered, and the practical capacitance value range can be obtained based on the theoretical value range of capacitance.

Supposing that in Equation (11), the capacitance at room temperature (25 °C) is C. The actual capacitance at a certain temperature is C'. The maximum and minimum values of the ratio C' to C is defined as A_T and B_T, respectively. Therefore, the maximum and minimum values of the actual capacitance considering the temperature characteristics can be expressed as A_TC and B_TC, respectively.

In order to ensure that the minimum capacitance within the given temperature range can still meet the requirements of output ripple voltage index, the minimum value of actual capacitance B_TC should be greater than Equation (11), the lower limit C_{min}. Similarly, the maximum capacitance must satisfy the requirements of output intrinsic safety, the maximum value of actual capacitance A_TC should be less than Equation (11), the upper limit C_{max}. According to Equation (11), the actual value range of capacitance considering temperature characteristics is

$$\frac{1}{4B_T mfR_{L,min}} \leq C \leq \frac{C_B}{A_T} \tag{12}$$

From Equation (12), the maximum and minimum values of the actual capacitance considering the temperature characteristics can be expressed as follows, respectively.

$$\begin{aligned} C_{\mathrm{T,max}} &= \tfrac{C_B}{A_T} \\ C_{\mathrm{T,min}} &= \tfrac{1}{4B_T m f R_{L,\min}} \end{aligned} \tag{13}$$

It can be seen that the minimum value $C_{\mathrm{T,min}}$ considering temperature characteristics of an intrinsic safe Buck converter increases with the decrease of switching frequency f, while the maximum value $C_{\mathrm{T,max}}$ is independent of f.

3. Design of Minimum Switching Frequency

Since the maximum capacitance $C_{\mathrm{T,max}}$ is independent of f considering temperature characteristics, while the minimum capacitance $C_{\mathrm{T,min}}$ increases with the decrease of f, it is pointed out that there must be a minimum switching frequency to make the capacitance range effective. When $C_{\mathrm{T,min}}$ equals to $C_{\mathrm{T,max}}$, the minimum switching frequency $f_{\min}$ is obtained as

$$f_{\min} = \frac{A_T}{4B_T m C_B R_{L,\min}} \tag{14}$$

The actual switching frequency f of the converter must be greater than $f_{\min}$.

Assuming that the expected value of capacitance C range from $C'_{\mathrm{T,min}}$ to $C_{\mathrm{T,max}}$, where $C'_{\mathrm{T,min}}$ is the minimum value of the actual capacitance; ΔC is the differential capacitance value of $C_{\mathrm{T,max}}$ and $C'_{\mathrm{T,min}}$. Then, the actual value range of C can be expressed as

$$\frac{C_B}{A_T} - \Delta C \le C \le \frac{C_B}{A_T} \tag{15}$$

When $C'_{\mathrm{T,min}} = C_{\mathrm{T,min}}$, the actual operating frequency f of the converter can be expressed as

$$f = \frac{C_{\mathrm{B}} f_{\min}}{C_{\mathrm{B}} - A_T \Delta C} \tag{16}$$

In Equation (16), it can be seen that the actual switching frequency f of the Buck converter is related to the temperature coefficient A_{T}, the minimum switching frequency $f_{\min}$ and the differential capacitance value ΔC. The actual switching frequency f is increases with the increase of ΔC.

4. Optimization Design Method and Design Example

4.1. Optimization Design Method

An optimal design method of output for an intrinsically safe Buck converter based on minimum switching frequency and considering temperature characteristics is proposed. The specific design process is shown in Figure 4.

First of all, the critical ignition capacitor C_{B} corresponding to the output voltage V_{o} can be determined according to the critical ignition curve of the simple capacitor circuit. A_{T} and B_{T} are obtained according to the temperature characteristic curve of the capacitor.

Then, the minimum switching frequency of the converter is calculated according to Equation (14). Based on the expected capacitance variation ΔC, the range of the actual capacitance C is determined according to Equation (15). The actual operating frequency is calculated according to Equation (16).

Finally, the value of inductance is determined according to the output ripple voltage index.

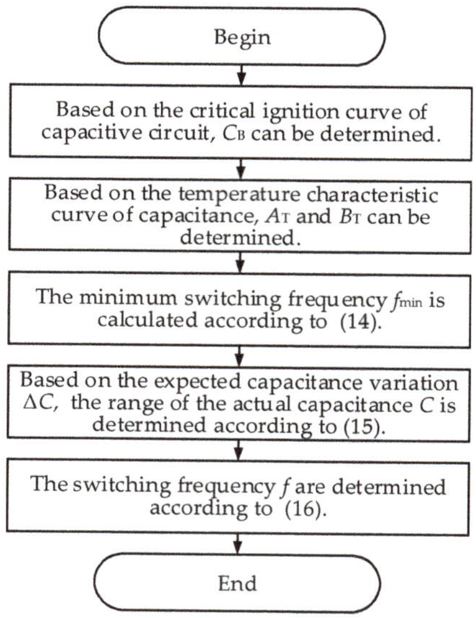

Figure 4. Proposed design method of an intrinsically safe Buck converter.

4.2. Design Examples

An output intrinsically safe Buck converter prototype was designed to verify the correctness of the above theoretical analysis and the proposed design method. We suppose that the specific indexes of the prototype are V_i = 20–24 V, V_o = 18 V, $V_{pp,max}$ = 1%V_o, R_L = 9–60 Ω. The ambient temperature ranges from −25 °C to 85 °C.

In order to verify the feasibility of the proposed design method, three groups of different parameters are set for experiments. According to the design process shown in Figure 4, the specific parameters are designed as follows:

(1) Determining the values of C_B, A_T and B_T.

When V_o = 18 V, the corresponding critical ignition capacitance C_B is 50 μF according to the critical ignition curve of a simple capacitor circuit.

Aluminum electrolytic capacitance is selected as the output filter capacitor. The temperature characteristics curve [27] is shown in Figure 5.

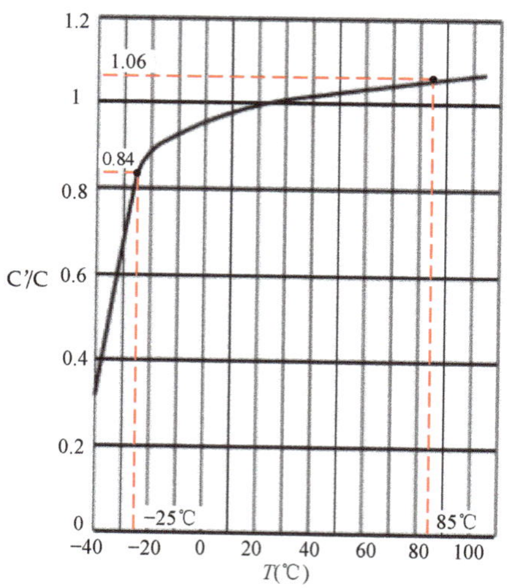

Figure 5. Temperature characteristics of aluminum electrolytic capacitor.

It can be seen from Figure 5 that A_T and B_T are about 1.055 and 0.84, respectively, within the temperature variation range (−25 °C to 85 °C).

(2) Solving the minimum switching frequency f_{min}.

According to Equation (13), the maximum capacitance $C_{T,max}$ is 47.4 μF. The fitting curve of the minimum actual capacitance $C_{T,min}$ versus frequency is shown in Figure 6.

Figure 6 shows that the minimum actual capacitance increases with the decrease of switching frequency. When the minimum actual capacitance is equal to the maximum actual capacitance, the minimum switching frequency considering the temperature characteristics is obtained. Substituting $R_{L,min}$ = 9 Ω, m = 0.01, A_T = 1.055, B_T = 0.84 and C_B = 50 μF into Equation (14), the minimum switching frequency f_{min} is 69.8 kHz.

(3) Determining the value range of the actual capacitance C and the switching frequency f.

We set the three different groups of ΔC as 5 μF, 20 μF and 30 μF, respectively. According to Equation (15), the calculated and actual values of capacitance C are shown in Table 1.

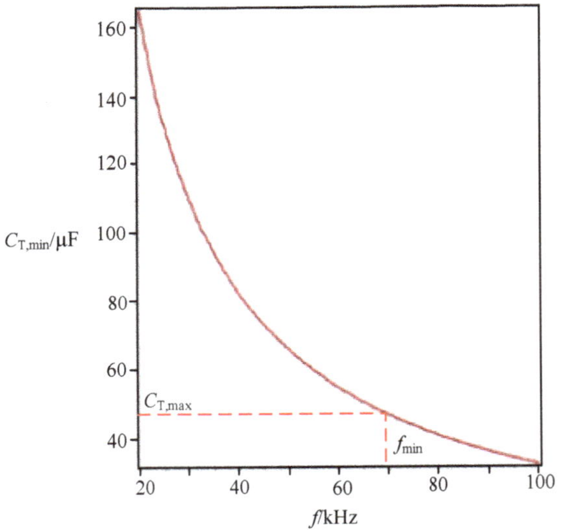

Figure 6. Fitting curve of the minimum actual capacitance versus frequency.

According to Equation (16), the calculated and actual values of the switching frequencies f are shown in Table 1, respectively.

Table 1. Calculated and actual values of the prototype.

Experimental Group	ΔC (μF)	Parameter	Calculated Value (Value Range)	Actual Value (Value Range)
Group 1	5	C (μF) f (kHz) L (μH)	(42.4, 47.4) 78 14.1	(43, 47) 80 15
Group 2	20	C (μF) f (kHz) L (μH)	(27.4, 47.4) 120.7 9.4	(28, 47) 120 10
Group 3	30	C (μF) f (kHz) L (μH)	(17.4, 47.4) 190.1 5.9	(18, 47) 190 7

(4) Determining the actual lower limit of inductance L.

When f is 80 kHz, 120 kHz and 190 kHz, the minimum values of the inductance L_{min} can be obtained as 14.1 μH, 9.4 μH and 5.9 μH.

It can be seen from Table 1 that the larger the capacitance variation range ΔC is, the larger the range of capacitance C is, the higher the switching frequency f is, and the smaller the required inductance L is. However, too large a frequency will lead to large switching loss and reduced efficiency. Therefore, it is not recommended to use excessively large ΔC in practical application.

(3) Analysis of experimental results. In order to verify the correctness of the above parameters, the output ripple voltage and safety spark experimental tests were carried out on the prototype, as shown in Figure 7.

Figure 7. The experimental prototype.

(1) Verification of output ripple voltage index.

At room temperature, the lower limits of the capacitor design range in Table 1 is taken as the filter capacitors of the converter. When the temperature drops to the lower limit of the given range (−25 °C), the peak-to-peak values of the output ripple voltage corresponding to the above three conditions are 173 mV, 159 mV and 152 mV, respectively. When the temperature rises to 85 °C, the output ripple voltage still meets the specified index requirements. The output ripple voltages under different temperatures are shown in Table 2 (considering temperature characteristics method) for the three groups.

Table 2. Output ripple voltage of the prototype.

Design Method	Parameter		Vpp/mV		
			−25 °C	25 °C	85 °C
Considering temperature characteristics	Group 1	f = 80 kHz C = 43 μF L = 15 μH	173	150	112
	Group 2	f = 120 kHz C = 28 μF L = 10 μH	159	142	108
	Group 3	f = 190 kHz C = 18 μF L = 7 μH	152	138	106
Without considering Temperature characteristics	Group 4	f = 80 kHz C = 35 μF L = 16 μH	191	172	166
	Group 5	f = 80 kHz C = 48 μF L = 16 μH	158	131	122

It can be seen from Table 2 that the output ripple voltage of the Buck converter increases with the decrease of temperature, when the lowest temperature is −25 °C, the output ripple voltage is the maximum value, and it is less than the given index. It indicates that even for the lower limit of capacitance value, the output ripple voltage still meets the design requirements when the lowest temperature is taken. With the increase of the capacitance or the temperature, the output ripple voltage of the converter is less than the limit value of the maximum output ripple voltage.

(2) Safety spark test and verification of intrinsic safety performance.

The intrinsically safe spark test device is shown in Figure 8.

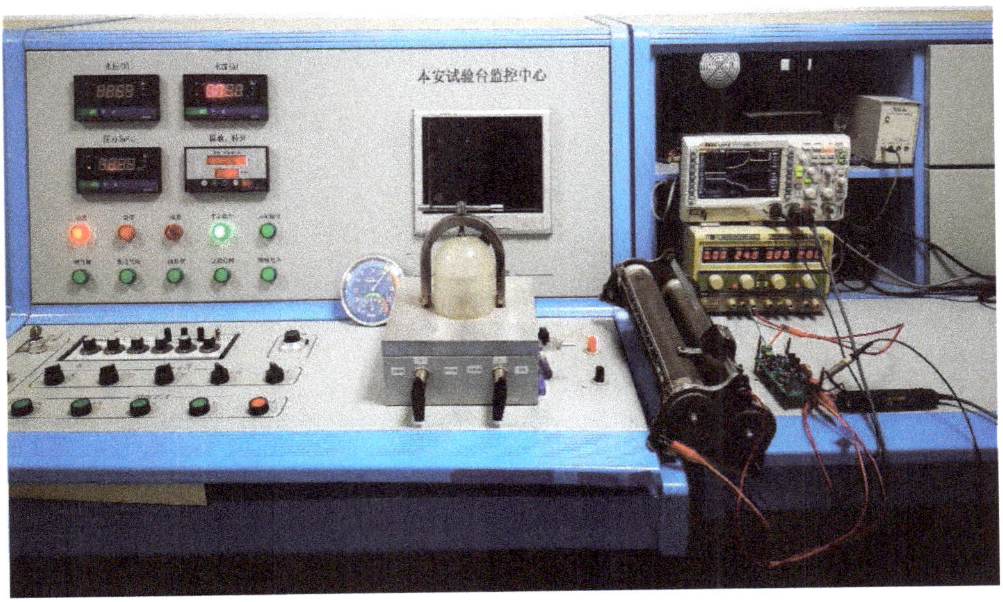

Figure 8. Intrinsically safe spark test device.

According to the curve of capacitance change rate shown in Figure 5, it is found that the capacitance increases gradually with the increase of temperature. Therefore, the upper limit of the capacitance design range at room temperature is taken as the filter of the converter considering the most dangerous situation. The prototype was placed in a humidifier to adjust its temperature to 85°C, and the safety spark test was carried out on the safety spark test device based on IEC standard.

The experimental results show that the specified explosive gas was not ignited. It indicates that the designed prototype of the Buck converter can meet the intrinsically safe requirements even at the specified highest temperature when the maximum capacitance is taken. Therefore, any capacitor in the range of capacitance value considering temperature characteristics obtained from Equation (15) can make the converter meet the output intrinsic safety requirements under a given temperature condition. The corresponding safety spark test results at different temperatures are shown in Table 3 (considering temperature characteristics method).

Table 3. Safety spark test results of the prototype.

Design Method	Parameter		Safety Spark Test		
			−25 °C	25 °C	85 °C
Considering temperature characteristics	Group 1	f = 80 kHz C = 47 μF L = 15 μH	safe	safe	safe
	Group 2	f = 120 kHz C = 47 μF L = 10 μH			
	Group 3	f = 190 kHz C = 47 μF L = 7 μH			
Without considering temperature characteristics	Group 4	f = 80 kHz C = 35 μF L = 16 μH	safe	safe	safe
	Group 5	f = 80 kHz C = 48 μF L = 16 μH	safe	safe	unsafe

(3) Comparative analysis and discussion.

In order to compare and analyze the influence of temperature on the electrical performance and intrinsic safety requirements of the converter, two groups of different parameters were set to test the Buck converter without considering the temperature characteristics of the capacitor. According to (11), the value range of capacitance is (34.7 μF, 50 μF) when f is 80 kHz. The minimum inductance L_{min} is 14 μH. The values of inductance, capacitance and their experimental results are shown in Table 3 (without considering temperature characteristics).

It can be seen from Tables 2 and 3 (without considering temperature characteristics) that the selected capacitance of Group 4 is close to the minimum value, and the capacitance of Group 5 is close to the maximum value. When the temperature is reduced to −25 °C, the output ripple voltage of Group 4 is greater than the specified requirement. However, when the temperature rises to 85 °C, Group 5 does not meet the output intrinsic safety requirements. It indicates that the value range of capacitance without considering the temperature characteristics is not reliable in a certain temperature range.

Therefore, in order to obtain a reliable converter parameter design range that meets the requirements of the specified output intrinsic safety and output ripple voltage indexes, the temperature characteristics of the capacitor must be considered within a given range of temperature when designing the intrinsically safe Buck converter.

5. Conclusions

A reliable design method of an output intrinsic safety Buck converter considering the temperature characteristics is proposed. The conclusions are as follows.

(1) The actual capacitance design range satisfying the output ripple voltage and output intrinsic safety requirement was obtained in a certain temperature variation range. It was concluded that the maximum actual capacitance allowed by the output intrinsic safety requirement is independent of operating frequency, and the minimum actual capacitance meeting output voltage ripple requirement increases with the decrease of operating frequency.

(2) It is pointed out that there must be a minimum switching frequency, which makes the converter meet both the output voltage ripple and output intrinsic safety requirement. The expression of a minimum switching frequency was derived through letting the minimum actual capacitance equal the maximum actual capacitance. Considering the influence of temperature on both the output voltage ripple and output intrinsic safety requirement, the actual operating frequency can be obtained.

(3) An optimal design method of an output intrinsically safe Buck converter considering temperature characteristics was proposed in a given temperature variation range.

This design method can also be applied to other switching converters. It plays an important role in the development and application of intrinsically safe converters.

Author Contributions: Project administration, Y.L.; supervision, S.L.; validation, Y.S.; writing—original draft, Y.L.; writing—review and editing, Y.L., Z.L. and Y.Z. All authors have read and agreed to the published version of the manuscript.

Funding: This work was supported by the National Natural Science Foundation of China under Grant 51777167.

Conflicts of Interest: The authors declare no conflict of interest.

References

1. Meng, Q.; Tian, Y. Analysis of Potential Hazards of Analog Energy Storage Components in the Intrinsic Safety Circuits and their Intrinsic Safety Criteria. *Trans. China Electrotech. Soc.* **2022**, *37*, 676–685.
2. Standardization Administration of China. *Explosive Atmospheres—Part 4: Equipment Protection by Intrinsic Safety "i":GB 3836.4-2010*; China Standards Press: Beijing, China, 2021.
3. Lu, W.; Li, Y. Research of the Key Technologies for Explosion-Proof and Intrinsically Safe High Voltage Frequency Converter. *Appl. Mech. Mater.* **2014**, *494*, 1552–1555. [CrossRef]
4. Liu, S.; Wu, H.; Wang, C.; Nie, S.; Qi, L. Inner Intrinsically Safe Criterion and Design Considerations of Buck Converter Based on Equivalent Inductance. *Electr. Power Compon. Syst.* **2019**, *47*, 248–260. [CrossRef]
5. Liu, S.; Wang, H.; Hu, C.; Zhang, H.; Wang, B. Analysis of the Optimal Operating Mode of Auxiliary Inductance and LC Parameter Optimization Design of Forward-Flyback Converter with Additional LC. *Trans. China Electrotech. Soc.* **2022**, *37*, 389–396.
6. Cao, X.; Chen, W.; Ning, G.; Zhang, K.; Song, Z.; Qiao, G.; Sun, L. Optimization Design of High-power High-frequency Transformer Based on Multi-objective Genetic Algorithm. *Proc. CSEE* **2018**, *38*, 1348–1355.
7. Liu, S.; Ma, Y.; Wen, X.; Qi, L. Research on the Most Dangerous Output Short-circuit Discharge Conditions of Output Intrinsic Safety Buck-Boost Converters. *Trans. China Electrotech. Soc.* **2015**, *30*, 253–260.
8. Liu, S.L.; Cui, Q.; Li, Y. Output short-circuit spark discharging energy and output intrinsic safety criterion of Buck converters. *Acta Phys. Sin.* **2013**, *62*, 438–447.
9. Shen, K.; Wang, J.; Ban, M.; Ji, Y.; Cai, X. Staircase modulation scheme based on PSO for modular multilevel converter. *Electr. Power Autom. Equip.* **2014**, *34*, 78–83, 89.
10. Huang, Y.; Zhang, X.; Wu, L. Optimization of BBMC main circuit parameters based on particle swarm optimization. *J. Hunan Univ. Sci. Technol. (Nat. Sci. Ed.)* **2014**, *29*, 89–92.
11. Zhao, B.; Wang, G.; Song, J.; Liu, Y. Optimal Design Method of the LCLC Resonant Converter Based on Particle-Swarm-Optimization Algorithm. *J. Electron. Inf. Technol.* **2021**, *43*, 1622–1629.
12. Wang, K.; Zang, Q.-J. Application of MOPSO Algorithm in Optimal Design for Boost Converter. *J. Yantai Univ. (Nat. Sci. Eng. Ed.)* **2017**, *30*, 317–322.
13. Nemec, M.; Ambrozic, V.; Rihar, A.; Zajec, P. DC Capacitor Testing Capability Intrinsic to Multi-leg Converters. In Proceedings of the 2020 IEEE 14th International Conference on Compatibility, Power Electronics and Power Engineering, Setubal, Portugal, 8–10 July 2020; Volume 22, pp. 94–99.
14. Huang, J.; Li, L.; Ren, S.; Liu, S. Analysis and Design of an Intrinsically Safe Buck-Boost Converter on Considering of the Filter Capacitor with Equivalent Series Resistance. *Trans. China Electrotech. Soc.* **2021**, *36*, 1658–1670.
15. Liu, S.; Hao, Y.; Li, Y.; You, M. Design Methods of Intrinsically Safe Buck Converter Based on the Maximum Output Power. *Trans. China Electrotech. Soc.* **2021**, *36*, 542–551.
16. Zhou, D.; Wang, H.; Blaabjerg, F. Mission profile based system-level reliability analysis of DC/DC converters for a backup power. *Appl. IEEE Trans. Power Electron.* **2018**, *33*, 8030–8039. [CrossRef]
17. Corti1, F.; Reatti, A.; Patrizi, G.; Ciani, L.; Catelani, M.; Kazimierczuk, M.K. Probabilistic evaluation of power converters as support in their design. *IET Power Electron.* **2020**, *13*, 4542–4550. [CrossRef]
18. Liu, Y.; Wang, H.; Huang, M.; Zha, X.; Gong, J.; Sun, J. Reliability-oriented optimization of theLC filter in a buck DC-DC converter. *IEEE Trans. Power Electron.* **2017**, *32*, 6323–6337. [CrossRef]
19. Owen, H.; Wilson, T.; Feng, S.; Lee, F. A computer-aided design procedure for flyback step-up DC-to-DC converters. *IEEE Trans. Magn.* **1972**, *8*, 289–291. [CrossRef]
20. Yang, Y.; Qi, L.; Li, L. Analysis of Output Ripple Voltage of Essential Safety for Interleaving Magnetics Buck Converter. *Trans. China Electrotech. Soc.* **2014**, *29*, 181–188.
21. Gizatullin, F.-A.; Salikhov, R.-M.; Efimenko, N.-V.; Karimova, A.G.; Demin, A.U. Instrumentation for modeling of discharge processes in ignition capacitive systems. *J. Theor. Appl. Phys.* **2019**, *13*, 263–267. [CrossRef]
22. Hasanpour, S.; Siwakoti, Y.; Blaabjerg, F. Analysis of a New Soft-Switched Step-Up Trans-Inverse DC/DC Converter Based on Three-Winding Coupled-Inductor. *IEEE Trans. Power Electron.* **2022**, *37*, 2203–2215. [CrossRef]

23. Liu, S.; Liu, J.; Kou, L.; Zhong, J. Analysis of Output Ripple Voltage of Buck DC/DC Converter and Its Application. *Trans. China Electrotech. Soc.* **2007**, *2*, 91–97.
24. Kang, Q.; Xu, C.; Tian, M.; Song, J. Design of high-power intrinsically safe power supply based on fault current change rate. *Ind. Mine Autom.* **2021**, *47*, 6–12.
25. Liu, J.; Liu, S.; Yang, Y.; Zhang, Y.-M. Output Intrinsically Safe Behavior of Buck Converters and Its Optimal Design. *Proc. CSEE* **2005**, *19*, 52–57.
26. Yu, Y.; Xie, D.Y.; Li, S.G.; Wu, X.J. Summary of Intrinsic Safety Electric Circuit Technology. *Coal Sci. Technol.* **2011**, *39*, 61–65.
27. Corporation TDK. Aluminum Electrolytic Capacitors General Technical Information. Tokyo, Japan. 2021. Available online: https://www.tdk-electronics.tdk.com.cn/download/540988/6ad5ed9e1ff4f727c328cb92da2adf2b/pdf-generaltechnicalinformation.pdf (accessed on 23 April 2021).

energies

MDPI

Article

Steady-State Analysis and Optimal Design of an LLC Resonant Converter Considering Internal Loss Resistance

Jeong-Sang Yoo [1,*], Yong-Man Gil [1] and Tae-Young Ahn [2]

1 Department of Electronics Engineering, Cheongju University, Cheongju 28503, Korea
2 Department of Electrical & Control Engineering, Cheongju University, Cheongju 28503, Korea
* Correspondence: jeongsangyoo@gmail.com

Abstract: In this paper, a steady-state model of an LLC resonant half-bridge converter with internal loss resistance is proposed, in order to maximize power conversion efficiency, and steady-state characteristic equations of DC voltage gain and input impedance are derived for the optimal design of the converter. First, to confirm the validity of the steady-state characteristic equation and the optimal design process, a prototype converter with a maximum output of 2 kW was designed. Through comparison of simulation, calculation, and experimental results obtained from the prototype test, it is shown that the calculation results proposed in this paper were closer to the experimental results than the calculation results obtained under the lossless condition. In addition, the relationship between the switching frequency and the load current of the prototype was compared, in order to determine the operating range of the switching frequency, which is important in the converter design stage. In this case, it was confirmed that the calculated value reflecting the internal loss showed a close result. In conclusion, we confirm the usefulness of the analysis results reflecting the internal loss resistance proposed in this paper and the optimal design process.

Keywords: LLC resonant converter; steady-state analysis; internal power losses; AC equivalent circuit; internal resistance; DC voltage gain

Citation: Yoo, J.-S.; Gil, Y.-M.; Ahn, T.-Y. Steady-State Analysis and Optimal Design of an LLC Resonant Converter Considering Internal Loss Resistance. *Energies* **2022**, *15*, 8144. https://doi.org/10.3390/en15218144

Academic Editor: Salvatore Musumeci

Received: 28 September 2022
Accepted: 26 October 2022
Published: 1 November 2022

1. Introduction

The climate crisis has been steadily emerging, where the biggest cause of climate change is known to be greenhouse gas and carbon emissions. There is an urgent need for low-loss power conversion technology that can reduce carbon emissions in the process of generating, distributing, and consuming power energy. Therefore, the technical demand for a high-efficiency power conversion device for the efficient use of limited electric energy is increasing. A switching power supply with AC input can be divided into an AC-DC rectifier and a DC-DC converter. In particular, the DC-DC converter uses various types of circuit methods, based on the position of the power semiconductor switch and the structure of the transformer. Among the insulated circuit methods using transformers, the LLC resonant converter, which is suitable for medium- to high-capacity scenarios, is a circuit method that has attracted attention in various industrial fields, due to its advantages such as low loss, high efficiency, and small number of parts needed [1–3]. The LLC resonant converter has low switching loss as well as low voltage and current surges, as both the main switch and the rectification switch operate in a soft-switching condition. In addition, it is known that the effective current value of the device is small, such that the conduction loss is small and, thus, high-efficiency power conversion is achievable [4,5].

Figure 1 shows the basic circuit structure of a half-bridge LLC resonant converter. In the figure, the main switches of the converter are S1 and S2, and the two switches are controlled with a fixed duty of 50% and a variable switching frequency. On the primary side of the transformer, there are three resonant components: The resonant capacitor C_R, resonant inductor L_R, and magnetizing inductor L_M. Meanwhile, the rectifying diode, output capacitor, and load resistor are shown on the secondary side. Figure 2 shows the operation waveform of the LLC resonant converter operating in steady-state. As can be seen from the figure, the primary side voltage v_p of the transformer switches its polarity with the same area size due to the secondary side output voltage V_o of the transformer, while the current i_{LM} of the magnetizing inductor, which is configured in parallel, operates as a triangular waveform with a constant slope [6–8]. As the current resonates with the two resonant components C_R and L_R, the main switch and the diode show zero current switching (ZCS) characteristics and so the LLC resonant converter has low switching loss and high-efficiency power conversion characteristics. Meanwhile, due to the resonance waveform having the same sized area, the magnetic flux density of the core uses both positive and negative polarities for the primary current of the transformer. Therefore, the core size and core loss may be reduced by increasing the core utilization [9–11].

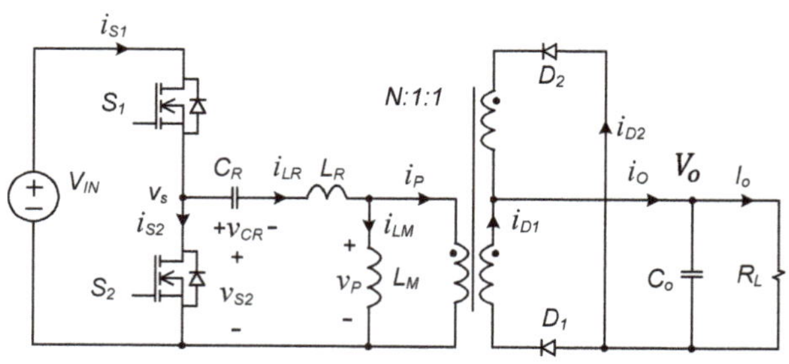

Figure 1. Basic circuit structure of half-bridge LLC resonant converter.

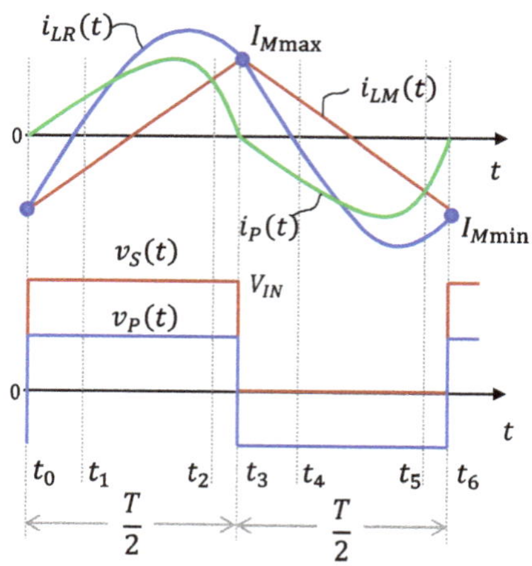

Figure 2. Operating waveform of LLC resonant converter in steady-state.

In general, the steady-state characteristic design formula used in the circuit design process of LLC resonant converters is based on an ideal equivalent model that does not reflect internal power loss. For this reason, it is well-known that there exists a difference between the design result of an LLC resonant converter and the operating characteristics of the actual circuit. In particular, errors occur in the operating switching frequency range, which is an important criterion when designing a converter [12,13]. In addition, these errors are the biggest obstacle to the optimal design of important devices, such as transformers and inductors, whose basic characteristics change depending on the operating frequency.

In this paper, an AC equivalent circuit model considering the internal loss resistance of the LLC resonant converter is proposed. From the proposed AC equivalent circuit, important steady-state characteristic equations necessary for converter design, such as

DC voltage gain and input impedance, are derived. In order to verify the validity of the results of the equivalent circuit and the steady-state characteristic equation, the calculation result of the steady-state characteristic, as well as those of the simulation and experimental circuit, were compared and confirmed [14–17]. In particular, the steady-state characteristic equation was applied to the optimal design process of the LLC resonant converter, in order to determine the main element values, and the results were used in an experimental circuit with a resonant frequency of 125 kHz and a maximum output of 2 kW.

2. Steady-State Analysis of LLC Resonant Converter

2.1. AC Equivalent Circuit

Figure 3 shows the equivalent circuit expressing the series parasitic resistances of each element in the LLC resonant converter of Figure 1. In the figure, the main switch is an ideal switch, the output capacitor and the load resistance are equivalent to a constant voltage, and the internal parasitic resistance is shown in series with the important components. In particular, the equivalent circuit in Figure 3 is divided into two equivalent circuits for each state, according to the operating state of switches S1 and S2 when operating in steady-state, as shown in Figure 4. In general, a non-switching linear equivalent circuit is required to derive the steady-state characteristic expression reflecting the internal loss resistance. In a previous study, the fundamental harmonic approximation (FHA) approximation method [18,19], which could approximate a square wave by a sine wave, has been used. In this paper, the internal power loss resistance is reflected in the circuit, and the input and output voltages are converted using an FHA approximation method to propose an equivalent circuit with loss resistance, as shown in Figure 5. At this time, existing parasitic resistors in each major element in Figure 3 could be simplified to three series resistors r_1, r_2, and r_3, as shown in the equivalent circuit of Figure 5. The load resistance R_L is expressed as the AC equivalent resistance R_{AC} converted to the primary side of the transformer, as in Equation (1):

$$R_{AC} = \frac{8N^2}{\pi^2} R_L. \tag{1}$$

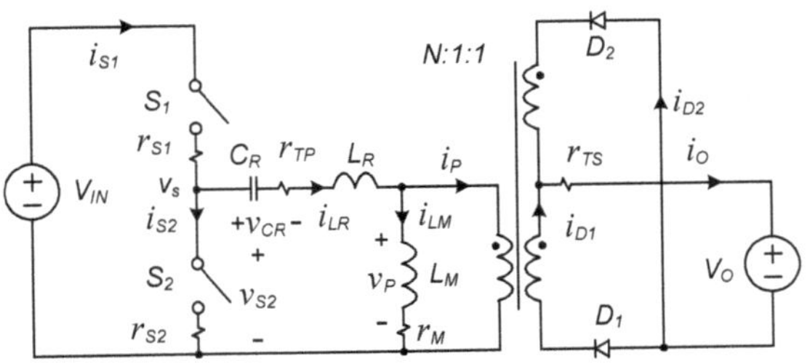

Figure 3. Equivalent circuit of LLC resonant converter with internal resistance.

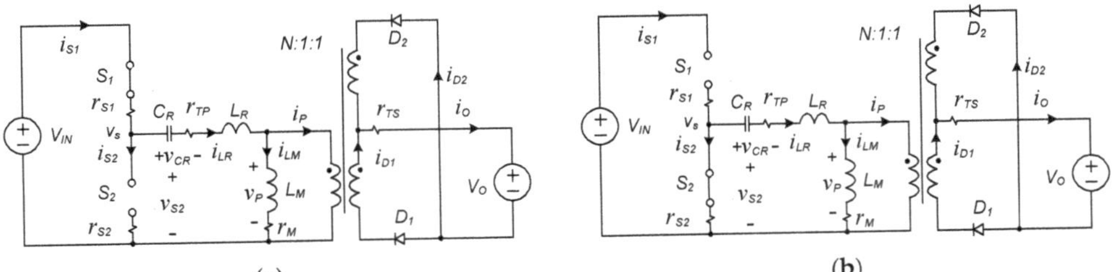

(a) (b)

Figure 4. Equivalent circuit for each state of the LLC resonant converter in steady-state: (**a**) When switch S1 is turned on; and (**b**) when switch S2 is turned on.

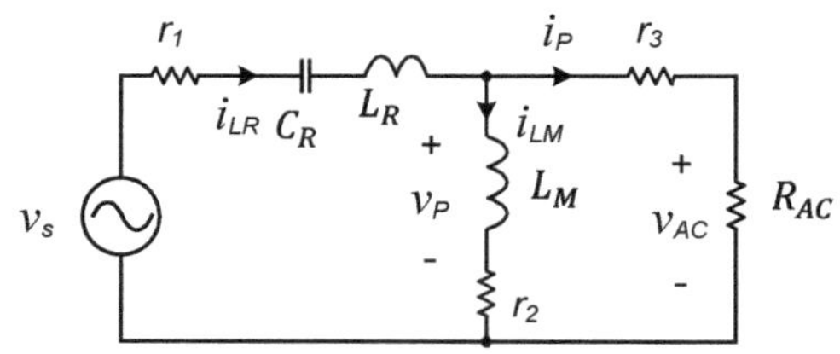

Figure 5. AC equivalent circuit of LLC resonant converter with internal loss resistance.

2.2. DC Conversion Ratio

First, Equations (2)–(6) are defined to derive the steady-state characteristic expression:

$$K_L = \frac{L_M}{L_R}, \tag{2}$$

$$Z_o = \sqrt{\frac{L_R}{C_R}}, \tag{3}$$

$$\omega_o = \frac{1}{\sqrt{L_R\, C_R}}, \tag{4}$$

$$\omega_n = \frac{\omega_f}{\omega_o}, \tag{5}$$

$$Q = \frac{Z_o}{R_{AC}}, \tag{6}$$

where

Z_o denotes the characteristic impedance;
L_M denotes the magnetizing inductance;
L_R denotes the resonant inductance;
K_L denotes the inductance ratio;
C_R denotes the resonant capacitor;
ω_o denotes the resonant angular frequency;
ω_n denotes the angular normalized frequency;
ω_f denotes the angular switching frequency.

In order to obtain the DC voltage gain in the steady-state for the AC equivalent circuit of Figure 5, the voltage gain M is defined as in Equation (7). At this time, the ratio between the input voltage $v_s(t)$ and the output voltage $v_{AC}(t)$ can be calculated through the impedance ratio of the equivalent circuit, and the result is shown as Equation (8), where the internal resistance r_K used in the equation is given in Equation (9).

$$M = \frac{v_{AC}(t)}{v_s(t)} \tag{7}$$

$$M = \frac{\left(\frac{r_2}{Z_o}\right) + j\omega_n K_L}{K_L\left(\frac{r_K}{Z_o^2 K_L} + 1 - \omega_n^2\right)Q + j\omega_n\left[\left(\frac{r_2+r_3}{R_{AC}} + 1\right) + \left(\frac{r_1+r_3}{R_{AC}} + 1\right)K_L - \frac{1}{\omega_n^2}\left(\frac{r_2+r_3}{R_{AC}} + 1\right)\right]} \tag{8}$$

$$r_K = r_1 r_2 + r_2 r_3 + r_3 r_1 + r_1 R_{AC} + r_2 R_{AC}, \tag{9}$$

In order to calculate only the magnitude of the input and output voltage gain M in Equation (8), the absolute values of both sides need to be obtained, and the result is shown in Equation (10). Meanwhile, it is expected that the products of the parasitic resistances $r_1 r_2$, $r_2 r_3$, and $r_3 r_1$ among the internal resistances of Equation (9) are very small; if the values of the products of the parasitic resistances are assumed to be zero, then Equation (11) can be derived. Assuming that each parasitic resistance is equal to the same resistance r, as in

Equation (12), it is arranged as Equation (13). At this time, the ratio between the internal parasitic resistance r and the characteristic impedance Z_o is defined as the internal loss equivalent resistance R_K, expressed as in Equation (14).

$$|M| = \frac{\sqrt{\left(\frac{r_2}{Z_o}\right)^2 + \omega_n^2 K_L^2}}{\sqrt{K_L^2\left(\frac{r_K}{Z_o^2 K_L} + 1 - \omega_n^2\right)^2 Q^2 + \omega_n^2\left[\left(\frac{r_2+r_3}{R_{AC}} + 1\right) + \left(\frac{r_1+r_3}{R_{AC}} + 1\right)K_L - \frac{1}{\omega_n^2}\left(\frac{r_2+r_3}{R_{AC}} + 1\right)\right]^2}}. \tag{10}$$

$$r_K \approx (r_1 + r_2)R_{AC} \tag{11}$$

$$r_1 \approx r_2 \approx r_3 \approx r \tag{12}$$

$$\frac{r_2 + r_3}{R_{AC}} \approx \frac{r_1 + r_3}{R_{AC}} \approx \frac{r_1 + r_2}{R_{AC}} \approx \frac{2r}{R_{AC}} = 2Q\left(\frac{r}{Z_o}\right) = 2QR_K \tag{13}$$

$$R_K = \frac{r}{Z_o} \tag{14}$$

From the above assumptions and approximations, the DC input and output voltage ratio M of Equation (10) can be arranged as Equations (15) and (16). If the internal loss equivalent resistance R_K is 0, it becomes Equation (17). Equation (17) is the same as the DC input and output voltage gain equation obtained from the existing lossless AC equivalent circuit. Therefore, Equation (16) can be used as the steady-state DC voltage gain characteristic expression of the LLC resonant converter reflecting the internal loss resistance depicted in Figure 3.

$$M = \frac{2NV_o}{V_{IN}} \tag{15}$$

$$M = \frac{\sqrt{1 + \left(\frac{R_K}{\omega_n K_L}\right)^2}}{\sqrt{(2QR_K + 1)^2\left[1 + \frac{1}{K_L}\left(1 - \frac{1}{\omega_n^2}\right)\right]^2 + \left(\frac{2R_K}{\omega_n Q K_L} + \frac{1}{\omega_n} - \omega_n\right)^2 Q^2}} \tag{16}$$

$$M = \frac{1}{\sqrt{\left[1 + \frac{1}{K_L}\left(1 - \frac{1}{\omega_n^2}\right)\right]^2 + \left(\frac{1}{\omega_n} - \omega_n\right)^2 Q^2}} \tag{17}$$

2.3. Frequency Characteristics of Voltage Gain

The frequency characteristics of the DC voltage gain in Equation (15) for the LLC resonant converter reflecting the internal loss resistance were analyzed. Table 1 shows each convergence value for the normalized angular velocity ω_n in Equation (15). First, when $Q = 0$, the load current is 0 from Equations (1) and (6) and in this case, the LLC resonant converter is in a no-load state. When the normalized angular velocity ω_n approaches 0, the voltage gain is 0; when the normalized angular velocity ω_n approaches infinity, the voltage gain converges to a constant value $\frac{K_L}{K_L+1}$, expressed in terms of the inductance ratio K_L. In addition, when the normalized angular velocity is 1, it is expressed as a value reflecting the internal loss equivalent resistance R_K; it can be seen that the voltage gain infinitely increases when the normalized angular velocity is a specific value $\frac{1}{\sqrt{K_L+1}}$. When the Q value is large, the voltage gain is 0 at a low normalized angular velocity; when the normalized angular velocity ω_n is 1, it can be seen that the voltage gain is expressed as a value reflecting the value of the internal loss resistance. In particular, when the normalized angular velocity is 1 and the internal loss resistance value is 0 in both cases, the voltage gain becomes 1. This result is the same as the frequency response characteristic of the existing voltage gain [20–22].

Table 1. DC voltage gain M of converter according to each normalized frequency ω_n.

Conditions	Q = 0	Q = ∞
$\omega_n = 0$	$M = 0$	$M = 0$
$\omega_n = \frac{1}{\sqrt{K_L + 1}}$	$M = \infty$	-
$\omega_n = 1$	$M = \sqrt{\left(\frac{R_K}{K_L}\right)^2 + 1}$	$M = \frac{\sqrt{1 + \left(\frac{R_K}{K_L}\right)^2}}{2QR_K + 1}$
$\omega_n = \infty$	$M = \frac{K_L}{K_L + 1}$	$M = 0$

Figure 6 shows the frequency characteristics of the voltage gain when the inductance ratio K_L is 4. Figure 6a shows when Q is 0, and Figure 6b shows when Q is 3. Figure 6a shows the frequency characteristics of the LLC resonant converter voltage gain under the no-load condition. As can be seen from the figure, as for the conventional voltage gain, when the normalized angular velocity is $\frac{1}{\sqrt{K_L+1}}$, the voltage gain increases infinitely; meanwhile, when the normalized angular velocity increases, the voltage gain converges to $\frac{K_L}{K_L+1}$. However, when ω_n is 1, M has a value greater than 1, due to internal loss. When Q is 3, unlike the existing frequency characteristics, when ω_n is 1, the voltage gain is not 1 but, instead, a peak value smaller than 1 reflecting the internal loss resistance [23,24]. The frequency characteristic in Figure 6 is the same as the convergence value in Table 1.

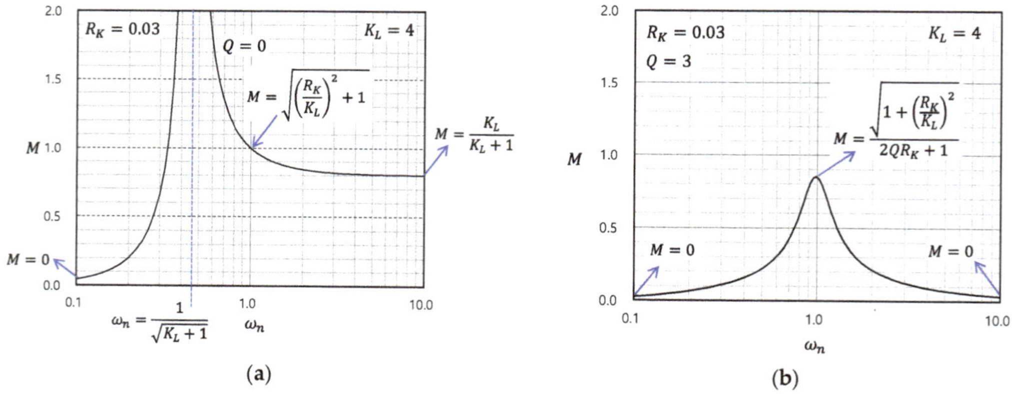

Figure 6. DC voltage gain M of LLC resonant converter according to each normalized frequency ω_n: (**a**) When Q is 0; and (**b**) when Q is large.

Figure 7 shows the voltage gain frequency characteristics for several Q values. Figure 7a shows a characteristic graph when the internal loss resistance R_K is 0, and Figure 7b shows a characteristic graph when the internal loss resistance R_K is 0.03. As can be seen from Figure 7b, when ω_n is 1, if there is an internal loss resistance, the voltage gain does not reach 1, and it can be seen that the peak value decreases as Q increases. Figure 8 shows the voltage gain obtained by comparing the calculated value and the experimental value when the inductance ratio K_L is 1.4, the transformer turn ratio N is 1.05, and the characteristic impedance Z_o is 61.24. In the figure, Q is set as 0.69, 1.37, or 6.85, and when ω_n is 1, it can be seen that the voltage gain does not reach 1 and the peak value decreases as Q increases. This result is consistent with the above description, and the fact that the experimental value and the calculated value were consistent (within a certain range) supports the validity of the steady-state characteristic expression of Equation (15).

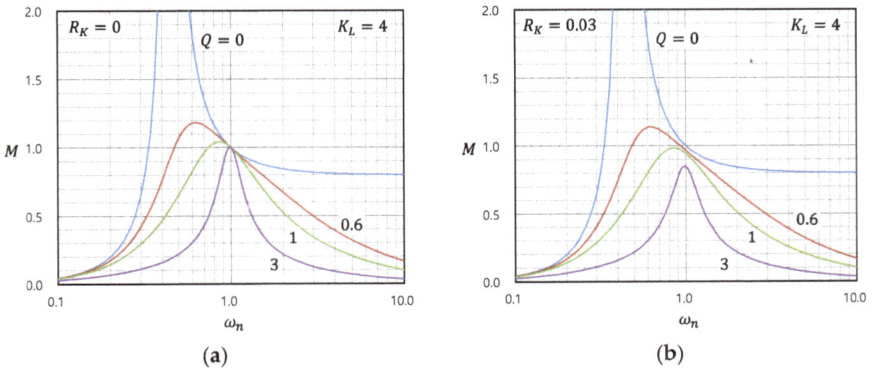

Figure 7. Frequency characteristics of DC voltage gain M for LLC resonant converter according to internal loss resistance R_K: (**a**) When the internal loss resistance R_K is 0; and (**b**) when R_K is 0.03.

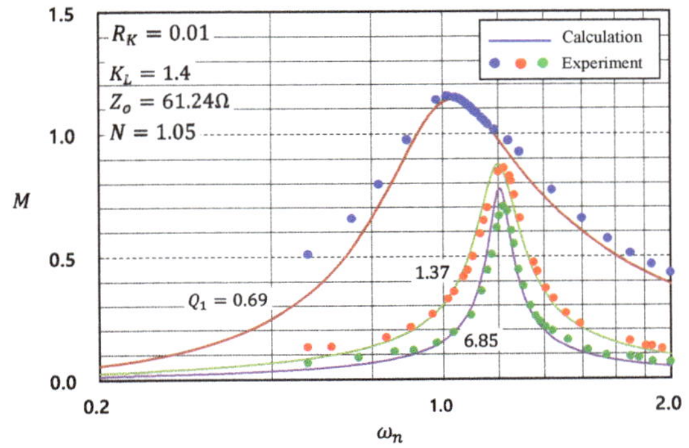

Figure 8. Frequency characteristics of the DC voltage gain M.

2.4. Input Impedance Characteristics

Figure 9 shows the equivalent circuits used to analyze the input impedance characteristics of the AC equivalent circuit with internal loss resistance. Figure 9a is the circuit used to find the input impedance for the AC equivalent circuit of Figure 5, and Figure 9b is the equivalent circuit in which the components in each branch are expressed by their respective impedances. Equation (18) was used to derive the normalized input impedance Z_N from Figure 9a. In this case, the normalized input impedance Z_N is the ratio of the input impedance Z_{IN} and the characteristic impedance Z_o. If Equation (18) is divided into real and imaginary terms and the absolute value is taken, Equation (19) is obtained [25,26]. If the internal resistance is set to 0 in Equation (19), Equation (20) is obtained, and it can be seen that this is the same as the existing lossless input impedance characteristic equation. In order to analyze the frequency characteristics of the normalized input impedance Z_N, Equations (21) and (22) were obtained by substituting zero and infinity for Q, respectively. Table 2 summarizes the convergence results of the normalized input impedance Z_N, according to the normalized angular velocity ω_n. It was found that the normalized angular velocity ω_n, which becomes the lowest value of the normalized input impedance Z_N, differs depending on the Q value; in particular, each lowest value corresponds to about twice the internal equivalent resistance R_K. The junction of the two conditions can be derived from the conditional expression of Equation (23), and the normalized angular velocity ω_n and the normalized input impedance Z_N at this time are expressed by Equations (24) and (25), respectively. Figure 10 shows the frequency characteristic graph of Equation (19), and the results of Table 2 are reflected in the graph [27].

$$Z_N = \frac{Z_{IN}}{Z_o} = \left(R_K + j\omega_n - j\frac{1}{\omega_n}\right) + \left(\frac{R_K + j\omega_n(R_K Q + 1)K_L}{1 + 2R_K Q + j\omega_n Q K_L}\right) \tag{18}$$

$$|Z_N| = \sqrt{\left[R_K + \frac{R_K + \omega_n^2(R_K Q + 1)QK_L^2}{(1 + 2R_K Q)^2 + \omega_n^2 Q^2 K_L^2}\right]^2 + \left[\omega_n - \frac{1}{\omega_n} + \frac{\omega_n K_L}{(1 + 2R_K Q)^2 + \omega_n^2 Q^2 K_L^2}\right]^2} \tag{19}$$

$$|Z_n| = \left|\frac{Z_{IN}}{Z_o}\right| = \sqrt{\left(\frac{\omega_n^2 K_L^2 Q}{1 + \omega_n^2 K_L^2 Q^2}\right)^2 + \left(\omega_n - \frac{1}{\omega_n} + \frac{\omega_n K_L}{1 + \omega_n^2 K_L^2 Q^2}\right)^2} \tag{20}$$

$$|Z_N|_{Q=0} = \sqrt{4R_K^2 + \left(\omega_n - \frac{1}{\omega_n} + \omega_n K_L\right)^2} \tag{21}$$

$$|Z_N|_{Q=\infty} = \sqrt{\left(R_K + \frac{R_K \omega_n^2 K_L^2}{4R_K^2 + \omega_n^2 K_L^2}\right)^2 + \left(\omega_n - \frac{1}{\omega_n}\right)^2} \tag{22}$$

$$|Z_N|_{Q=0} = |Z_N|_{Q=\infty} \tag{23}$$

$$\omega_n = \sqrt{\frac{2}{K_L + 2}} \tag{24}$$

$$|Z_N| = \sqrt{4R_K^2 + \frac{K_L^2}{2K_L + 4}} \tag{25}$$

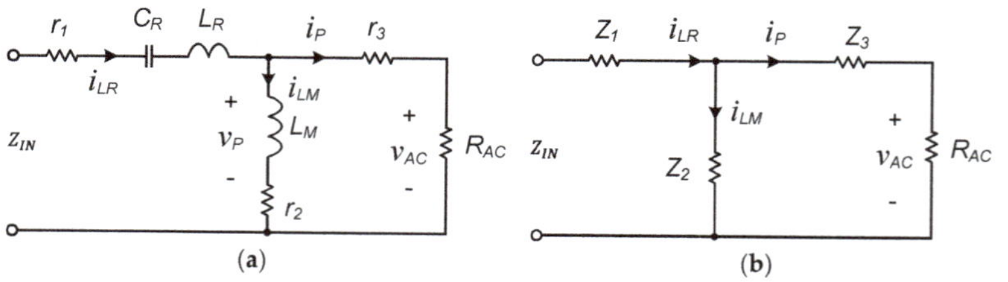

Figure 9. The input impedance Z_N of the AC equivalent circuits: (**a**) Equivalent circuit of input impedance Z_N; and (**b**) simplified equivalent circuit of input impedance Z_N.

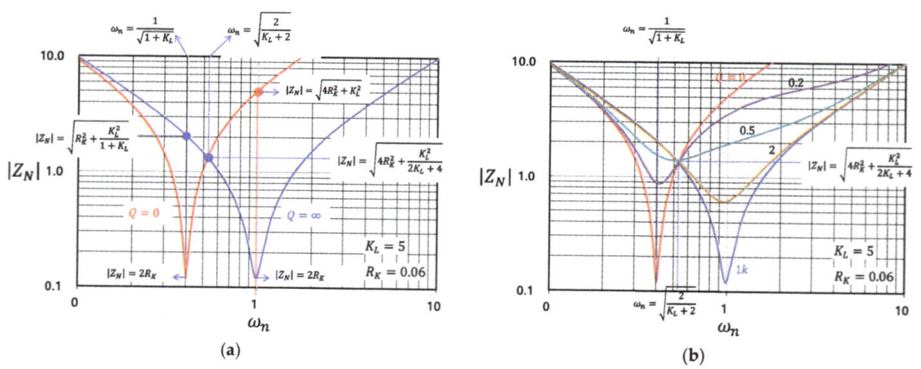

Figure 10. The frequency characteristic of the normalized input impedance $|Z_N|$: (**a**) At $Q = 0$ and infinity; and (**b**) for various values of Q.

Table 2. Normalized input impedance Z_N of the converter, according to each normalized frequency ω_n.

Conditions	Q = 0	Q = ∞
$\omega_n = 0$	$\lvert Z_N \rvert = \infty$	$\lvert Z_N \rvert = \infty$
$\omega_n = \frac{1}{\sqrt{K_L+1}}$	$\lvert Z_N \rvert = 2R_K$	$\lvert Z_N \rvert = \sqrt{R_K^2 + \frac{K_L^2}{1+K_L}}$
$\omega_n = 1$	$\lvert Z_N \rvert = \sqrt{4R_K^2 + K_L^2}$	$\lvert Z_N \rvert = 2R_K$
$\omega_n = \infty$	$\lvert Z_N \rvert = \infty$	$\lvert Z_N \rvert = \infty$

3. Optimal Design of LLC Resonant Converter

3.1. Optimal Design Process

The design process of an LLC resonant converter with internal loss, using the steady-state analysis results mentioned in the previous section, is detailed here. First of all, it is necessary to determine the maximum Q value in the DC voltage gain range required for design, as in the previous research. The peak value of M appearing at a specific Q value occurs at the lowest point of the normalized input impedance of Equation (26) and, in this case, it can be obtained under the condition that the imaginary term is 0, as in Equation (27). Equation (28), which summarizes Equation (27), is the normalized angular velocity $\omega_{n\ max}$ at which M becomes maximum for a specific Q value [28].

$$Z_N = \left[R_K + \frac{R_K + \omega_n^2(R_K Q + 1)QK_L^2}{(1+2QR_K)^2 + \omega_n^2 Q^2 K_L^2}\right] + j\left[\omega_n - \frac{1}{\omega_n} + \frac{\omega_n K_L}{(1+2QR_K)^2 + \omega_n^2 Q^2 K_L^2}\right] \tag{26}$$

$$\omega_{n\ max} - \frac{1}{\omega_{n\ max}} + \frac{\omega_{n\ max} K_L}{(1+2QR_K)^2 + \omega_{n\ max}^2 Q^2 K_L^2} = 0 \tag{27}$$

$$\omega_{n\ max} = \left(\sqrt{(A^2+B)} - A\right)^{\frac{1}{2}} \tag{28}$$

$$A = \frac{K_L + B}{2Q_{max}^2 K_L^2} - \frac{1}{2}, \quad B = (1 + 2R_K Q_{max})^2$$

Substituting the result of Equation (28) into Equation (29) for DC voltage gain, an expression satisfying Equation (30) is obtained, where Q is proportional to the load current as in Equation (31). Figure 11 is a graph of the results obtained from Equations (27) and (30). Figure 11a is the graph from Equation (30) for each K_L value, and Figure 11b is the graph from Equation (27) for each K_L value. Figure 11c is a Q–M graph according to the internal equivalent resistance, from which it can be seen that the peak value of M is relatively low when there is an internal equivalent resistance in the figure [29,30]. Figure 11d shows the peak value of M at a specific value of Q. In general, this graph serves as a criterion for finding an important Q value in the design process of an LLC resonant converter under a given DC voltage gain.

$$M_{peak} = \frac{\sqrt{1 + \left(\frac{R_K}{\omega_n K_L}\right)^2}}{\sqrt{(2Q_{max}R_K + 1)^2\left[1 + \frac{1}{K_L}\left(1 - \frac{1}{\omega_n^2}\right)\right]^2 + \left(\frac{2R_K}{\omega_n Q_{max} K_L} + \frac{1}{\omega_n} - \omega_n\right)^2 Q_{max}^2}} \tag{29}$$

$$(2QR_K+1)^2\left[1+\frac{1}{K_L}\left(1-\frac{1}{\omega_n^2}\right)\right]^2+\left(\frac{2R_K}{\omega_n QK_L}+\frac{1}{\omega_n}-\omega_n\right)^2Q^2-\frac{1+\left(\frac{R_K}{\omega_n K_L}\right)^2}{M^2}=0 \tag{30}$$

$$Q=\left(\frac{Z_o\pi^2}{8N^2V_o}\right)I_o \tag{31}$$

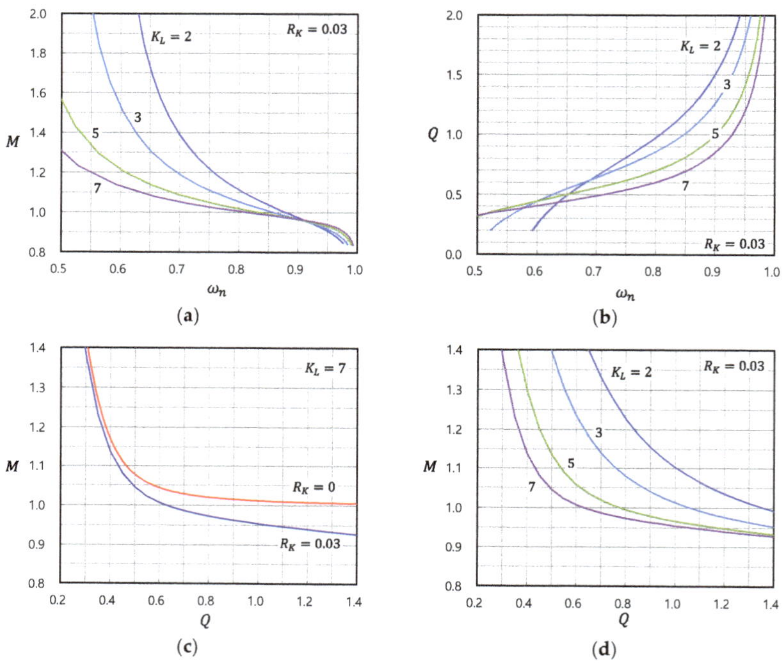

Figure 11. Characteristic graph of peak DC voltage gain M: (**a**) Peak DC voltage gain M vs. normalized frequency ω_n; (**b**) Q vs. Normalized frequency ω_n; (**c**) DC voltage gain M in accordance with internal loss resistance R_K; and (**d**) DC voltage gain M in accordance with inductance ratio K_L.

Figure 12 below shows the operation range of the LLC resonant converter, partially enlarged in the DC voltage gain characteristic graph of Figure 7. In the figure, M takes the maximum and minimum values of the normalized angular velocity ω_n, according to the input voltage range, and the operation area is formed as a closed loop connecting the points A, B, C, and D. At this time, within a specific Q value range proportional to the load current, the peak value of M can be determined as in Equation (27). In order to design a stable control system in the LLC resonant converter, it is necessary to ensure that the peak value of M is on the outer left side of the operating range, as shown in the figure [31].

Table 3 shows the design specifications of the LLC resonant converter designed in this paper. The input voltage range is 360–400 V, the maximum output is about 2 kW, the output voltage is 54 V, the resonance frequency is 125 kHz, and the inductance ratio is 8. At the normal input voltage of 390 V in Table 3, the transformer turn ratio is given as in Equation (32). Although the maximum output is 2 kW when the maximum output including the margin is 2.250 kW, the AC resistance R_{AC} is as given in Equation (33). The maximum and minimum voltage gain, with respect to the input voltage, are shown in Equations (34) and (35), respectively. If the peak value of M for Q is obtained from Equation (30), it can be obtained from two points, as shown in Figure 13. Point A is the case with internal equivalent resistance, and point B is the lossless case [32]. From the figure, it can be seen that Q is about 0.44 at point A, and about 0.52 at B. In general, the difference in

Q value affects the operating switching frequency range. The characteristic impedance is expressed in Equation (37), and the resonant components that play an important role in the operation of the converter are expressed in Equations (38)–(40).

$$N = \frac{V_{IN(nom)}}{2V_o} = \frac{380}{2 \times 54} = 3.5 \tag{32}$$

$$R_{AC} = \frac{8N^2V_o^2}{\pi^2 P_{o(max)}} = \frac{8 \times 3.5^2 \times 54}{\pi^2 \times 2250} = 12.9\ \Omega \tag{33}$$

$$M_{min} = \frac{2NV_o}{V_{IN(max)}} = \frac{2 \times 3.5 \times 54}{400} = 0.95 \tag{34}$$

$$M_{max} = \frac{2NV_o}{V_{IN(min)}} = \frac{2 \times 3.5 \times 54}{360} = 1.05 \tag{35}$$

$$Q_{max} = 0.44 \tag{36}$$

$$Z_o = QR_{AC} = 5.7\ \Omega \tag{37}$$

$$C_R = \frac{1}{2\pi f_o Z_o} = 226\ \text{nF} \tag{38}$$

$$L_R = \frac{Z_o}{2\pi f_o} = 7.3\ \mu\text{H} \tag{39}$$

$$L_M = K_L L_R = 58.4\ \mu\text{H} \tag{40}$$

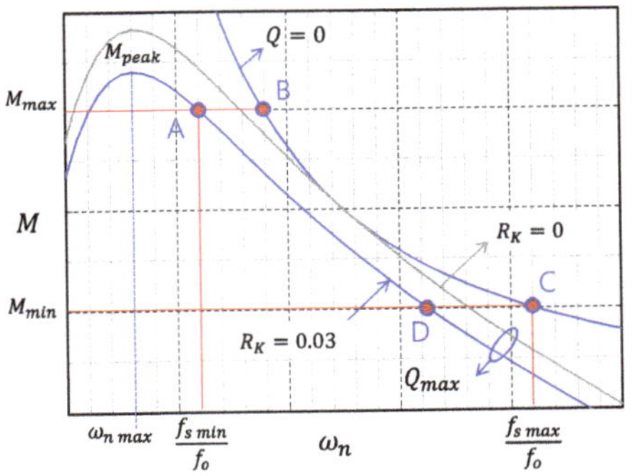

Figure 12. Operating range of LLC converter.

Table 3. Design Conditions for LLC Resonant Converters.

Parameter	Symbol	Value	Unit
Input voltage range	V_{IN}	360–400	V
Nominal input voltage	$V_{IN\ (nom)}$	380	V
Output voltage	V_o	54	V
Maximum output power	$P_{o\ max}$	2.0	kW
Maximum output current	$I_{o\ max}$	36	A
Resonant frequency	f_o	125	kHz
Inductance ratio	K_L	8	-

3.2. Circuit Simulation Results and Comparison

To verify the LLC resonant converter design result in the previous section, a circuit simulation was performed in this paper. We used the simulation software PSIM 11.0, and the circuit diagram is shown in Figure 14. In the figure, the internal resistances of the converter reflect the internal losses. A photocoupler was applied to isolate the control signal, and a frequency-limited voltage-controlled oscillator (VCO) circuit was used to stabilize the output voltage. Figure 15 shows the steady-state result waveform of the PSIM simulation obtained under two load conditions. In Figure 15a, the load current is about 4 A, while in Figure 15b the load current is about 36 A—the maximum output. At a typical input voltage of 390 V, the switching frequency was close to the resonance frequency, and both the magnetization current and the resonance current showed stable waveforms. This result is considered to indirectly confirm that the previously designed main resonant components were designed relatively appropriately [33–35].

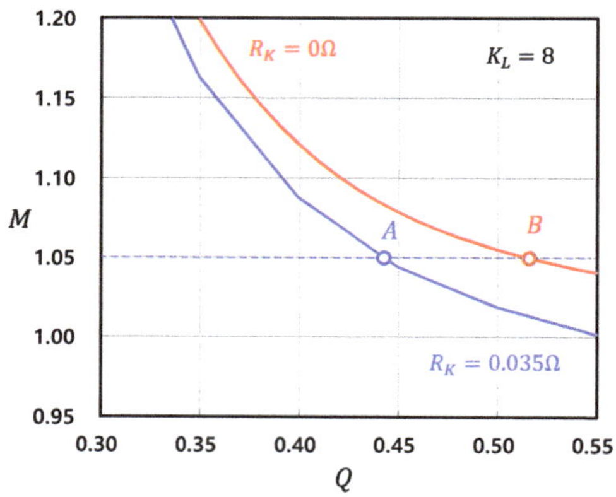

Figure 13. Internal loss resistance and change in Q value.

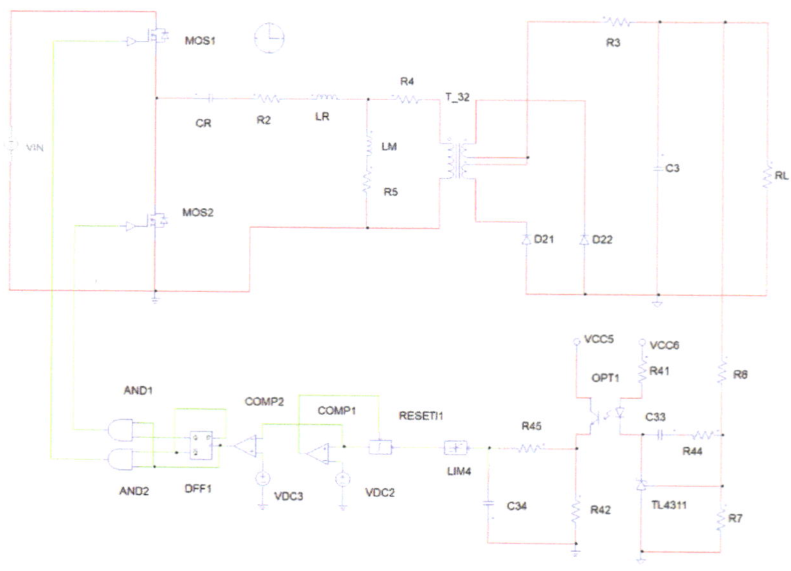

Figure 14. PSIM simulation schematic of a half-bridge LLC resonant converter.

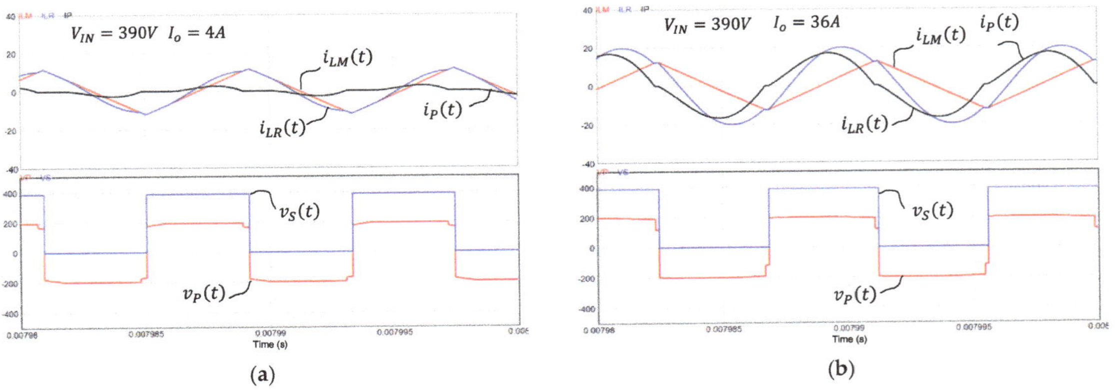

Figure 15. Steady-state waveform result of half-bridge LLC resonant converter with PSIM simulation: (**a**) At low load power; and (**b**) at full load power.

4. Experimental Results

In order to verify the design result of the LLC resonant converter in the previous section, an experimental circuit was constructed and its characteristics were analyzed. Table 4 shows the models and electrical specifications of the main components used in the experimental circuit. In Figure 16, Vishay's 650 V class MOSFET was used for the main switch and the gate on-resistance was 39 mΩ. Six resonant capacitors were used in parallel to increase the current rating, and a ferrite core (PQ3552 model) was used for the transformer. The rectifying diode on the secondary side of the transformer applied a synchronous rectifier circuit to reduce the rectification loss, and the on-resistance of the rectifier switch was 5.1 mΩ. Figure 17 provides a photograph of the experimental configuration, which shows the experimental circuit and measuring instruments used for the test. The power supply used in the experiment was a 61,505 from Chroma (Foothill Ranch, CA, USA), the electronic load was a PLA5K-600-300 from Amrel (San Diego, CA, USA), the power analyzer was a WT1802E from YOKOGAWA (Tokyo, Japan), and the oscilloscope was a 44MXS-B from LeCroy (Chestnut Ridge, NY, USA). Figure 18 shows the steady-state operation waveform of the experimental circuit. Figure 18a,b shows the results when the input voltage was 390 V whereas Figure 18c,d shows the results when the input voltage was 400 V. Figure 18a,c shows when the load current was at 4 A whereas Figure 18b,d shows when the load current was at 36 A. In the figure, the waveforms represent the resonance current i_{LR}, magnetizing current i_{LM}, the transformer primary current i_p, and the transformer primary voltage v_p, from the top. In the figure, each current and voltage waveform maintained a stable state, with respect to load change and input voltage change, and was controlled at an appropriate switching frequency state. In particular, when the input voltage was 390 V, the waveforms of the experimental result and the PSIM simulation result were similar.

Table 4. Main components and electrical ratings used in the experimental circuit.

Parameter	Symbol	Value	Specifications
Main switch	$S_{1,2}$	SiHG73N60E	650 V, 73 A, 39 mΩ
Resonant capacitor	C_R	234 nF	39 nF × 6
Resonant inductor	L_R	7 µH	PQ3535
Resonant inductor	L_M	58 µH	PQ3535
Transformer turns ratio	N	35:10:10	PQ3552
Rectifier switch	$D_{1,2}$	IPP051N15N5	150 V, 120 A, 5.1 mΩ
Output capacitor	C_o	960 µF	120 µF × 8

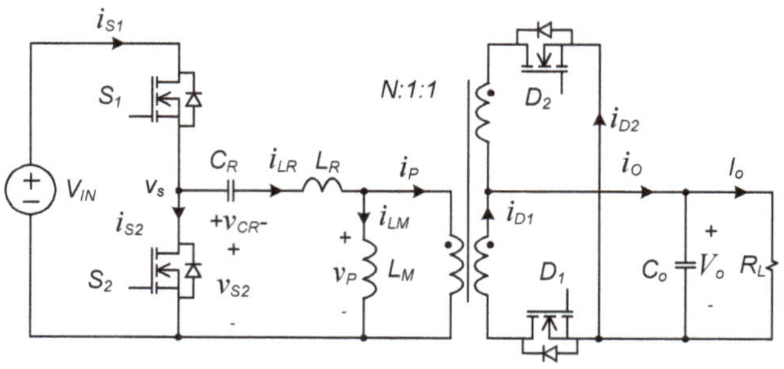

Figure 16. Basic schematic for experimental circuit.

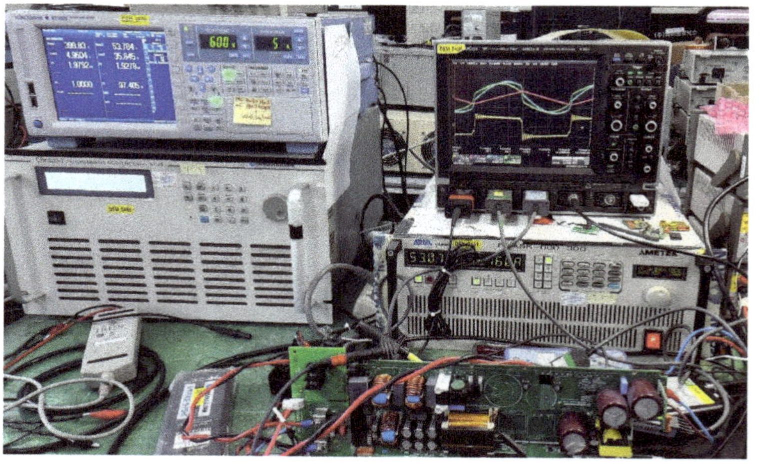

Figure 17. Experimental circuit and experimental configuration.

Figure 19 shows the relationship between the DC voltage gain and frequency measured in the experimental circuit. Figure 19a shows the comparison of experimental values, calculated values, and simulation results of input voltage and switching frequency when the load current was maintained at 4 A. Figure 19b shows the results for the voltage gain M versus the normalized angular velocity ω_n. When the load was small, the difference between the case with and without the internal equivalent resistance was not large. The difference between the experimental value and the calculated value in the figure is thought to be due to the fundamental harmonic approximation (FHA) approximation method, which increases the error at no load, being used for the voltage gain equation derived from the AC equivalent circuit. Figure 20 shows the measured voltage gain characteristics when the load current was 36 A. Figure 20a shows the results for the input voltage versus the switching frequency, while Figure 20b shows the results for the voltage gain M versus the normalized angular velocity ω_n. As can be seen from the figure, as the load power increases, the calculated, simulation, and experimental values showed similar trends; however, it can be seen that the lossless conditions and the experimental values showed a large difference. Therefore, the steady-state characteristic equation and design process derived in this paper are considered to be valid. Figure 21 compares the relationship between the switching frequency and the load current of the experimental circuit. Figure 21a,b shows the results under input voltages of 390 V and 400 V, respectively. The load current was measured up to 35 A. In the figure, the solid line denotes the calculated value and the dotted line is the simulation value. In the figure, when the internal equivalent resistance is $R_K = 0$, the switching frequency decreases according to the load current, but the decrement is significantly smaller than the experimental value. On the other hand, when the internal equivalent resistance is $R_K = 0.025$, it presented more similar characteristics to the experimental value. Therefore, the AC equivalent circuit, steady-state characteristic equation, and optimal design proposed in this paper are effective in the actual design, in terms of the amount of change in switching frequency with respect to input voltage and load current change. Figure 22 shows the power conversion efficiency of the experimental circuit. When the input voltage was a typical voltage of 390 V, the maximum efficiency was 97.6%; even at an input voltage of 420 V, it was 97.3%. As previously mentioned through equivalent circuits and derived equations in Sections 2 and 3, the parasitic resistance of the main components is required to be considered for a more precise and optimal design of the LLC resonant converter. In particular, since switching frequency and resonant frequency are generally from hundreds of kHz to MHz, most of the power loss occurs from switching loss in main switches, conduction loss in transformer winding, and core loss in the core of transformer and inductors [36].

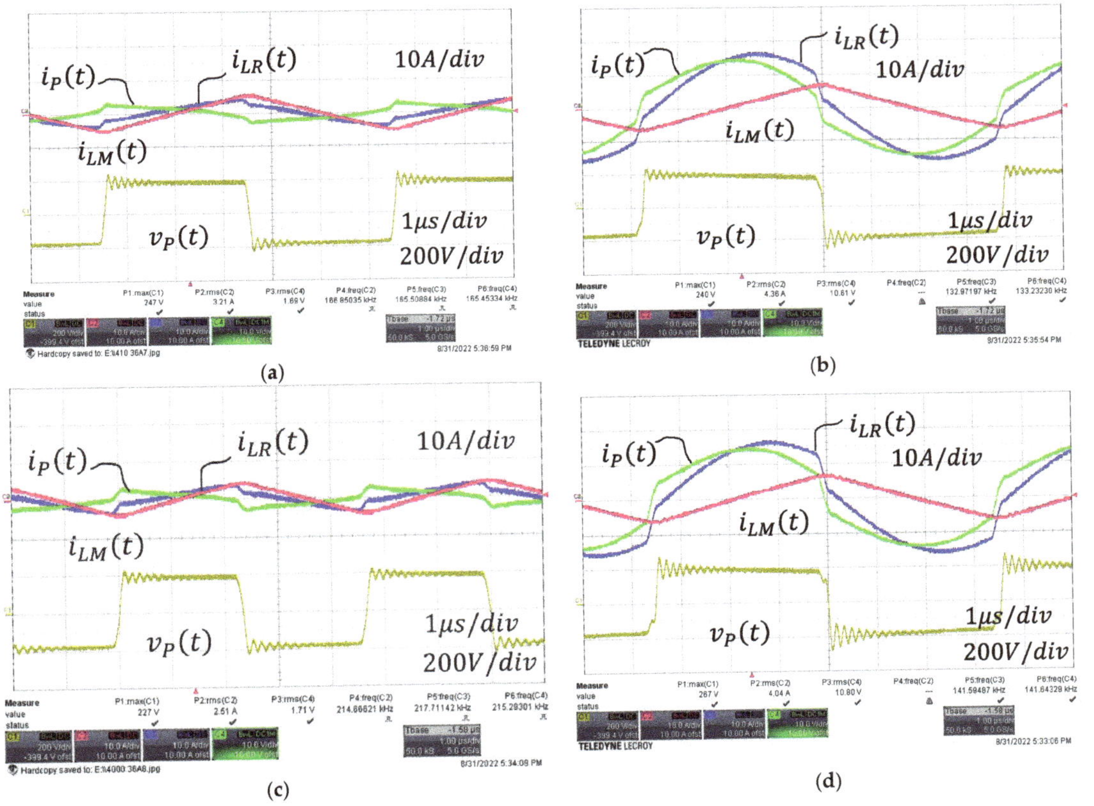

Figure 18. Experimental waveform in steady-state: (**a**) At an input voltage of 390 V and load current of 4 A; (**b**) at an input voltage of 390 V and load current of 36 A; (**c**) at an input voltage of 400 V and load current of 4 A; and (**d**) at an input voltage of 400 V and load current of 36 A.

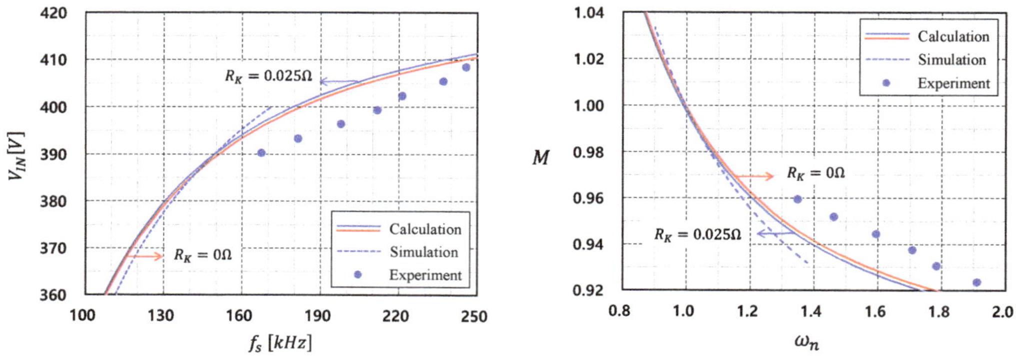

(a) (b)

Figure 19. Frequency characteristics of DC voltage gain M at low load current: (**a**) Input voltage characteristics in accordance with switching frequency variation; and (**b**) DC voltage gain M characteristic in accordance with normalized frequency variation.

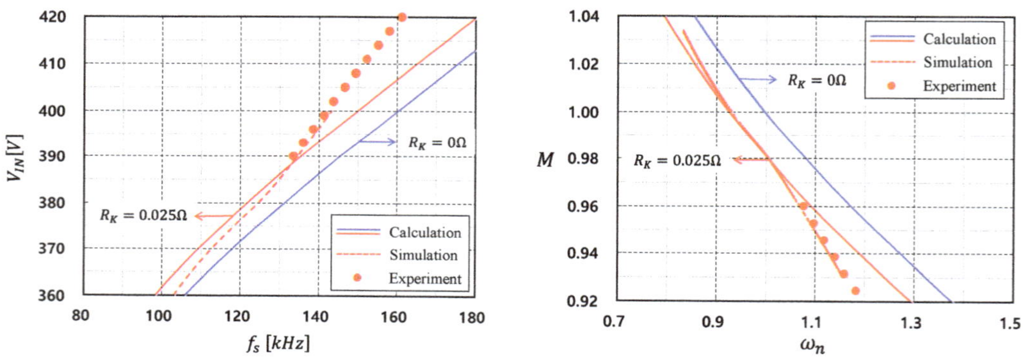

(a) (b)

Figure 20. Frequency characteristics of DC voltage gain M at high load current: (**a**) Input voltage characteristics in accordance with switching frequency variation; and (**b**) DC voltage gain M characteristic in accordance with normalized frequency variation.

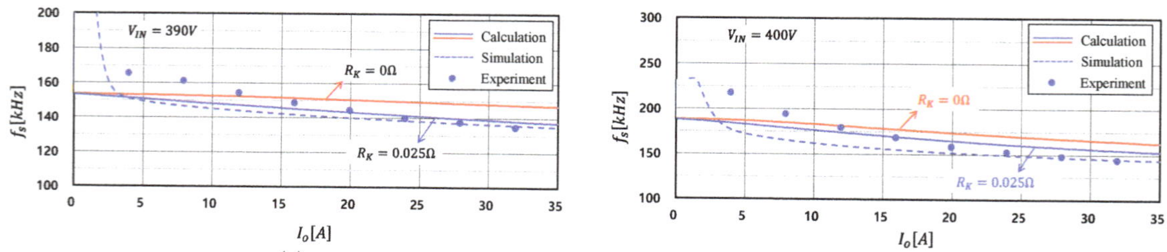

(a) (b)

Figure 21. Load characteristics of switching frequency: (**a**) Input voltage of 390 V; and (**b**) input voltage of 400 V.

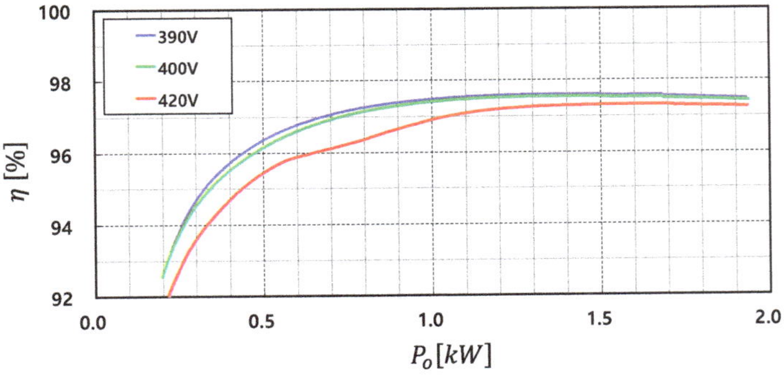

Figure 22. Measured power conversion efficiency of the experimental circuit.

5. Discussion

In order to confirm the validity of the steady-state characteristic equation and the optimal design process, a converter with a maximum output of 2 kW, an input voltage of 390 V, an output voltage of 54 V, and a resonance frequency of 125 kHz was designed, and we compared the simulation, calculation, and experimental results. First, at low output current, there was a difference between the experimental value and the calculated value due to the fundamental harmonic approximation (FHA) method, which increases the error at no load, being used for the voltage gain equation. As the load power increased, the calculation, simulation, and experimental results showed similar trends. In particular, it could be seen that the calculation results proposed in this paper were closer to the experimental results than the calculation results under the lossless condition. From this, it can be seen that the steady-state characteristic equation reflecting the internal loss resistance derived in this paper and the optimal design process are helpful in actual design.

Next, the relationship between the switching frequency and the load current of the experimental circuit was compared, in order to determine the operating range of the switching frequency, which is important in the converter design stage. The load current was measured up to 35 A. When the internal equivalent resistance was calculated as 0, the switching frequency decreased according to the load current, but the amount of decrement was significantly smaller than the experimental value. When the internal equivalent resistance was considered, the switching frequency characteristics were similar to the experimental results. The experimental and calculation results were compared in the switching frequency characteristic graph when the converter operated under light load and maximum load conditions. In this case, it was confirmed that the calculated value reflecting the internal loss also showed a close result.

6. Conclusions

In this paper, a steady-state model of an LLC resonant half-bridge converter with internal loss resistance was proposed, and steady-state characteristic equations of DC voltage gain and input impedance were derived. From the frequency characteristic graph of DC voltage gain, it was revealed that the voltage gain was reduced, compared to the previous case where the internal equivalent circuit was not considered. Additionally, the optimal design process of important components required for the design of LLC resonant converters was shown, using the characteristic equations and graphs of the input impedance.

The change of the switching frequency with respect to the load current at each input voltage was compared with the calculated value and the experimental result. As a result, it was possible to confirm the usefulness of the analysis results, reflecting the internal loss resistance proposed in this paper and the process of optimal design. The power conversion efficiency of the experimental circuit showed a maximum of 97.6% when the input voltage was 390 V, and the maximum efficiency remained as high as 97.3% when the input voltage was 420 V.

Author Contributions: Conceptualization, T.-Y.A.; methodology, J.-S.Y., Y.-M.G., and T.-Y.A.; validation, J.-S.Y., Y.-M.G. and T.-Y.A.; formal analysis, T.-Y.A.; investigation, J.-S.Y. and Y.-M.G.; resources, T.-Y.A.; data curation, J.-S.Y. and T.-Y.A.; writing—original draft preparation, J.-S.Y. and T.-Y.A.; writing—review and editing, J.-S.Y., Y.-M.G. and T.-Y.A.; visualization, J.-S.Y. and T.-Y.A.; supervision, T.-Y.A.; project administration, Y.-M.G. and T.-Y.A.; funding acquisition, T.-Y.A. All authors have read and agreed to the published version of the manuscript.

Funding: This work was supported by the Industrial Strategic Technology Development Program-Automobile industry technology development (1415180408, Development of common component technology for power steering and braking systems for hydrogen electric commercial vehicles) funded by the Ministry of Trade, Industry & Energy (MOTIE, Korea).

Data Availability Statement: Not applicable.

Conflicts of Interest: The authors declare no conflict of interest.

References

1. Khoobroo, E.; Ashrafinia, B.; Akhbari, M. Optimal design of LLC resonant converter with improved controllability. In Proceedings of the 5th Annual International Power Electronics, Drive Systems and Technologies Conference (PEDSTC 2014), Tehran, Iran, 5–6 February 2014; pp. 396–401. [CrossRef]
2. Beiranvand, R.; Rashidian, B.; Zolghadri, M.R.; Hossein Alavi, S.M. A Design Procedure for Optimizing the LLC Resonant Converter as a Wide Output Range Voltage Source. *IEEE Trans. Power Electron.* **2012**, *27*, 3749–3763. [CrossRef]
3. Zheng, R.; Liu, B.; Duan, S. Analysis and Parameter Optimization of Start-Up Process for LLC Resonant Converter. *IEEE Trans. Power Electron.* **2015**, *30*, 7113–7122. [CrossRef]
4. Lee, I.O.; Moon, G.W. The *k*-Q Analysis for an LLC Series Resonant Converter. *IEEE Trans. Power Electron.* **2013**, *29*, 13–16. [CrossRef]
5. Tan, X.; Ruan, X. Equivalence Relations of Resonant Tanks: A New Perspective for Selection and Design of Resonant Converters. *IEEE Trans. Ind. Electron.* **2016**, *63*, 2111–2123. [CrossRef]
6. Hu, Z.; Wang, L.; Wang, H.; Liu, Y.F.; Sen, P.C. An Accurate Design Algorithm for LLC Resonant Converters—Part I. *IEEE Trans. Power Electron.* **2016**, *31*, 5435–5447. [CrossRef]
7. Hu, Z.; Wang, L.; Qiu, Y.; Liu, Y.F.; Sen, P.C. An Accurate Design Algorithm for LLC Resonant Converters—Part II. *IEEE Trans. Power Electron.* **2016**, *31*, 5448–5460. [CrossRef]
8. Fang, X.; Hu, H.; Chen, F.; Somani, U.; Auadisian, E.; Shen, J.; Batarseh, I. Efficiency-Oriented Optimal Design of the LLC Resonant Converter Based on Peak Gain Placement. *IEEE Trans. Power Electron.* **2013**, *28*, 2285–2296. [CrossRef]
9. Beiranvand, R.; Zolghadri, M.R.; Rashidian, B.; Alavi, S.M.H. Optimizing the LLC–LC Resonant Converter Topology for Wide-Output-Voltage and Wide-Output-Load Applications. *IEEE Trans. Power Electron.* **2011**, *26*, 3192–3204. [CrossRef]
10. Fang, Z.; Cai, T.; Duan, S.; Chen, C. Optimal Design Methodology for LLC Resonant Converter in Battery Charging Applications Based on Time-Weighted Average Efficiency. *IEEE Trans. Power Electron.* **2015**, *30*, 5469–5483. [CrossRef]
11. Yu, R.; Ho, G.K.Y.; Pong, B.M.H.; Ling, B.W.K.; Lam, J. Computer-Aided Design and Optimization of High-Efficiency LLC Series Resonant Converter. *IEEE Trans. Power Electron.* **2012**, *27*, 3243–3256. [CrossRef]
12. Li, G.; Xia, J.; Wang, K.; Deng, Y.; He, X.; Wang, Y. Hybrid Modulating of Parallel-series LLC Resonant Converter and Phase Shift Full-Bridege Converter for a Dual-output DC-DC Converter. *IEEE Trans. Power Electron.* **2019**, *7*, 833–842. [CrossRef]
13. Qian, T.; Qian, C. An Adaptive Frequency Optimization Scheme for LLC Converter with Adjustable Energy Transferring Time. *IEEE Trans. Power Electron.* **2019**, *34*, 2018–2024. [CrossRef]
14. Da Silva, R.L.; Borges, V.L.F.; Possamai, C.E.; Barbi, I. Solid-State Transformer for Power Distribution Grid Based on a Hybrid Switched-Capacitor LLC-SRC Converter: Analysis, Design, and Experimentation. *IEEE Access* **2020**, *8*, 141182–141207. [CrossRef]
15. Shafaei, R.; Perez, M.C.G.; Ordonez, M. Planar Transformers in LLC Resonant Converters: High-Frequency Fringing Losses Modeling. *IEEE Trans. Power Electron.* **2020**, *35*, 9632–9649. [CrossRef]
16. Duan, J.; Zhang, D.; Gu, R. Partial-Power Post-Regulated LLC Resonant DC Transformer. *IEEE Trans. Ind. Electron.* **2022**, *69*, 7909–7919. [CrossRef]
17. Glitz, E.S.; Ordonez, M. MOSFET Power Loss Estimation in LLC Resonant Converters: Time Interval Analysis. *IEEE Trans. Power Electron.* **2019**, *34*, 11964–11980. [CrossRef]
18. Liu, J.; Zhang, J.; Zheng, T.Q.; Yang, J. A Modified Gain Model and the Corresponding Design Method for an LLC Resonant Converter. *IEEE Trans. Power Electron.* **2017**, *32*, 6716–6727. [CrossRef]
19. Kim, D.K.; Moon, S.; Yeon, C.O.; Moon, G.W. High-Efficiency LLC Resonant Converter with High Voltage Gain Using an Auxiliary LC Resonant Circuit. *IEEE Trans. Power Electron.* **2016**, *31*, 6901–6909. [CrossRef]

20. Xu, H.; Yin, Z.; Zhao, Y.; Huang, Y. Accurate Design of High-Efficiency LLC Resonant Converter with Wide Output Voltage. *IEEE Access* **2017**, *5*, 26653–26665. [CrossRef]
21. Chen, S.Y.; Li, Z.R.; Chen, C.L. Analysis and Design of Single-Stage AC/DC LLC Resonant Converter. *IEEE Trans. Ind. Electron.* **2012**, *59*, 1538–1544. [CrossRef]
22. Fang, X.; Hu, H.; Shen, Z.J.; Batarseh, I. Operation Mode Analysis and Peak Gain Approximation of the LLC Resonant Converter. *IEEE Trans. Power Electron.* **2012**, *27*, 1985–1995. [CrossRef]
23. Ivensky, G.; Bronshtein, S.; Abramovitz, A. Approximate Analysis of Resonant LLC DC-DC Converter. *IEEE Trans. Power Electron.* **2011**, *26*, 3274–3284. [CrossRef]
24. Duan, F.; Xu, M.; Yang, X.; Yao, Y. Canonical Model and Design Methodology for LLC DC/DC Converter with Constant Current Operation Capability under Shorted Load. *IEEE Trans. Power Electron.* **2016**, *31*, 6870–6883. [CrossRef]
25. Jung, J.H.; Kim, H.S.; Ryu, M.H.; Baek, J.W. Design Methodology of Bidirectional CLLC Resonant Converter for High-Frequency Isolation of DC Distribution Systems. *IEEE Trans. Power Electron.* **2013**, *28*, 1741–1755. [CrossRef]
26. Hu, S.; Deng, J.; Mi, C.; Zhang, M. Optimal design of line level control resonant converters in plug-in hybrid electric vehicle battery chargers. *IET Electr. Syst. Transp.* **2014**, *4*, 21–28. [CrossRef]
27. Yang, Y.; Zhang, L.; Ma, T. Conduction Loss Analysis and Optimization Design of Full Bridge LLC Resonant Converter. In Proceedings of the 2018 International Power Electronics Conference (IPEC-Niigata 2018-ECCE Asia), Niigata, Japan, 20–24 May 2018; pp. 2703–2707. [CrossRef]
28. Yang, C.H.; Liang, T.J.; Chen, K.H.; Li, J.S.; Lee, J.S. Loss analysis of half-bridge LLC resonant converter. In Proceedings of the 2013 1st International Future Energy Electronics Conference (IFEEC), Tainan, Taiwan, 3–6 November 2013; pp. 155–160. [CrossRef]
29. Zhao, C.; Xinke, W.; Meng, P.; Qian, Z. Optimum design consideration and implementation of a novel synchronous rectified soft-switched phase-shift full-bridge converter for low-output-voltage high-output-current applications. *IEEE Trans. Power Electron.* **2009**, *24*, 388–397. [CrossRef]
30. Beiranvand, R.; Rashidian, B.; Zolghadri, M.; Alavi, S.M. Using LLC resonant converter for designing wide-range voltage source. *IEEE Trans. Ind. Electron.* **2011**, *58*, 1746–1756. [CrossRef]
31. Kim, J.; Moon, G. A new LLC natural resonant converter with narrow switching frequency variation and reduced conduction losses. *IEEE Trans. Power Electron.* **2014**, *29*, 4278–4287. [CrossRef]
32. Jami, M.; Beiranvand, R.; Mohamadian, M.; Ghasemi, M. Optimization the LLC resonant converter for achieving maximum efficiency at a predetermined load value. In Proceedings of the 6th Power Electronics, Drive Systems & Technologies Conference (PEDSTC2015), Tehran, Iran, 3–4 February 2015; pp. 149–155. [CrossRef]
33. Chen, W.; Xu, M.; Ge, Y.; Guo, Q.; Jiang, T.; Che, X. Parameter Design of High Power LLC Resonant Converter. In Proceedings of the 2022 IEEE 5th International Conference on Electronics Technology (ICET), Chengdu, China, 13–16 May 2022; pp. 421–425. [CrossRef]
34. Ahn, T.Y. A Study on Steady State Characteristics of LLC Resonant Half Bridge Converter Considering Internal Losses. *IKEEE* **2018**, *22*, 985–991. [CrossRef]
35. Yoon, K.H.; Noh, Y.J.; Phum, S.; Meas, S.; Jang, S.H.; Kim, E.S. LLC resonant converter with wide input voltage and load range at fixed switching frequency. In Proceedings of the 2012 Twenty-Seventh Annual IEEE Applied Power Electronics Conference and Exposition (APEC), Orlando, FL, USA, 5–9 February 2012; pp. 1338–1342. [CrossRef]
36. Yang, B.; Lee, F.C.; Zhang, A.J.; Huang, G. LLC resonant converter for front end DC/DC conversion. In Proceedings of the APEC, Seventeenth Annual IEEE Applied Power Electronics Conference and Exposition (Cat. No.02CH37335), Dallas, TX, USA, 10–14 March 2002; Volume 2, pp. 1108–1112. [CrossRef]

energies

Article

A High Frequency Multiphase Modular Hybrid Transformerless DC/DC Converter for High-Voltage-Gain High-Current Applications

Hu Xiong [1], Jiayuan Li [1], Bin Xiang [1], Xiaoguang Jiang [1] and Yuan Mao [2,*]

1 State Grid Hubei Electric Power Research Institute, No.227 Xudong Avenue, Wuhan 430077, China
2 College of Information Engineering, The Zhejiang University of Technology, Hangzhou 310023, China
* Correspondence: maoyuan@zjut.edu.cn; Tel.: +86-17133387919

Abstract: In order to meet the demands of desirable efficiency, transformerless DC/DC equipment with great voltage step-down are inevitable needed. This research offers a unique type of high-frequency, high-voltage-gain DC/DC converter, which comprises a switched capacitor (SC) converter and a buck converter. Thanks to the transformation of a two-stage converter to a single-stage converter, it has a considerable ratio of step-down voltage transformation and a reasonable duty cycle. In addition, it can permit low voltage stress on the switches. The simple control method and easy driving circuit implementation makes it scalable for high-power-level devices. Low cost can be realized as fewer components are needed. Under all operational circumstances, total soft-charging and low equipment voltage stresses are accomplished. Compared to those classic high-voltage-gain converters, the proposed converter exhibits merits of higher efficiency, higher flexibility, lower ripples, and lower costs. A comprehensive analysis is carried out for the converter's steady-state operations. With a 1 MHz switching frequency, a 900 W prototype of a 20-time converter is constructed, with a peak efficiency of 92.5%. Simulations and experiments verify the effectiveness of the theoretical investigation of the converter's operation.

Keywords: high-frequency hybrid converter; DC microgrids; high efficiency; low-voltage-stress; high-voltage conversion ratio; cost-effective

Citation: Xiong, H.; Li, J.; Xiang, B.; Jiang, X.; Mao, Y. A High Frequency Multiphase Modular Hybrid Transformerless DC/DC Converter for High-Voltage-Gain High-Current Applications. *Energies* **2023**, *16*, 2518. https://doi.org/10.3390/en16062518

Academic Editor: Salvatore Musumeci

Received: 20 January 2023
Revised: 25 February 2023
Accepted: 2 March 2023
Published: 7 March 2023

1. Introduction

In an inter-grid scenario with a lot of distributed generation (DG), the traditional power grid architecture, which is made up of numbers of power stations, transmission, and distribution networks, will be displaced by a number of microgrids [1]. For instance, DC microgrids are well used in car, airplane, data center, and boat systems [1–7]. In many applications, a DC bus is used to distribute power so that lighting systems, motor drives, and devices that store energy can work together. A high-gain converter is needed to ensure that the DC voltage buses feed low-voltage loads, such as those in records centers, in a way that is both efficient and good for the power quality.

Step-down conversion is often carried out using DC/DC buck converters since they have fewer active switches and passive components. However, for high-duty cycle operation, the energy efficiency of the converter decreases substantially. Traditional buck DC/DC converters have a limited voltage gain as they lose a lot of power when the voltage goes very high [4,5]. They are not good for applications that need a lot of voltage gain. To get a huge voltage gain, a two-stage converter consisting of two cascaded buck converters has been suggested [8]. To stop the beat-frequency effect, a set of controllers that work together would have to be used to control the active switches. The controller's design would become more challenging as a result [8]. Additionally, if the input voltage and load value vary, instability is quite probable [9].

The switched capacitor (SC) converter belongs to a non-isolated DC/DC converter that may boost voltage while preserving efficiency [10–18]. SC converters without magnetic components would be tiny and powerful. Dickson, Fibonacci (FIB), series–parallel, and other high-voltage-gain SC converters have been studied previously [19–27]. Generally, the SC converter cannot achieve both high efficiency and sufficient line and load regulation [28–31]. Efficiency will suffer greatly if precision control is sought [28–30]. A two-stage DC/DC converter may improve control and voltage gain. The standard buck converter is utilized in the second stage, owing to regulatory considerations and the necessity for high-voltage-gain with few components. Additionally, multi-phase buck converters may provide a significant current capacity for high-power applications. The regulation issue is often solved and voltage gain is increased by using two-stage DC/DC conversion [32]. Figure 1 depicts the system diagram for this two-stage converter topology. A multiphase buck converter with regulation makes up the second stage, whereas the first stage is made up of an uncontrolled SC converter. Most of the high-voltage-gain is produced by the first-stage converter, while precise control is produced by the second stage. The design has two downsides, despite its advantages: (a) the two-stage design has a larger bill of materials (BOM) cost since it has more circuit components; and (b) the efficiency may drastically decrease with two switching stages and large switching losses by the relative components.

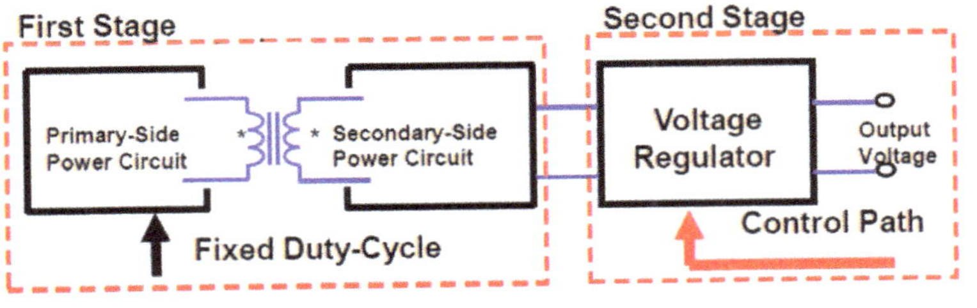

Figure 1. Conventional two-stage DC/DC converter with high-voltage-gain.

For high-voltage and high-current devices, a hybrid transformerless DC/DC converter with great efficiency is discussed in this article. With substantially less switches as well as y control. It has a reasonable duty cycle, and low cost because it combines a two-stage into one stage, which minimizes the number of components needed. To reduce current ripple, the recommended converter employs interleaving control. The principle of the suggested converter is validated, and the improved performance is shown in both simulations and experiments. Fair comparisons among the proposed converter and other popular single-stage converters are given in Table 1. Here, the star symbols (i.e., "*") indicate the polarities of the transformer. It is exhibited that even with a larger voltage gain, the suggested circuit can implement a higher efficiency.

Table 1. Comparisons among different single-stage converters.

The Converter Topology in	Voltage Gain (Times)	Output Power (W)	Switching Frequency (kHz)	Peak Efficiency (%)
[33]	15	2000	100	90
[34]	15	2000	100	92
[35]	16.6	100	100	90
[36]	16.6	300	100	92.5
Proposed	20	900	100	92.5

2. Operation Principle and Configuration for the Designed DC/DC Converter

This section describes the suggested DC/DC SC-buck converter's system architecture and modular design. The suggested converter's operating theory is then described.

2.1. System Topology and Modular Scheme

The suggested SC-buck converter is depicted in Figure 2. On the basis of the aforementioned theories, some assumptions are completed first.

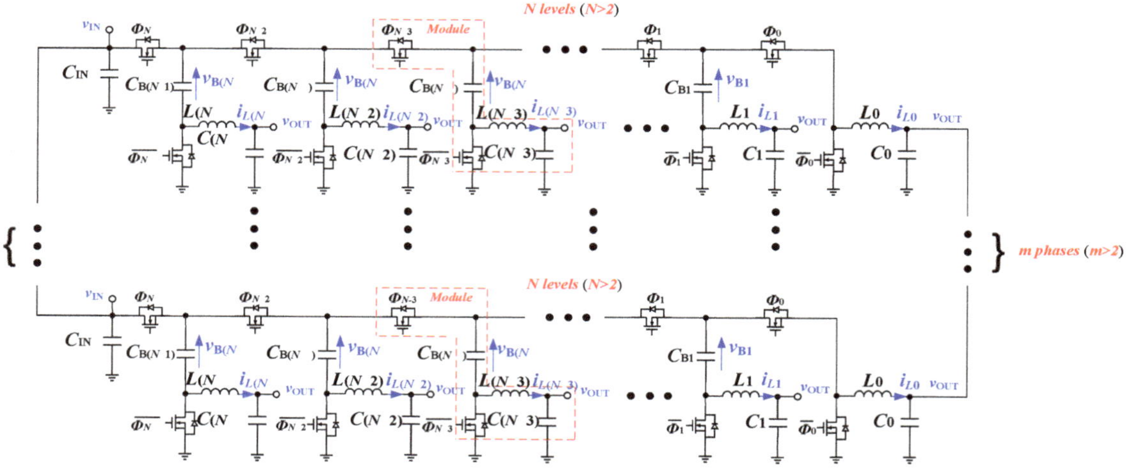

Figure 2. Configuration of the suggested converter at the system level.

a. Each switching component in the recommended converter is perfect.
b. Voltage variation between the capacitors is ignored; therefore, all capacitors have ideal values to maintain virtually constant holding voltages throughout operation. As a result, the capacitors are assumed as ideal voltage sources.
c. The dead-time between activating one switch and deactivating a complimentary switch is minimal compared to the conduction time of each switch, and may therefore be ignored. The dead-time is excluded from the examination of circuit structure in order to make it simpler.
d. Every switch within the modules has equal switching frequency while the suggested converter is working in steady-state.

The suggested *M*-phase converter contains *N* programmable modules of SC-buck circuits in each of its phases. With the exception of the first module, which is devoid of a flying capacitor, C_{B0}, each phase is made up of two complementary MOSFETs, a flying MOSFET Φ_i $(i = 0, 1, 2, \ldots, -1)$ and a bottom MOSFET $\overline{\Phi_i}$ $(i = 0, 1, 2, \ldots, N-1)$, a flying capacitor $C_{B(i)}$ $(i = 1, 2, \ldots, N-1)$, and an input capacitor C_i $(i = 0, 1, 2, \ldots, N-1)$. The design of each module in the proposed converter's switched-capacitor-buck circuit is shown in Figure 3.

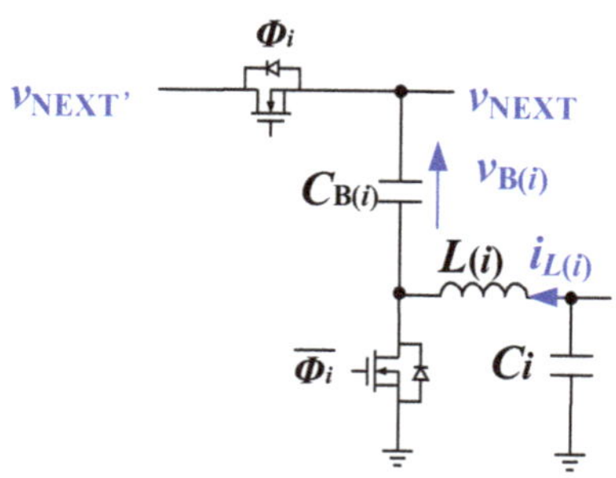

Figure 3. Configuration of the SC-circuit model.

The proposed converter's input and output voltages are represented by v_{IN} and v_{OUT}, respectively; the inductor currents on L_i (i = 0, 1,2, ..., $N-1$) and the voltages acros $C_{\mathrm{B}(i)}$ (i = 1, 2, ..., $N-1$) are represented by i_{Li} (i = 0, 1, 2, ..., $N-1$) and $v_{\mathrm{B}(i)}$ (i = 1, 2, ..., $N-1$). Additionally, $\overline{d_i}$ (i = 0, 1, 2, ..., $N-1$) is used to denote the duty ratios of the MOSFETs. The bottom MOSFETs' complementary duty ratios d_i (i = 0, 1, 2, ..., $N-1$) are shown as Φ_i (i = 0, 1, 2, ..., $N-1$). Evidently, $d_i + \overline{d_i} = 1$.

The steady-state functioning of L_i (i = 0, 1 ,2 , ..., $N-1$) is given by the following expressions to achieve voltage-second balance.

$$\begin{cases} V_{\mathrm{OUT}} = \left[V_{\mathrm{B}(i+1)} - V_{\mathrm{B}(i)}\right] \\ V_{\mathrm{B}(N)} = V_{\mathrm{IN}} \\ V_{\mathrm{B0}} = 0 \end{cases} \quad (i = 0,1,2,\ldots,N-1) \tag{1}$$

On the basis of (1), the voltage gain M can be obtained as

$$M = \frac{V_{\mathrm{OUT}}}{V_{\mathrm{IN}}} (i = 1,2,3,\ldots,N) \tag{2}$$

To achieve charge equilibrium across all $C_{\mathrm{B}(i)}$, the following expression is derived:

$$I_{L0}D_0 = I_{L1}D_1 = I_{L2}D_2 = I_{L(N-1)}\, D_{N-1} \tag{3}$$

According to (3), each inductor's steady-state current will be the same because the flying MOSFET's duty ratios are equal.

The benefits that the modular design can reduce the intricacy of the power converter system and provide uniform thermal distribution are realized in the majority of applications by spreading the inductor currents evenly. It is desirable for the converter to maintain a constant temperature throughout with no hot areas. This means that all devices will experience the same power stress distribution. It allows the converter to provide the most power at the required temperature.

Consequently, the duty ratios, D_i, of each flying MOSFET are all adjusted to the same value.

$$D_0 = D_1 = D_2 = \ldots = D_{N-1} = D \tag{4}$$

where D stands for the flying MOSFETs' consistent duty ratio to ensure equitable current sharing.

Substitute (4) into (1) and (2),

$$M = \frac{V_{\mathrm{OUT}}}{V_{\mathrm{IN}}} = \frac{D}{N} (i = 1,2,3,\ldots,N) \tag{5}$$

The M in (5) is equivalent to that of an N-phase buck converter with a N:1 ratio SC converter that is duty-cycle-regulated. It should be noted that the duty ratios are designed for the proposed circuit operating at the continuous current mode (CCM).

2.2. Operating Principle

The duty cycle should be near to 1/2 for optimal efficiency in typical buck converters [4]. Based on the design procedure in [33], both the maximum and lowest duty cycles must be constrained to optimize the converter in terms of efficiency, cost, and size. Consequently, the value of N can be obtained from (5). In this research, a three-phase converter with four modules in each phase is used to show the working concept.

Figure 4a depicts the converter in steady-state operation with three phases and four modules working in each phase. The flying MOSFET 0 and 2 are controlled by signal PWM-1, while MOSFET 3 and MOSFET 4 are controlled by PWM-2 that is in reverse phase with PWM-1. The converter's matching timing diagram is depicted in Figure 4b. In Figure 4a, the converter is shown operating in three phases, with four modules operating in each phase as it would in a typical situation. Signal PWM-1 is in charge of controlling the flying MOSFETs 0 and 2. Signals in PWM-2 that are out of phase with PWM-1 are used to control components 3 and 4. Figure 4 displays the suggested converter's timing diagram, b.

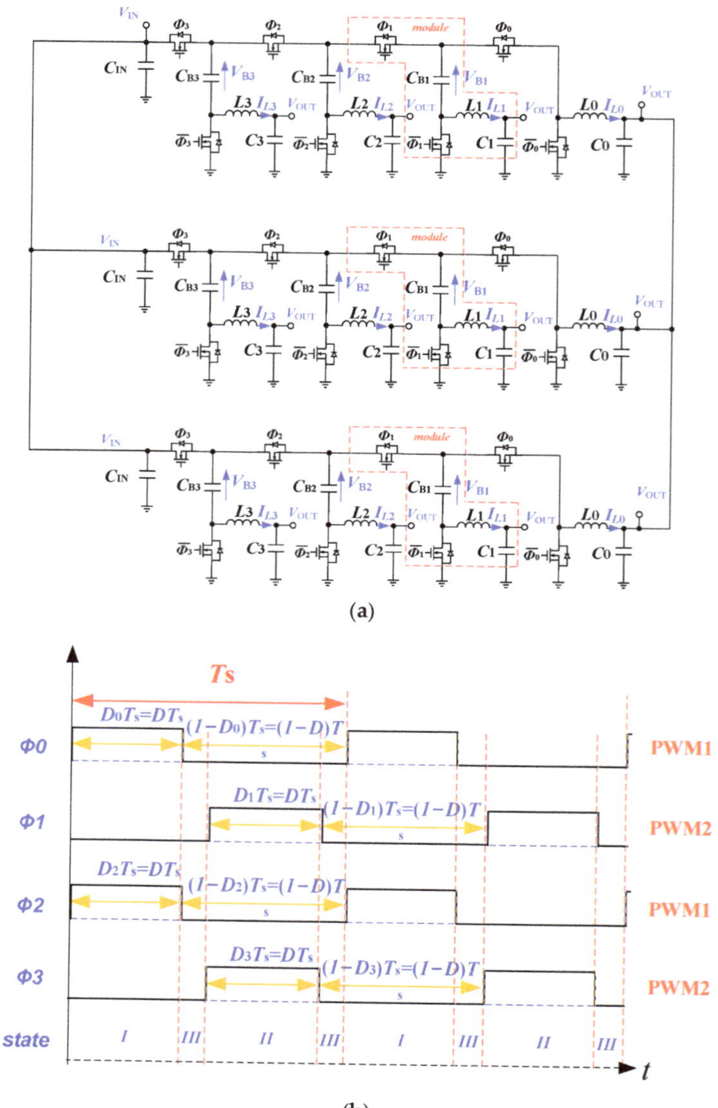

Figure 4. (**a**) Schematic diagram of the suggested converter; and (**b**) the corresponding timing schematic.

The switches controlled through PWM-1 will "ON", whereas the switches controlled through PWM-2 will "OFF" in state I. Consequently, the switches controlled through the complement of PWM-1 will "OFF" and the switches controlled through the complement of PWM-2 will "ON". According to Figure 5a, V_{IN} charges the C_{OUT} and the C_{B2} through L_1 and L_3, respectively. For C_{B2} and C_{OUT}, respectively, C_{B1} and C_{B3} are simultaneously discharged. Meanwhile, V_{IN} charges L_0 and L_2, storing energy in L_0 and L_2. The voltage on C_{OUT} powers the load.

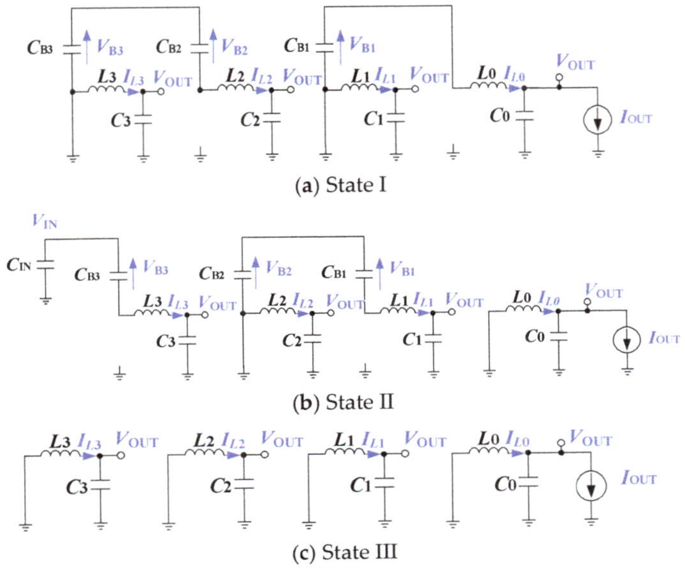

(a) State I

(b) State II

(c) State III

Figure 5. Equivalent circuit diagrams for the three distinct operational states for each phase.

Every switch controlled through PWM-1 will "OFF" and every switch controlled through PWM-2 will "ON" in state II. Consequently, the switches operated by PWM-1's complement will "ON," while the switches operated by PWM-2's complement will "OFF." According to Figure 5b, C_{B1} and C_{B3} are charged by the V_{IN} via L_0 and L_2. The load and C_{B3} are being concurrently discharged by C_{OUT} and C_{B2}, respectively. While this is happening, V_{IN} charges L_1 and L_3 and stores energy in them.

All switches under PWM-1 control will "OFF", while the switches under PWM-2 control will "OFF" in state III. The switches that are controlled by PWM-1 and PWM-2's complementary are therefore "ON". The flying capacitors do not charge or discharge, since all of the flying MOSFETs are deactivated. V_{IN} charges and stores energy in inductors L_0–L_3. The simultaneous supply of the load by the voltage on C_{OUT} is shown in Figure 5c.

The following equations are deduced based on (4) and the assumption that all flying MOSFETs' duty ratios D_i ($i = 1, 2, 3$) are set to be the same to D in order to realize an equitable current sharing condition across L_0–L_3.

$$\begin{cases} V_{B1}D = V_{OUT} \\ (V_{B2} - V_{B1})D = V_{OUT} \\ (V_{B3} - V_{B2})D = V_{OUT} \\ (V_{IN} - V_{B3})D = V_{OUT} \end{cases} \tag{6}$$

Therefore, (6) can be rewritten as

$$\begin{cases} V_{OUT} = \frac{D}{4}V_{IN} \\ V_{B1} = \frac{1}{4}V_{IN} \\ V_{B2} = \frac{2}{4}V_{IN} \\ V_{B3} = \frac{3}{4}V_{IN} \end{cases} \tag{7}$$

Based on (7), $C_{B(i)}$ ($i = 1,2,3$) throughout this converter owns an offset voltage $V_{B(i)}$ ($i = 1,2,3$) that resembles that of conventional converters.

According to (7), the steady-state voltage gain M with $N = 4$ is obtained as

$$M = \frac{V_{OUT}}{V_{IN}} = \frac{D}{4} \quad (8)$$

which matches (2) when N equals to 4.

In line with the operating principle illustrated above, each flying capacitor is continually softly charged/discharged, efficiently mitigating the loss made by the flying capacitors' voltage ripple owning to the converter's hybrid architecture, which interconnects the buck inductor with the SC stage. In conventional SC-based converters, this crucial function will prevent the inrush current. This eliminates the charge-sharing losses that normally occur during charging. Because of this, the hybrid converter proposed here will always be gently charged, no matter the tolerance of the flying capacitors.

2.3. Interleaving Procedure

With the right gate driving signal control, 360/4 interleaving between each module of the suggested converter might be put into practice. The timing diagram of the gate signals of Φ_i ($i = 0, 1, 2, 3$) is shown in Figure 6. This circuit combines a four-module switching capacitor converter with a standard four-phase interleaved buck converter. This may lead to the cancellation of four-phase current ripple and a significant decrease in current ripple. As a result, the current stress on the capacitors may be mitigated, and it would be possible to avoid using the enormous capacitor bank that is often needed to buffer the substantial current ripple. The corresponding current and voltage stresses for the switches are provided in Table 2.

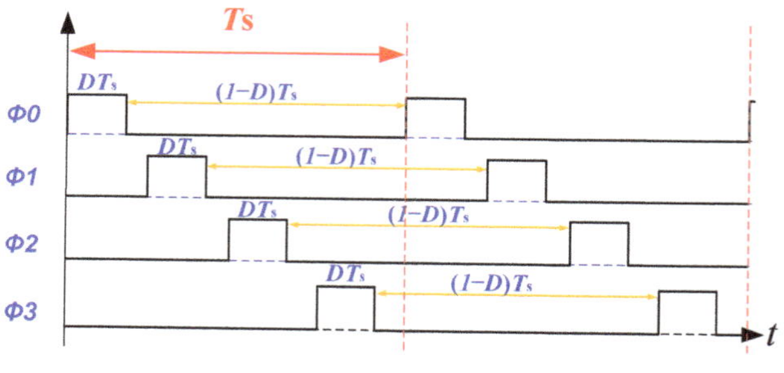

Figure 6. Timing schematic of the gate signals for the MOSFET Φ_i ($i = 0, 1, 2, \ldots, N-1$) with flawless interleaving operation under four modules in each phase for the suggested converter.

Table 2. Voltage and current stresses of the switches.

Power Switches	Voltage Stress	Current Stresss
Φ_{N-1}	(V_{OUT}/N)	Peak of $I_{L(N-1)}$
Φ_i ($i = 0, 1, \ldots, N-2$)	$2(V_{OUT}/N)$	Peak of I_{Li} ($i = 0, 1, \ldots, N-2$)
$\overline{\Phi_i}$ ($i = 0, 1, \ldots, N-1$)	(V_{OUT}/N)	Peak of I_{Li} ($i = 0, 1, \ldots, N-1$)

2.4. Circuit Design and Optimization

Switching frequency, the number of modules in the converter, maximum output current, input voltage, and output voltage are the key parameters that need to be specially designed for optimizing the layout of the circuit.

The inductance of the inductor for each module can be calculated using

$$L = \frac{V_{IN} \cdot (1 - D)}{\Delta I_L \cdot f_s} \quad (9)$$

where the inductor current ripple is $\Delta I_L = \alpha \cdot I_{O_{MAX}}$ and α is typically 0.2 or 0.3.

The capacitance of the flying capacitor for each module is derived using

$$C = \frac{I_{O_{MAX}} \cdot D}{\Delta V_C \cdot N \cdot f_s} \tag{10}$$

where ΔV_C is the voltage ripple across the flying capacitors. The capacitances are determined by the switching frequency, load current, duty cycles of the flying active switches, number of modules, and desired voltage ripples. The voltage ripples are required to be controlled at the lowest level for high efficiency.

For the power switches, the maximum voltages are obtained through a scaling function of N. All switches' current rating are obtained by the peak value of their inductor currents.

3. Simulation and Experimental Verifications

First, simulation was finished to confirm the facticity of the suggested converter in interleaved operation. Table 3 displays the simulated parameters. The proposed converter is regulated in the reverse mode. The input voltage of the prototype is 2.4 V, while the output voltage is 48 V, which is a typical DC bus voltage for low-voltage DC microgrids. Figure 7a represents the full-load current waveforms of L_1–L_4. The voltage waveforms of the capacitors C_{B1}–C_{B3} are shown in Figure 7b. The average voltages on C_{B1}–C_{B3} are 12 V, 24 V, and 36 V, respectively, with an output voltage of 48 V, as anticipated in (7). Figure 7c shows the input current and output voltage at full load. The voltage across FET Φ_1 and $\overline{\Phi_1}$ is also processed, as shown in Figure 7d. Based on Table 3, the highest voltage stress for FET Φ_i (i = 0, 1, 2) is 24 V. The highest voltage stress for FET $\overline{\Phi_i}$ (i = 0, 1, 2, 3) and Φ_3 is 12 V.

Table 3. Main parameters of the simulation as well as hardware.

Depiction	Symbols	Values
Inductor	L_0–L_3	0.2 µH
Flying Capacitor	C_{B1}–C_{B3}	47 µF
Resistor Load	R_{OUT}	6.95 mOhm
Output Voltage	V_{OUT}	2.4 V
Output Power	P_{OUT}	900 W
Input Voltage	V_{IN}	48 V

A prototype of the suggested hybrid DC/DC converter was built with M = 3, meaning three phases were active, and N = 4, meaning four modules were activated in each phase (see Figure 8). The hardware prototype consists of three stages. The power rating of each phase equals to 300 W, and V_{OUT} equals to 2.4 V; thus, the output current of each phase equals to 125 A. Each step of the prototype model consists of four parts. With an output voltage of 2.4 V, each module's flying capacitors and power inductor will carry around 31 A of current, while each module's power inductor will carry about 31 A of current. Three different types of components are used: (a) power MOSFETs, (b) inductors, and (c) capacitors. Three PCB boards, such as a DSP control board, a power stage board, and a signal processing board, are used in each step of the setup. The power stage board also comprises six components. The board can function in a four-module configuration when four modules are active and two are deactivated. A complete list of the parts used in the setup is provided in Table 4. The DSP TMS320F28335 is used to build the digital closed-loop controller. BSC009NE2LS5 MOSFETs are being used. The MOSFET has a turn-on resistance of less than 1 mΩ and a rated voltage of 30 V. The drivers and control logic circuitry of the MOSFET would be further manufactured and combined onto a semiconductor chip to provide a more desirable way. The required ripple voltage, the flying active switches' duty cycle, the number of modules, and the output current at full-load frequency all affect the flying capacitor's capacitance. Voltage ripple should be as little as feasible to maximize efficiency. The efficiency of the flying capacitor will be enhanced by its own value. The capacitances of the flying capacitors are 47 µF, with the ripple of the peak-to-peak capacitor voltage being lower than 1 V. Flying capacitors may be thought of as a continuous voltage source/sink because of their low voltage ripple in comparison to their DC value. Electrical properties like input impedance, power losses, etc., are often connected with capacitance levels in the most of DC-DC converters. To produce less voltage ripple than DC flying capacitors, Class-II multilayer ceramic capacitors (MLCCs) with low equivalent series resistance (ESR) would be adopted. Additionally, to achieve a higher power density, class-II MLCCs sometimes provide a high capacitance per unit area. The flying capacitors employed in this setup are Class-II (e.g., X7R, X6S, etc.) MLCC capacitors. More capacitors should be added in parallel if the current rating of the selected capacitor is insufficient. For this setup, ten 4.7 μF MLCCs in parallel connection are adopted for each module, and the total capacitance is 47 μF. The current grade for every MLCC equals 4 A. The size of the MLCC piece is 0.6 mm × 0.3 mm. According to the operating principle of our suggested design, because of the hybrid design of the suggested converter, which connects the buck inductor to the SC stage, each flying capacitor can be softly charged/discharged during all operating periods. In conventional switched capacitor-based converters, this crucial characteristic will prevent the inrush current. Therefore, complete soft-charging functioning and low device strains are accomplished under all working circumstances.

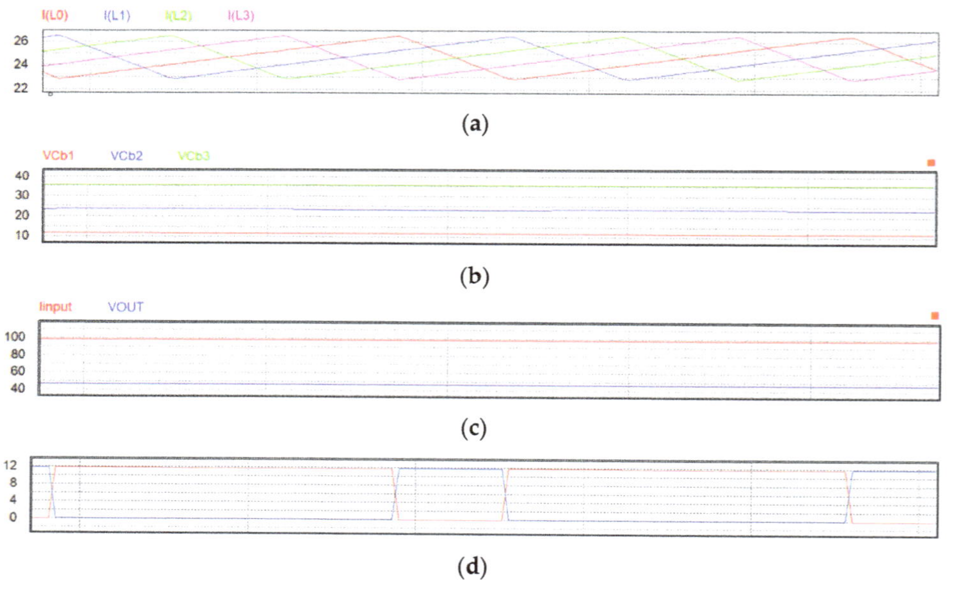

Figure 7. Simulated results of the suggested converter in the reverse operation mode with (**a**) inductor L_0–L_3 current (A), (**b**) voltage for the converter capacitor C_{B1}–C_{B3} (V), (**c**) input current (A) and output voltage (V); and (**d**) device Φ_3 and $\overline{\Phi_3}$ drain-to-source voltage (V).

In general, the current ratings of all power MOSFET switches may be derived from the peak inductor current. In reality, all switch current ratings are identical to power inductor current values. The power MOSFET utilized in the experimental setup is manufactured by Infineon and has the component number BSC009NE2LS5. The device's current flow under full load is substantially less than the maximum continuous drain current of 100 A. Full load current may flow via this MOSFET. This power MOSFET is put in a 4 mm × 4 mm package to power both the bottom and flying MOSFETs. The size of power MOSFETs may be decreased via system integration by further combining a pair of flying MOSFETs and a bottom MOSFET into a single package (which is not shown here to illustrate the concept). The prototype power inductor is a Wurth part with part number 744323020, based on the current that each module's power inductor conducts. The saturation current of the inductor equals 52 A, which is a lot larger than its peak current. Its dimensions are 10.2 mm by 10.2 mm (with a height equal to 5 mm). The test condition and the simulation condition are the same, as shown in Table 3.

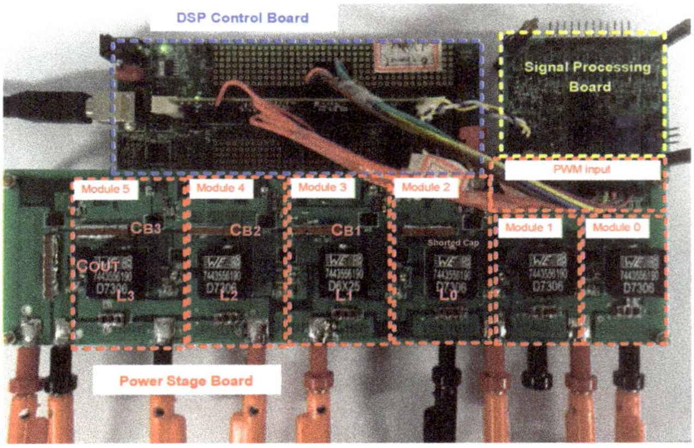

(**a**)

(**b**)

Figure 8. Experimental setup of the suggested converter (**a**) single-phase for a power rating of 300 W; (**b**) three-phase for a power rating of 900 W.

Table 4. Components for the prototype.

Depiction	Part#
Inductor	744323020 (0.2 µH)
Level Shifter	ADUM5240
Switching Device	BSC009NE2LS5
Gate Driver	LTC4440
Flying Capacitor *CB*1	C3216X7R1H475K160AC
Flying Capacitor *CB*2	C3216X7R1H475K160AC
Flying Capacitor *CB*3	C3216X7R1H475K160AC
Digital Controller	TMS320F28335

Figure 9 illustrates the observed waveforms of the current and voltage. Figure 9a illustrates the observed PWM signals of Φ_i $(i = 0, 1, 2, 3)$, and Figure 9b represents the observed waveforms of the output voltage and capacitor voltages C_{B1}–C_{B3}. As demonstrated, the voltage on C_{B1} is 12 V, C_{B2} is 24 V, and C_{B3} is 36 V, which corresponds to the previous analysis. Figure 9c illustrates the current waveforms flow through L_0–L_3 under full load conditions (i.e., 900 W for 3 phases). The observed drain-to-source voltage's waveforms for Φ_1 and $\overline{\Phi_1}$ are shown in Figure 9d. The aforementioned findings indicate that the suggested converter can realize a four-phase interleaved operation.

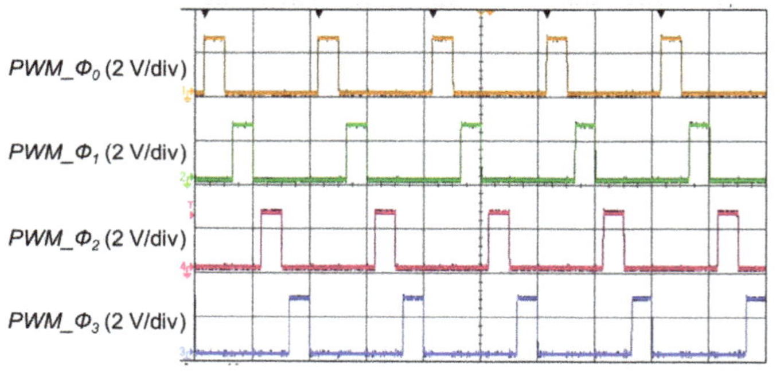

(a)

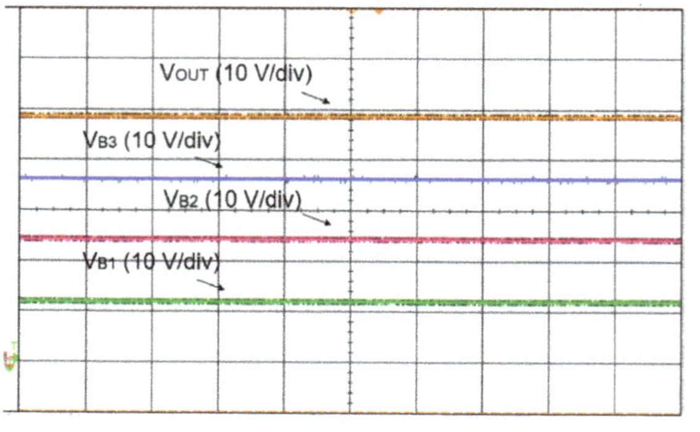

(b)

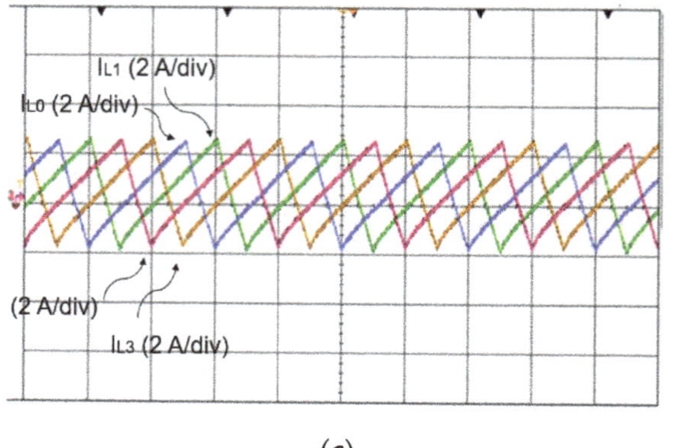

(c)

Figure 9. *Cont.*

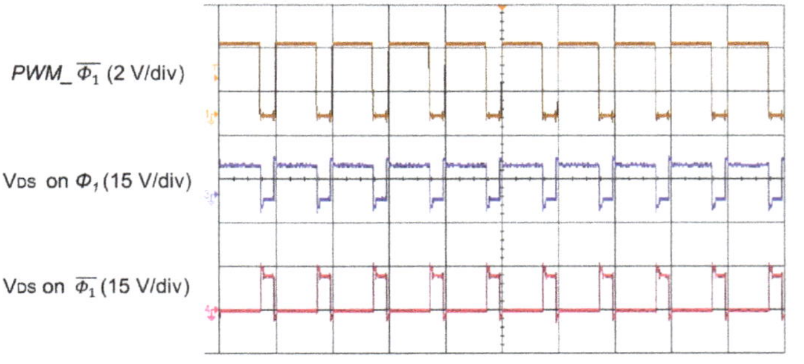

(d)

Figure 9. (**a**) PWM signals on MOSFETs Φ_i ($i = 0, 1, 2, 3$) (**b**) voltages cross C_{B1}–C_{B3} and output voltage (**c**) currents for L_0–L_3 (**d**) drain-to-source voltage on Φ_1 and $\overline{\Phi_1}$.

The observed efficiency of the suggested converter switching at 1 MHz while operating in interleaved mode is shown in Figure 10. The input/output voltage are observed through the fluke multimeters under the high-resolution mode to achieve the most accurate results. The output current is tested with programmable chroma loads. The converter's measured maximum efficiency is 92.5%, as depicted in Figure 10.

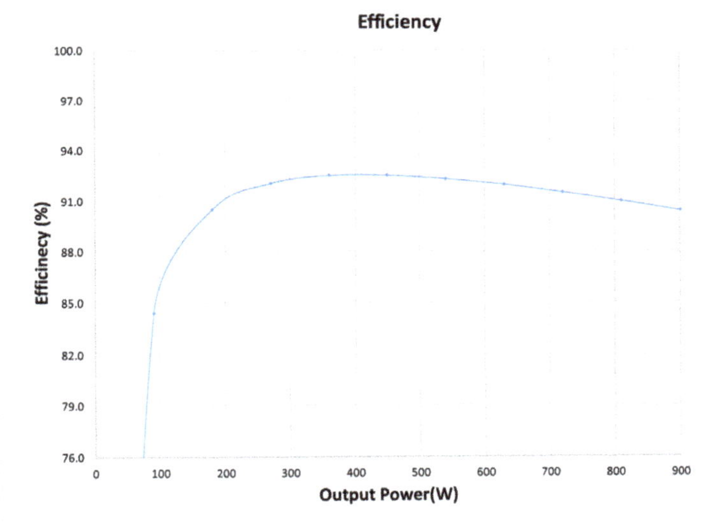

Figure 10. Observed efficiency for the converter under 2.4 V output voltage and 900 W output power.

Similar investigations were also conducted for the suggested circuit with $N = 5$. Four modules are with the load conditions in Table 3, and one module is load-free. The switching signals of the four modules with loads are shown in Figure 11a, while the switching signals of the rest module are shown in Figure 11b. The corresponding waveforms of V_{OUT}, I_{OUT}, and V_{DS} for Φ_1 are presented in Figure 12.

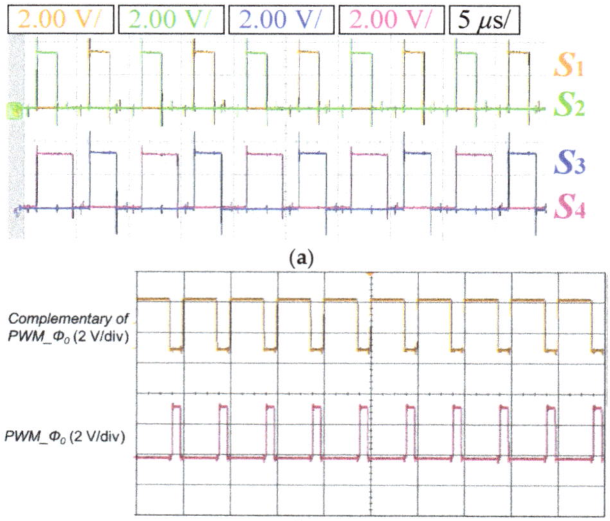

Figure 11. (**a**) PWM signals on MOSFETs Φ_i $(i = 1, 2, 3, 4)$ and (**b**) PWM signals of MOSFETs Φ_0 and $\overline{\Phi_0}$ of the studied circuit with $N = 5$.

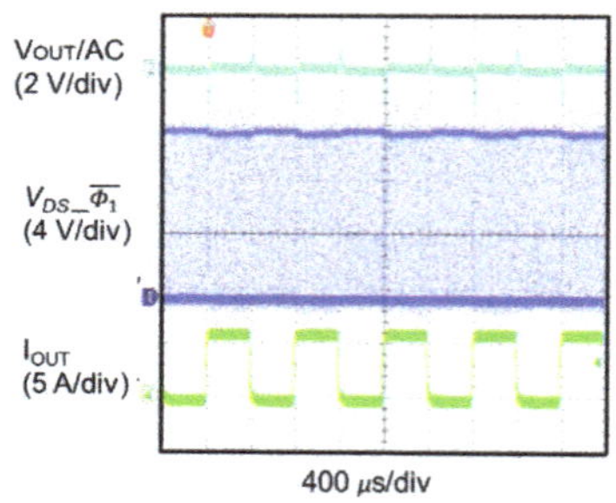

Figure 12. Waveforms of V_{out}, I_{out}, and V_{DS} for Φ_1 for the proposed circuit with $N = 5$.

4. A Comparison for Several Types of High-Voltage-Gain-Converters

Adopting two-stage DC/DC conversion is one common method for achieving high-current capacity, high-voltage-gain, and appropriate regulation all at the same time, as was previously described. Figure 13a shows a typical two-stage system for high-current and -power applications. A four-phase buck converter is formed as the first stage, while the second is a switched-capacitor converter with a 4:1 ratio. For applications requiring high-voltage-gain, this two-stage converter is often employed. Compared to single-phase buck operation, four-phase buck operation can achieve precise regulation with rapid reaction and readily manage high-power and high-current strains. Figure 13b shows the suggested converter acquiring the same voltage gain and power rating as the two-stage method to allow for a fair comparison. Given the same voltage gain and power rating, the two-stage system has eighteen active switches in each phase, while the suggested hybrid converter has eight active switches in each phase. All switches from the SC converter stage can be saved, resulting in a low BOM cost. With just one switching stage and fewer switching devices needed, the suggested converter may also achieve substantially greater efficiency with only one set of active switches and lower switching losses. In addition, Table 5 compares the performance of the suggested converter to converters from earlier stages. The recommended converter is proven to have a higher efficiency with the same voltage gain.

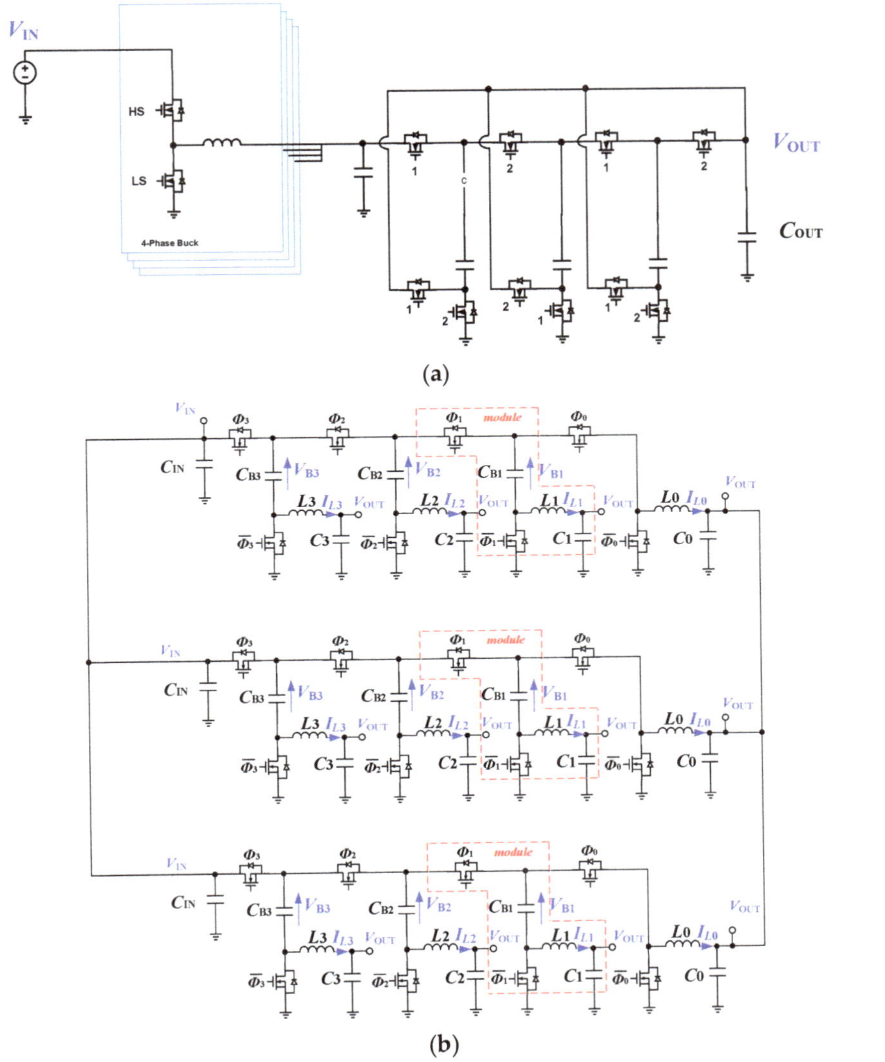

Figure 13. (**a**) Traditional two-stage solution with high-voltage-gain step-down conversion for high-current and -power devices. (**b**) Suggested conveter.

Table 5. Comparisons for the high-voltage-gain converters.

Converter in	Voltage Gain (Times)	Scalability	Switching Frequency (kHz)	Peak Efficiency (%)
[33]	20	Poor	1000	90%
Cascaded multi-phase buck	20	Medium	1000	91.7%
Proposed	20	Very good	1000	92.5%

5. Conclusions

In this work, a novel class of high-frequency transformerless converters is presented for high-efficiency applications that need high-voltage step-down ratios. The characteristics can be summarized as follows.

a. A high-voltage step-down ratio that is adjustable and has a medium duty cycle.
b. High efficiency due to the employment of low-voltage, high-powered switching devices and the smaller number of MOSFETs in the single merging stage, making it suitable for high-frequency operation.
c. Due to the interleaved operation, there is no pulsing current and minimal current ripple.
d. The total cost is relatively low because of the hybrid design using less MOSFETs and the integration of two-stage converters to single-stage converters.
e. Inherent modularity and scalability for high-power applications.
f. Mitigation current and voltage spike issues and electro-magnetic interference (EMI) concerns as a result of the flying capacitors' soft-charging action.

The research analyzes the recommended converter's steady-state performance. A 48 V to 2.4 V, 900 W converter system was created to illustrate the benefits of the recommended topology. The highest efficiency was 92.5 percent. The validity of the theoretical analysis was determined by simulation and experimentation. Applications needing a high-current, high-voltage-gain, and high-power hold a lot of potential for the recommended converter. The suggested hybrid converter may therefore be used in power conversion applications.

Author Contributions: Conceptualization, H.X. and Y.M.; methodology, H.X.; software, J.L.; validation, H.X., B.X. and X.J.; formal analysis, H.X.; investigation, H.X. and J.L.; resources, Y.M.; data curation, H.X.; writing—original draft preparation, H.X.; writing—review and editing, Y.M.; visualization, J.L., B.X. and X.J.; supervision, Y.M.; project administration, Y.M. All authors have read and agreed to the published version of the manuscript.

Funding: This research received no external funding.

Data Availability Statement: Data is unavailable due to ethical restrictions.

Conflicts of Interest: The authors declare no conflict of interest.

References

1. Guerrro, J.M.; Vasquez, J.; Matas, J.; de Vicuna, L.; Castilla, M. Hierarchical control of droop-controlled AC and DC microgrids—A general approach toward standardization. *IEEE Tran. Ind. Electron.* **2011**, *58*, 158–172. [CrossRef]
2. Yang, Y.; Mok, K.; Tan, S.; Hui, S.Y.R. Nonlinear dynamic power tracking of low-power wind energy conversion system. *IEEE Tran. Power Electron.* **2015**, *30*, 5223–5236. [CrossRef]
3. Huang, Y.; Mei, Y.; Xiong, S.; Tan, S.; Tang, C.; Hui, S.Y.R. Reverse electrodialysis energy harvesting system using high-gain step-up dc/dc converter. *IEEE Trans. Sustain. Energy* **2018**, *9*, 1578–1587. [CrossRef]
4. Huang, Y.; Xiong, S.; Tan, S.; Hui, S.Y. Non-isolated harmonics-boosted resonant DC/DC converter with high-step-up gain. *IEEE Trans. Power Electron.* **2017**, *9*, 7770–7781.
5. Yang, Y.; Tan, S.; Hui, S.Y.R. Mitigating distribution power loss of DC microgrids with DC electric springs. *IEEE Trans. Smart Grid* **2018**, *9*, 5897–5906. [CrossRef]
6. Pan, C.T.; Lai, C.M. A high-efficiency high step-up converter with low switch voltage stress for fuel-cell system applications. *IEEE Trans. Ind. Electron.* **2010**, *57*, 1998–2006.
7. Lineykin, S.; Ben-Yaakov, S. Modeling and analysis of thermoelectric modules. *IEEE Trans. Ind. Electron.* **2007**, *43*, 505–512. [CrossRef]
8. Huber, L.; Jovanovic, M.M. A design approach for server power supplies for networking applications. In Proceedings of the Fifteenth Annual IEEE Applied Power Electronics Conference and Exposition, New Orleans, LA, USA, 6–10 February 2000; pp. 1163–1169.
9. Feng, X.G.; Liu, J.; Lee, F.C. Impedance specifications for stable dc distributed power systems. *IEEE Trans. Power Electron.* **2002**, *17*, 157–162. [CrossRef]
10. Schaef, C.; Rentmeister, J.; Stauth, J.T. Multimode Operation of Resonant and Hybrid Switched-Capacitor Topologies. *IEEE Trans. Power Electron.* **2018**, *33*, 10512–10523. [CrossRef]
11. Schaef, C.; Din, E.; Stauth, J.T. A Hybrid Switched-Capacitor Battery Management IC With Embedded Diagnostics for Series-Stacked Li–Ion Arrays. *IEEE J. Solid-State Circuits* **2017**, *52*, 3142–3154. [CrossRef]

12. Le, H.; Crossley, J.; Sanders, S.; Alon, E. A sub-ns response fully integrated battery-connected switched-capacitor voltage regulator delivering 0.19W/mm2 at 73% efficiency. In Proceedings of the 2013 IEEE International Solid-State Circuits Conference Digest of Technical Papers, San Francisco, CA, USA, 17–21 February 2013; pp. 372–373.
13. Le, H.; Sanders, S.; Alon, E. Design Techniques for Fully Integrated Switched-Capacitor DC-DC Converters. *IEEE J. Solid-State Circuits* **2011**, *46*, 2120–2131. [CrossRef]
14. Li, K.; Hu, Y.; Ioinovici, A. Generation of the large DC gain step-up non-isolated converters in conjunction with renewable energy sources starting from a proposed geometric structure. *IEEE Trans. Power Electron.* **2017**, *32*, 5323–5340. [CrossRef]
15. Wu, G.; Ruan, X.; Ye, Z. Non-isolated high step-up DC–DC converters adopting switched-capacitor cell. *IEEE Trans. Ind. Electron.* **2015**, *62*, 383–393. [CrossRef]
16. Dickson, J. On-chip high-voltage generation in mnos integrated circuits using an improved voltage multiplier technique. *IEEE J. Solid-State Circuits* **1976**, *11*, 374–378. [CrossRef]
17. Chen, M.; Li, K.; Hu, J.; Ioinovici, A. Generation of a family of very high DC gain power electronics circuits based on switched-capacitor-inductor cells starting from a simple graph. *IEEE Trans. Circuits Syst.* **2016**, *63*, 2381–2392. [CrossRef]
18. Eguchi, K.; Ueno, F.; Zhu, H.; Tabata, T.; Fujiyoshi, A. An IC chip of a Dickson-type DC-DC converter with bootstrapped gate transfer switches. In Proceedings of the IEEE Region 10 Conference TENCON 2004, Chiang Mai, Thailand, 24 November 2004; pp. 77–80.
19. Kiratipongvoot, S.; Tan, S.; Ioinovici, A. Phase-shift interleaving control of variable-phase switched-capacitor converters. *IEEE Trans. Ind. Electron.* **2013**, *60*, 5575–5584. [CrossRef]
20. Cheung, C.K.; Tan, S.; Tse, C.; Ioinovici, A. On energy efficiency of switched-capacitor converters. *IEEE Trans. Power Electron.* **2013**, *28*, 862–876. [CrossRef]
21. Law, K.K.; Cheng, K.; Yeung, Y.P.B. Design and analysis of switched-capacitor-based step-up resonant converters. *IEEE Trans. Circuits Syst.* **2005**, *52*, 943–948. [CrossRef]
22. Ye, Y.; Cheng, K.E.; Liu, J.; Xu, C. A family of dual-phase-combined zero-current switching switched-capacitor converters. *IEEE Trans. Power Electron.* **2014**, *29*, 4209–4218. [CrossRef]
23. Ben-Yaakov, S. Behavioral average modeling and equivalent circuit simulation of switched capacitors converters. *IEEE Trans. Power Electron.* **2012**, *27*, 632–636. [CrossRef]
24. Rentmeister, J.S.; Stauth, J.T. Modeling the Dynamic Behavior of Hybrid-Switched-Capacitor Converters in State Space. In Proceedings of the 2018 IEEE 19th Workshop on Control and Modeling for Power Electronics (COMPEL), Padova, Italy, 25–28 June 2018; pp. 1–7.
25. Schaef, C.; Stauth, J.T. A 12-volt-input hybrid switched capacitor voltage regulator based on a modified series-parallel topology. In Proceedings of the 2017 IEEE Applied Power Electronics Conference and Exposition (APEC), Tampa, FL, USA, 26–30 March 2017; pp. 2453–2458.
26. Farth, H.M.H.; Othman, M.; Eltamaly, A.; Al-Saud, M.S. Maximum power extraction from a partially shaded PV system using an interleaved boost converter. *Energies* **2018**, *11*, 2543. [CrossRef]
27. Farth, H.M.H.; Eltamaly, A.; Al-Saud, M.S. Interleaved boost converter for global maximum power extraction from the photovoltaic system under partial shading. *IET Renew. Power Gener.* **2019**, *13*, 1232–1238. [CrossRef]
28. Cervera, A.; Ben-Yaakov, S.; Peretz, M. Single-stage switched-resonator converter topology with wide conversion ratio for volume-sensitive applications. In Proceedings of the 2017 IEEE Applied Power Electronics Conference and Exposition (APEC), Tampa, FL, USA, 26–30 March 2017; pp. 1706–1712.
29. Schaef, C.; Din, E.; Stauth, J.T. 10.2 A digitally controlled 94.8%-peak-efficiency hybrid switched-capacitor converter for bidirectional balancing and impedance-based diagnostics of lithium-ion battery arrays. In Proceedings of the 2017 IEEE International Solid-State Circuits Conference (ISSCC), San Francisco, CA, USA, 5–9 February 2017; pp. 180–181.
30. Schaef, C.; Reese, B.; Sullivan, C.; Stauth, J.T. Design aspects of multi-phase interleaved resonant switched-capacitor converters with mm-scale air-core inductors. In Proceedings of the 2015 IEEE 16th Workshop on Control and Modeling for Power Electronics (COMPEL), Vancouver, BC, Canada, 12–15 July 2015; pp. 1–5.
31. Schaef, C.; Kesarwani, K.; Stauth, J.T. 20.2 A variable-conversion-ratio 3-phase resonant switched capacitor converter with 85% efficiency at 0.91W/mm^2 using 1.1nH PCB-trace inductors. In Proceedings of the 2015 IEEE International Solid-State Circuits Conference—(ISSCC) Digest of Technical Papers, San Francisco, CA, USA, 22–26 February 2015; pp. 1–3.
32. Cao, D.; Peng, F.Z. A family of zero current switching switched-capacitor dc-dc converters. In Proceedings of the 2010 Twenty-Fifth Annual IEEE Applied Power Electronics Conference and Exposition (APEC), Palm Springs, CA, USA, 21–25 February 2010; pp. 1365–1372.
33. Krismer, F.; Biela, J.; Kolar, J.W. A comparative evaluation of isolated bi-directional DC/DC converters with wide input and output voltage range. In Proceedings of the Fourtieth IAS Annual Meeting, Conference Record of the 2005 Industry Applications Conference, Hong Kong, China, 2–6 October 2005; pp. 599–606.
34. Krismer, F.; Kolar, J.W. Closed form solution for minimum conduction loss modulation of DAB converters. *IEEE Trans. Power Electron.* **2012**, *27*, 174–188. [CrossRef]

35. Shang, F.; Krishnamurthy, M.; Isurin, A. A novel high gain step-up resonant DC-DC converter for automotive application. In Proceedings of the 2016 IEEE Applied Power Electronics Conference and Exposition (APEC), Long Beach, CA, USA, 20–24 March 2016; pp. 880–885.
36. Hwu, K.I.; Jiang, W.; Yau, Y.T. An isolated high step-up converter with continuous input current and LC snubber. In Proceedings of the 2016 IEEE Applied Power Electronics Conference and Exposition (APEC), Long Beach, CA, USA, 20–24 March 2016; pp. 2415–2421.

energies

MDPI

Article

Quasi-Resonant Single-Switch High-Voltage-Gain DC-DC Converter with Coupled Inductor and Voltage Multiplier Cell

Giordano Luigi Schiavon, Eloi Agostini, Jr. * and Claudinor Bitencourt Nascimento

Department of Electrical Engineering, Federal University of Technology–Parana–UTFPR, Ponta Grossa 84017-220, PR, Brazil
* Correspondence: eloiagostini@utfpr.edu.br

Abstract: This paper introduces a quasi-resonant high-efficiency high-step-up DC–DC converter requiring a reduced number of components. The proposed circuit uses a coupled inductor associated with voltage multiplier cells to ensure high-voltage-gain operation without the necessity of an extremely high number of turns ratio. Quasi-resonant operation guarantees zero current switching (ZCS) for some diodes of the converter. A detailed steady-state analysis is carried out aiming at the adequate design of the circuit. Experimental results taken from the testing of a 400 W prototype operating in closed loop with an input voltage range of 25–48 V, output voltage of 400 V and switching frequency of 100 kHz validate the analysis carried out and demonstrate the feasibility of the proposed converter.

Keywords: DC–DC converter; high voltage gain; coupled inductor; voltage multiplier cell

Citation: Schiavon, G.L.; Agostini, E., Jr.; Nascimento, C.B. Quasi-Resonant Single-Switch High-Voltage-Gain DC-DC Converter with Coupled Inductor and Voltage Multiplier Cell. *Energies* **2023**, *16*, 3874. https://doi.org/10.3390/en16093874

Academic Editor: Salvatore Musumeci

Received: 6 March 2023
Revised: 26 April 2023
Accepted: 27 April 2023
Published: 3 May 2023

1. Introduction

Techniques associated with the generation of electrical energy have been improved to address the challenges associated with climate change and contribute to sustainable development. In this sense, production of electrical energy from renewable resources, such as wind, solar and biomass, is not an alternative anymore but a necessity [1,2]. Considering photovoltaic (PV) and fuel cell (FC) systems, the voltage levels are usually low and in the form of direct current (DC), diverging from the usual alternate current (AC) system that prevails in the distribution of electricity in most segments [3]. Thus, in order to adapt the voltage levels while taking into account technical and economic aspects, electronic energy processing is mandatory, from which low cost and high efficiency are expected. In this sense, high-step-up DC–DC converters are widely employed in applications where the primary energy source is characterized by a voltage level much lower than that required for the end use [4–8].

Providing energy to a DC distribution system or conditioning the voltage as an intermediate stage of a DC–AC conversion system are some applications of DC–DC converters. One of the main challenges in the use of such converters is to avoid expressive switching and conduction losses. Therefore, using circuits with a reduced number of components, featuring low-voltage stress on the semiconductors and with the ability of recycling the leakage energy are key to improving the overall system efficiency. However, achieving all these characteristics is not an easy task. One classic example is the boost converter, which is indeed a simple circuit but exhibits poor performance under high-voltage-gain conditions [9]. Addressing the limitations of the boost converter by the use of techniques to improve the voltage gain has been extensively investigated in the literature [10–12]. Among the several techniques discussed in [12], one can be highlighted: boost-based solutions using a single active switch, a single coupled inductor and voltage multiplier cells (VMCs), such as the boost-flyback and the boost associated with VMCs introduced in [13] and [14], respectively, in addition to other solutions recently reported in the literature [15–19]. In all these examples, voltage gain is dependent on the duty cycle (D) and on the number of turns ratio (n) of the coupled inductor. As a result, higher voltage conversion ratios can be achieved without extreme duty cycle values, and therefore a higher efficiency is expected, since conduction and switching losses are lower compared to the conventional step-up converters.

Converters employing a single active switch and reduced number of diodes and capacitors are usually not suited for applications where the power processed is higher than a few hundred watts. However, these circuits are interesting for processing energy from PV modules rated in the range of 200–600 W. In this sense, this work introduces a high-voltage-gain DC–DC converter based on the boost configuration using coupled inductor and voltage multiplier cells, as depicted in Figure 1. The key features of the proposed circuit that make it suitable for PV applications are: high efficiency; reduced number of components; low voltage stress on the semiconductors; and voltage distribution among the output capacitors. In addition, quasi-resonant characteristics guarantee ZCS operation for diodes D_2 and D_3, thus contributing to improving the system efficiency [20–22]. In addition, a high voltage conversion ratio can be achieved if compared with similar topologies, even though a relatively low number of turns ratio of the coupled inductor is adopted. Therefore, weight, volume and magnetic losses can be minimized.

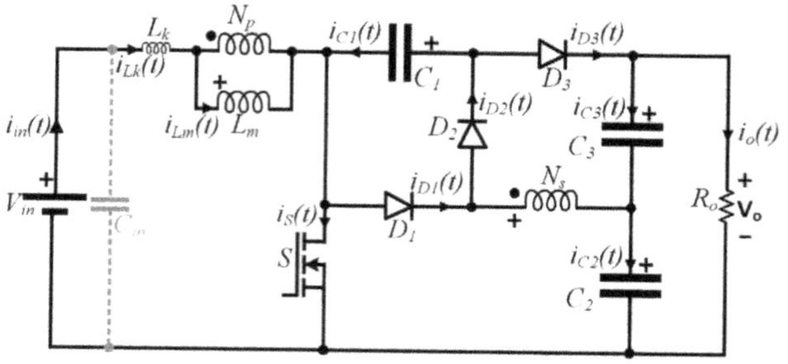

Figure 1. Proposed high-voltage-gain DC–DC converter employing coupled inductor and voltage multiplier cells.

2. Principle of Operation in Steady State

The proposed high-step-up DC–DC converter depicted in Figure 1 contains a VMC composed of the winding N_s of the coupled inductor associated with the pairs D_2/C_2 and D_3/C_3. Since this topology is able to impose high voltage levels on the capacitors, adopting a number of turns ratio close to one is possible without the need for overly high values of duty cycle. The steady-state analysis of the proposed converter is carried out for deriving a mathematical model that allows the proper choice of every component contained in the circuit. In this sense, the following assumptions are made:

- All elements are considered ideal, except for the leakage inductance of the coupled inductor;
- The voltage on the capacitors and the magnetizing current are assumed to be ripple free;
- The analysis is carried out within one period of the switching frequency f_s.

Six operating stages are verified from the analysis of the converter during one switching period, as depicted in Figure 2. From these six stages, only four are relevant in terms of the energy processed by the system. Hence, transition stages regarding the intervals Δt_1 and Δt_4 will be neglected from the mathematical analysis.

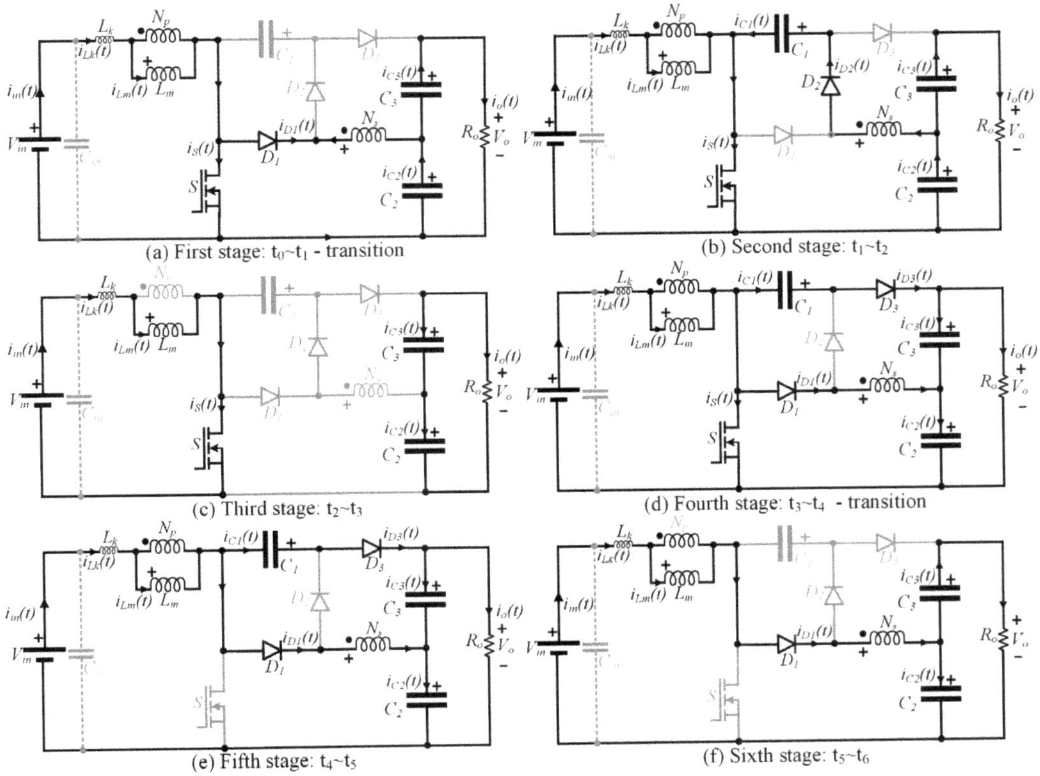

(a) First stage: t_0~t_1 - transition (b) Second stage: t_1~t_2 (c) Third stage: t_2~t_3 (d) Fourth stage: t_3~t_4 - transition (e) Fifth stage: t_4~t_5 (f) Sixth stage: t_5~t_6

Figure 2. Operating stages in steady state. First and fourth stages correspond to brief transition intervals that play minor roles in the energy processed by the system.

During the second and third stages, which correspond to the intervals Δt_2 and Δt_3, respectively, switch S is turned on, the magnetizing inductance L_m stores energy, and capacitors C_2 and C_3 provide energy that has been stored during the previous stages to the system. These intervals are characterized as linear stages, since resonance between the leakage inductance L_k and the capacitors contained in the circuit is not expressive. This operational condition is finished when S is turned off.

The fifth and sixth stages, defined by the intervals Δt_5 and Δt_6, occur when S remains in the off state. During these stages, energy stored in L_m is provided to the circuit. Resonance between L_k, C_1 and C_3 becomes expressive during the fifth stage. In this sense, if it is guaranteed that the resonant frequency is higher than the switching frequency, D_2 operates with ZCS, thus contributing to increasing the system efficiency. It is noteworthy that the fifth stage plays a crucial role in defining the voltage gain of the converter. Regarding the sixth stage, the converter returns to a linear operating characteristic.

Figure 3 presents the main theoretical waveforms regarding the converter operation in steady state. As can be seen, an abrupt change in i_{Lk} occurs at the end of interval Δt_3, which is a simplifying assumption justified by the fact that the fourth stage corresponds to a very brief transition state, during which the energy processed can be neglected.

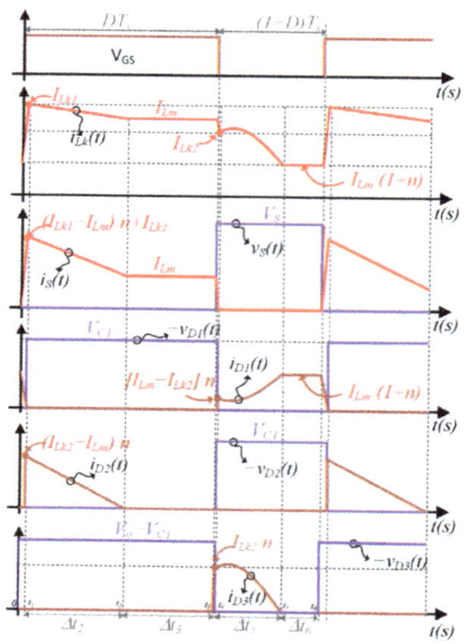

Figure 3. Main theoretical waveforms within one switching period for steady-state operation.

3. Mathematical Model

Fundamental information regarding the operation of the proposed converter can be extracted from the mathematical model of the circuit, such as the voltage conversion ratio and voltage stress on the semiconductor devices. Such knowledge is key to designing the circuit properly, aiming at reduced conduction and switching losses to ensure high-efficiency operation.

3.1. Steady-State Analysis

The proposed mathematical model takes into consideration the following definitions:

$$\begin{aligned} &\Delta t_i = t_i - t_{i-1};\ T_s = \tfrac{1}{f_s};\ n = \tfrac{N_s}{N_p};\ \lambda = \tfrac{L_k}{L_m};\ k = \tfrac{L_m}{L_k+L_m}; \\ &M = \tfrac{V_o}{V_{in}};\ M_{C1} = \tfrac{V_{C1}}{V_{in}};\ M_{C2} = \tfrac{V_{C2}}{V_{in}};\ M_{C3} = \tfrac{V_{C3}}{V_{in}}, \end{aligned} \tag{1}$$

where i defines the interval of each stage beginning at t_{i-1} and ending at t_i; T_s corresponds to the switching period; f_s is the switching frequency; n defines the number of turns ratio of the coupled inductor; the factors λ and k relate the magnetizing and leakage inductances; and M, M_{C1}, M_{C2} e M_{C3} correspond to the total and partial static gains of the circuit.

The fundamental equation for $i_{Lk}(t)$ valid for the interval Δt_2 is given by (2).

$$I_{Lm} = I_{Lk1} + \left(1 - \frac{M_{C1} - M_{C2}}{n}\right)\frac{\Delta t_2}{\lambda L_m}V_{in} \tag{2}$$

The energy balance of the algebraic sum of the leakage and magnetizing inductances yields (3).

$$M_{C2} + M_{C3} - M_{C1} = \frac{1}{1-D} \tag{3}$$

Equation (4) can be determined from the analysis of the equivalent circuit of the second stage depicted in Figure 2b.

$$M_{C1} - M_{C2} = nk \tag{4}$$

As previously mentioned, resonance between L_k and the pair C_1/C_3 is expressive during the fifth stage of the proposed high-step-up DC–DC converter. Thus, the analysis of this stage can be carried out using the current and voltage on these elements.

Currents on C_1 and C_3 can be written in terms of $i_{Lk}(t)$ as given by (5) and (6), respectively.

$$i_{C1}(t) = \frac{I_{Lm}}{n} - \frac{(1+n)}{n} i_{Lk}(t) \tag{5}$$

$$i_{C3}(t) = -\left(\frac{I_{Lm}}{n} + I_o\right) + \frac{(1+n)}{n} i_{Lk}(t) \tag{6}$$

Equation (7) provides the voltage on the leakage inductance during the resonant stage.

$$v_{Lk}(t) = V_{in} - V_o + \frac{(1+n)}{n} v_{C1}(t) - \frac{1}{n} v_{C3}(t) \tag{7}$$

Differentiating (7) results in

$$\frac{dv_{Lk}(t)}{dt} = \frac{(1+n)}{n}\frac{dv_{C1}(t)}{dt} - \frac{1}{n}\frac{dv_{C3}(t)}{dt}. \tag{8}$$

Equation (8) can be rewritten in terms of the currents on C_1 and C_3, as given by

$$L_k \frac{d^2 i_{Lk}(t)}{dt^2} = \frac{(1+n)}{nC_1} i_{C1}(t) - \frac{1}{nC_3} i_{C3}(t). \tag{9}$$

Substituting (5) and (6) into (9) results in the differential equation

$$L_k C_{eq} \frac{d^2 i_{Lk}(t)}{dt^2} + i_{Lk}(t) = A, \tag{10}$$

where

$$C_{eq} = \frac{n^2}{(1+n)}\left[\frac{C_1 C_3}{C_1 + (1+n)C_3}\right], \tag{11}$$

$$A = \frac{1}{(1+n)}\left[I_{Lm} + \frac{nC_1}{C_1 + (1+n)C_3} I_o\right]. \tag{12}$$

Applying the Laplace transform in (10) results in (13), which represents the current $i_{Lk}(t)$ in the s domain.

$$I_{Lk}(s) = \frac{\frac{A\omega_o^2}{s} + sI_{Lk2}}{s^2 + \omega_o^2} \tag{13}$$

where

$$\omega_o = \frac{1}{\sqrt{L_k C_{eq}}}. \tag{14}$$

At this time, current $i_{Lk}(t)$ during the fifth stage can be determined by applying the inverse Laplace transform in (13), as given by (15). It is noteworthy that the resonant frequency f_o must be higher than the switching frequency f_s to guarantee ZCS for D_3.

$$i_{Lk}(t) = \left(V_o - \frac{1+nD}{1-D} V_{in} - V_{C3}\right)\frac{\sin(\omega_o t)}{nL_k\omega_o} + A + (I_{Lk2} - A)\cos(\omega_o t) \tag{15}$$

Differentiating (15) and multiplying the result by L_k provides the value of v_{Lk}, as given by

$$v_{Lk}(t) = \frac{1}{n}\left[V_o - V_{C3} - \frac{(1+nD)V_{in}}{(1-D)}\right]\cos(\omega_o t) + \omega_o L_k (I_{Lk3} - A)\sin(\omega_o t). \tag{16}$$

Assuming that at $t = 0$ v_{Lk} is approximately equal to 0, one can derive the partial static gain M_{C3} as

$$M_{C3} = M - \frac{1+nD}{1-D}. \tag{17}$$

The circuit analysis reveals that the static gain M corresponds to the sum of M_{C2} and M_{C3}, as given by (18), which allows computing M_{C2} as (19).

$$M = M_{C2} + M_{C3} \tag{18}$$

$$M_{C2} = \frac{1 + nD}{1 - D} \tag{19}$$

Substituting (19) into (4) yields the value of M_{C1} given by

$$M_{C1} = \frac{1 + n}{1 - D} + n\,k\,. \tag{20}$$

Lastly, the total voltage gain M can be determined by substituting (4) and (17) into (3).

$$M = \frac{2 + n\,D}{1 - D} + n\,k \tag{21}$$

It can be concluded from (21) that lower values of k yield lower voltage conversion ratios. On the other hand, if the leakage inductance is low, the factor k tends toward one, and consequently the voltage gain converges to an ideal operating condition. In addition, the impact of k is minimized when the number of turns ratio is lower.

Figure 4 presents several curves of the voltage gain M as a function of the duty cycle D for different values of n and k. For the condition $n = 1$, it is evident that the influence of k is neglectable. On the other hand, as n increases and k decreases, the voltage gain deviates from the ideal case.

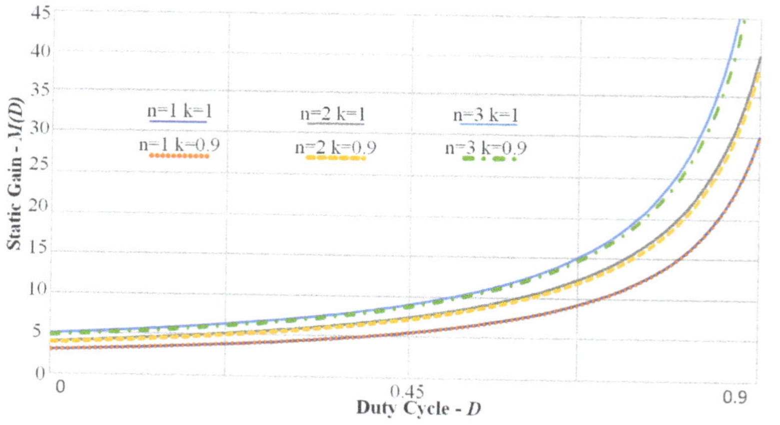

Figure 4. Voltage gain M as a function of the duty cycle D for different values of n and k.

The unknowns I_{Lk1} and I_{Lk2}, as well as the intervals Δt_2 and Δt_5, can be determined using (2) and (15) along with the average current on D_2 e D_3, which in turn are equivalent to the average output current I_o and are given by (22) and (23), respectively.

$$\frac{1}{T_s}\int_0^{\Delta t_5} i_{C1}(t)\,dt = I_o \tag{22}$$

$$\left(\frac{I_{Lk1} - I_{Lm}}{2n}\right)\frac{\Delta t_2}{T_s} = I_o \tag{23}$$

To complete the mathematical model of the proposed converter, it is fundamental that the equations defining the values of C_1, C_2 and C_3, the magnetizing inductance L_m and the voltage stress on the semiconductor devices are derived.

Based on the operating stages depicted in Figure 2, it is possible to conclude that the values of C_1, C_2 and C_3 can be determined by (24), (25) and (26), respectively. However, C_1 and C_3 can also be determined using the resonance criterion given by (14) and (11), which is used in this work to calculate these capacitances.

$$C_1 = \frac{1}{\Delta V_{C1}} \int_0^{\Delta t_5} i_{C1}(t)\, dt \tag{24}$$

$$C_2 = \frac{1}{\Delta V_{C2}} \left[\int_0^{\Delta t_5} \left(\frac{i_{Lk}(t)}{n} - I_o \right) dt + \frac{I_{Lm}}{1+n} \Delta t_6 \right] \tag{25}$$

$$C_3 = \frac{D T_s I_o}{\Delta V_{C3}} \tag{26}$$

where ΔV_{C1}, ΔV_{C2}, ΔV_{C3} are the voltage ripples on C_1, C_2 and C_3, respectively.

The magnetizing inductance L_m can be computed using

$$L_m = \frac{(1-D)(V_{C1} - V_{C3}) T_s}{n \Delta I_{Lm}}, \tag{27}$$

where ΔI_{Lm} is the current ripple on L_m.

It can be concluded from Figure 2b,e that the voltage stress on D_1 and D_2 is equal to the voltage on C_1 given by (20).

$$V_{D1} = V_{D2} = \left(\frac{1+n}{1-D} + n\, k \right) V_{in} \tag{28}$$

The voltage stress on S and D_3 can be determined from the analysis of the second and fifth operating stages, resulting in

$$V_S = V_{D3} = \frac{V_{in}}{1-D}. \tag{29}$$

The proposed converter can be designed and validated in the laboratory based on the mathematical analysis detailed so far. Moreover, a detailed efficiency analysis can be performed to determine the advantages of the proposed circuit for the intended applications, especially PV and FC systems.

3.2. Control Strategy

As known, PV and FC systems require a proper control system to ensure maximum power point tracking (MPPT) operation. In such a system, the DC–DC converter stage is directly responsible for MPPT realization by means of controlling the input voltage or current. In this paper, a control strategy aiming at controlling the input current is adopted, in accordance with the schematic depicted in Figure 5.

In the proposed scheme, input voltage and current are measured using the digital signal processor TMS320F28377S, in which an input current compensator is implemented. First-order filters with a cutoff frequency of 1 kHz are used for signal conditioning and analog-to-digital conversion is performed with a sampling frequency equal to 100 kHz. In this work, a proportional–integral (PI) controller with proportional and integral gains of 0.005 and 15, respectively, is used to adjust the input current, which in turn allows MPPT to be realized. It is noteworthy that the small-signal analysis of the proposed circuit is very extensive due to the high number of operating stages and quasi-resonant operation, and therefore it will not be detailed in this paper. Nevertheless, closed-loop operation will be investigated to demonstrate the suitability of the proposed converter for PV and FC applications.

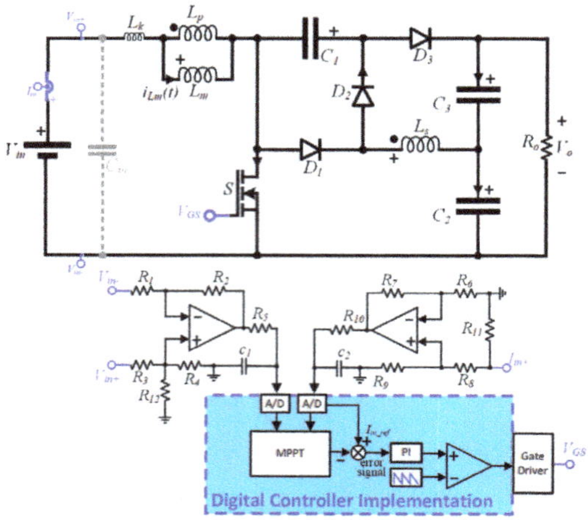

Figure 5. Control strategy for the proposed converter.

4. Experimental Results

The experimental verification of the proposed converter is carried out using a 400 W prototype designed in consideration of the specifications shown in Table 1.

Table 1. Design specifications for a 400 W prototype.

Symbol	Description	Value
V_{in}	Input voltage	48 V
f_s	Switching frequency	100 kHz
P_o	Output power	400 W
V_o	Output voltage	400 V
n	Number of turns ratio	1
k	Inductance factor	95%
L_k	Leakage inductance	2 µH
f_o	Resonant frequency	$1.6{\cdot}f_s$
ΔI_{Lm}	Current ripple on L_m	45% of I_{in}
ΔV_{C3}	Voltage ripple on C_3	1% of V_{C3}

As previously mentioned, the resonance criterion was adopted to determine the capacitances C_1, C_2 and C_3. In this sense, the leakage inductance was initially estimated as $L_k = 2$ µH. Next, Equation (14) was used to determine C_{eq}, which in turn can be used to calculate C_1 and C_3 using (11). It is noteworthy that f_o was chosen to be higher than f_s, and thus ZCS is guaranteed for D_3. Based on the knowledge that the most expressive resonance occurs in the fifth stage, which corresponds to the interval Δt_5 that can last as long as $(1 - D) \cdot T_s$, it was decided that $f_o = 1.6{\cdot}f_s$. The main results obtained from the design of the proposed converter are listed in Table 2 and a picture of the prototype is shown in Figure 6.

Table 2. Main design results.

Parameter	Value
D	0.644
L_m	80 µH, E42/15, $N_p = N_s = 16$
C_1	3 µF/400 V
C_2	3 µF/400 V
C_3	3 µF/400 V
D_1, D_2 and D_3	MUR 840
S	IRFP4668PBF

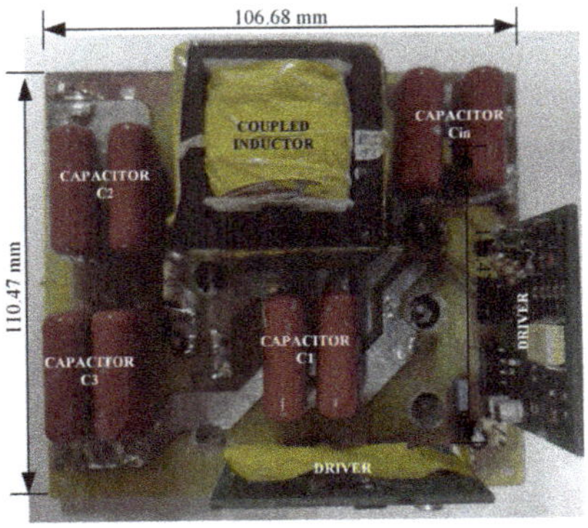

Figure 6. Picture of the 400 W prototype of the proposed converter.

Voltage and current at the input and output are depicted in Figure 7a. Input current was measured before the decoupling capacitor C_{in}, resulting in a filtered low-current ripple waveform. Figure 7b shows the voltage on capacitors C_1, C_2 and C_3. The results are in accordance with the theoretical predictions given by (20), (19) and (17), respectively. The voltage and current on the switch S are depicted in Figure 8a. The waveforms indicate dissipative commutation, but it is evident that the drain-to-source voltage is much lower than the output voltage. Consequently, a low-R_{DSon} MOSFET can be used, thus minimizing the conduction and switching losses on this device. Voltage and current on D_1, D_2 and D_3 are shown in Figures 8b and 9a,b, respectively. These results confirm that the commutation of D_1 is dissipative, as a considerable reverse recovery current is observed. However, this reverse current does not appear on the switch current (c.f., Figure 8a), and therefore its impact on the switching losses becomes limited. It can be also verified that D_2 and D_3 operate with ZCS as a result of the resonant characteristic of the current on these devices. The most expressive resonance occurring during the fifth stage and predicted in the theoretical analysis is evident on the waveform of the current on D_3. The measured resonant frequency was approximately 190 kHz, deviating from the 160 kHz ($1.6{\cdot}f_s$) defined in Table 1. This difference was expected, because the actual leakage inductance of the coupled inductor measured in laboratory was L_k = 1.45 μH, a value lower than the 2 μH estimated during design.

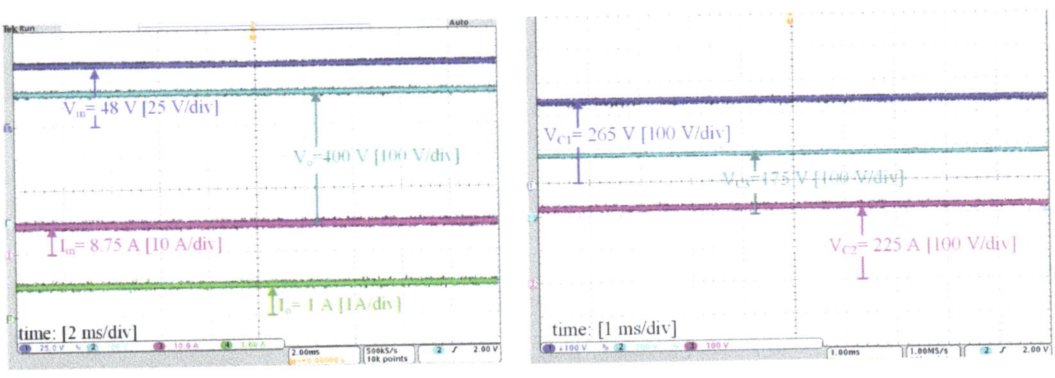

(a) (b)

Figure 7. (**a**) Voltages and currents on the input and output; (**b**) Voltages on capacitors C_1, C_2 and C_3.

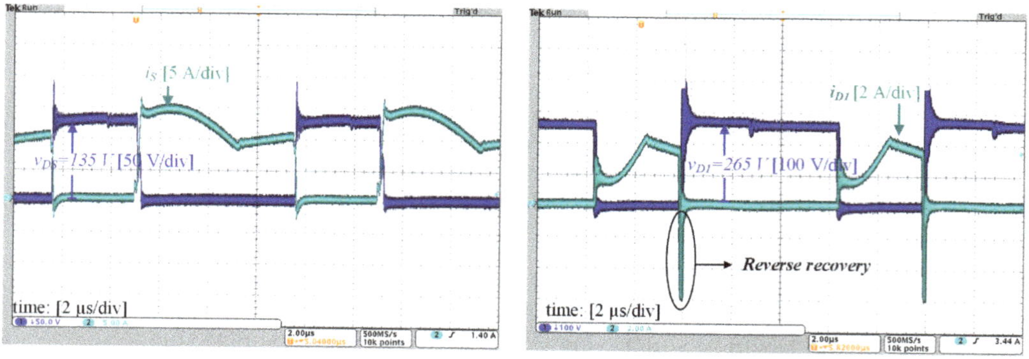

(a) **(b)**

Figure 8. Voltage and current on (**a**) switch S; (**b**) diode D_1.

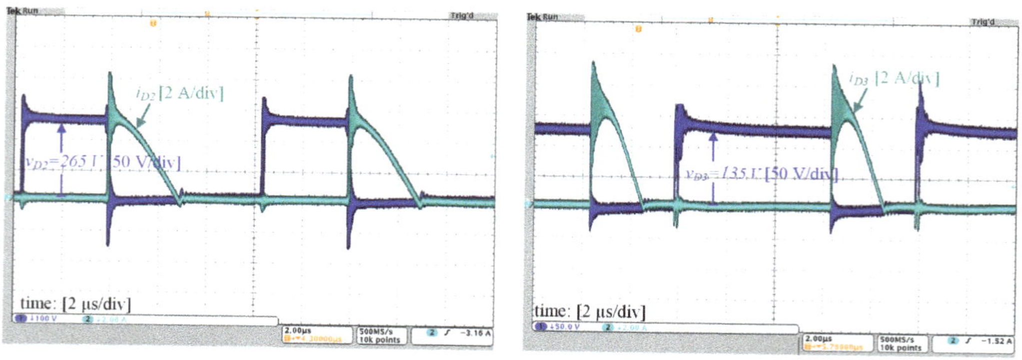

(a) **(b)**

Figure 9. Voltage and current on (**a**) diode D_2; (**b**) diode D_3.

Efficiency measurements were taken using the power precision analyzer Yokogawa WT500 and are presented in Figure 10. Tests under three different situations were performed to create conditions in accordance with the specifications of 250–400 W solar modules. Figure 10a presents the efficiency curve versus output power considering input and output voltages fixed at 48 V and 400 V, respectively. A maximum efficiency of 96.9% was measured at 60% of rated output power, while the full-load efficiency was 96.46%. Efficiency versus input voltage variation, maintaining the output power fixed at 400 W, is shown in Figure 10b. It is evident that the efficiency is considerably reduced at low input voltage due to increased current levels. A more realistic situation is when the output power decreases as the input voltage is reduced. Figure 10c shows the efficiency versus input voltage considering an output power variation of 250–400 W. Although an efficiency reduction is also observed as the input voltage decreases, the impact is not as severe as that verified in the test with P_o fixed at 400 W.

Finally, closed-loop operation with the output voltage fixed at 400 V is demonstrated in Figure 11. Steps in the input current reference from 4.25 A to 8.75 A and from 8.75 A to 4.25 A were applied, and the response demonstrates the proper operation of the converter using the control strategy depicted in Figure 5. In the tests, the output voltage was regulated at 400 V by the electronic load NHR9430 operating in the constant voltage mode to emulate the behavior of a grid-tied inverter stage. It is also noteworthy that the input current was measured before the filter capacitor C_{in}, and hence its ripple is lower than the theoretical prediction depicted in the waveform of i_{Lk} (c.f., Figure 3).

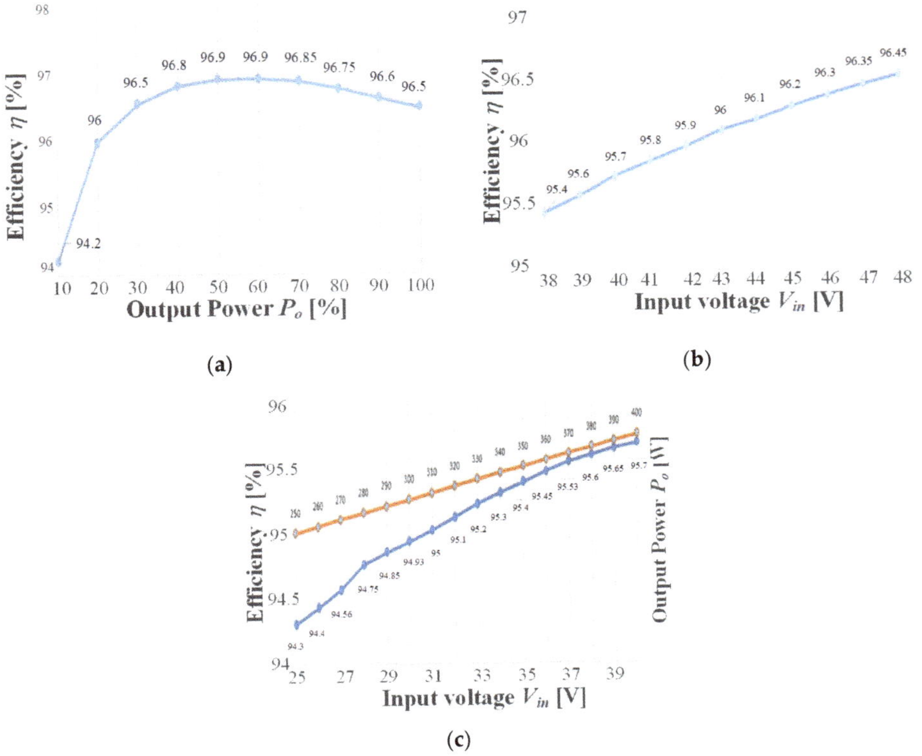

Figure 10. Efficiency curves: (**a**) as a function of the output power with V_{in} = 48 V and V_o = 400 V; (**b**) as a function of the input voltage (V_{in} = 38–48 V) with P_o = 400 W; and (**c**) as a function of the input voltage (V_{in} = 25–40 V) with variable output power (P_o = 250–400 W).

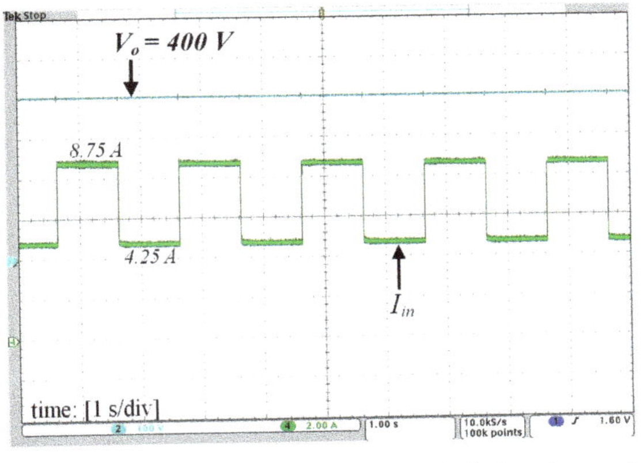

Figure 11. Closed-loop response for steps in the input current reference from 4.25 A to 8.75 A and from 8.75 A to 4.25 A. The output voltage is imposed by the electronic load NHR9430 operating in the constant voltage mode to emulate the behavior of a grid-tied inverter stage.

Comparative Analysis

The proposed converter is compared with the high-voltage-gain topologies proposed in [5,6,13–17], as summarized in Table 3. In this comparative analysis, only circuits containing a single active switch and one coupled inductor with two windings are considered. In addition, it is assumed that the leakage inductance of the coupled inductor is much smaller than its magnetizing inductance. As can be concluded from Table 3, both the voltage gain M and the voltage stresses on the semiconductors are usually related to the number of diodes and capacitors contained in the circuits. The circuits proposed in [5,13] contain only two diodes and two capacitors, although in [5] the voltage stress on the switch and on the diode D_2 is equal to the entire output voltage level. As a result, higher conduction and switching losses are expected. The converters introduced in [15,17] exhibit the higher voltage gains, but more diodes and capacitors are employed. The boost-flyback converter proposed in [13] presents the lower voltage conversion ratio, and also requires the use of an auxiliar snubber to prevent potentially destructive voltage spikes on D_3. As a consequence, if the same duty cycle is adopted, a higher number of turns ratio is required to reach the same voltage gain, and therefore weight and cost of the coupled inductor are also increased. It is also noteworthy that the snubber required for proper operation increases the losses of the circuit, which has an impact on the overall system efficiency. The converters proposed in [6,14,16] present similar constructive and operational characteristics. The highest efficiencies are verified in the proposed converter and in the circuit introduced in [6]. However, the proposed converter has a higher voltage gain and was tested with a higher switching frequency.

Table 3. Comparison between boost-based high-step-up DC–DC converters using a single active switch and one coupled inductor with only two windings.

Reference	Voltage Gain (V_o/V_{in})	Voltage Stress on the Switch	Voltage Stress on the Diodes			Number of Capacitors	Number of Diodes	Efficiency η
			V_{D1}	V_{D2}	V_{D3}			
Proposed	$\frac{2+n}{1-D}$	$\frac{V_{in}}{1-D}$	$\left(\frac{1+n}{1-D}+n\right)V_{in}$	$\left(\frac{1+n}{1-D}+n\right)V_{in}$	$\frac{V_{in}}{1-D}$	3	3	96.5% @400 W, 100 kHz, $n = 1$
[5]	$\frac{1+n}{1-D}$	$\frac{1+n}{1-D}V_{in}$	$\left(\frac{1+n}{1-D}-n+1\right)V_{in}$	$\frac{1+n}{1-D}V_{in}$	-	2	2	94% @300 W, 100 kHz, $n = 3$
[6]	$\frac{1+n}{1-D}$	$\frac{V_{in}}{1-D}$	$\frac{V_{in}}{1-D}$	$\frac{nV_{in}}{1-D}$	$\frac{nV_{in}}{1-D}$	3	3	96.5% @400 W, 90 kHz, $n = 1.72$
[13]	$\frac{1+nD}{1-D}$	$\frac{V_{in}}{1-D}$	$\frac{V_{in}}{1-D}$	Limited by the RCD snubber [1]	Voltage stress on the snubber diode [1]	3	3	93% @35 W, 38 kHz, $n = 2$
[14]	$\frac{1+n}{1-D}$	$\frac{V_{in}}{1-D}$	$\frac{V_{in}}{1-D}$	$\frac{nV_{in}}{1-D}$	$\frac{nV_{in}}{1-D}$	3	3	Not reported, 20 kHz, $n = 1$
[15] [2]	$\frac{1+n}{1-D}+n$	$\frac{V_{in}}{1-D}$	$\frac{V_{in}}{1-D}$	$\frac{nV_{in}}{1-D}$	$\frac{nV_{in}}{1-D}$	4	4	94.3% @400 W, 50 kHz, $n = 1$
[16]	$\frac{1+n}{1-D}$	$\frac{V_{in}}{1-D}$	$\frac{V_{in}}{1-D}$	$\frac{nV_{in}}{1-D}$	$\frac{nV_{in}}{1-D}$	3	3	94% @300 W, 100 kHz, $n = 6$
[17] [3]	$\frac{2+n}{(1-D)^2}$	$\frac{V_{in}}{(1-D)^2}$	$\frac{V_{in}}{1-D}$	$\frac{DV_{in}}{(1-D)^2}$	$\frac{V_{in}}{(1-D)^2}$	4	5	91.1% @500 W, 40 kHz, $n = 1.86$

[1] Voltage stresses were not reported in the paper. [2] Voltage stress on the fourth diode is similar to D_2 and D_3. [3] Voltage stresses on D_{r1} and D_o are not included.

5. Conclusions

A novel high-step-up DC–DC converter based on the boost converter employing one coupled inductor and voltage multiplier cells was proposed in this paper. The circuit contains a single active switch, three diodes and three capacitors, and is capable of achieving a high conversion ratio even when a low number of turns ratio is adopted. It was demonstrated that using a low number of turns ratio minimizes the influence of the leakage inductance on the voltage gain. In addition, the resonant characteristic of the circuit provides ZCS operation for two of its diodes, thus reducing the switching losses and improving the system efficiency. A detailed efficiency analysis was performed in the laboratory, and the results demonstrate that the proposed converter is a viable solution for applications with a power rating on the order of a few hundred watts, since it is capable of providing high voltage gain with a reduced number of components.

Author Contributions: Conceptualization, G.L.S. and C.B.N.; software, G.L.S.; validation, G.L.S., E.A.J. and C.B.N.; formal analysis, G.L.S., E.A.J. and C.B.N.; investigation, G.L.S.; data curation, G.L.S.; writing—original draft preparation, C.B.N.; writing—review and editing, E.A.J.; supervision, C.B.N.; project administration, C.B.N.; funding acquisition, E.A.J. and C.B.N. All authors have read and agreed to the published version of the manuscript.

Funding: This research was funded in part by the Coordination for the Improvement of Higher Education Personnel (CAPES)–Financing Code 001.

Data Availability Statement: No new data were created in this work.

Conflicts of Interest: The authors declare no conflict of interest.

Nomenclature

Abbreviations	
AC	Alternate Current
DC	Direct Current
FC	Fuel Cell
MPPT	Maximum Power Point Tracking
PI	Proportional–Integral
PV	Photovoltaic
VMC	Voltage Multiplier Cell
ZCS	Zero Current Switching
Symbols	
ΔI_{Lm}	Current ripple on L_m
Δt_i	Interval of the i-th operating stage
$\Delta V_{C1}, \Delta V_{C2}, \Delta V_{C3}$	Voltage ripple on C_1, C_2 and C_3
λ	Inductance factor 1
ω_o	Resonant angular frequency
A	Simplifying term
C_{eq1}	Equivalent capacitance
D	Duty cycle
f_o	Resonant frequency
f_s	Switching frequency
i_{C1}, i_{C2}, i_{C3}	Instantaneous current on C_1, C_2 and C_3
i_{in}	Instantaneous input current
i_{Lk}	Instantaneous current on L_k
$I_{Lk}(s)$	Laplace transform of i_{Lk}
I_{Lk1}	Current on L_k at $t = t_1$
I_{Lk2}	Current on L_k at $t = t_4$
I_{Lm}	Average value of the current on L_m
I_o	Average value of the output current
k	Inductance factor 2
M	Voltage gain
M_{C1}, M_{C2}, M_{C3}	Partial voltage gains on C_1, C_2 and C_3
n	Number of turns ratio
R_{DSon}	On-resistance of the MOSFET
T_s	Switching period
v_{C1}, v_{C2}, v_{C3}	Instantaneous voltage on C_1, C_2 and C_3
V_{D1}, V_{D2}, V_{D3}	Maximum voltage stress on diodes D_1, D_2 and D_3
V_{in}	Input voltage
V_{GS}	Gate-to-source voltage on the MOSFET
v_{Lk}	Instantaneous voltage on L_k
V_o	Output voltage
V_S	Maximum voltage stress on switch S

References

1. Shubbak, M.H. Advances in solar photovoltaics: Technology review and patent trends. *Renew. Sustain. Energy Rev.* **2019**, *115*, 109383. [CrossRef]
2. Li, Z.; Ma, T.; Yang, H.; Lu, L.; Wang, R. Transparent and colored solar photovoltaics for building integration. *Solar RRL* **2021**, *5*, 2000614. [CrossRef]
3. Premkumar, M.; Subramaniam, U.; Haes Alhelou, H.; Siano, P. Design and Development of Non-Isolated Modified SEPIC DC-DC Converter Topology for High-Step-Up Applications: Investigation and Hardware Implementation. *Energies* **2020**, *13*, 3960. [CrossRef]
4. Zheng, Y.; Smedley, K.M. Analysis and design of a single-switch high step-up coupled-inductor boost converter. *IEEE Trans. Power Electron.* **2020**, *35*, 535–545. [CrossRef]
5. Hass, E.S.; Nascimento, C.B. A simple self-clamped high step-up DC-DC converter employing coupled inductor. *Braz. Power Electron. J.* **2019**, *24*, 204–213.
6. Rosas, I.P.; Agostini, E.; Nascimento, C.B. Single-switch high-step-up DC-DC converter employing coupled inductor and voltage multiplier cell. *IEEE Access* **2022**, *10*, 82626–82635. [CrossRef]
7. Wu, Y.-E.; Wu, Y.-L. Design and Implementation of a High Efficiency, Low Component Voltage Stress, Single-Switch High Step-Up Voltage Converter for Vehicular Green Energy Systems. *Energies* **2016**, *9*, 772. [CrossRef]
8. Rahimi, T.; Islam, M.R.; Gholizadeh, H.; Mahdizadeh, S.; Afjei, E. Design and Implementation of a High Step-Up DC-DC Converter Based on the Conventional Boost and Buck-Boost Converters with High Value of the Efficiency Suitable for Renewable Application. *Sustainability* **2021**, *13*, 10699. [CrossRef]
9. Pereira, A.V.C.; Cavalcanti, M.C.; Azevedo, G.M.; Bradaschia, F.; Neto, R.C.; Santos de Carvalho, M.R. A Novel Single-Switch High Step-Up DC–DC Converter with Three-Winding Coupled Inductor. *Energies* **2021**, *14*, 6288. [CrossRef]
10. Li, W.; He, X. Review of nonisolated high-step-up DC/DC converters in photovoltaic grid-connected applications. *IEEE Trans. Ind. Electron.* **2011**, *58*, 1239–1250. [CrossRef]
11. Liu, H.; Hu, H.; Wu, H.; Xing, Y.; Batarseh, I. Overview of high-step-up coupled-inductor boost converters. *IEEE J. Emerg. Sel. Top. Power Electron.* **2016**, *4*, 689–704. [CrossRef]
12. Forouzesh, M.; Siwakoti, Y.P.; Gorji, S.A.; Blaabjerg, F.; Lehman, B. Step-up DC–DC converters: A comprehensive review of voltage-boosting techniques, topologies, and applications. *IEEE Trans. Power Electron.* **2017**, *32*, 9143–9178. [CrossRef]
13. Tseng, K.C.; Liang, T.J. Novel high-efficiency step-up converter. *IEE Proc. Electr. Power Appl.* **2004**, *151*, 182–190. [CrossRef]
14. Baek, J.-W.; Ryoo, M.-H.; Kim, T.-J.; Yoo, D.-W.; Kim, J.-S. High Boost Converter Using Voltage Multiplier. In Proceedings of the 31st Annual Conference of IEEE Industrial Electronics Society (IECON), Raleigh, NC, USA, 6–10 November 2005.
15. Hsieh, Y.-P.; Chen, J.-F.; Liang, T.-J.; Yang, L.-S. A novel high step-up DC–DC converter for a microgrid system. *IEEE Trans. Power Electron.* **2011**, *26*, 1127–1136. [CrossRef]
16. Wai, R.-J.; Duan, R.-Y. High step-up converter with coupled inductor. *IEEE Trans. Power Electron.* **2005**, *20*, 1025–1035. [CrossRef]
17. Hu, X.; Gong, C. A High Voltage Gain DC–DC Converter Integrating Coupled-Inductor and Diode–Capacitor Techniques. *IEEE Trans. Power Electron.* **2014**, *29*, 789–800.
18. Abbasian, S.; Farsijani, M. A single-switch high step-up Zero Current Switching DC-DC converter based on three-winding coupled inductor and voltage multiplier cells with quasi resonant operation. *Int. J. Circ. Theory Appl.* **2022**, *50*, 4419–4433. [CrossRef]
19. Schmitz, L.; Martins, D.C.; Coelho, R.F. Generalized high step-up DC-DC boost-based converter with gain cell. *IEEE Trans. Circ. Syst. I Reg. Papers* **2017**, *64*, 480–493. [CrossRef]
20. Lee, H.-S.; Yun, J.-J. Quasi-resonant voltage doubler with snubber capacitor for boost half-bridge DC–DC converter in photovoltaic micro-inverter. *IEEE Trans. Power Electron.* **2019**, *34*, 8377–8388. [CrossRef]
21. Hasanpour, S.; Baghramian, A.; Mojallali, H. A modified SEPIC-based high step-up DC–DC converter with quasi-resonant operation for renewable energy applications. *IEEE Trans. Ind. Electron.* **2019**, *66*, 3539–3549. [CrossRef]
22. Roshandel, E.; Mahmoudi, A.; Kahourzade, S.; Emami, H.D. DC-DC High-Step-Up Quasi-Resonant Converter to Drive Acoustic Transmitters. *Energies* **2022**, *15*, 5745. [CrossRef]

energies MDPI

Perspective

Gallium Nitride Power Devices in Power Electronics Applications: State of Art and Perspectives

Salvatore Musumeci * and Vincenzo Barba

Dipartimento Energia "Galileo Ferraris", Politecnico di Torino, 10129 Torino, Italy; vincenzo.barba@polito.it
* Correspondence: salvatore.musumeci@polito.it

Abstract: High-electron-mobility transistors based on gallium nitride technology are the most recently developed power electronics devices involved in power electronics applications. This article critically overviews the advantages and drawbacks of these enhanced, wide-bandgap devices compared with the silicon and silicon carbide MOSFETs used in power converters. High-voltage and low-voltage device applications are discussed to indicate the most suitable area of use for these innovative power switches and to provide perspective for the future. A general survey on the applications of gallium nitride technology in DC-DC and DC-AC converters is carried out, considering the improvements and the issues expected for the higher switching transient speed achievable.

Keywords: GaN FET; HEMT; enhancement mode; depletion mode; WBG; SiC MOSFET; super-junction MOSFET; cascode; LIDAR

Citation: Musumeci, S.; Barba, V. Gallium Nitride Power Devices in Power Electronics Applications: State of Art and Perspectives. *Energies* **2023**, *16*, 3894. https://doi.org/10.3390/en16093894

Academic Editor: K. T. Chau

Received: 13 April 2023
Revised: 28 April 2023
Accepted: 3 May 2023
Published: 4 May 2023

1. Introduction

Wide-bandgap materials have begun to replace silicon in several power electronics applications. At present, gallium nitride (GaN) is probably the most challenging technology in the field of power electronics, allowing for the development of attractive devices with increased power density, reduced on resistance, and very-high-frequency switching. The wide bandgap of the semiconductor material results in a high critical electric field, which can lead to designs of electronic devices with a shorter drift region and therefore a lower on-state resistance when compared to a silicon-based device featuring the same voltage rating [1]. Due to the high switching speed, GaN power electronic devices require a careful design of the power loop layout in converter applications [2]. Furthermore, appropriate packaging is necessary to both reduce the stray inductances and to dissipate the heat due to the device's high energy density. The substantial reduction in chip size compared to silicon power devices with the same current and voltage rate and the high switching frequency that is currently attainable (up to tens of MHz) [3] allows for a decrease in the power converter's global volume. The reduction in the power converter size is a key point for the integration of converters and actuators such as integrated modular motor drives (IMMDs) [4]. Furthermore, the reduction in switching power losses provides increasingly efficient solutions for converter applications featuring emerging opportunities for the expanding power electronics markets. There are several fields of power electronics applications benefiting from the technological advancements of GaN devices, especially in the low-voltage (<200 V) power supply areas, which are, at present, having a growing impact on modern society, such as in telecom/datacom, server SMPS, and wireless charging. Furthermore, it also is an enabling technology for the integration of renewable energy with power converters, electric vehicles, industrial automation, and modular battery management systems (BMSs) [5], generally at a higher voltage rate (currently, 650 V is the standard for high-voltage GaN, with some power device fabrications achieving a maximum voltage of up to 1200 V). The wide-bandgap semiconductors in all mentioned applications improve the power converters' features, reducing their weight, volume, and lifecycle cost [6]. Furthermore, the increasing demand for compact and long, autonomous, battery-powered portable devices has led to the proposal of GaN technology as a possible attractive response for near-future applications. In China, which has a very representative global market index, consumer electronics applications are projected to grow from USD 79.6 million in 2021 to USD 964.7 million in 2027 in the power GaN device market (source: Power GaN report, Yole intelligence 2023). Figure 1 describes the evolution of the penetration of GaN devices into the market of power electronics, highlighting their applications [7]. Currently, cost is one of the significant limits, while the reliability and robustness are gradually improving as new generations of devices arrive on the market, thanks to the continuous effort of the application designers and engineers who highlight their limits and application opportunities.

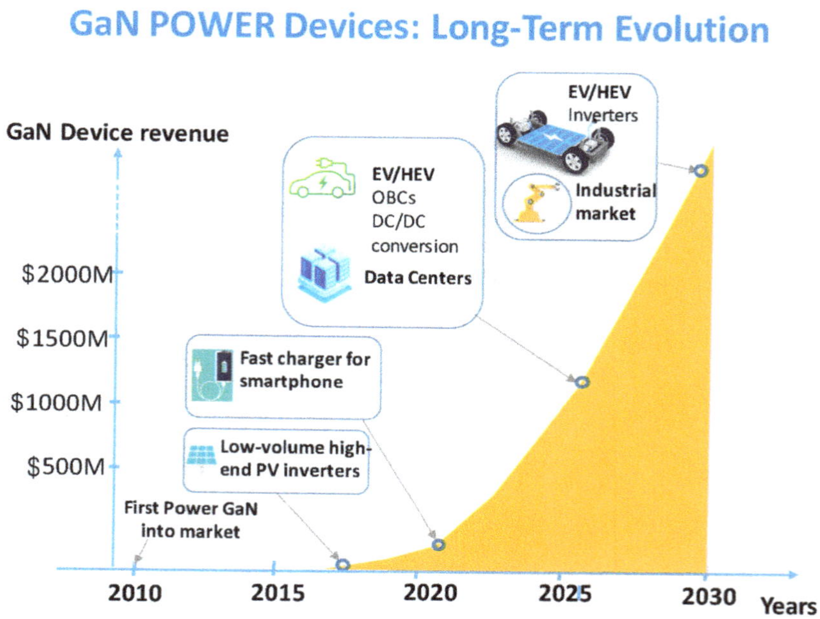

Figure 1. GaN power electronic devices in long-term evolution for the Chinese market (Source Power GaN report, Yole intelligence 2023 [7]).

In this article, the impact of the GaN technology in power electronics applications is described and discussed, considering the state of the art and a feasible outlook and perspectives. The article begins with an overview of gallium nitride technology, considering the advantages and drawbacks and examining possible developments. In the next sections, a general survey on the applications of GaN-based devices and their impact in improving the overall performance of the converters compared to competing power devices are described. The promising evolution of the future is also evaluated.

2. GaN Technology: Overview and Development

The first gallium nitride devices appeared recently, around 2004, as depletion-mode radio frequency (RF) transistors made in Japan by the Eudyna Corporation. The first semiconductor RF transistor used GaN on silicon carbide (SiC) substrates [8]. After a few years, the GaN-based technology began to be used in power electronics switches in different structural arrangements [9]. In 2009, Efficient Power Conversion (EPC) developed the first enhancement-mode GaN FET. Since this date, the evolution of GaN devices has carried on without stopping for increasingly ambitious targets.

The timeline of the development of GaN-based power devices and the main industrial players involved are depicted in Figure 2.

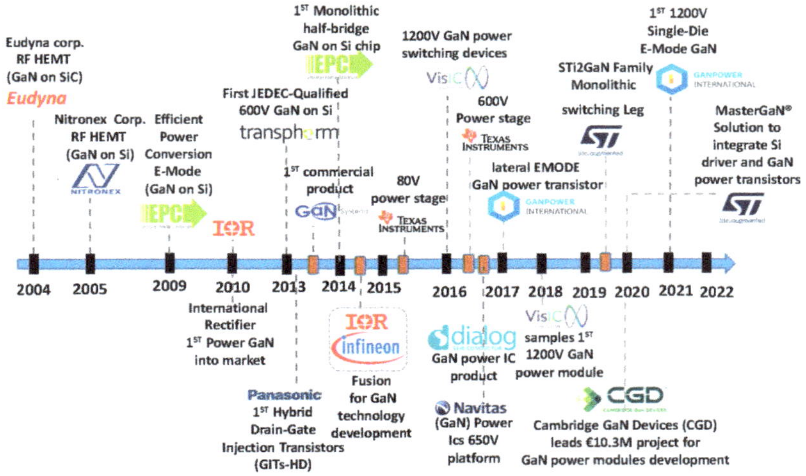

Figure 2. GaN power electronics devices milestones and main industrial players.

GaN power devices are wide-bandgap materials (WBG) belonging to the class of high-electron-mobility transistors (HEMTs). These semiconductor electronic transistors feature a two-dimensional electron gas (2DEG) created by a junction between two crystalline materials featuring diverse atomic spacing and band gaps. The polarization effect induces the 2DEG phenomena in the heterojunction (AlGaN/GaN), causing the electron's high mobility [1]. Several factors have led power switch designers to put significant effort into making devices with WBG materials. To better recognize the causes behind the advanced performance of WBG power devices, a comparison of the significant material properties of silicon (Si), silicon carbide (SiC), and GaN is graphically depicted in Figure 3a. The SiC devices considered in Figure 3 are 4H-SiC technology. Power electronics device manufacturers use this technology for its isotropic structure [10]. From Figure 3a arises better WBG material properties compared with Si, such as a high thermal conductivity and electron mobility and the breakdown electrical field. GaN and SiC have different behaviors with advantages and drawbacks, making one or the other preferable based on the type of application.

GaN has a lower thermal conductivity than the SIC; indeed, the conductivity of SiC is 2:5 times higher than Si; this increased thermal conductivity feature transfers heat from the semiconductor device junction to the case in an improved way. Better thermal conductivity enables higher current/power densities for SiC devices. Instead, the main GaN peculiarity is the HEMT property, which means that the mobility in the transistor channel is very high. The electron mobility is typically around 2000 cm^2/Vs, almost 100 times higher than SiC MOSFETs. This is a noticeable advantage regarding low ON resistance values. In this way is possible to obtain devices with a much smaller area [11]. Moreover, the switching transients are very short, achieving higher switching frequencies than the SiC devices. After the comparison of the device technologies discussed above, it may be asserted that SiC devices offer a higher power/current density. In contrast, GaN devices provide a lower conduction loss and a higher switching frequency. The device's applications cover different areas of technological development with some overlays, as summarized in Figure 3b. From the point of view of the cost, the SiC material currently has a lower cost than the GaN.

Most manufacturers achieve GaN on Silicon structures on which large-size Si substrates are used for the GaN epitaxial layer growth to exploit the existing facilities and know-how and decrease fabrication costs [12]. The current technological approach led to the development of GaN semiconductor power devices with a lateral structure. In the lateral structure, the electron flow between the source and drain terminals features an inhomogeneous distribution of the electric field in the device, showing a peak in specific device areas; this phenomenon limits its full voltage-blocking technological potential [13]. Vertical structures for GaN power devices are under development [14]. As in silicon and SiC devices, the vertical design allows for an increase in the breakdown voltage by increasing the thickness of the voltage-blocking layer. The vertical structure is a promising solution for growing penetration in the high-voltage markets dominated by SiC and silicon devices. To develop competitive vertical GaN device structures, some issues must be overcome, such as ion implantation for p-GaN and long-term reliability [15]. Reliability is a crucial constraint for the HEMT devices to be released into extended-power applications for the mass market. The device's reliability is studied by manufacturers in depth to achieve and fulfil standards to keep the GaN structure safe. In particular, the fabrication of the AlGaN crystal can be controlled accurately, and an imperfection or unexpected defects lead to failure in the switching operation [3]. Another aspect to consider related to the ruggedness of GaN devices is the conduction resistance R_{DSon} during switching events.

GaN power transistors show a dynamic on resistance (Dynamic-R_{DSon}) increase in switching-mode power converters [16]. In the lateral AlGaN/GaN device structure, the amount of the electric field in the gate edge adjacent to the drain side region strongly influences the Dynamic-R_{DSon} variation. Furthermore, the adjacent semiconductor states and the injection and trapping of electrons affect the regions' degradation, leading to an increase in device loss and temperature (undesired features in the GaN HEMT structure) [17,18].

In recent years, GaN power transistor designers have worked hard on controlling and reducing this phenomenon which leads to serious operating problems during switching in the power converters, degrading its performance with a decrease in the drain current (I_D) during switching (called a current collapse) [19].

One example is the advent of the high-voltage, normally off GaN devices of the gate injection transistor (GIT) structure which effectively acts on the degradation of the Dynamic-R_{DSon}. The device injects holes from an additional p-AlGaN layer at the AlGaN/GaN heterojunction to release the trapped electrons and increase the electron density in the channel by drastically reducing the on-state resistance [20].

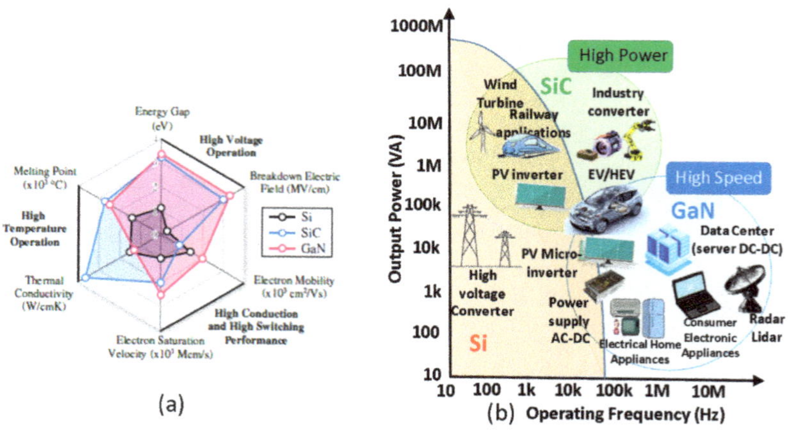

Figure 3. (**a**) Radar chart of the key material properties of Si, 4H-SiC, and GaN devices at ambient temperature (25 °C). Source [21]. (**b**) Si, SiC, and GaN power electronics switch applications areas.

2.1. *GaN Power Devices: Classifications and Operation*

The transistors based on GaN technology are lateral structure power devices belonging to the field effect transistor (FET) family, with a current conduction channel between the drain and source terminals. A gate voltage modulates the conduction current. The natural operation, similar to FET devices, is in depletion mode, from which the simplest structure of a GaN FET is a normally on switch. A simplified structure of a depletion GaN FET is reported in Figure 4a. Generally, the gate electrode is made via a Schottky contact on the surface of the layer. The Schottky barrier becomes inversely polarized through the application of a negative voltage to this electrode with respect to the source, bringing the device to the OFF state. Figure 4a highlights the 2DEG created in the heterojunction (AlGaN/GaN). However, the GaN power transistor designer has developed typically off structures to make the HEMT power electronics switch increasingly competitive with SiC- and Si-based GaN FET device structures.

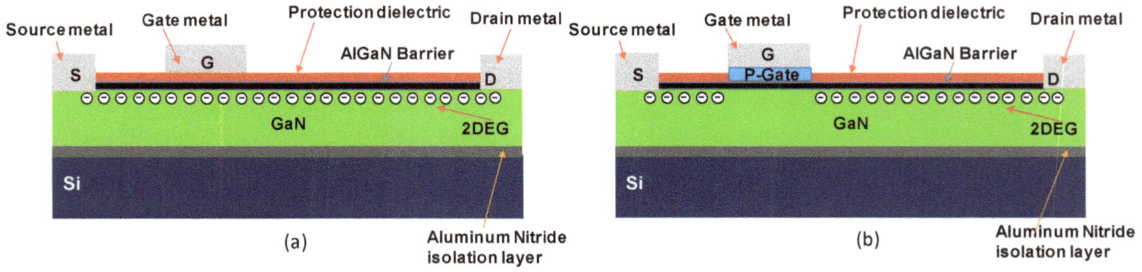

Figure 4. (**a**) Normally on GaN (d-mode): simplified structure. (**b**) Normally off GaN (e-mode): simplified structure.

The normally off GaN FET is obtained via an enhanced structure (e-mode), as depicted in the simplified structure of Figure 4b (considering the device in its off-state). The positively charged (p-type) GaN gate obtains the e-mode GaN (this is the first commercial device structural arrangement [8]). The layer is grown on top of the AlGaN barrier. The p-type layer effectively depletes the two-dimensional electron gas with $V_{GS} = 0$, achieving a normally off device. The 2DEG is fully restored through a suitable positive voltage (greater than the threshold voltage) between the gate and the source.

If the gate voltage is under the device threshold voltage, an equivalent diode behavior appears, and reverse conduction can happen. The virtual equivalent diode in GaN features a voltage drop that is higher than the body diode of a Si MOSFET. It does not show a reverse recovery charge, Q_{rr} (no minority carriers are involved in the conduction), at the turn-off [21].

Currently, the following GaN power transistor families are available in the HEMT device market (Figure 5).

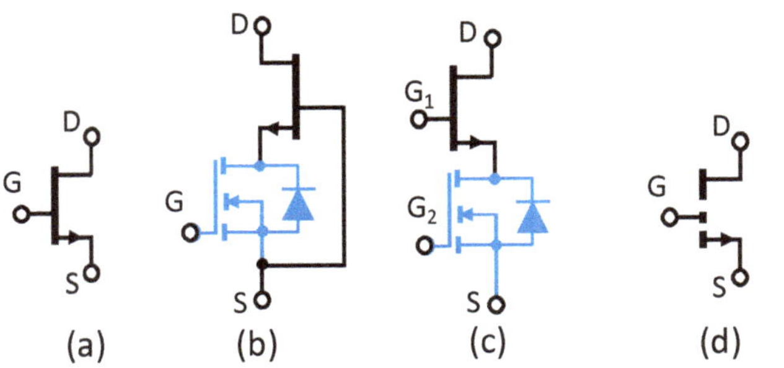

Figure 5. (**a**) Normally on GaN (d-mode). (**b**) Normally off Cascode GaN (d-mode). (**c**) Direct-drive GaN (d-mode). (**d**) Normally off GaN (e-mode).

The normally on GaN shown in Figure 5a is the basic brick of HEMT GaN devices. The structure is an FET operating in depletion mode (d-mode). It features the lowest R_{DSon} due to having the simplest structure [21]. The GaN FET is in an on state with a gate voltage (V_{GS}) of 0 V, while it is turned off with a V_{GS} equal to −15 V. Like the FET, the d-mode GaN is naturally bidirectional if the gate voltage polarization enables the conduction channel between the source and drain. Currently, this structure is not on the market as a single power switch for the normally on state conditions, but it is used to achieve a cascode structure.

The cascode structure reported in Figure 5b attains a switch normally off, adding a low voltage (LV) MOSFET in series with the GaN source. The cascode solution combines the advantaged switching /conduction capability and the structural simplicity of d-mode GaN technology with the friendly gate-driving technique of the Si MOSFET. The GaN FET gate is connected to the MOSFET source to maintain the HEMT device in an on state, while the MOSFET is used for the switching transients. The reverse cascode operation appears when the GaN switch is turned off, the LV Si MOSFET is turned off. In this condition, a reverse voltage is applied to the GaN switch; a current flows through the body diode of the LV Si MOSFET and the channel of the normally on GaN FET with a reduced voltage drop (as it is a naturally bidirectional device). However, an overall lower off state reverse voltage (i.e., <1 V) during dead times is obtained. The cascode configuration may be used in low-voltage applications (<200 V); nevertheless, in high-voltage applications (from 650 V up to 1200), it features a viable usage for higher-current switches [22].

The critical issues of the cascode arrangement are related to the following points:

- The complexity of the package solution;
- The increase in the stray inductances in the power loop;
- The presence of the reverse recovery of the MOSFET body diode;
- The quasi-uncontrolled GaN switching due to a lack of control drive of the HEMT gate.

Figure 5c shows a modified arrangement of the cascode configuration called a direct-drive GaN power transistor. The d-mode GaN gate terminal is not connected to the low-voltage Si MOSFET source in the direct-drive operation. The Si MOSFET is used as a protection device to prevent series shoot through (i.e., requiring an enable gate signal after the converter start-up). The d-mode GaN FET can be directly driven with a negative unipolar voltage (i.e., −15 V, 0 V). The main advantage of this direct-drive arrangement approach is the capability to drive the d-mode GaN FET, exploiting its switching properties and avoiding the uncontrolled commutation of the conventional cascode implementation [21]. Compared to the previous cascode solution, the direct-drive connection increases stray inductances. Furthermore, the two gates available increase the device pin to four.

The enhancement-mode (e-mode) GaN power transistor (Figure 5d) is achieved when the depletion-mode device modifies the gate structure to shift the threshold voltage positively to create a conductive channel [8]. The switching HEMT device is similar to an enhancement-mode MOSFET. The driver circuit must supply a gate voltage from 0 V to +6 V. While the threshold voltage V_{GSth} is low, in the range of 1–2 V, the reverse conduction during the dead time is characterized by an equivalent diode conduction mechanism that causes a higher voltage drop than a MOSFET body diode, increasing the reverse conduction power losses. Currently, the e-mode GaN power transistors are the most widespread family of GaN FET devices and are the group to which the major efforts of academic and industrial designers are directed to optimize performance, as in the case of the high-voltage GIT structure (from Panasonic) already discussed, which was developed to reduce the degradation effect of the dynamic R_{DSon} [20].

2.2. Integration of GaN Structures

The first monolithic half-bridge in a single chip appeared in the GaN market arena in 2014 from EPC [23]. The half-bridge integration reduces the power loop inductances, decreasing the drain voltage peak. Furthermore, the power device's circuit space in the board layout is strongly reduced. In the DC-DC converter for synchronous Buck topology, non-symmetric-area devices are used to optimize the figure of merit (FOM) of the high-side and low-side devices [24]. This monolithic half-bridge solution was the crucial step that led to the integration of driving, control, and sensing with protection functions together in a power stage in the following years. Several GaN power device manufacturers have taken up these technological challenges by providing the market with various integrated solutions with power stages and multiple functions that can be used in typical applications of both DC-DC and DC-AC converters. Figure 6 depicts some of the most attractive solutions developed in recent years. In Figure 6a, the low-voltage half-bridge monolithic arrangement (EPC 2152 −80 V, 15 A half-bridge as a power stage, from EPC) with two symmetric e-mode GaN transistors integrating all drive circuitry and a level shift with a bootstrap function is depicted [25]. All the power integrated circuits (ICs), power stages, and signal circuits are developed in GaN technology. The GaN FET symbol for EPC maintains the MOSFET drawing arrangement. This integrated half-bridge can be used in DC-DC converters as single switching legs and in inter-leaving arrangements and inverters for low-power brushless DC (BLDC) motors, such as in e-bike applications [26].

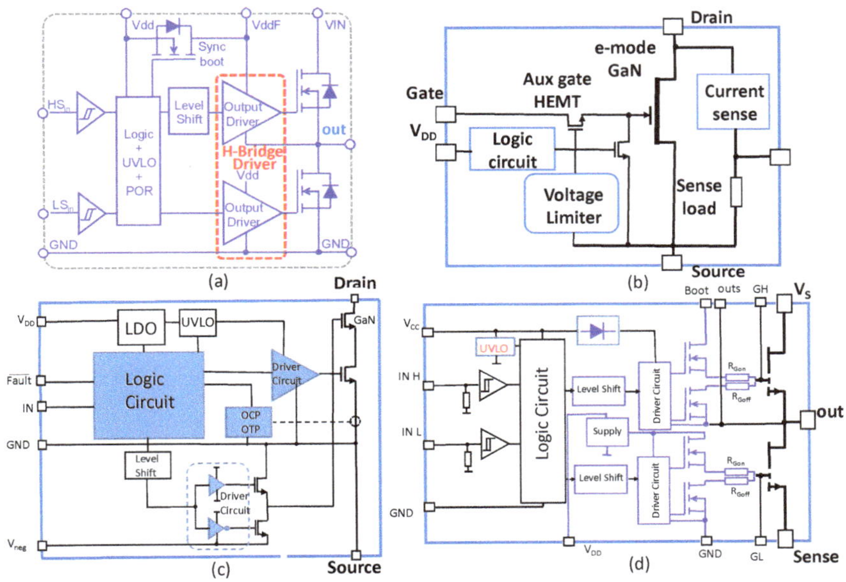

Figure 6. (**a**). Low-voltage switching leg power stage integrated with driver and protection circuit, source [25]. (**b**) High-voltage normally- off e-mode GaN FET integrated with driver and protection circuit from CGD, source [27]. (**c**) High-voltage direct-drive GaN FET (d-mode). From Texas Instruments, source [28]. (**d**) High-voltage normally-off (e-mode) GaN FET half bridge in SiP arrangement from STMicroelectronics [29].

Cambridge GaN Devices (CGD) offers several integrated circuit solutions (at 650 V). In Figure 6b, the single e-mode high-voltage monolithic GaN power stage, ICeGaN™, with smart sensing and protection and controller and driver function is reported. The integrated solution is to drive it the same as a Si MOSFET. ICeGaN™ accepts an input voltage V_{in} from 0 to 20 V, and the threshold voltage is augmented to 3 V for better driving control [27]. The application is oriented to a DC-DC single-switch converter, such as a quasi-resonant Flyback converter, for consumer power supply. The Texas instrument provides the LMG341x family's integrated cascode solution (power stage: 600 V; 40 A at 25 °C) with a direct-drive implementation and driver and protection circuits [28]. The schematic of the power integrated circuit (IC) is shown in Figure 6c. A key feature of the gate driver is the control of the slew rate during hard-switching transients. The protection circuits associated with the function direct-drive solution allow for an increase in the reliability of the integrated GaN-based cascode device.

STMicroelectronics has developed "MasterGaN" devices, an integrated platform (system-in-package, SiP) embedding a silicon technology half-bridge driver with two e-mode GaN transistors in a switching-leg configuration (Figure 6d). Currently, the integrated switching-leg MasterGan1 features 600 V with 10 A at 25 °C of drain current [29]. As shown in Figure 6, the GaN devices' circuit symbol is still non-uniform; every manufacturer draws the GaN power transistor in a different mode. It is necessary to define a standard symbol in the future.

In this field of power, GaN transistor integration highlights the monolithic solution for two GIT GaN devices arranged in a single chip to obtain a bidirectional switch, which has been realized in recent years for applications in protection power switches for voltages up to 400 V and hybrid relays [30].

A system-on-chip (SoC) circuit is much more attractive for the reduced number of components and space occupied by the power converter achieved. The target of future integrated structures based on HEMT GaN technology is to achieve a single-power IC that is driven by a microcontroller with a simple digital signal, high switching performances, and adequate current density [31].

The power modules are another crucial target for obtaining high-current power switches to compete with SiC MOSFET devices in high-power applications. In [32], a prototype of a 650 V/60 A GaN power module in an SP1 package is described. The basic power stage is a cascode configuration (as reported in Figure 5b). In the power module, a switching leg with two cascode devices in parallel connection in the high-side and low-side positions, respectively, is implemented. Furthermore, an RC snubber is integrated to control the drain voltage slope. The power module production for high-current devices in e-mode GaN power transistors is related to the development of the vertical HEMT structure. It is a perspective to consider in the coming years.

2.3. *Package Solutions*

The package solution plays a crucial role in reducing parasitic inductances to allow for an increase in switching frequencies without high ringing in the converter circuit waveforms and while avoiding drain voltage overshoot. A low-inductance package design is crucial. A suitable package material and layout decrease heat dissipation and limit electromagnetic interference (EMI). A GaN FET features a die size smaller than a MOSFET of an equivalent current density and breakdown voltage characteristics, allowing for a noticeable decrease in the package volume. Furthermore, the GaN FET case must achieve efficient cooling paths on the top and bottom sides.

Furthermore, electrical insulation is requested for a ground-referenced heat sink for a half-bridge topology [33]. An illustrative example regarding low-voltage e-mode GaN devices relating to EPC manufacture is the chip-scale package (CSP) arrangement. This package layout has "solderable bars" on the device surface. The developed solution allows a direct solder onto a printed circuit board (PCB). The CSP achieves a noticeable decrease in stray inductance.

Furthermore, efficient cooling can be obtained. The CSP solution is illustrated in Figure 7. The solderable bars are depicted in Figure 7a with the pin connections. In Figure 7b, the package assembly on the PCB is described. Finally, a photo of a half-bridge circuit in a CSP arrangement on the PCB top layer is shown in Figure 7c.

For high-voltage, the Discretes Flat No-leads (DFN) package is used by several manufacturers for an achievable high-voltage (up to 1200 V) that exhibits few stray inductances. For integrated GaN technology, the Quad Flat No-Lead (QFN) package is preferred, with some improvement as described in [34]. The growth of GaN devices in the semiconductor market goes hand in hand with the development of packages that can exploit the advantage of the switching characteristics of HEMT transistors, featuring a compact size to reduce parasitic inductance and the isolation strength capability to achieve the high-voltage switches that are increasingly demanded. Furthermore, the package solutions must exhibit a relevant heat dissipation capacity on both the bottom and high sides of the case.

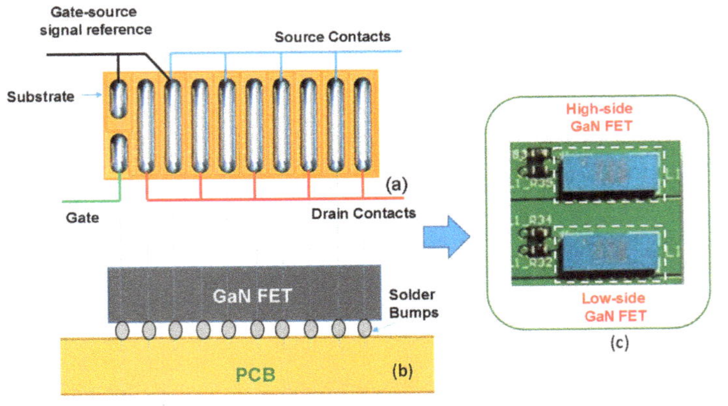

Figure 7. (**a**). Package description of solderable bars with source, drain, gate, and substrate contacts. (**b**) Package on the PCB arrangement. (**c**) Picture of a top layer of a half-bridge PCB with two devices in the CSP package.

3. GaN in Power Electronics Applications

E-mode GaN FETs are already used in several power electronics application areas. At a low voltage, the superior characteristics of HEMT devices allow for the increasing use of GaN FETs over silicon MOSFETs. There are electrical benefits and a reduction in the space occupied, making HEMT components very attractive for producing compact and efficient electronic equipment. As mentioned above, the currently higher cost than the MOSFETs is the principal limit for the even more widespread use of GaN FETs. In high-voltage power electronics systems, other established devices in these fields include WBG SiC MOSFETs and the more technologically mature silicon super-junction (SJ) MOSFETs and IGBT devices. This condition causes GaN power transistors to be used for a restricted but crucial area of application with a need for higher switching frequencies. In [35], a comparison of power losses among several SiC MOSFETs, SJ silicon MOSFETs, and GaN FETs (in d-mode cascode switching devices) at 400 V of DC bus and a different load current amplitude performed on inductive switching (Figure 8a) versus the switching frequency, show clearly the area of the best application of the tested silicon and WBG devices, as reported in Figure 8b. However, the GaN power transistors in high-voltage applications are increasingly experimented with in various fields of power converter applications, such as battery chargers in the automotive area or photovoltaic inverters [36,37], to recognize the advantages and disadvantages of the GaN FETs applications (in d-mode cascode and e-mode configurations) in these critical areas of the sustainable development in energy conversion. Several power electronics applications are discussed in the following to overview of the GaN FETs' impact.

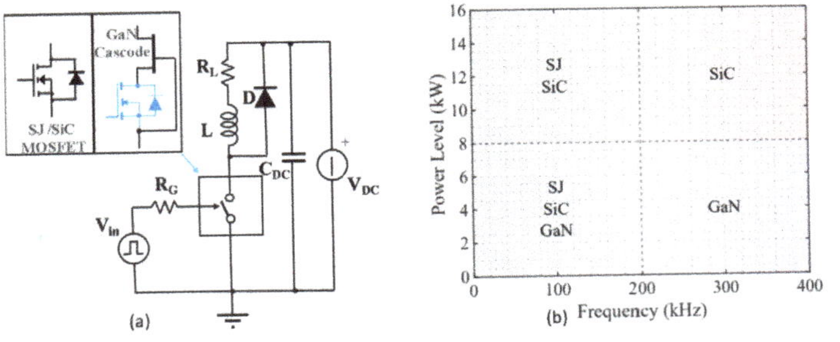

Figure 8. (**a**) Inductive-load-switching test circuit schematic. (**b**) Application areas trend for SiC, SJ, and GaN technologies. Drain-source voltage equal to 400 V, power levels in the range of 1 kW to 16 kW (current amplitude: 1 A–40 A), and switching frequencies from 1 kHz to 500 kHz [35].

3.1. GaN for DC-DC Power Converters

The power brick DC-DC converter is strongly diffused in power supply arrangements for industrial and domestic electronic equipment such as game consoles, laptops, computers, servers, communication networking, and storage systems. The power brick that complies with the standard sizes and footprints settled by the Distributed-power Open Systems Alliance (DOSA) is the solution for a flexible modular power supply. The advantage of using the power brick is that the equipment design engineers do not need to be experts in power supply design [38]. The power brick comprises the switching topology (a synchronous buck or boost topology) and a controller. For example, a typical 1/16th brick has a 48 V input, an output voltage of 12 V, and 25 A of output current (300 W). The power density per unit area is a crucial challenge for the brick arrangement. HEMT devices are excellent candidates for compact and high-efficiency modular power supply systems. The low-voltage GaN FETs feature a compact footprint with an outstanding FOM. The size and shape of the magnetic components can be improved via the increasing switching frequency achievable by GaN FETs. Furthermore, using high-switching HEMT devices reduces dead time in the implementation of a synchronous buck converter [39]. Choosing a GaN-compatible controller is a challenge for obtaining a high-performance power brick.

The advantages in terms of the main electrical parameters of a low-voltage e-mode GaN FET (GS61004B from GaN System in GaNPX® packaging) for a power brick application compared with an equivalent application trench-gate MOSFET (ISZ230N10NM6 from Infineon in a PG-TSDSON-8FL package) are reported in Table 1. The trench-gate-technology MOSFET device used has a very advantageous FOM (achieved by the product of R_{DSon} and Q_G) when compared to MOSFET devices with an equivalent R_{DSon} resistance. It features a very low gate charge Q_G with respect to most MOSFETs available on the market. Despite the small Q_G of the chosen MOSFET, the FOM is very favorable to the GaN FET device, as shown in the graph of Figure 9a, related to the results in Table 1. The output capacitor C_{oss} is a parameter related to the switching losses [21,40]. The MOSFET shows a higher C_{oss} than the GaN FET. In reverse conduction, the e-mode GaN power transistor shows an equivalent diode behavior with Q_{rr} = 0 but with a higher voltage drop. Generally, the dead-time reverse-conduction losses are lower for the MOSFETs, but the dead-time duration is strongly reduced for GaN FET devices [41]. The switching characteristics are strongly advantageous for GaN FETs, as demonstrated by the FOM features. However, the high frequency that can be reached brings a crucial cure in the layout arrangement to reduce the stray inductances in the power loop and gate circuit. Furthermore, the GaN FET cost currently is higher than the cost of a Si MOSFET.

Table 1. Comparison of low-voltage (100 V) main electrical parameters between a commercial e-GaN and an equivalent application trench-gate MOSFET device. The reference temperature is 25 °C.

LV Device	I_D (A)	V_{DS} (V)	V_{GS} (V)	R_{Dson} (mΩ)	C_{oss} (pF)	Q_G (nC)	Q_{rr} (nC)	V_F (V)
e-mode GaN	38	100	−10 to +7	16	110	3.3	0	2.3
MOSFET	31	100	±20	23	150	9.3	23	1

Table 2. Comparison of high-voltage (650 V) main electrical parameters comparison among a commercial e-GaN and equivalent application SiC MOSFET and SJ MOSFET devices. The reference temperature is 25 °C.

HV Device	I_D (A)	V_{DS} (V)	V_{GS} (V)	R_{DSon} (mΩ)	Coss (pF)	Q_G (nC)	Q_{rr} (nC)	V_F (V)
e-mode GaN	30	650	−10 to +7	50	65	6.1	0	3.3
SiC MOSFET	35	650	−5 to 18	58	73	29	30	3
SJ MOSFET	30	650	±20	86	84	33	5100	1.2

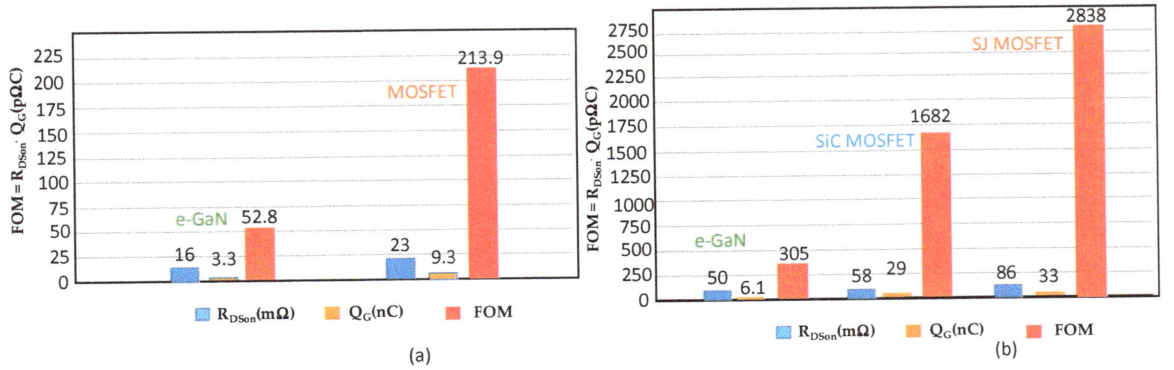

Figure 9. (**a**) R_{DSon}, Q_G, and FOM comparison between a low-voltage (100 V) e-mode GaN FET and Si MOSFET related to Table 1. (**b**) R_{DSon}, Q_G, and FOM comparison among a high-voltage e-mode GaN FET with a SiC and Si MOSFETs related to Table 2.

The LLC resonant converter is one of the topologies in which the use of e-mode GaN FETs allows for the achievement of the best power density [42]. The zero voltage switching achieved in an LLC converter permits an increase in efficiency, and the high switching frequency reachable (from 1 to tens of MHz) dramatically reduces the resonant tank's size. The LLC converter is one of the much-used converter topologies for inductive wireless power transfer (WPT) systems [43]. Compared with traditional Si-based MOSFETs, the compact size obtained with HEMT devices is a crucial factor for wireless charging applications, especially in drone applications [44] where the available space for the charger is limited. Another relevant field of application is the lighting of LEDs for driving. GaN devices, with their high switching frequency and high-power density characteristics, enable compact and efficient driving systems that can be inserted into the lamp, reducing the overall size [45].

In high-current applications such as in the battery interface (for example, in a 48 V mild-hybrid system), the DC-DC converter is arranged with bidirectional switching legs in an interleaved solution [46]. The GaN FET used must be automotive-qualified. The interleaved solution is associated with a high switching frequency, allowing for a substantial reduction in the current ripple and a reduction in the size of the inductor. Furthermore, the monolithic solution described above (Section 2.2) permits a compact size layout that is crucial for automotive applications.

In high-voltage e-mode GaN FET applications, the battery charger is one of the most diffused power electronic systems. The HEMT device may make a solid effort to increase the switching performance and reduce the occupied converter volume. Considering a typical voltage of 400 V for the DC bus, the maximum rated voltage for the switching device in a common half-bridge topology used for these kinds of applications V_{DSmax} is 650 V. Using this design constraint, three different technology-based devices are compared. In Table 2, the main electrical parameters for a high-voltage e-mode GaN FET (GS66508B from GaN System in e GaNPX® package) are compared with an equivalent application SiC MOSFET (SCT055HU65G3AG from STM, in U3PAK package) and a silicon super junction (SJ) MOSFET (SiHB100N60E from Vishay in D2PAK—TO-263). From Table 2, the better switching characteristics of the e-mode GaN FET arise.

Figure 9b depicts the graphics of the FOM comparison in relation to the Table 2 results. In this range of voltage, the e-mode GaN FET can be an effective alternative at SiC and Si MOSFETs. Over the cost, the maximum voltage currently available due to the lateral physical structure is the main drawback limiting market penetration. One of the critical disadvantages of GaN-HEMTs in DC-DC converter applications is their reduced capability to withstand unclamped inductive switching (UIS) [47], which is caused by the non-removable structure of holes generated by the avalanche breakdown. The UIS ruggedness and capability of GaN FET HEMTs can be improved via a technological design approach that implements a hole removal structure and optimized package thermal resistance [48].

Furthermore, short-circuit robustness (SC) has been studied recently. The experimental investigations in [49,50] show that high voltage (until 650 V) GaN FET e-mode devices can sustain the SC for the standard 10 μs with a drain–source voltage V_{DS} lower than about half the maximum expected voltage. This crucial fault condition must be further investigated to furnish detailed information to the GaN FET designers to improve the switch SC ruggedness.

Recently, developers of HEMT devices have also been experimenting with multilevel converters, such as flying capacitor inverter topology, to exploit the compactness of GaN transistors and the high switching frequencies in these power circuits [51]. The investigation into the application of the DC-DC modular multilevel converter (MMC) for solar electric propulsion in spacecraft power systems is described in [52]. For high voltages, d-mode GaN Power transistors in cascode configuration are currently used.

3.2. GaN in Motor Drive Applications

In recent years, the use of GaN FETs for low voltages has experienced remarkable development in applications for inverter drives for brushless direct current (BLDC) and AC motors (mostly for voltages < 200 V) due to the possibility of reaching high switching frequencies ($\geq$100 kHz), reducing the overall dimensions of the inverter, and therefore allowing for the integration of the motor, inverter, and control in a single compact system obtaining an efficient and integrated modular motor drive [53]. Light e-mobility devices, such as E-kick scooters, e-Bikes, skateboards, hoverboards, low-speed EVs, segways, mopeds, and e-scooters, benefit from the advantages of the HEMT GaN technology [54]. In low-voltage motor drives, the DC bus voltage V_{DC} is in the range of 24–96 V. The switching frequency for a MOSFET-based inverter is equal to or below 40 kHz. The dead time with MOSFET switching legs is maintained from 200 ns to 500 ns. The dead-time length impacts the generation of a sixth harmonic on the electrical frequency of the generated torque. This harmonic amplitude influences the motor efficiency performance, increasing the mechanical vibration and the winding temperature [55]. Dead time in e-mode GaN FETs can be reduced to tens of ns, improving the quality of the waveform and drastically decreasing the sixth harmonic effect [56]. From the point of view of power losses, the reverse conduction in GaN FETs must be maintained as low as possible to avoid cross-conduction in the inverter legs as well as to reduce the power losses caused by the higher voltage drop of the equivalent diode conduction [57].

An e-mode GaN FET inverter easily reaches 100 kHz of switching frequency. By increasing the PWM frequency, a double-positive effect can be achieved. The first advantage is the current ripple reduction. The second advantage is the input voltage ripple decrease that consequently permits a DC link capacitor reduction. Furthermore, at a conventional motor drive switching frequency (maximum 40 kHz with Si MOSFET in low voltage applications), the DC link capacitor is generally arranged with the electrolytic capacitor. When increasing the switching frequency, the capacitance values decrease, allowing for the use of ceramic capacitors. The non-polarized capacitors feature a lower series resistance (ESR) with a minimum in the range of 100–200 kHz, showing good temperature stability and reliability. The substantial reduction in the inverter input filter also influences the inverter's efficiency and compactness [26]. From the above consideration, the low-voltage DC battery-powered motor is moving from a Si-MOSFET-based inverter to HEMT e-mode GaN technology in the coming years. In high-voltage inverter applications, there are several experimental prototypes and investigations from academic and industrial researchers. In [53], a three-phase inverter based on the e-mode GaN transistor GS66516T produced by GaN Systems is experimentally evaluated (V_{DC} = 350 V, an output rated current of 90 A for phase, at 20 kHz of switching frequency). Furthermore, an IMMD design approach using GaN power transistors for inverter-driving permanent magnet synchronous motors (PMSMs) is presented in [58]. The application of high-voltage GaN power transistors in two-level and especially in multi-level inverters are in development, and exciting results are expected in the following years.

3.3. GaN for LIDAR Application

Light detection and ranging (LIDAR) identifies the technology that measures the distance to an object by illuminating it with laser light. Lidar is currently used as a remote recognition device for developing self-driving vehicles and for ADASs (advanced driver assistance systems).

Two modes of operation are prevalent in LIDAR applications: direct time of flight (DToF) and indirect time of flight [23]. The first LIDAR form sends individual pulses and measures the time of reflection to calculate the distances. The IToF method compares the phase of the transmitted and reflected signal pulse to calculate the distance of the target. In recent years, the IToF solution has grown for its versatility, simple design, and low cost. IToF is applied mainly in the medium range of distances (from 1 m to 10 m). It is primarily used in robots, unmanned aerial vehicles (UAV), and autonomous drive vehicles. The principle of operation of the IToF is shown in Figure 10a.

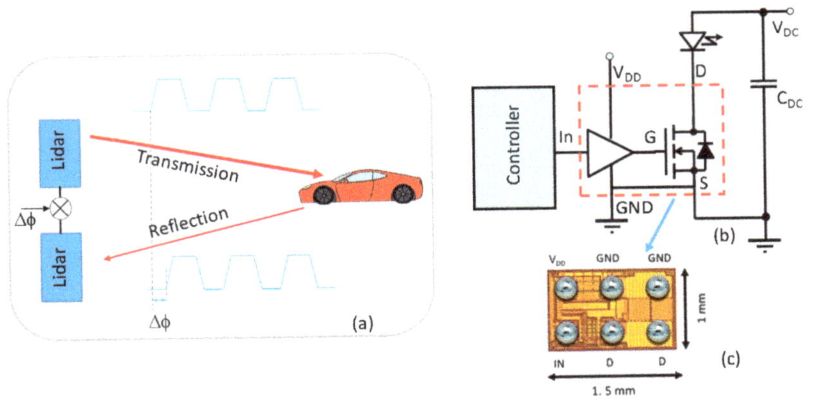

Figure 10. (**a**) Indirect time of flight system principle. (**b**) Simplified power circuit for the laser-pulse current generation with an e-mode GaN FET power stage and integrated driver circuit(EPC21601). (**c**) Picture of the die where the GaN FET power stage and driver circuit are integrated.

The LIDAR laser requires a specific pulse current for operation in the ToF system. The current pulse must be obtained through a low-voltage circuit based on a switch capable of quickly managing high currents. The e-mode GaN FET is an excellent candidate for this kind of application. In this direction, the EPC has developed an integrated power stage and driver circuit with a voltage switching time of 750 ps at a typical voltage supply of 30 V with 15 A of the peak current that is suitable for LIDAR operation and an operative frequency up to 200 MHz [59]. One of the most-used circuit topologies for implementing such a pulse laser current is shown in Figure 10b, and the layout of the IC, with the power stage and control circuit, is depicted in Figure 10c. The benefit of HEMT GaN technology is the compactness of the integration solution, which reduces the circuit part count associated with the reachable high-switching transients.

4. Discussion

The advent of GaN HEMT technology for power electronics has brought a novel breath of fresh air to the semiconductor power device arena. Gallium nitride, which is grown as a thin film on top of a standard silicon substrate, creates new horizons for developing fast and compact devices for increasing efficiency and performing applications. The integration capability in the lateral GaN devices developed by several manufacturers allows for a reduction in the count-part of components for developing effective and innovative high-frequency power circuits. HEMT characteristics lead to a revolutionary layout arrangement to avoid ringing voltage due to stray inductances on the power loop. The power stage and driver circuit integration (SoC) are moving toward this target. A crucial point in the challenge of the massive application of HEMT GaN power transistors is the development of packages able to reduce the parasitic inductances and optimize the thermal management in the bottom and upper sides of the case. The GaN FET in modules arrangement is a further challenge. The e-mode GaN FET in a vertical structure is a future target for increasing the modular arrangement capability and maximum voltage available. Currently, the maximum high-current and high-voltage technologies are obtained in the cascode configuration of d-mode GaN devices. To increase the voltage managed, a multilevel converter will be investigated extensively in terms of advanced topology, innovative layout, and thermal management. Moreover, the technology development for high-voltage devices is moving on an innovative concept, the multi-channel monolithic-cascode high-electron-mobility transistor (MC^2 -HEMT), in which the low-voltage device is an e-mode GaN FET. Instead, the high-voltage device is a d-mode HEMT GaN FET. In [60], the voltage of the first devices developed was 3 kV, and the maximum voltage forecast for this new concept of GaN-based devices may go up to 10 kV.

In low-voltage applications, the e-mode GaN FETs feature broad application fields for superior performance compared to the equivalent pure silicon MOSFET devices. Si MOSFETs are still widely used due to the high knowledge of their static and dynamic characteristics in the various fields of power electronics and the maturity of the vertical structure technology. Meanwhile, the applications of lateral e-mode GaN FET devices are still under study, and the development of specific power converters is required. For example, Si MOSFETs have been investigated extensively in a parallel connection to obtain a fast and high current switch, while the behavior of GaN FETs is under study to optimize the parallel assembly [61,62]. Furthermore, another important limit for market penetration in consumer power electronics is the high cost. Cost reduction is a crucial future target for the extensive use of GaN FETs. GaN technology-manufacturing devices are working hard to obtain increasingly competitive construction processes. However, there are application fields, such as the case of LIDAR, where GaN FETs have become significantly necessary despite the high cost due to their high switching speed characteristics.

5. Conclusions

In this paper, an overview of HEMT GaN FET devices for low-voltage and high-voltage applications in the power electronics area are discussed, considering the state of the art and perspective. A GaN technology overview is presented, and the electrical characteristics are compared with pure Silicon and SiC MOSFETs. The high switching speed of the GaN power transistors allows for a compact design and more efficient power converters than Si and SiC MOSFETs. In the paper, the low-voltage and high-voltage device features and applications are evaluated to highlight their benefits and limits. From the outcome reported, the capability arises to use much more extensive e-mode GaN FETs in low-voltage power converters. The trend of growth of the e-mode GaN FETs shows that HEMT devices are moving on to substitute Si MOSFETs in several applications in the near future. In the high-voltage field, the use of GaN FETs in e-mode and d-mode in cascode configurations suffers from several competitors, such as Si IGBTs and SJ MOSFETs or WBG SiC MOSFETs. The lateral GaN FETs structure still does not reach the high current delivered by Si IGBT or WBG SiC modules. Still, for a lower current, the power converter compactness achievable with the GaN FETs is an enabling factor for its use in different types of applications, from lower power to medium power applications that require high frequencies of operation. The vertical structure of the GaN FETs would allow for the achievement of higher maximum voltages and modules that may be implemented more easily. Currently, vertical structure arrangement is under investigation, and significant results are expected in the next few years. Finally, GaN FETs are promising devices for improving the efficiency, compactness, and high-frequency operation of battery-powered equipment, e-mobility devices, wireless power transfer chargers, etc. The benefits of the advantages of HEMT technology and its further improvements are foreseen for the near future.

Shortly, the main advantages and drawbacks of GaN FET power transistors applied in the power electronics converters investigated are summarized in Table 3.

Table 3. Main advantages and drawbacks of GaN FET in the investigated power electronics applications.

Application	Advantages	Drawbacks
DC-DC power converters for power supply or battery charger applications	• Higher power density per unit area; • High switching frequency achievable (tens of MHz); • Size of magnetic components reduction; • Dead time reduction.	• Higher cost than SIC MOSFET and Si MOSFET; • Special cure in the PCB layout arrangement; • Limits in maximum voltage reachable for the current physical lateral structure adopted; • Critical UIS and short circuit management.
DC-AC Power converters for motor drive applications	• Decreasing the motor's sixth harmonic torque by reducing dead time; • Output current ripple reduction thanks to a higher switching frequency; • Input voltage ripple decrease and DC link capacitor reduction.	• Device cost; • Special cure in the PCB layout arrangement; • Both high dv/dt and voltage-ringing control to avoid motor insulation damage; • Critical short-circuit fault condition management.
LIDAR	• Lower R_{Dson}, which achieves lower conduction losses; • Lower parasitic capacitances, decreasing in power losses during charging/discharging with the higher switching frequency; • Smaller devices with less overheating, lower board-space requirements, and higher reliability.	• Higher cost than Si MOSFET; • Special cure in the management of stray inductances with a design approach typical of RF circuits.

To continue this overview work, the next evolution of the investigation into the e-mode GaN FET as a switch in power converter applications will regard two key points. The analysis of the monolithic structures' impact on the ruggedness and compactness of the power converter system and the realization of a high-current switch through an optimized parallel connection.

Author Contributions: Conceptualization, S.M. and V.B.; methodology, S.M.; software, V.B.; validation, S.M. and V.B.; formal analysis, S.M.; investigation, S.M. and V.B.; resources, V.B.; data curation, S.M. and V.B.; writing—original draft preparation, S.M.; writing—review and editing, S.M.; and V.B.; visualization, V.B.; supervision, S.M.; project administration, S.M.; funding acquisition, S.M. All authors have read and agreed to the published version of the manuscript.

Funding: This research was funded by Ministero (MIUR) PRIN, Innovative Solutions for Renewables in Energy Communities (ISoREC): 202054TZLF.

Conflicts of Interest: The authors declare no conflict of interest.

References

1. Meneghini, M.; De Santi, C.; Abid, I.; Buffolo, M.; Cioni, M.; Khadar, R.A.; Nela, L.; Zagni, N.; Chini, A.; Medjdoub, F.; et al. GaN-based power devices: Physics, reliability, and perspectives. *J. Appl. Phys.* **2021**, *130*, 181101. [CrossRef]
2. Sun, B.; Jorgensen, K.L.; Zhang, Z.; Andersen, M.A. Research of Power Loop Layout and Parasitic Inductance in GaN Transistor Implementation. *IEEE Trans. Ind. Appl.* **2020**, *57*, 1677–1687. [CrossRef]
3. Zhong, Y.; Zhang, J.; Wu, S.; Jia, L.; Yang, X.; Liu, Y.; Zhang, Y.; Sun, Q. A review on the GaN-on-Si power electronic devices. *Fundam. Res.* **2022**, *2*, 462–475. [CrossRef]
4. Wang, J.; Li, Y.; Han, Y. Integrated Modular Motor Drive Design with GaN Power FETs. *IEEE Trans. Ind. Appl.* **2015**, *51*, 3198–3207. [CrossRef]
5. Lipu, M.S.H.; Miah, S.; Ansari, S.; Meraj, S.T.; Hasan, K.; Elavarasan, R.M.; Al Mamun, A.; Zainuri, M.A.A.M.; Hussain, A. Power Electronics Converter Technology Integrated Energy Storage Management in Electric Vehicles: Emerging Trends, Analytical Assessment and Future Research Opportunities. *Electronics* **2022**, *11*, 562. [CrossRef]
6. Kumar, A.; Moradpour, M.; Losito, M.; Franke, W.-T.; Ramasamy, S.; Baccoli, R.; Gatto, G. Wide Band Gap Devices and Their Application in Power Electronics. *Energies* **2022**, *15*, 9172. [CrossRef]

7. China's Lead in Power GaN Consumer Devices Drives the Future Use in Telecom/Datacom and Automotive. Yole Group. 2 March 2023. Available online: https://www.yolegroup.com/strategy-insights/chinas-lead-in-power-gan-consumer-devices-drives-the-future-use-in-telecom-datacom-and-automotive/ (accessed on 5 April 2023).
8. Lidow, A.; De Rooij, M.; Strydom, J.; Reusch, D.; Glaser, J. *GaN Transistors for Efficient Power Conversion*, 3rd ed.; John Wiley & Sons: Hoboken, NJ, USA, 2019.
9. Liu, A.-C.; Tu, P.-T.; Langpoklakpam, C.; Huang, Y.-W.; Chang, Y.-T.; Tzou, A.-J.; Hsu, L.-H.; Lin, C.-H.; Kuo, H.-C.; Chang, E.Y. The Evolution of Manufacturing Technology for GaN Electronic Devices. *Micromachines* **2021**, *12*, 737. [CrossRef]
10. Liu, Y.; Lin, Z.; Cui, P.; Fu, C.; Lv, Y.; Cheng, Z. Influence of polarization Coulomb field scattering on high-temperature electron mobility in AlGaN/AlN/GaN heterostructure field-effect transistors. *Superlattices Microstruct.* **2018**, *120*, 389–394. [CrossRef]
11. Dimitrijev, S.; Han, J.; Haasmann, D.; Moghadam, H.A.; Aminbeidokhti, A. Power-switching applications beyond silicon: The status and future prospects of SiC and GaN devices. In Proceedings of the 2014 29th International Conference on Microelectronics Proceedings—MIEL 2014, Belgrade, Serbia, 12–14 May 2014; pp. 43–46. [CrossRef]
12. Chen, K.J.; Häberlen, O.; Lidow, A.; lin Tsai, C.; Ueda, T.; Uemoto, Y.; Wu, Y. GaN-on-Si Power Technology: Devices and Applications. *IEEE Trans. Electron Devices* **2017**, *64*, 779–795. [CrossRef]
13. De Santi, C.; Buffolo, M.; Rossetto, I.; Bordignon, T.; Brusaterra, E.; Caria, A.; Meneghesso, G. Review on the degradation of GaN-based lateral power transistors. *E-Prime—Adv. Electr. Eng. Electron. Energy* **2021**, *1*, 100018. [CrossRef]
14. Meneghesso, G.; Meneghini, M.; De Santi, C.; Zanoni, E. GaN-based lateral and vertical devices: Physical mechanisms limiting stability and reliability. In Proceedings of the 2019 Electron Devices Technology and Manufacturing Conference (EDTM), Singapore, 12–15 March 2019; pp. 68–70. [CrossRef]
15. Oka, T. Recent development of vertical GaN power devices. *Jpn. J. Appl. Phys.* **2019**, *58*, SB0805. [CrossRef]
16. Kuring, C.; Tannhaeuser, M.; Dieckerhoff, S. Improvements on Dynamic On-State Resistance in Normally-off GaN HEMTs. In Proceedings of the PCIM Europe 2019, International Exhibition and Conference for Power Electronics, Intelligent Motion, Renewable Energy and Energy Management, Nuremberg, Germany, 7–9 May 2019; pp. 1–8.
17. Yang, S.; Han, S.; Sheng, K.; Chen, K.J. Dynamic On-Resistance in GaN Power Devices: Mechanisms, Characterizations, and Modeling. *IEEE J. Emerg. Sel. Top. Power Electron.* **2019**, *7*, 1425–1439. [CrossRef]
18. Wang, W.; Liang, Y.; Zhang, M.; Lin, F.; Wen, F.; Wang, H. Mechanism Analysis of Dynamic On-State Resistance Degradation for a Commercial GaN HEMT Using Double Pulse Test. *Electronics* **2021**, *10*, 1202. [CrossRef]
19. Ho, S.-Y.; Lee, C.-H.; Tzou, A.-J.; Kuo, H.-C.; Wu, Y.-R.; Huang, J. Suppression of Current Collapse in Enhancement Mode GaN-Based HEMTs Using an AlGaN/GaN/AlGaN Double Heterostructure. *IEEE Trans. Electron Devices* **2017**, *64*, 1505–1510. [CrossRef]
20. Uemoto, Y.; Hikita, M.; Ueno, H.; Matsuo, H.; Ishida, H.; Yanagihara, M.; Ueda, T.; Tanaka, T.; Ueda, D. Gate Injection Transistor (GIT)—A Normally-Off AlGaN/GaN Power Transistor Using Conductivity Modulation. *IEEE Trans. Electron Devices* **2007**, *54*, 3393–3399. [CrossRef]
21. Cittanti, D.; Vico, E.; Bojoi, I.R. New FOM-Based Performance Evaluation of 600/650 V SiC and GaN Semiconductors for Next-Generation EV Drives. *IEEE Access* **2022**, *10*, 51693–51707. [CrossRef]
22. Huang, X.; Liu, Z.; Li, Q.; Lee, F.C. Evaluation and Application of 600 V GaN HEMT in Cascode Structure. *IEEE Trans. Power Electron.* **2014**, *29*, 2453–2461. [CrossRef]
23. Lidow, A. (Ed.) *GaN Power Devices and Applications*; John Wiley & Sons Ltd.: Hoboken, NJ, USA, 2021; ISBN 0996649220.
24. Musumeci, S.; Armando, E.; Mandrile, F.; Scrimizzi, F.; Longo, G.; Mistretta, C. Experimental Evaluation of an Enhanced GaN-Based Non-Symmetric Switching Leg Integrated Module for Synchronous Buck Converter Applications. In Proceedings of the 2021 23rd European Conference on Power Electronics and Applications (EPE'21 ECCE Europe), Ghent, Belgium, 6–10 September 2021; pp. 1–10. [CrossRef]
25. EPC2152 Datasheet and Application Note. Available online: https://epc-co.com/epc/products/gan-fets-and-ics/epc2152 (accessed on 5 April 2023).
26. Musumeci, S.; Mandrile, F.; Barba, V.; Palma, M. Low-Voltage GaN FETs in Motor Control Application; Issues and Advantages: A Review. *Energies* **2021**, *14*, 6378. [CrossRef]
27. Cambridge GaN Devices CGD65A055S2 Datasheet. Available online: https://cms.camgandevices.com/uploads/CGD_65_A055_S2_8f7276942a.pdf (accessed on 5 April 2023).
28. Direct-Drive Configuration for GaN Devices. Available online: https://www.ti.com/lit/wp/slpy008a/slpy008a.pdf?ts=1680655995471 (accessed on 5 April 2023).
29. MASTERGAN1 High Power Density 600V Half-Bridge Driver with Two Enhancement-Mode GaN HEMT. Available online: https://www.st.com/resource/en/datasheet/mastergan1.pdf (accessed on 5 April 2023).
30. Musumeci, S.; Panizza, M.; Stella, F.; Perraud, F. Monolithic Bidirectional Switch Based on GaN Gate Injection Transistors. In Proceedings of the 2020 IEEE 29th International Symposium on Industrial Electronics (ISIE), Delft, The Netherlands, 17–19 June 2020; pp. 1045–1050. [CrossRef]
31. GaN Power ICs. Available online: https://navitassemi.com/gan-power-ics/ (accessed on 5 April 2023).
32. Doumergue, P.-L.; Bontemps, S. First 650V 60A GaN Power Module in a Compact, Package. In Proceedings of the PCIM Europe 2022, International Exhibition and Conference for Power Electronics, Intelligent Motion, Renewable Energy and Energy Management, Nuremberg, Germany, 10–12 May 2022; pp. 1–8. [CrossRef]

33. Musumeci, S.; Palma, M.; Mandrile, F.; Barba, V. Low-Voltage GaN Based Inverter for Power Steering Application. In Proceedings of the 2021 AEIT International Conference on Electrical and Electronic Technologies for Automotive (AEIT AUTOMOTIVE), Torino, Italy, 17–19 November 2021; pp. 1–6. [CrossRef]
34. Mishra, D.; Arora, V.; Nguyen, L.; Iriguchi, S.; Sada, H.; Clemente, L.; Spencer, S. Packaging Innovations for High Voltage (HV) GaN Technology. In Proceedings of the 2017 IEEE 67th Electronic Components and Technology Conference (ECTC), Orlando, FL, USA, 30 May–2 June 2017; pp. 1480–1484. [CrossRef]
35. Prado, E.O.; Bolsi, P.C.; Sartori, H.C.; Pinheiro, J.R. An Overview about Si, Superjunction, SiC and GaN Power MOSFET Technologies in Power Electronics Applications. *Energies* **2022**, *15*, 5244. [CrossRef]
36. Das, M.; Bertilsson, K. GaN Based Converters for Battery Charging Application of Electric Vehicle. In Proceedings of the IECON 2019—45th Annual Conference of the IEEE Industrial Electronics Society, Lisbon, Portugal, 14–17 October 2019; pp. 5134–5139. [CrossRef]
37. Stubbe, T.; Mallwitz, R.; Rupp, R.; Pozzovivo, G.; Bergner, W.; Haeberlen, O.; Kunze, M. GaN Power Semiconductors for PV Inverter Applications—Opportunities and Risks. In Proceedings of the CIPS 2014, 8th International Conference on Integrated Power Electronics Systems, Nuremberg, Germany, 25–27 February 2014; pp. 1–6.
38. Power Brick. Available online: https://www.sunpower-uk.com/glossary/power-brick/ (accessed on 5 April 2023).
39. Mauromicale, G.; Raciti, A.; Rizzo, S.A.; Susinni, G.; Fusillo, F.; Palermo, A.; Scrimizzi, F. Efficiency of state-of-the-art GaN devices in a synchronous-rectifier buck converter. In Proceedings of the IECON 2019—45th Annual Conference of the IEEE Industrial Electronics Society, Lisbon, Portugal, 14–17 October 2019; pp. 1726–1731. [CrossRef]
40. Armando, E.; Fusillo, F.; Musumeci, S.; Scrimizzi, F. Low Voltage Trench-Gate MOSFETs for High Efficiency Auxiliary Power Supply Applications. In Proceedings of the 2019 International Conference on Clean Electrical Power (ICCEP), Otranto, Italy, 2–4 July 2019; pp. 165–170. [CrossRef]
41. Han, D.; Sarlioglu, B. Deadtime Effect on GaN-Based Synchronous Boost Converter and Analytical Model for Optimal Deadtime Selection. *IEEE Trans. Power Electron.* **2016**, *31*, 601–612. [CrossRef]
42. Mortazavizadeh, S.A.; Palazzo, S.; Amendola, A.; De Santis, E.; Di Ruzza, D.; Panariello, G.; Sanseverino, A.; Velardi, F.; Busatto, G. High Frequency, High Efficiency, and High Power Density GaN-Based LLC Resonant Converter: State-of-the-Art and Perspectives. *Appl. Sci.* **2021**, *11*, 11350. [CrossRef]
43. Choi, J.; Tsukiyama, D.; Tsuruda, Y.; Rivas, J. 13.56 MHz 1.3 kW resonant converter with GaN FET for wireless power transfer. In Proceedings of the 2015 IEEE Wireless Power Transfer Conference (WPTC), Boulder, CO, USA, 13–15 May 2015; pp. 1–4. [CrossRef]
44. Faraci, G.; Rizzo, S.A.; Schembra, G. Unmanned aerial vehicles and wind generation serving isolated areas. In Proceedings of the 2020 International Symposium on Power Electronics, Electrical Drives, Automation and Motion (SPEEDAM), Sorrento, Italy, 24–26 June 2020; pp. 138–143. [CrossRef]
45. Barba, V.; Solimene, L.; Palma, M.; Musumeci, S.; Ragusa, C.S.; Bojoi, R. Modelling and Experimental Validation of GaN Based Power Converter for LED Driver. In Proceedings of the 2022 IEEE International Conference on Environment and Electrical Engineering and 2022 IEEE Industrial and Commercial Power Systems Europe (EEEIC/I&CPS Europe), Prague, Czech Republic, 28 June—1 July 2022; pp. 1–6. [CrossRef]
46. Alharbi, S.S.; Alharbi, S.S.; Matin, M. Effects of Gallium Nitride (GaN) Cascode FETs on the Efficiency of an Interleaved DC-DC Converter. In Proceedings of the 2018 IEEE International Conference on Electro/Information Technology (EIT), Rochester, MI, USA, 3–5 May 2018; pp. 384–389. [CrossRef]
47. Raciti, A.; Musumeci, S.; Chimento, F.; Privitera, G. A new thermal model for power MOSFET devices accounting for the behavior in unclamped inductive switching. *Microelectron. Reliab.* **2016**, *58*, 3–11. [CrossRef]
48. Saito, W.; Naka, T. UIS test of high-voltage GaN-HEMTs with p-type gate structure. *Microelectron. Reliab.* **2016**, *64*, 552–555. [CrossRef]
49. Li, H.; Li, X.; Wang, X.; Wang, J.; Alsmadi, Y.; Liu, L.; Bala, S. E-mode GaN HEMT short circuit robustness and degradation. In Proceedings of the 2017 IEEE Energy Conversion Congress and Exposition (ECCE), Cincinnati, OH, USA, 1–5 October 2017; pp. 1995–2002. [CrossRef]
50. Landel, M.; Gautier, C.; Lefebvre, S. Study of short-circuit robustness of p-GaN and cascode transistors. *Microelectron. Reliab.* **2022**, *138*, 114695. [CrossRef]
51. Barth, C.B.; Assem, P.; Foulkes, T.; Chung, W.H.; Modeer, T.; Lei, Y.; Pilawa-Podgurski, R.C. Design and control of a GaN-based, 13-level, flying capacitor multilevel inverter. In Proceedings of the 2016 IEEE 17th Workshop on Control and Modeling for Power Electronics (COMPEL), Trondheim, Norway, 27–30 June 2016; pp. 1–6. [CrossRef]
52. Barchowsky, A.; Kozak, J.P.; Grainger, B.M.; Stanchina, W.E.; Reed, G.F. A GaN-based modular multilevel DC-DC converter for high-density anode discharge power modules. In Proceedings of the 2017 IEEE Aerospace Conference, Big Sky, MT, USA, 4–11 March 2017; pp. 1–10. [CrossRef]
53. Nakayama, Y.; Kanazawa, Y.; Kondo, F.; Inoue, M.; Ohta, K.; Doki, S.; Shiozaki, K. Efficiency Improvement of Motor Drive System by using a GaN Three Phase Inverter. In Proceedings of the 2019 IEEE International Conference on Industrial Technology (ICIT), Melbourne, VIC, Australia, 8 July 2019; pp. 1599–1604. [CrossRef]
54. Musumeci, S.; Barba, V.; Palma, M. GaN-Based Low-Voltage Inverter for Electric Scooter Drive System. In Proceedings of the 2022 AEIT International Annual Conference (AEIT), Rome, Italy, 3–5 October 2022; pp. 1–6. [CrossRef]

55. Ermolaev, A.; Erofeev, V.; Plekhov, A.; Titov, D. Magnetic Vibration in Induction Motor Caused by Supply Voltage Distortion. *Energies* **2022**, *15*, 9600. [CrossRef]
56. Mandrile, F.; Musumeci, S.; Palma, M. Dead Time Management in GaN Based Three-Phase Motor Drives. In Proceedings of the 2021 23rd European Conference on Power Electronics and Applications (EPE'21 ECCE Europe), Ghent, Belgium, 6–10 September 2021; pp. 1–10. [CrossRef]
57. Musumeci, S.; Barba, V.; Mandrile, F.; Palma, M.; Bojoi, R.I. Dead Time Reverse Conduction Investigation in GaN-Based Inverter for Motor Drives. In Proceedings of the IECON 2022—48th Annual Conference of the IEEE Industrial Electronics Society, Brussels, Belgium, 17–20 October 2022; pp. 1–6. [CrossRef]
58. Uğur, M.; Keysan, O. Multi-physics design optimisation of a GaN-based integrated modular motor drive system. *J. Eng.* **2019**, *2019*, 3900–3905. [CrossRef]
59. EPC21601data Sheet. Available online: https://epc-co.com/epc/products/gan-fets-and-ics/epc21601 (accessed on 5 April 2023).
60. Xiao, M.; Ma, Y.; Du, Z.; Pathirana, V.; Cheng, K.; Xie, A.; Zhang, Y. Multi-Channel Monolithic-Cascode HEMT (MC2-HEMT): A New GaN Power Switch up to 10 kV. In Proceedings of the 2021 IEEE Interntional Electron Devices Meeting (IEDM), San Francisco, CA, USA, 9 March 2022; pp. 5.5.1–5.5.4. [CrossRef]
61. Barba, V.; Musumeci, S.; Palma, M.; Bojoi, R. Maximum Peak Current and Junction-to-ambient Delta-temperature Investigation in GaN FETs Parallel Connection. *Power Electron. Devices Compon.* **2023**, *5*, 100035. [CrossRef]
62. Wattenberg, M.; Lorenz, O.; Sanchez, J. Efficiently Paralleling GaN-Transistors for High Current and High Frequency Applications Using a Butterfly Layout. In Proceedings of the 2022 24th European Conference on Power Electronics and Applications (EPE'22 ECCE Europe), Hanover, Germany, 5–9 September 2022; pp. 1–10.

energies

MDPI

Article

A Family of Zero-Voltage-Transition Magnetic Coupling Bidirectional DC/DC Converters †

Shanshan Wang, Ming Gao and Jianjiang Shi *

College of Electrical Engineering, Zhejiang University, Hangzhou 310007, China; 11710055@zju.edu.cn (S.W.); 11810059@zju.edu.cn (M.G.)

* Correspondence: jianjiang@zju.edu.cn

† This paper is an extended version of our paper published in 2022 12th International Conference on Power, Energy and Electrical Engineering (CPEEE), Shiga, Japan, 25–27 February 2022; pp. 180–184.

Abstract: In this paper, a family of zero-voltage-transition (ZVT) magnetic coupling bidirectional DC/DC converters (BDCs) is proposed. The coupling inductor has two functions: to serve as a filter; and to provide the auxiliary current. In addition, the phase-shifting control method is used to reduce the conduction loss of the auxiliary circuit. The auxiliary switches are all working under zero-current-switching (ZCS) conditions; thus, all the switches have no switching loss. Furthermore, compared with the traditional ZVT implementation method based on the coupled inductor, this topology avoids the input source current notch and has a smaller ripple. Finally, a prototype of 800 W BDC is built to verify the feasibility of the proposed topology.

Keywords: zero-voltage-transition; zero-current-switching; buck/boost converter; coupled inductor

Citation: Wang, S.; Gao, M.; Shi, J. A Family of Zero-Voltage-Transition Magnetic Coupling Bidirectional DC/DC Converters. *Energies* **2023**, *16*, 4760. https://doi.org/10.3390/en16124760

Academic Editor: Salvatore Musumeci

Received: 10 April 2023
Revised: 7 June 2023
Accepted: 12 June 2023
Published: 16 June 2023

1. Introduction

With the continuous growth of systems with energy storage devices, research on bidirectional DC–DC converters has been greatly promoted. Bidirectional DC–DC converters are widely used in photovoltaic and energy storage systems [1,2], electric/hybrid vehicles [3,4], uninterrupted power supply systems [5], fuel cell power systems [6,7], etc. In these applications, converters are required to have low current ripple requirements at the energy storage port. If electrical isolation is not required between the high-voltage port and low-voltage port, the BDC topology is usually chosen due to its low cost and simple structure. For BDC, switching loss is the main reason for limiting frequency increases. Therefore, in order to achieve a high-efficiency and high-frequency power conversion of BDC, it is necessary to introduce soft-switching technology.

One effective method to achieve the purpose of soft-switching is to make the BDC work in near-critical conduction modes [8–10]. However, the input current has a larger ripple, which leads to a high conduction loss and easy core saturation.

Quasi-resonant technology realizes the zero-voltage-switching (ZVS) of the main switch by using *LC* resonance to generate zero voltage when the main switch is turned on [11,12]. However, a large circulating current exists in the circuit, which leads to an increase in conduction loss. In addition, the switching frequency changes in a wide range when the load changes greatly, which will make the design of passive devices more difficult.

Active clamp ZVS technology also needs to add a resonant inductor in the main circuit to realize the ZVS of the main switch [13,14]. The circulating current of the auxiliary circuit is large, which increases the conduction loss. Moreover, the main switch has high voltage stress.

As the auxiliary circuit is removed from the main circuit and the auxiliary switch operates only during the period before and after the main switch is turned on, the conduction loss in the ZVT converter is reduced, while the voltage stress of the main switch is lower than that of the quasi-resonant converter and active clamp converter [15–18]. These ZVT converters all need additional inductors as resonant inductors, which are disadvantageous to the improvement of power density because of the large volume and weight of magnetic components. Several soft-switching technologies based on coupled inductors have been proposed in [19–28].

In [19], a coupled winding is added to the same core of the main inductor to realize the soft switching of the main switch. In the ZVT BDC proposed in [20], a coupled inductor and two unidirectional auxiliary switches are used in the auxiliary circuit. A ZVT BDC is described in [21], while the auxiliary circuit consists of a coupled inductor and two auxiliary switches. A family of ZVS Buck, ZVS Boost, and ZVS BDC topologies is presented in [22], and the auxiliary circuit consists of a coupled inductor and a diode. In [23], a family of ZVS magnetic coupling BDC topologies is proposed. In those BDCs, the additional circuits contain two auxiliary switches. Two auxiliary voltage sources composed of passive components are used in the auxiliary circuit of the ZVS BDC in [24]. A ZVT BDC is proposed in [25], in which the auxiliary circuit consists of a coupled inductor, two unidirectional switches, and an auxiliary capacitor. In [26], a ZVT BDC is studied, in which all the switches implement ZVS or ZCS conditions. In [27], several ZVS DC–DC converters based on coupled inductors are proposed by changing the connection position of the coupling branch. A ZVS magnetic coupling BDC with no current notch at the low-voltage port is proposed in [28]. In [29], a ZVT PWM BDC based on coupled inductors is proposed. This topology needs two cores, increasing the complexity of the system. An overview of soft-switching technologies for BDC is given in [30].

As shown in Figure 1, there is an input current notch in the ZVS BDCs proposed in [20,21,23,24,26], which would hurt the life of storage equipment. In this paper, a family of ZVT magnetic coupling BDCs is proposed. The coupling inductor not only serves as a filter but also provides the auxiliary current. Compared with the traditional ZVS implementation method, this topology avoids the input source current notch and has a smaller ripple.

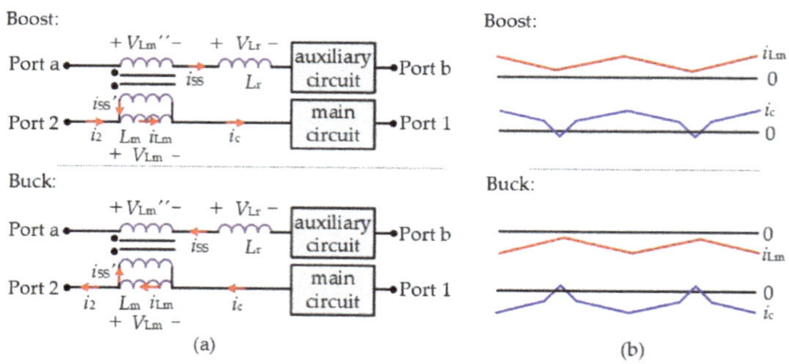

Figure 1. General circuit of ZVS BDCs with coupled filter inductors in [20,21,23,24,26]. (**a**) General circuit; (**b**) key waveforms of the input currents.

In Section 2, the topologies and operation process are demonstrated. In Section 3, the ZVT conditions of the main switches, the loss analysis, and the comparison between the proposed ZVT BDCs and the existing ZVT BDCs are given in detail. Moreover, the current frequency spectrum of the low-voltage port comparison between the proposed ZVT BDC and the traditional ZVT BDC is provided. In Section 4, the experimental results verify the validity of the topology. Finally, the conclusions are summarized.

2. Operation Process Analysis

2.1. Proposed ZVT BDCs

The ZVT magnetic coupling BDCs are shown in Figure 2. S_1 and S_2 are the switches of the main power circuit. S_{SS1} and S_{SS2} are the switches of the auxiliary circuit. C_{S1}, C_{S2}, and C_a are the three auxiliary capacitors. The coupled inductor is equivalent to an ideal transformer T, L_m (the excitation inductor), and L_r (the leakage inductor). C_1 and C_2 are filter capacitors of Port 1 and Port 2. The difference between Topology 1 (as shown in Figure 2a) and Topology 2 (as shown in Figure 2b) is that the auxiliary circuit of Topology 1 is connected between Port 2 and the ground, and the auxiliary circuit of Topology 2 is connected between Port 1 and Port 2.

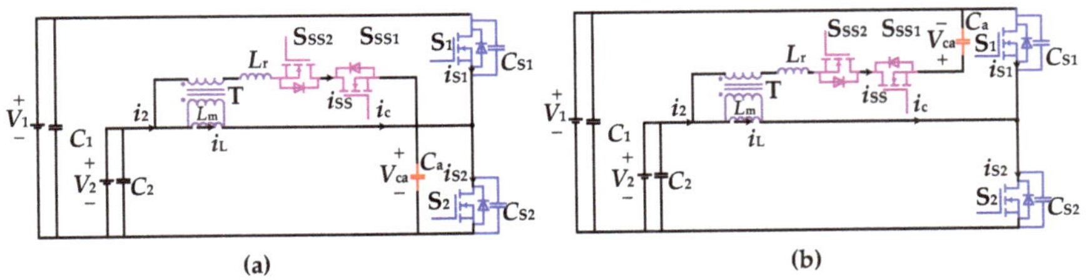

Figure 2. The proposed ZVS BDCs. (**a**) Topology 1; (**b**) Topology 2.

2.2. Operation Analysis of the Proposed ZVS BDCs

2.2.1. Operation Analysis of Topology 1

In Topology 1, as shown in Figure 2a, the switching process of each switch is shown in Figure 3. Taking the ZVS implementation of S_2 as an example, the switching process is analyzed, and the equivalent circuit of each stage is shown in Figure 4.

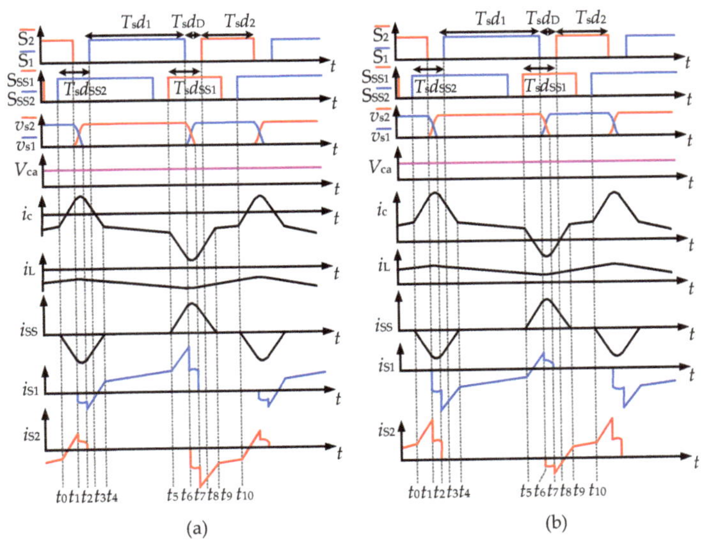

Figure 3. ZVS implementation of the main switch for topology 1. (**a**) ZVS implementation of S_1; (**b**) ZVS implementation of S_2.

Interval 1 [t_0–t_1]: At t_0, S_{SS2} is turned under the ZCS condition. The current i_{SS} increases because of the voltage clamp action of C_a, and the equivalent circuit is shown in Figure 4a.

Interval 2 [t_1–t_2]: At t_1, S_2 is turned off under the ZVS condition because of C_{S1} and C_{S2}. During this interval, v_{S1} decreases, v_{S2} increases, C_{S1} discharges, C_{S2} charges, and the equivalent circuit is shown in Figure 4b. The current i_{SS} at the resonance stage is:

$$i_{SS} = i_L(t_1) + 2A_{11}C_S\omega_{11}\sin(\omega_{11}(t - t_1) + \theta_{11}) \tag{1}$$

$$\omega_{11} = \frac{1}{\sqrt{2L_rC_S}} \tag{2}$$

$$A_{11} = \sqrt{V_{ca}^2 + \left[\frac{i_{SS}(t_1) - i_L(t_1)}{2C_S\omega_{11}}\right]^2} \tag{3}$$

$$\theta_{11} = \arctan\frac{i_{SS}(t_1) - i_L(t_1)}{-2C_S\omega_{11}V_{ca}} + \pi \tag{4}$$

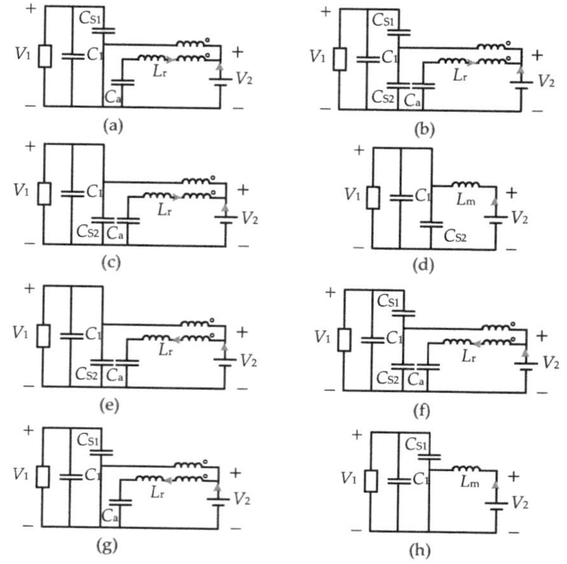

Figure 4. The equivalent circuit for each stage of Topology 1. (**a**) t_0–t_1; (**b**) t_1–t_2; (**c**) t_2–t_4; (**d**) t_4–t_5; (**e**) t_5–t_6; (**f**) t_6–t_7; (**g**) t_7–t_9; (**h**) t_9–t_{10}.

Interval 3 [t_2–t_3]: At t_2, v_{S1} decreases to 0, v_{S2} increases to V_1, and i_{S1} is a negative value, the equivalent circuit for this stage is shown in Figure 4c.

Interval 4 [t_3–t_4]: At t_3, S_1 is turned on under the ZVS condition.

Interval 5 [t_4–t_5]: The current i_{SS} is reduced to 0 at t_4, and the equivalent circuit for this stage is shown in Figure 4d.

Interval 6 [t_5–t_6]: S_{SS1} is turned on at t_5. The current i_{SS} increases because of the voltage clamp action of the capacitor C_a, and the equivalent circuit is shown in Figure 4e.

Interval 7 [t_6–t_7]: S_1 is turned off under ZVS condition at t_6 because of C_{S1} and C_{S2}. During this period, v_{S1} increases, v_{S2} decreases, C_{S1} charges, C_{S2} discharges, and the equivalent circuit is shown in Figure 4f. The auxiliary current i_{SS} at the resonance stage is:

$$i_{SS} = i_L(t_6) + 2A_{12}C_S\omega_{12}\sin(\omega_{12}(t - t_6) + \theta_{12}) \tag{5}$$

$$\omega_{12} = \frac{1}{\sqrt{2L_rC_S}} \tag{6}$$

$$A_{12} = \sqrt{(V_1 - V_{ca})^2 + \left[\frac{i_{SS}(t_6) - i_L(t_6)}{2C_S\omega_{12}}\right]^2} \tag{7}$$

$$\theta_{12} = \arctan\frac{i_{SS}(t_6) - i_L(t_6)}{2C_S\omega_{12}(V_1 - V_{ca})} \tag{8}$$

Interval 8 [t_7–t_8]: At t_7, v_{S1} increases to V_1, v_{S2} decreases to 0, and i_{S2} is a negative value. The equivalent circuit for this stage is shown in Figure 4g.

Interval 9 [t_8–t_9]: S_2 is turned on under the ZVS condition at t_8.

Interval 10 [t_9–t_{10}]: The current i_{SS} is reduced to 0 at t_9, and the equivalent circuit for this stage is shown in Figure 4h.

2.2.2. Operation Analysis of Topology 2

In Topology 2, as shown in Figure 2b, the switching process of each switch is shown in Figure 5. The switching process of S_1 is analyzed, and the equivalent circuit of each stage is shown in Figure 6.

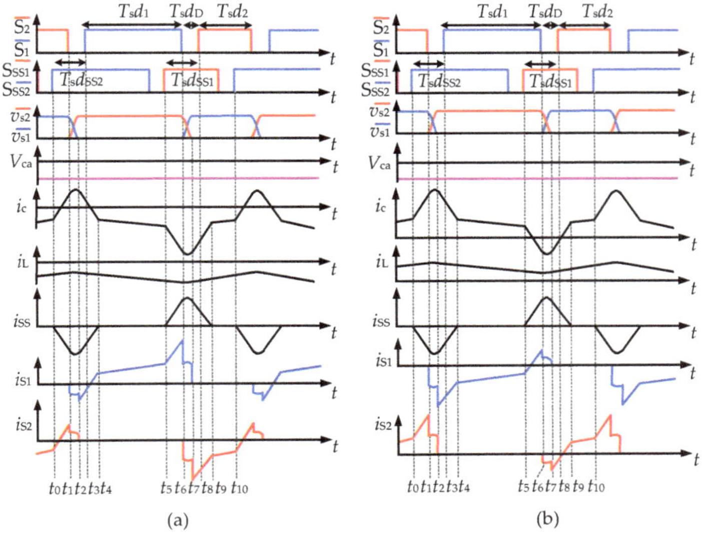

Figure 5. ZVS implementation of the main switch for Topology 2. (**a**) ZVS implementation of S_1; (**b**) ZVS implementation of S_2.

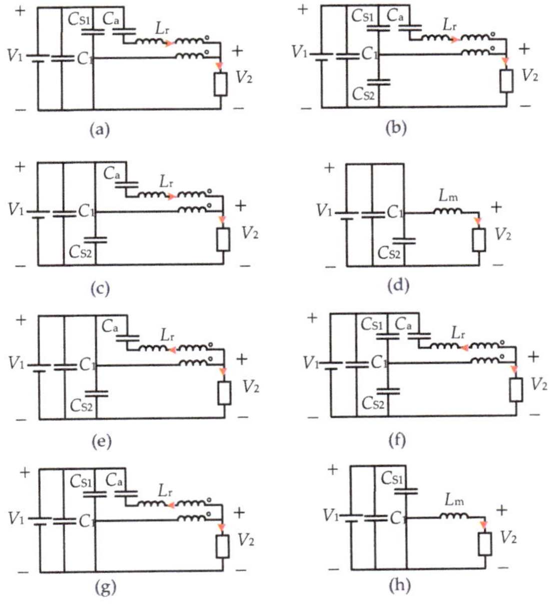

Figure 6. The equivalent circuit for each stage of Topology 2. (**a**) t_0–t_1; (**b**) t_1–t_2; (**c**) t_2–t_4; (**d**) t_4–t_5; (**e**) t_5–t_6; (**f**) t_6–t_7; (**g**) t_7–t_9; (**h**) t_9–t_{10}.

Interval 1 [t_0–t_1]: At t_0, S_{SS2} is turned on under the ZCS condition.

Interval 2 [t_1–t_2]: At t_1, S_2 is turned off under the ZVS condition. The current i_{SS} at the resonance stage is:

$$i_{SS} = i_L(t_1) + 2A_{21}C_S\omega_{21}\sin(\omega_{21}(t - t_1) + \theta_{21}) \tag{9}$$

$$\omega_{21} = \frac{1}{\sqrt{2L_rC_S}} \tag{10}$$

$$A_{21} = \sqrt{(V_1 + V_{ca})^2 + \left[\frac{i_{SS}(t_1) - i_L(t_1)}{2C_S\omega_{21}}\right]^2} \tag{11}$$

$$\theta_{21} = \arctan\frac{i_{SS}(t_1) - i_L(t_1)}{2C_S\omega_{21}(-V_1 - V_{ca})} + \pi \tag{12}$$

Interval 3 [t_2–t_3]: At t_2, v_{S1} decreases to 0, v_{S2} increases to V_1, and i_{S1} is a negative value.

Interval 4 [t_3–t_4]: At t_3, S_1 is turned on under the ZVS condition.

Interval 5 [t_4–t_5]: S_1 is on during this period.

Interval 6 [t_5–t_6]: S_{SS1} is turned on under the ZCS condition at t_5.

Interval 7 [t_6–t_7]: S_1 is turned off under the ZVS condition at t_6. The current i_{SS} at the resonance stage is:

$$i_{SS} = i_L(t_6) + 2A_{22}C_S\omega_{22}\sin(\omega_{22}(t - t_6) + \theta_{22}) \tag{13}$$

$$\omega_{22} = \frac{1}{\sqrt{2L_rC_S}} \tag{14}$$

$$A_{22} = \sqrt{{V_{ca}}^2 + \left[\frac{i_{SS}(t_6) - i_L(t_6)}{2C_S\omega_{22}}\right]^2} \tag{15}$$

$$\theta_{22} = \arctan\frac{i_{SS}(t_6) - i_L(t_6)}{-2C_S\omega_{22}V_{ca}} \tag{16}$$

Interval 8 [t_7–t_8]: At t_7, i_{S2} is a negative value.

Interval 9 [t_8–t_9]: S_2 is turned on under the ZVS condition at t_8.

Interval 10 [t_9–t_{10}]: S_2 is on during this period.

2.3. Other ZVT BDC Topologies

In addition to the topologies shown in Figure 2, other ZVT BDCs based on coupled inductor are shown in Figure 7.

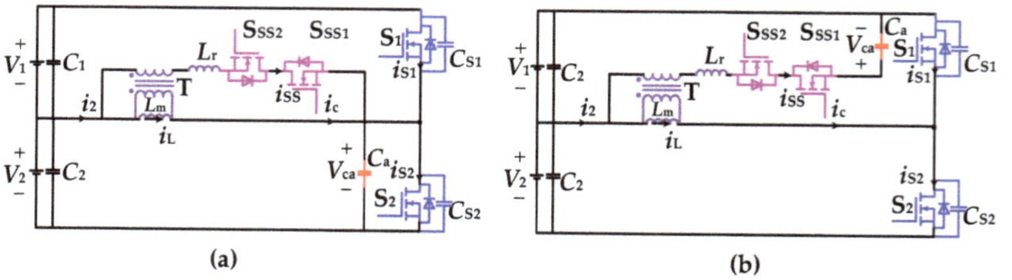

Figure 7. Other ZVT BDCs. (**a**) Topology 1; (**b**) Topology 2.

3. Performance Analysis

3.1. Analysis of Input Current Ripple

As shown in Figures 3 and 5, i_L is nearly constant during the resonance stage (t_1–t_2) and (t_6–t_7), and the ripple of i_L is:

$$\Delta I_L \approx \frac{V_2}{L_m} d_2 T_S \tag{17}$$

To further illustrate the advantages of the proposed ZVT BDCs, the harmonic comparison of i_L is compared between the ZVT BDC (shown in Figure 2a) and the traditional ZVS BDC (represented in [23]). Those two simulation models are built, with the simulation performed at the same voltage and power levels.

The simulation results are shown in Figures 8 and 9. Figure 8 shows the simulation results in Boost mode and Figure 9 in Buck mode. Figure 8a,b show the current i_2 and its harmonic analysis of ZVS BDC in [23], and when P_1 = 800 W, the harmonic amplitude of i_2 is 3.12 dBA (100 kHz). Figure 8c,d present the simulation results of the proposed ZVT BDC; the harmonic amplitude of i_2 is 0.63 dBA (100 kHz). As shown in Figure 9, when P_2 = 800 W, the proposed ZVT BDC has a 1.89 dBA (100 kHz) decrease in the harmonic amplitude of i_2 compared to ZVS BDC in [23].

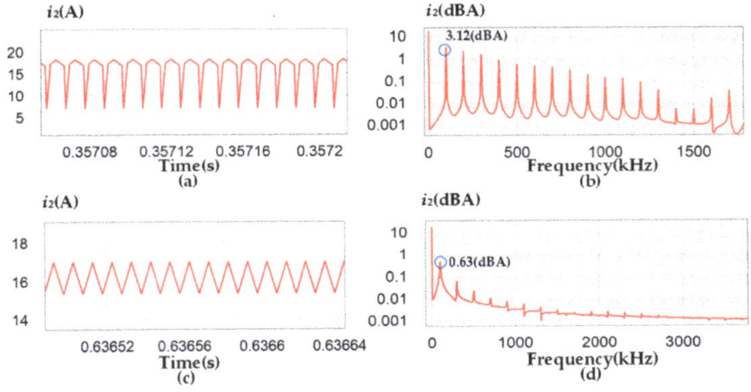

Figure 8. Harmonic comparison in Boost mode. (**a**) Input current of ZVS BDC in [23]; (**b**) harmonic analysis of input current of ZVS BDC in [23]; (**c**) input current of ZVT BDC (shown in Figure 2a); (**d**) harmonic analysis of input current of ZVT BDC (shown in Figure 2a).

3.2. Loss Analysis

In order to theoretically analyze the loss comparison between the traditional BDC and the proposed ZVT BDC, two simulation models are established. These work at 100 kHz and 500 kHz, respectively. The parameters are shown in Table 1, the proposed ZVT BDC is shown in Figure 2a, and the loss estimation is shown in Table 2.

Table 1. Detail parameters of 800 W ZVS Boost converter.

f_S = 100 kHz				f_S = 500 kHz			
Parameter	**Value**	**Parameter**	**Value**	**Parameter**	**Value**	**Parameter**	**Value**
V_2/V	50	V_1/V	100	V_2/V	50	V_1/V	100
P_{1max}/W	800	n	1	P_{1max}/W	800	n	1
L_c/μH	3.8	L_m/μH	130	L_c/μH	0.85	L_m/μH	160
C_1/μF	1000	C_2/μF	1000	C_1/μF	1000	C_2/μF	1000
C_S/nF	2.2	C_a/μF	200	C_S/nF	1	C_a/μF	80

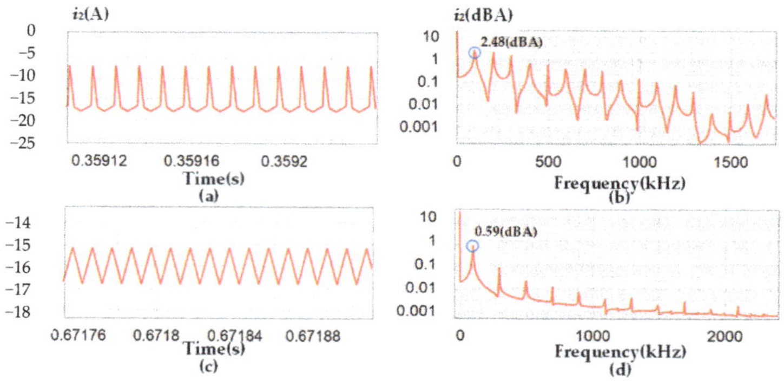

Figure 9. Harmonic comparison in Buck mode. (**a**) Output current of ZVS BDC in [23]; (**b**) harmonic analysis of output current of ZVS BDC in [23]; (**c**) output current of ZVT BDC (shown in Figure 2a); (**d**) harmonic analysis of output current of ZVT BDC (shown in Figure 2a).

Table 2. Loss comparison.

Loss Source	f_S = 100 kHz		f_S = 500 kHz	
	Traditional BDC	Proposed BDC	Traditional BDC	Proposed BDC
P_{c_S}	2.1 W	2.5 W	2.1 W	3.1 W
P_{c_SS}	---	5.4 W	---	10.4 W
P_{s_S}	19.1 W	---	95.5 W	---
P_{RR}	---	0.8 W	---	4.2 W
Core loss	1.9 W	2.1 W	0.3 W	0.4 W
Copper loss	2.7 W	3.3 W	2.9 W	4.3 W

For the proposed ZVT BDC, MOSFET IRFP4668 is applied as the main switch, and MOSFET SUP90142E is utilized as the auxiliary switch, respectively. The on-resistance of the main switch is 0.008 Ω, the on-resistance of the auxiliary switch is 0.017 Ω, and the diode forward voltage of the auxiliary switch is 0.85 V (the junction temperature (Tj) is 75°). The core of the coupled inductor is E220–18. In the traditional BDC model, the filter inductor adopts the magnetic core KS25060A, and the main switch adopts MOSFET IRFP4668.

For the traditional BDC, the main loss includes P_{c_S} (conduction loss of S_1 and S_2), P_{s_S} (switching loss of S_1 and S_2), core loss, and copper loss. For the proposed ZVS BDC, the main loss includes P_{c_S} (conduction loss of S_1 and S_2), P_{c_SS} (conduction loss of S_{SS1} and S_{SS2}), P_{RR} (Reverse recovery loss of S_{SS1} and S_{SS2}), core loss, and copper loss.

When the switching frequency of the converter is increased to more than 500 kHz, the switching loss of the conventional BDC will greatly reduce the efficiency of the converter; the introduction of soft-switching measures is especially necessary.

3.3. Topology Comparison

The ZVS conditions can be realized when the circuit is operating under Buck mode or Boost mode in the ZVS BDC in [19,22,27]. Table 3 shows the topology comparison between the ZVS BDC in the existing literature and the proposed ZVS BDC (as shown in Figure 2a).

Table 3. Topology comparison.

Topology	Number of Auxiliary Switches	Number of Auxiliary Diodes	Current Notch	Voltage Stress of Auxiliary Switches
ZVT BDC [17]	2	2	exist	$\max\{DV_1, (1-D)V_1\}$
ZVS BDC [18]	2	0	exist	Buck: V_2 Boost: $V_1 - V_2$
ZVT BDC [20]	2	2	exist	$(1 + nD)V_2$
ZVT BDC [21]	2	0	exist	nDV_1
ZVS BDC [23]	2	0	exist	S_{a1}: $(n+1)V_2$ S_{a2}: $(n+1)(V_1 - V_2)$
ZVS BDC [24]	2	2	exist	Buck: $\max\{\lvert(n-1)(V_1-V_2)\rvert, V_2 + n(V_1 - V_2)\}$ Boost: $\max\{\lvert(n-1)V_2\rvert, V_1 + (n-1)V_2\}$
ZVT BDC [25]	2	2	no	Buck: >0.5 V_2 Boost: >0.5 V_2
ZVT BDC [26]	2	0	exist	S_{a1}: $(n+1)(V_1 - V_2)$ S_{a2}: $(n+1)V_2$
ZVS BDC [28]	2	2	no	S_{SS1}: $\frac{2}{3}\left[\frac{L_1(V_1-V_2)}{L_1+L_c-2M} + V_2\right] - \frac{1}{3}V_1$ S_{SS2}: $-\frac{2}{3}\frac{(L_c-M)V_2}{L_1+L_c-2M} + \frac{1}{3}V_1$
Proposed ZVT BDCs	2	0	no	S_{SS1}: $V_1 - V_{ca}$ S_{SS2}: V_{ca}

The topologies in [17,18] construct auxiliary circuits by adding auxiliary inductors, and the topology in [24] constitutes an auxiliary circuit by adding an auxiliary inductor and introducing a coupling inductor. The topologies in [19–23,25–28] and this paper use coupling inductors to construct auxiliary branches, which reduce the number of inductors. In order to realize soft switching in both operating modes (Buck and Boost), two auxiliary switches are required. To reduce the loss of auxiliary circuits, the diodes in [20,25,28] need to have good reverse recovery characteristics and low–forward voltage drops, which will increase the cost. In the topologies in [17–24,26], to realize the ZVT of main switches, notches exist in the current at Port 2. When Port 2 is connected to a battery, the current notch can adversely affect the battery life. However, the proposed ZVT BDC can avoid this notch phenomenon. For the topologies with no current notch in [25,28] and the proposed ZVT BDC, the current stress of the auxiliary switch depends on the current and ripple at Port 2, and the current stress is similar when the power and voltage levels are the same. In terms of voltage stress, the topology of the proposed ZVT BDC is less than that of [25].

4. System Design

4.1. Design of Initial Value V_{ca}

4.1.1. Design of Initial Value V_{ca} in Topology 1

For Topology 1, shown in Figure 2a, in the switching process shown in Figure 3, the resonance process should be ignored to simplify the analysis. The voltage V_{ca} of the auxiliary capacitor C_a during the period (t_0–t_1) is:

$$V_{ca} = \frac{I_{SS}L_r}{t_1 - t_0} \tag{18}$$

During the period (t_5–t_6), the voltage V_{ca} of C_a is:

$$V_{ca} = V_1 - \frac{I_{SS}L_r}{t_6 - t_5} \tag{19}$$

When the period (t_0–t_1) and (t_5–t_6) are the same, the phase-shifting duty cycles of the auxiliary switch S_{SS1} and S_{SS2} are the same. According to (18) and (19), the initial value V_{ca} = 0.5 V_1 can be obtained. Considering the resonant process in the dead time, the actual V_{ca} is different from 0.5 V_1.

4.1.2. Design of Initial Value V_{ca} in Topology 2

For Topology 2, shown in Figure 2b, the switching process is shown in Figure 5. During the period (t_0–t_1), V_{ca} is:

$$V_{ca} = -V_1 + \frac{I_{SS}L_r}{t_1 - t_2} \tag{20}$$

During the period (t_5–t_6), V_{ca} is:

$$V_{ca} = -\frac{I_{SS}L_r}{t_6 - t_5} \tag{21}$$

When the periods (t_0–t_1) and (t_5–t_6) are the same, the initial value V_{ca} = −0.5 V_1 can be obtained according to (20) and (21). Considering the resonant process in the dead time, the actual V_{ca} is different from −0.5 V_1.

4.2. Design of the Coupled Inductor

In the topology shown in Figure 2, the turn ratio of transformer T is n, the primary side is connected to the main circuit, the excitation inductor acts as a filter, and the secondary side is connected to the auxiliary circuit. i_L is the current flowing through the excitation inductor, i_{SS} is the current of the auxiliary circuit, i_{SS}' is the current converted to the primary side, and $i_{SS}' = i_{SS}/n$. The current of Port 2 is:

$$i_2 = i_{SS} + i_c \tag{22}$$

The current i_c is:

$$i_c = i_L - i_{SS}' \tag{23}$$

According to (22) and (23), the current of port 2 is $i_2 = i_{SS} + i_L - i_{SS}'$. Therefore, when $n = 1$, $i_2 = i_L$.

In the topology shown in Figure 2, L_m is the filter inductor for the BDC, duty cycle d_2 is $(1 - V_2/V_1)$, and the current ripple for i_L is:

$$\Delta i_L \approx \frac{V_2}{L_m} d_2 T_S \tag{24}$$

The ripple $\Delta i_2 = \Delta i_L$, taking $\Delta i_L = x_1 i_L$, x_1 is the ripple coefficient, and the value of L_m can be obtained according to the requirement of ripple.

4.2.1. Design of L_r in Topology 1

In Topology 1, the maximum value of auxiliary current i_{SS} is:

$$\Delta i_{SSmax\_1} = \frac{V_1 - V_{ca}}{L_r} T_S d_{SS1} \tag{25}$$

For Buck mode, the following condition should be met to realize the ZVS of S_1:

$$I_L + \frac{1}{2}\Delta i_L + \Delta i_{SSmax\_1} > 0 \tag{26}$$

For Boost mode, the following condition should be met:

$$I_L - \frac{1}{2}\Delta i_L - \Delta i_{SSmax\_1} < 0 \tag{27}$$

From (26) and (27), the value of L_r can be obtained.

4.2.2. Design of L_r in Topology 2

In topology 2, the maximum value of auxiliary current i_{SS} is:

$$\Delta i_{SSmax\_2} = \frac{-V_{ca}}{L_r} T_S d_{SS2} \quad (28)$$

To realize the ZVS of S_1 and S_2, the following conditions should be satisfied:

$$I_L + \frac{1}{2}\Delta i_L + \Delta i_{SSmax\_2} > 0 \quad (29)$$

$$I_L - \frac{1}{2}\Delta i_L - \Delta i_{SSmax\_2} < 0 \quad (30)$$

From (29) and (30), the value of L_r can be obtained.

4.3. Design of C_S

4.3.1. Design of C_S in Topology 1

For Topology 1, according to the ZVS implementation process as shown in Figure 3b, dead time needs to satisfy:

$$\Delta t_{12} = t_2 - t_1 = \frac{1}{\omega_{11}}(\arccos\frac{-V_{ca}}{A_{11}} - \arccos\frac{V_1 - V_{ca}}{A_{11}}) < t_D \quad (31)$$

$$\Delta t_{67} = t_7 - t_6 = \frac{1}{\omega_{12}}(\arccos\frac{-V_{ca}}{A_{12}} - \arccos\frac{V_1 - V_{ca}}{A_{12}}) < t_D \quad (32)$$

At time t_7, the values of i_c and i_{S2} are:

$$i_c(t_7) = -2A_{12}C_S\omega_{12}\sin(\omega_{12}(t_7 - t_6) + \theta_{12}) \quad (33)$$

$$i_{S2}(t_7) = -A_{12}C_S\omega_{12}\sin(\omega_{12}(t_7 - t_6) + \theta_{12}) \quad (34)$$

When S_2 is turned on, i_{S2} needs to meet:

$$i_{S2}(t_{S2\_on}) = i_{S2}(t_7) + \left(\frac{V_{ca}}{L_r} + \frac{V_2}{L_m}\right) \times (t_D - \Delta t_{67}) < 0 \quad (35)$$

According to (31)–(35),the value of C_S can be obtained.

4.3.2. Design of C_S in Topology 2

For Topology 2, according to the ZVS implementation process as shown in Figure 5a, in order to realize the ZVS of S_1, the following conditions should be met:

$$\Delta t_{12} = t_2 - t_1 = \frac{1}{\omega_{21}}(\arccos\frac{-V_1 - V_{ca}}{A_{21}} - \arccos\frac{-V_{ca}}{A_{21}}) < t_D \quad (36)$$

$$\Delta t_{67} = t_7 - t_6 = \frac{1}{\omega_{22}}(\arccos\frac{-V_1 - V_{ca}}{A_{22}} - \arccos\frac{-V_{ca}}{A_{22}}) < t_D \quad (37)$$

$$i_{S1}(t_{S1\_on}) = i_{S1}(t_2) + \left(\frac{-V_{ca}}{L_r} + \frac{V_1 - V_2}{L_m}\right) \times (t_D - \Delta t_{12}) < 0 \quad (38)$$

According to (36)–(38), the value of C_S can be obtained.

4.4. Design of C_a

4.4.1. Design of C_a in Topology 1

As the voltage source of the auxiliary circuit, the voltage V_{ca} of the auxiliary capacitor C_a can be determined according to the duration of (t_0–t_1) and (t_5–t_6), and the voltage ripple ΔV_{ca} is:

$$\Delta V_{ca} = \frac{1}{C_a}\int_{t_0}^{\frac{T_S}{2}} i_{SS}(t)dt \tag{39}$$

In order to simplify the analysis, ignoring the resonance process and according to (39) and Figure 3, ΔV_{ca} can be obtained:

$$\Delta V_{ca} \approx \frac{V_1 V_{ca} T_S^2 d_{SS2}^2}{2C_a L_r (V_1 - V_{ca})} \tag{40}$$

Taking $\Delta V_{ca} = x_2 V_{ca}$, x_2 is the voltage ripple coefficient, and the value of auxiliary capacitance C_a can be obtained according to the requirement of voltage ripple.

4.4.2. Design of C_a in Topology 2

According to (39), and Figure 5, ΔV_{ca} can be obtained:

$$\Delta V_{ca} \approx -\frac{V_1 (V_1 + V_{ca}) T_S^2 d_{SS2}^2}{2C_a L_r V_{ca}} \tag{41}$$

Similarly, according to the ripple requirements, the value of the auxiliary capacitor C_a can be obtained.

Taking the ZVT BDC (as shown in Figure 2a) design process as an example, a step-by-step design methodology working in Boost mode is given in Table 4. The other topologies and the design process in Buck mode are similar and will not be repeated.

4.5. Design of the Control Method

Figure 10 shows the implementation of the control method, which is divided into two cases (Buck mode and Boost mode). The PI regulator of the output voltage (V_2 under Buck, V_1 in Boost) generates the modulation signal v_{ctrl}, and the phase difference between the carrier signals v_{triS1} and v_{triS2} is 180°, generating the drive signals of S_1 and S_2 by comparators. To obtain the driving signals of S_{SS1} and S_{SS2}, firstly, judge the working mode of the system (Buck mode or Boost mode) according to the voltages and currents of Ports 1 and 2. The phase-shifting duty cycles d_{SS1} and d_{SS2} are calculated according to the working mode. In order to ensure that the auxiliary current is large enough, d_{SS1} and d_{SS2} require a certain amount of margin Δd. The phase-shifting controller shifts the carrier signals v_{triS1} and v_{triS2} according to d_{SS1} and d_{SS2}, and the carrier signals of S_{SS1} and S_{SS2}, v_{triSS1} and v_{triSS2}, are obtained. The driving signals of S_{SS1} and S_{SS2} are generated by comparators.

Table 4. Step-by-step design methodology.

Step 1	Determine the voltages (V_1 and V_2) and power level according to the system requirements.
Step 2	Ignore the resonance process and select initial value of V_{ca}, $V_{ca} = 0.5V_1$.
Step 3	Design L_m of T according to the requirements of current ripple, and $L_m \geq \frac{V_2 d_2 T_S}{x_1 I_L}$.
Step 4	Select $n = 1$.
Step 5	Design L_r of T according to $I_L - 0.5\Delta i_L - \Delta i_{SS} < 0$, in which $\Delta i_{SS} = \frac{V_1 - V_{ca}}{L_r} T_S d_{SS1}$.
Step 6	Determine value range of (C_S, t_d) according to $\begin{cases} t_{12} < t_d \\ t_{67} < t_d \\ i_{S2}(t_{S2\_on}) < 0 \end{cases}$, and select C_S and t_d.
Step 7	Design C_a according to the requirements of voltage ripple, and $C_a \geq \frac{1}{x_2 V_{ca}}\int_{t_0}^{\frac{T_S}{2}} i_{SS}(t)dt$.
Step 8	Simulation and experimental verification.

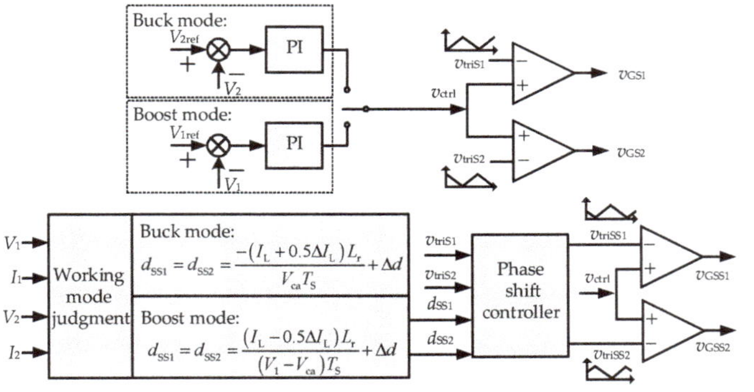

Figure 10. Control schematic diagram.

5. Experimental Verification

Using DSP TMS320F28377SPTPT as the controller, MOSFET IRFP4668 as the main switch, and MOSFET SUP90142E as the auxiliary switch, a BDC prototype is built, as shown in Figure 11. Detailed experimental parameters are shown in Table 5.

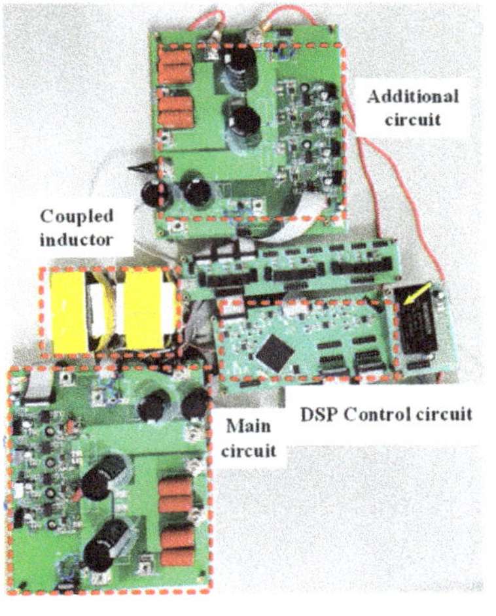

Figure 11. Experimental platform.

Table 5. Detail parameters of experimental platform.

Parameter	Value	Parameter	Value
V_2/V	40~60	V_1/V	100
f_S/kHz	100	n	1
L_c/μH	3.8	L_m/μH	130
C_1 and C_2/μF	1000	C_S/nF	2.2
P_{1max} and P_{2max}/W	800	C_a/μF	200

To verify the validity of the two ZVT BDCs shown in Figure 2, two sets of experiments were conducted, in each set of experiments; the power level is 800 W. The experimental waveforms of these two topologies operating in two modes (Buck and Boost) and two duty cycles (d_1 = 0.4 and d_1 = 0.6) are given. The control method utilized in the experiments is shown in Figure 10.

Figures 12–15 show the experimental waveforms of the topology shown in Figure 2a.

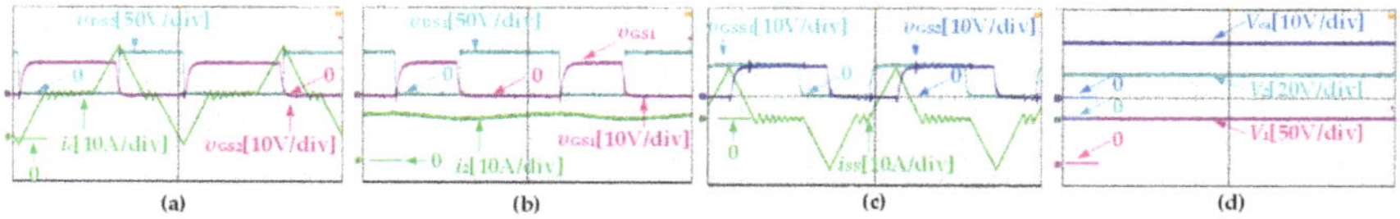

Figure 12. Experimental waveforms of BDC Topology 1 when P_1 = 800 W, V_1 = 100 V, d_1 = 0.4 (t − 2 μs/div). (**a**) v_{DS2}, v_{GS2}, and i_c; (**b**) v_{DS1}, v_{GS1}, and i_2; (**c**) v_{GSS1}, v_{GS2}, and i_{SS}; (**d**) V_1, V_2, and V_{ca}.

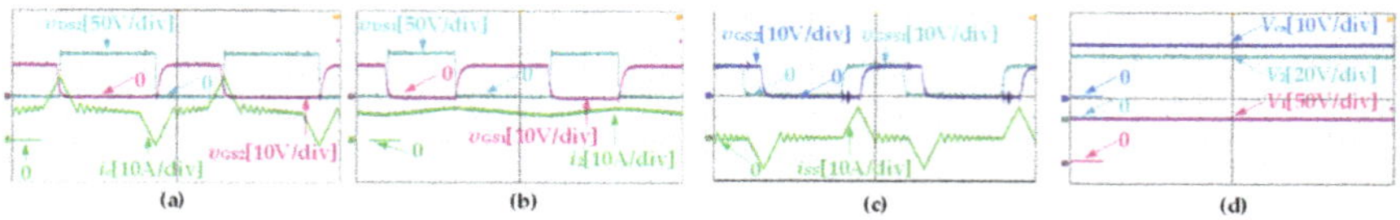

Figure 13. Experimental waveforms of BDC Topology 1 when P_1 = 800 W, V_1 = 100 V, d_1 = 0.6 (t − 2 μs/div). (**a**) v_{DS2}, v_{GS2}, and i_c; (**b**) v_{DS1}, v_{GS1}, and i_2; (**c**) v_{GSS1}, v_{GS2}, and i_{SS}; (**d**) V_1, V_2, and V_{ca}.

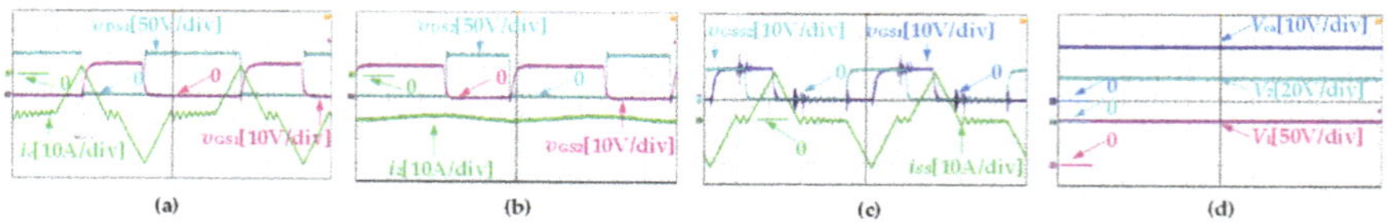

Figure 14. Experimental waveforms of BDC Topology 1 when P_2 = 800 W, V_1 = 100 V, d_1 = 0.4 (t − 2 μs/div). (**a**) v_{DS1}, v_{GS1}, and i_c; (**b**) v_{DS2}, v_{GS2}, and i_2; (**c**) v_{GSS2}, v_{GS1}, and i_{SS}; (**d**) V_1, V_2, and V_{ca}.

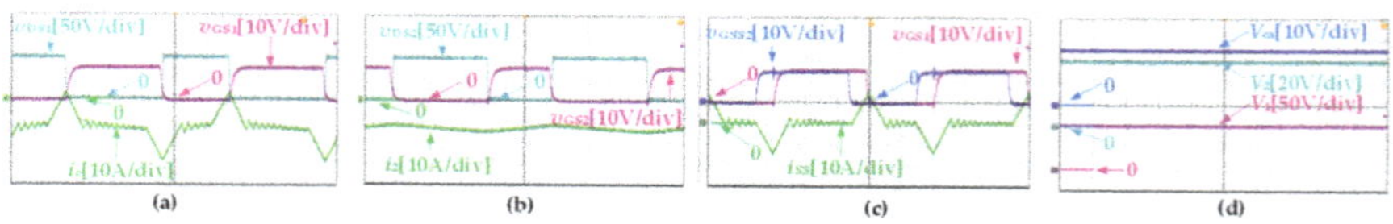

Figure 15. Experimental waveforms of BDC Topology 1 when P_2 = 800 W, V_1 = 100 V, d_1 = 0.6 (t − 2 μs/div). (**a**) v_{DS1}, v_{GS1}, and i_c; (**b**) v_{DS2}, v_{GS2}, and i_2; (**c**) v_{GSS2}, v_{GS1}, and i_{SS}; (**d**) V_1, V_2, and V_{ca}.

When ZVS BDC operates in Boost mode, P_1 = 800 W, V_1 = 100 V, Figures 12 and 13 are the experimental waveforms under the condition of V_2 = 40 V and V_2 = 60 V, respectively. The main switches S_1 and S_2 are turned on and off under ZVS conditions, and the auxiliary switches S_{SS1} and S_{SS2} are turned on and off under ZCS conditions. When V_2 is 40 V and 60 V, in order to achieve the ZVS of S_2, the phase shift angles between the drive signals v_{GSS1} and v_{G2} can be 60° and 45°, respectively.

When ZVS BDC operates in Buck mode, P_2 = 800 W, V_1 = 100 V, Figures 14 and 15 are the experimental waveforms under the condition of V_2 = 40 V and V_2 = 60 V, respectively. The main switches S_1 and S_2, are turned on and off under ZVS conditions, and the auxiliary switches S_{SS1} and S_{SS2} are turned on and off under ZCS conditions. When V_2 is 40 V and 60 V, in order to achieve the ZVS of S_1, the phase shift angles between the drive signals v_{GSS2} and v_{G1} can be 60° and 45°, respectively.

Figures 16–19 show the experimental waveforms of the topology shown in Figure 2b.

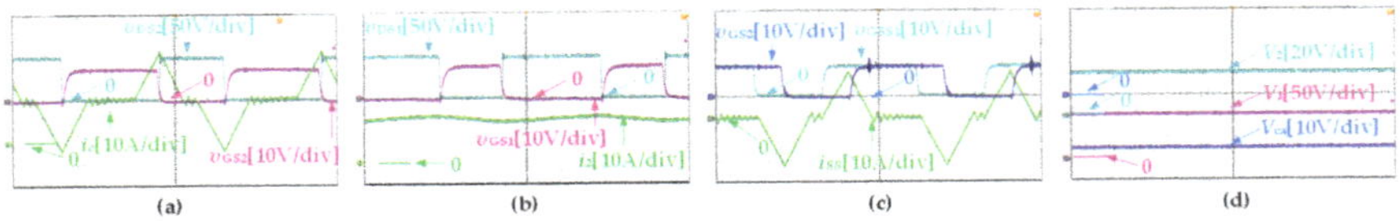

Figure 16. Experimental waveforms of BDC Topology 2 when P_1 = 800 W, V_1 = 100 V, d_1 = 0.4 (t – 2 μs/div). (**a**) v_{DS2}, v_{GS2}, and i_c; (**b**) v_{DS1}, v_{GS1}, and i_2; (**c**) v_{GSS1}, v_{GS2}, and i_{SS}; (**d**) V_1, V_2, and V_{ca}.

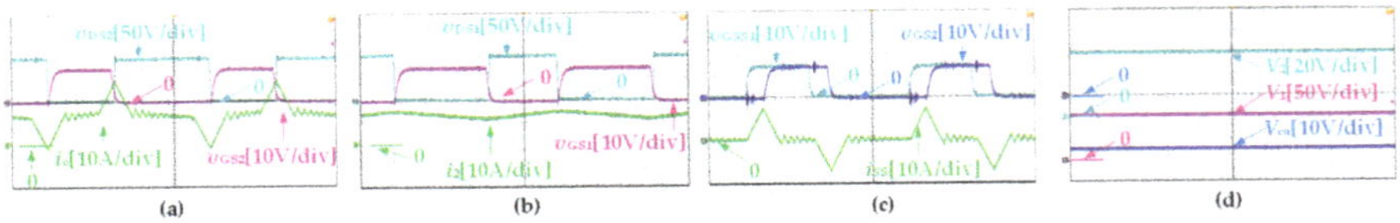

Figure 17. Experimental waveforms of BDC Topology 2 when P_1 = 800 W, V_1 = 100 V, d_1 = 0.6 (t – 2 μs/div). (**a**) v_{DS2}, v_{GS2}, and i_c; (**b**) v_{DS1}, v_{GS1}, and i_2; (**c**) v_{GSS1}, v_{GS2}, and i_{SS}; (**d**) V_1, V_2, and V_{ca}.

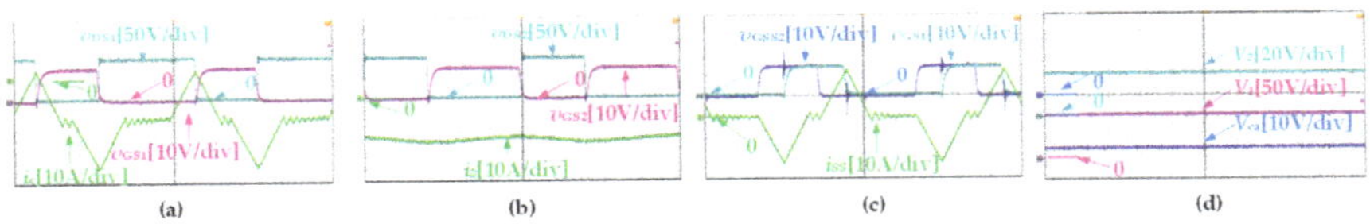

Figure 18. Experimental waveforms of BDC Topology 2 when P_2 = 800 W, V_1 = 100 V, d_1 = 0.4 (t – 2 μs/div). (**a**) v_{DS1}, v_{GS1}, and i_c; (**b**) v_{DS2}, v_{GS2}, and i_2; (**c**) v_{GSS2}, v_{GS1}, and i_{SS}; (**d**) V_1, V_2, and V_{ca}.

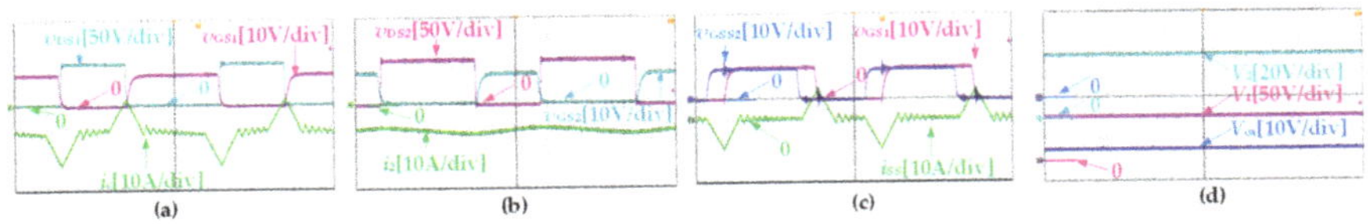

Figure 19. Experimental waveforms of BDC Topology 2 when P_2 = 800 W, V_1 = 100 V, d_1 = 0.6 (t – 2 μs/div). (**a**) v_{DS1}, v_{GS1}, and i_c; (**b**) v_{DS2}, v_{GS2}, and i_2; (**c**) v_{GSS2}, v_{GS1}, and i_{SS}; (**d**) V_1, V_2, and V_{ca}.

When ZVS BDC operates in Boost mode, P_1 = 800 W, V_1 = 100 V, Figures 16 and 17 are the experimental waveforms under the condition of V_2 = 40 V and V_2 = 60 V, respectively. The main switches S_1 and S_2 are turned on and off under ZVS conditions, and the auxiliary switches S_{SS1} and S_{SS2} are turned on and off under ZCS conditions. When V_2 is 40 V and 60 V, in order to achieve the ZVS of S_2, the phase shift angles between the drive signals v_{GSS1} and v_{G2} can be 60° and 45°, respectively.

When ZVS BDC operates in Buck mode, P_2 = 800 W, V_1 = 100 V, Figures 18 and 19 are the experimental waveforms under the condition of V_2 = 40 V and V_2 = 60 V, respectively. The main switches S_1 and S_2 are turned on and off under ZVS conditions, and the auxiliary switches S_{SS1} and S_{SS2} are turned on and off under ZCS conditions. When V_2 is 40 V and 60 V, in order to achieve the ZVS of S_1, the phase shift angles between the drive signals v_{GSS2} and v_{G1} can be 60° and 45°, respectively.

As shown in Figures 12–19, the current at Port 2 has no notch when these two proposed ZVT BDCs operate in both two modes (Buck and Boost). In Topology 1, $V_{ca} > 0$, and in Topology 2, $V_{ca} < 0$. When the power circuit operates in Buck mode, the current i_c flowing through the bridge arm (S_1 and S_2) changes from negative to positive before S_1 is turned on due to the auxiliary current. When the power circuit operates in Boost mode, i_c changes from positive to negative before S_2 is turned on, realizing ZVT conditions for S_1 and S_2.

When the average value of i_2 changes, the peak of i_{SS} has different requirements. Thus, d_{SS1} and d_{SS2} need to be adjusted. In addition, auxiliary switches are turned on and off under ZCS conditions.

As shown in Figure 20, the main loss sources of a traditional BDC include conduction loss of the main switch, switching loss of the main switch, copper loss and core loss of the inductor. The main loss sources of the proposed ZVS BDC include the conduction loss of the main switch, conduction loss of the auxiliary switch, RR loss of the auxiliary switch, copper loss, and core loss of inductor. For the traditional BDC topology, switching loss accounts for the largest proportion of the total loss, while for the proposed BDC, the largest source is conduction loss. As shown in Figure 21, compared to the traditional BDC, the maximum efficiency of the proposed BDC (shown in Figure 2a) can be improved by 1.2%.

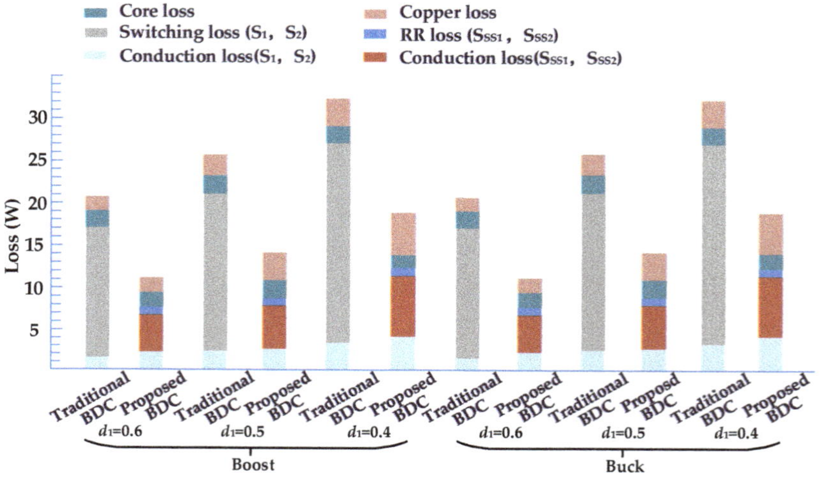

Figure 20. Loss analysis.

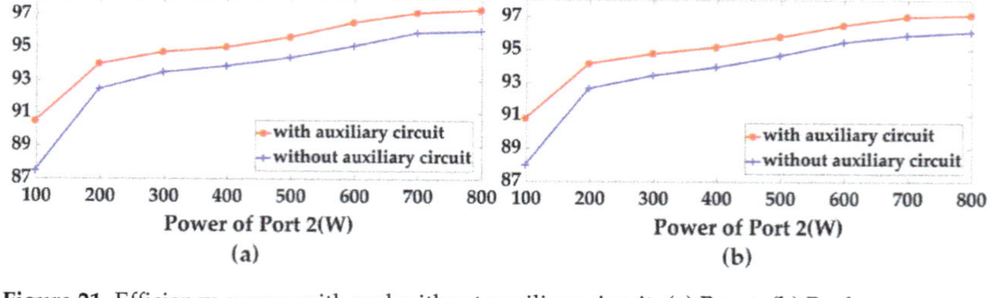

Figure 21. Efficiency curves with and without auxiliary circuit. (**a**) Boost; (**b**) Buck.

Through the theoretical analysis of the loss of traditional BDC, when the switching frequency is increased to more than 500 kHz, it is necessary to introduce the ZVT implementation measures. According to the efficiency curve of the existing topology, the loss analysis is carried out when the switching frequency is greater than 500 kHz, and the comparison of the theoretical efficiency values among different topologies under full load when the switching frequency $f_S \geq 500$ kHz is shown in Table 6.

Table 6. Efficiency comparison when the switching frequency $f_S \geq 500$ kHz.

Topology	Boost	Buck
ZVS BDC [17]	96.7%	95.2%
ZVS BDC [18]	97.2%	96.5%
ZVT BDC [20]	95.7%	95.6%
ZVT BDC [21]	97.7%	97.2%
ZVS BDC [23]	97.5%	97.5%
ZVS BDC [24]	96.3%	96.5%
ZVT BDC [25]	95.3%	95.4%
ZVT BDC [26]	92.2%	93.3%
ZVS BDC [28]	96.4%	96.2%
The proposed ZVT BDCs	96.7%	96.7%

6. Conclusions

In this paper, several ZVT magnetic coupling BDCs are proposed. In these topologies, the excitation inductor of the transformer acts as a filter, and the leakage inductor is used to generate auxiliary current. When the turn ratio is 1:1, the current of the original side and the secondary side is the same but in the opposite direction. Thus, the current notch at the low voltage port can be eliminated. The main switches can achieve ZVT conditions, and the auxiliary switches can achieve ZCS conditions. In order to verify the feasibility of these ZVT BDCs, 100-kHz and 800-W prototypes are built, and the experimental results of two working modes (Boost and Buck) are shown. Through detailed theoretical analysis and experimental verification, all switches of the converter realize soft switching, and no current notch exists at the low-voltage port. Compared with the traditional BDC, the maximum efficiency can be increased by 1.2%.

Author Contributions: Conceptualization, S.W.; methodology, S.W., M.G. and J.S.; validation, S.W. and M.G. All authors have read and agreed to the published version of the manuscript.

Funding: This work is supported by the National Natural Science Foundation of China (52077199).

Data Availability Statement: Not applicable.

Conflicts of Interest: The authors declare no conflict of interest.

References

1. Lee, H.; Yun, J. High-Efficiency Bidirectional Buck–Boost Converter for Photovoltaic and Energy Storage Systems in a Smart Grid. *IEEE Trans. Power Electron.* **2019**, *34*, 4316–4328. [CrossRef]
2. Mirzaei, A.; Forooghi, M.; Ghadimi, A.A.; Abolmasoumi, A.H.; Riahi, M.R. Design and construction of a charge controller for stand-alone PV/battery hybrid system by using a new control strategy and power management. *Sol. Energy* **2017**, *149*, 132–144. [CrossRef]
3. Dimitrov, B.; Hayatleh, K.; Barker, S.; Collier, G.; Sharkh, S.; Cruden, A. A Buck–Boost Transformerless DC–DC Converter Based on IGBT Modules for Fast Charge of Electric Vehicles. *Electronics* **2020**, *9*, 397. [CrossRef]
4. Park, D.-R.; Kim, Y. Design and Implementation of Improved High Step-Down DC–DC Converter for Electric Vehicles. *Energies* **2021**, *14*, 4206. [CrossRef]
5. Abusara, M.A.; Guerrero, J.M.; Sharkh, S.M. Line-Interactive UPS for Microgrids. *IEEE Trans. Ind. Electron.* **2013**, *61*, 1292–1300. [CrossRef]
6. Wu, J.; Wen, P.; Sun, X.; Yan, X. Reactive Power Optimization Control for Bidirectional Dual-Tank Resonant DC–DC Converters for Fuel Cells Systems. *IEEE Trans. Power Electron.* **2020**, *35*, 9202–9214. [CrossRef]
7. Ramirez-Murillo, H.; Restrepo, C.; Konjedic, T.; Calvente, J.; Romero, A.; Baier, C.R.; Giral, R. An Efficiency Comparison of Fuel-Cell Hybrid Systems Based on the Versatile Buck–Boost Converter. *IEEE Trans. Power Electron.* **2018**, *33*, 1237–1246. [CrossRef]
8. Huang, X.; Lee, F.C.; Li, Q.; Du, W. High-Frequency High-Efficiency GaN-Based Interleaved CRM Bidirectional Buck/Boost Converter with Inverse Coupled Inductor. *IEEE Trans. Power Electron.* **2016**, *31*, 4343–4352. [CrossRef]
9. Yao, Z.; Lu, S. A Simple Approach to Enhance the Effectiveness of Passive Currents Balancing in an Interleaved Multiphase Bidirectional DC–DC Converter. *IEEE Trans. Power Electron.* **2019**, *34*, 7242–7255. [CrossRef]
10. Ahmed, H.F.; Cha, H.; Kim, S.; Kim, D.; Kim, H. Wide Load Range Efficiency Improvement of a High-Power-Density Bidirectional DC–DC Converter Using an MR Fluid-Gap Inductor. *IEEE Trans. Ind. Appl.* **2015**, *51*, 3216–3226. [CrossRef]

11. Jayashree, E.; Uma, G. Design and implementation of zero-voltage switching quasi-resonant positive-output Luo converter using analog resonant controller UC3861. *IET Power Electron.* **2011**, *4*, 81–88. [CrossRef]
12. Moradisizkoohi, H.; Elsayad, N.; Mohammed, O.A. Experimental Demonstration of a Modular, Quasi-Resonant Bidirectional DC-DC Converter Using GaN Switches for Electric Vehicles. *IEEE Trans. Ind. Appl.* **2019**, *55*, 7787–7803. [CrossRef]
13. Liu, S.; Wu, X. A 1-MHz ZVS Boost DC–DC converter with active clamping using GaN power transistors. In Proceedings of the 2016 IEEE 8th International Power Electronics and Motion Control Conference (IPEMC-ECCE Asia), Hefei, China, 22–26 May 2016; pp. 557–562.
14. Lee, S.S.; Choi, S.W.; Moon, G.W. High efficiency active clamp forward converter with synchronous switch controlled ZVS operation. *J. Power Electron.* **2006**, *6*, 131–138.
15. Lee, I.; Kim, J.; Lee, T.; Jung, Y.; Won, C. A new bidirectional DC–DC converter with ZVT switching. In Proceedings of the 2012 IEEE Vehicle Power and Propulsion Conference (VPPC), Seoul, Republic of Korea, 9–12 October 2012; pp. 684–689.
16. Yang, J.; Do, H. High-Efficiency Bidirectional DC–DC Converter With Low Circulating Current and ZVS Characteristic Throughout a Full Range of Loads. *IEEE Trans. Ind. Electron.* **2014**, *61*, 3248–3256. [CrossRef]
17. Mohammadi, M.R.; Farzanehfard, H. Analysis of Diode Reverse Recovery Effect on the Improvement of Soft-Switching Range in Zero-Voltage-Transition Bidirectional Converters. *IEEE Trans. Ind. Electron.* **2015**, *62*, 1471–1479. [CrossRef]
18. Lee, J.H.; Yu, D.H.; Kim, J.G.; Kim, Y.H.; Won, C.Y. Auxiliary Switch Control of a Bidirectional Soft-Switching DC/DC Converter. *IEEE Trans. Power Electron.* **2013**, *28*, 5446–5457. [CrossRef]
19. Zhang, Y.; Sen, P.C. A new soft-switching technique for buck, boost, and buck–boost converters. *IEEE Trans. Ind. Appl.* **2003**, *39*, 1775–1782. [CrossRef]
20. Mohammadi, M.R.; Farzanehfard, H. New Family of Zero-Voltage-Transition PWM Bidirectional Converters with Coupled Inductors. *IEEE Trans. Ind. Electron.* **2012**, *59*, 912–919. [CrossRef]
21. Mohammadi, M.R.; Farzanehfard, H. A New Family of Zero-Voltage-Transition Nonisolated Bidirectional Converters with Simple Auxiliary Circuit. *IEEE Trans. Ind. Electron.* **2016**, *63*, 1519–1527. [CrossRef]
22. Chen, G.; Deng, Y.; Tao, Y.; He, X.; Wang, Y.; Hu, Y. Topology Derivation and Generalized Analysis of Zero-Voltage-Switching Synchronous DC–DC Converters With Coupled Inductors. *IEEE Trans. Ind. Electron.* **2016**, *63*, 4805–4815. [CrossRef]
23. Chen, G.; Deng, Y.; Chen, L.; Hu, Y.; Jiang, L.; He, X.; Wang, Y. A Family of Zero-Voltage-Switching Magnetic Coupling Nonisolated Bidirectional DC–DC Converters. *IEEE Trans. Ind. Electron.* **2017**, *64*, 6223–6233. [CrossRef]
24. Mohammadi, M.R.; Farzanehfard, H. Family of Soft-Switching Bidirectional Converters With Extended ZVS Range. *IEEE Trans. Ind. Electron.* **2017**, *64*, 7000–7008. [CrossRef]
25. Mohammadi, M.R.; Peyman, H.; Yazdani, M.R.; Mirtalaei, S.M.M. A ZVT Bidirectional Converter With Coupled-Filter-Inductor and Elimination of Input Current Notches. *IEEE Trans. Ind. Electron.* **2020**, *67*, 7461–7469. [CrossRef]
26. Nan, C.; Ayyanar, R. A 1 MHz Bi-directional Soft-switching DC–DC Converter with Planar Coupled Inductor for Dual Voltage Automotive Systems. In Proceedings of the 2016 IEEE Applied Power Electronics Conference and Exposition (APEC), Long Beach, CA, USA, 20–24 March 2016; pp. 432–439.
27. Cheng, X.; Zhang, Y.; Yin, C. A Family of Coupled-Inductor-Based Soft-Switching DC–DC Converter With Double Synchronous Rectification. *IEEE Trans. Ind. Electron.* **2019**, *66*, 6936–6946. [CrossRef]
28. Wang, S.; Hu, Y.; Gao, M.; Shi, J. Coupled inductor based zero-voltage-switching buck/boost converter. *J. Power Electron.* **2022**, *22*, 1059–1072. [CrossRef]
29. Rezvanyvardom, M.; Mirzaei, A. Zero-Voltage Transition Nonisolated Bidirectional Buck–Boost DC–DC Converter With Coupled Inductors. *IEEE J. Emerg. Sel. Top. Power Electron.* **2021**, *9*, 3266–3275. [CrossRef]
30. Cheng, X.; Liu, C.; Wang, D.; Zhang, Y. State-of-the-Art Review on Soft-Switching Technologies for Non-Isolated DC-DC Converters. *IEEE Access* **2021**, *9*, 119235–119249. [CrossRef]

energies

MDPI

Article

Implementation of an Adaptive Method for Changing the Frequency Division of the Counter Clock Signal in a Frequency-to-Code Converter

Dariusz Świsulski [1] and Piotr Warda [2,*]

1 Department of Metrology and Information Systems, Gdansk University of Technology, 80-233 Gdańsk, Poland; dariusz.swisulski@pg.edu.pl
2 Department of Automation and Metrology, Lublin University of Technology, 20-618 Lublin, Poland
* Correspondence: p.warda@pollub.pl; Tel.: +48-081-5384315

Abstract: Processing physical quantities into an indirect signal is a standard method of transferring information about the measured quantity to the master system, which analyzes the data obtained from the acquisition system. The intermediate signal is very often the voltage, but another transmission medium can be the frequency of the output signal of the "physical quantity-to-frequency" converter. The article presents the implementation of the adaptive method of selecting the clock signal frequency of the counter working in the converter. The issue of selecting the clock signal frequency for the required processing range of the transducer is discussed in detail. The application of the method using the STM32L476RG microcontroller is presented. The principle of checking the processing range of the developed transducer model is discussed. The algorithms of transducer operation in basic and adaptive modes of measuring the period of the variable frequency signal are proposed. The results of operation, in both modes, of the transducer model of frequency processing are presented, along with the metrological analysis of the results. The influence of selected approximations used to reconstruct the measured quantity on the final presentation of the measurement result is discussed.

Keywords: instantaneous frequency; frequency measurement; frequency-to-code converter

Citation: Świsulski, D.; Warda, P. Implementation of an Adaptive Method for Changing the Frequency Division of the Counter Clock Signal in a Frequency-to-Code Converter. *Energies* **2023**, *16*, 5399. https://doi.org/10.3390/en16145399

Academic Editors: Salvatore Musumeci and Andrea Mariscotti

Received: 9 May 2023
Revised: 27 June 2023
Accepted: 13 July 2023
Published: 15 July 2023

1. Introduction

The conversion of physical quantities into an indirect signal is a constantly developing field of metrology [1–5]. It is used for data transmission and processing in master systems, which can be, for example, PCs or other microprocessor systems with sufficient computing performance for a given application. Very often, the intermediate signal is the voltage [6,7], which can be easily converted into a digital form using an analog-to-digital converter. Another possibility is to use the frequency of the transducer's output signal as an intermediate carrier of information [8]. The frequency itself can also provide information about the examined phenomenon, such as, for instance, the stability of power grid operation, the variability of the voltage generated by photovoltaic inverters, power plant frequency control capabilities [9], etc. The frequency of the signal can be directly processed into numerical values [6,10], but in the case of time-varying values, this issue is still the subject of many publications [1,11–16] and research on new solutions [17–24]. The intensity of the research is favored by: high availability of converters of physical quantities into frequency (X/f) [25], low sensitivity of the information transmission channel using the frequency signal to electromagnetic interference and amplitude attenuation [26,27], high accuracy of constant frequency measurement, and widely available programmable systems containing counters enabling high accuracy frequency measurement [28,29].

The increasing offer of X/f converters and the continuous development and creation of new innovative constructions enabling the measurement of time-varying parameters have entailed the need to develop methods of variable-frequency processing [30].

Constant-frequency measurement is most often carried out using the digital method. Systems for digital measurement of constant frequency and period are well-known and widely described in the literature on metrology [6]. The frequency meter usually includes a counter system, a standard frequency generator, an input signal conditioning system, and a decoder to convert the information about the measured frequency read from the counter into readable form. The result of this decoding is presented on the display or passed to another part of the system. The operation of the frequency counter is based on the principle of summing up the pulses signaling the end of the current period and the beginning of the next period of the measured frequency signal [25]. The quotient of the number of summed pulses and the measurement time is an approximation of the measured frequency. In the case of period measurement, the pulses from the reference generator are summed up during the full period of the processed signal [17].

The length of the period is determined by the product of the counter state and the length of the clock signal period. Provided that the design of the instrument for measuring frequency and period has been developed correctly, i.e., a Schmitt trigger [25] is used in the input channel of the meter, and the conditions for stable operation of the reference signal generator have been ensured, the main source of the measurement error is the quantization error [31]. The constant frequency measurement result is stable and can be presented in a readable form on the instrument's display.

Unlike a constant frequency meter, processing a variable frequency signal into a clear presentation of the measurement data requires a more complex measurement system, which will allow for more advanced data processing. Although counter systems are also often used in this case [6,25], the further data processing procedure is more complicated. The variable frequency of the signal makes a direct presentation of the result on the display impossible. In addition, due to its variability, it seems reasonable to measure the frequency indirectly, through digital measurement of the period, or more precisely, subsequent periods of the processed signal. Knowledge of the length of successive periods will enable the most complete mapping of changes in the quantity processed by the X/f converter. The consequence of the proposed method of frequency measurement is the need to collect the measurement data in order to present it in a clear form on a graph of the measurand $x(t)$ as a function of time. Typically, this data is later used to create other, more complex reports. Due to the nature of the operation of the system converting frequency into numerical values, this type of device is called a "frequency-to-code" converter and is marked with the symbol f/N [10]. An example of the structure of the f/N converter that allows continuous processing of the signal period using a single counter is shown in Figure 1.

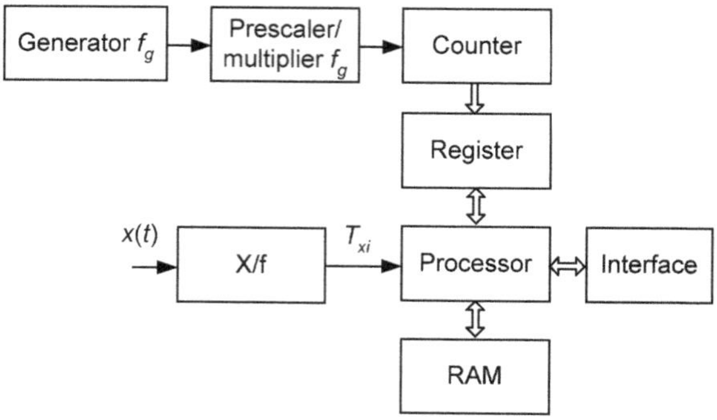

Figure 1. Sample structure of the f/N converter with single counter.

It is obvious that the acquisition of a volume-relevant set of numerical values for subsequent visualization, analysis, and archiving requires an adequately capacious memory in the system [32] and an interface that will allow for the effective transfer of the measurement data from the f/N converter to the master computer. An alternative to storing the data in the RAM of the f/N converter may be the transmission of consecutive numerical values obtained during the measurement directly to the master computer. However, this requires a careful analysis of the possibilities of data acquisition and transmission by the f/N converter and the possibility of receiving and archiving the measurement data sent by the computer controlling the measurement process [33]. As a consequence, taking into account the need to store a large amount of data, regardless of the storage method, it becomes reasonable to develop methods to reduce the memory occupied by the stored data. One of the possibilities is to limit the transmitted single numerical values to a variable with an acceptable number of bits, corresponding to the number of bits in the counter register used. Usually, 8- and 16-bit counters are available, but some microprocessor systems also offer 32-bit counters. Often, universal counters in microprocessor structures can be configured to create a counter structure with a larger capacity. Figures 2–4 show three graphs of theoretically achievable frequency measurement range for different sizes of the counter register. Each measurement range is limited by the assumption that the relative value of the quantization error will not exceed 1%, i.e., the minimum number of pulses registered by the counter of the f/N converter is 100. The quantization error is calculated according to the formula:

$$\delta_k = \frac{T_g}{T_{xi}} 100\%, \tag{1}$$

where: T_g—is the period of the clock generator signal with frequency f_g, and T_{xi}—is the period of the processed signal.

The lower limit of each of the presented ranges is the product of the maximum meter state and the length of the period T_g.

Figure 2 shows the measurement ranges for a meter with an 8-bit register, which gives a data set consisting of numbers with the least use of the measurement system memory. It can be seen that for the clock signal used by the counter with the frequency f_g = 80 MHz, it is theoretically possible to measure the maximum frequency of 800 kHz, while the lowest measurable frequency for f_g = 1 MHz and the maximum state of the counter 255 is about 3921.57 Hz.

Unfortunately, it can be seen that the three selected frequencies f_g allow processing only in a discontinuous frequency range of the processed signal. In order to enable measurement in a continuous range, a more frequent change of the f_g frequency division degree would be required, which would generate additional data informing about the set frequency division degree [31]. As a consequence, the effect of memory saving would be minimal, while continuous changes in the frequency division could lead, in extreme cases, to unstable operation of the measurement system.

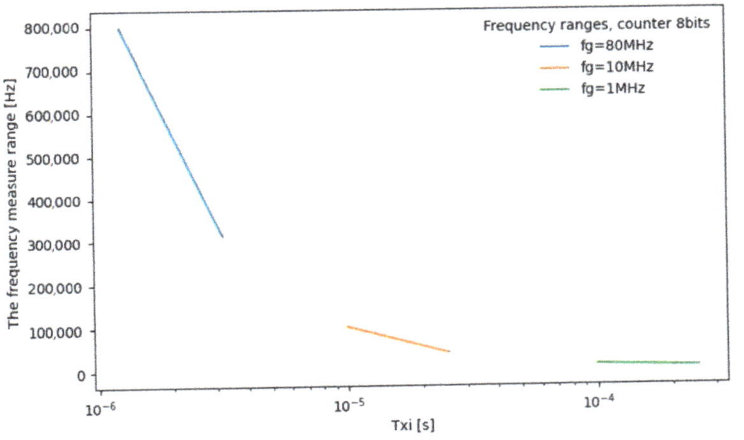

Figure 2. Measuring ranges for a counter with an 8-bit summing register.

Figure 3 shows the measurement ranges for a meter using a 16-bit register. The proposed frequencies of the clock signal allow obtaining a continuous measuring range, whereby for f_g = 80 MHz, the maximum range of 800 kHz is obtained, as before, while the minimum frequency for f_g = 1 MHz and the maximum state of the counter 65535 is about 15.26 Hz.

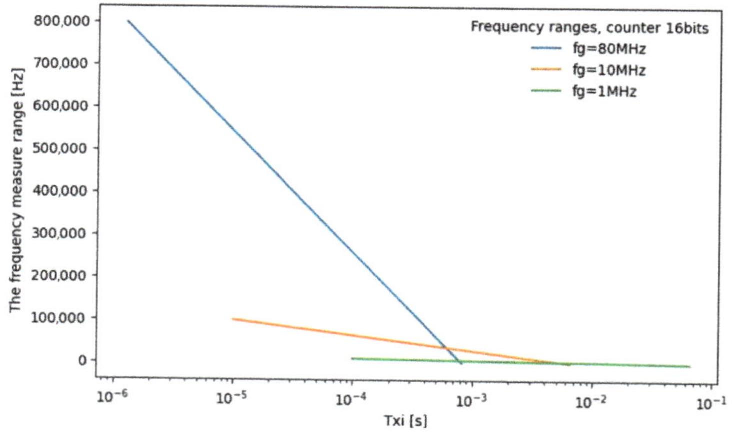

Figure 3. Measuring ranges for a counter with a 16-bit summing register.

Figure 4 shows the measurement ranges for a counter with a 32-bit register. In this case, it can be seen that the measuring range is the widest. Like for the previous counters, for f_g = 80 MHz, the limit value of the 800 kHz range is obtained, as imposed by the value of the minimum state of the counter register. In the range of the lowest frequencies, for f_g = 1 MHz and the maximum counter value of 4,294,967,295, a very significant reduction of the measurable values to the level of about 232 μHz is obtained. However, the wide measurement range of the 32-bit counter requires twice as much memory to store the measurement data as the 16-bit counter.

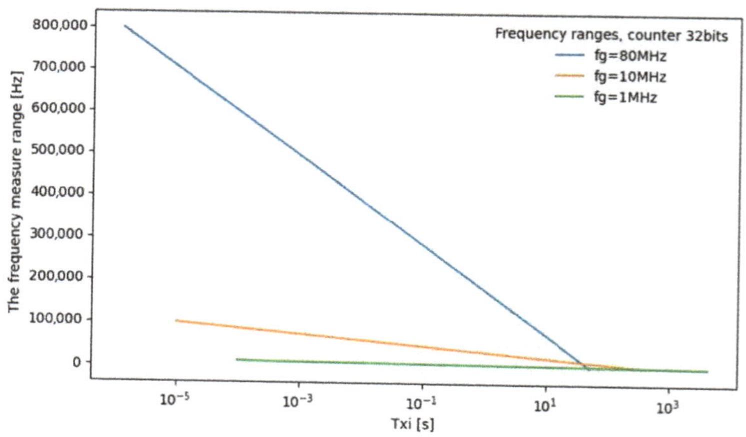

Figure 4. Measuring ranges for a counter with a 32-bit summing register.

In summary, it is reasonable to say that the f/N converter with a 32-bit counter can be used when it is necessary to measure very low frequencies, while in other cases it seems better to use a 16-bit counter with switching the division degree of the clock signal frequency f_g in the prescaler, depending on the instantaneous value of the measured frequency.

Taking into account the above considerations, it was assumed in the further part of the work that a 16-bit counter will be used in the tests, which will allow counting in the range from 0 to 65,535 and saving the instantaneous value of the summing register in an additional register.

The aim of the article is to present an example of the implementation of the method to change adaptively the frequency division of the counter clock signal working in the f/N converter and thus extend the range for low frequencies processed by the f/N converter with a 16-bit counter.

2. Processing of Successive Periods of the Signal

The test signal frequency f_{xi} corresponding to each successive period of T_{xi} is determined indirectly. The value of f_{xi} is calculated from the formula:

$$f_{xi} = \frac{1}{T_{xi}} \tag{2}$$

Neglecting the quantization error (1), the approximate value of T_{xi} is calculated as the product of the period T_g and the state N_{xi} of the f/N converter counter:

$$T_{xi} \approx T_g N_{xi} \tag{3}$$

One of the possible implementations of the functionality of the f/N converter is the use of a microprocessor system [34], which typically includes counter circuits, memory, interfaces, and a reference frequency generator. Its structure enables the design of a system that implements the functionality of digital signal processing [33].

The principle of continuous processing of the period T_{xi} of a variable frequency signal (hereinafter referred to as the frequency signal) by means of a system containing one counter (Figure 1) is shown in Figure 5 [31]. The first graph from the top presents the quantity $x(t)$ processed in the physical quantity-frequency converter, while the next graph shows the signal of the output voltage $U_{Xf}(t)$ of the X/f converter. The period of this signal is proportional to the instantaneous value of quantity $x(t)$. The third graph shows the voltage $U_g(t)$ generated at the output of the clock signal generator, and finally, the last graph presents the time-varying state $N(t)$ of the counter of the f/N converter working continuously. The instantaneous values of the counter state N_i are read at times t_i of the occurrence of pulses constituting the boundary of successive periods T_{xi}.

$$x(t) = \frac{1}{ST_{xi}} = \frac{1}{ST_g(N_{i+1} + N_{\max} * O - N_i)}, \tag{4}$$

where O is the number of recorded meter overflows and S is the sensitivity of the X/f converter.

In normal practice, if only hardware processing of successive N_{xi} values is required, only one case of overflow can be recorded. Storing the information about successive overflows requires the use of RAM to create an additional software counter. In applications requiring the fastest possible processing of subsequent T_{xi} periods, a much better solution will be to use another counter that counts overflows or to replace the used counter with another unit with a larger capacity.

The absolute value of the total absolute error of processing the information about the quantity $x(t)$ carried by the frequency of the input signal by the f/N converter is the sum of two errors. The first is the absolute frequency measurement error Δ_{fT} resulting from the formation of the quantization error in the T_{xi} measurement. The frequency f_{xi} is calculated indirectly from the measured values of T_{xi}; hence, this error should be calculated using the total differential method:

$$\Delta_{fT} = \left|\frac{\partial f_{xi}}{\partial T_{xi}}\right| |\Delta T_{xi}|. \tag{5}$$

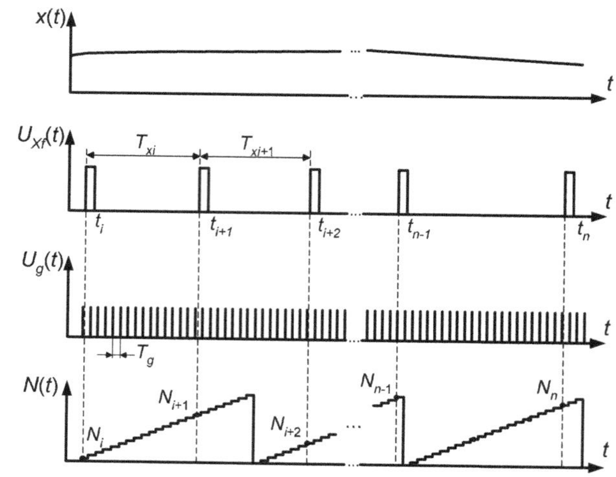

Figure 5. The principle of processing the period of variable frequency signal using a system with one counter [31].

Taking into account relations (2) and (4) and that the absolute quantization error of the T_{xi} measurement is $\pm T_{xi}$, after relevant substitutions the relation for Δ_{fT} takes the form:

$$\Delta_{fT} = \left| \frac{\partial \left(S T_{xi}^{-1} \right)}{\partial T_{xi}} \right| T_g. \tag{6}$$

Determining the derivative in the above relation gives the final formula for the error Δ_{fT}:

$$\Delta_{fT} = \frac{S T_g}{T_{xi}^2}. \tag{7}$$

The other component of the total error is the absolute averaging error [31]. Finally, for a sinusoidal test signal:

$$x(t) = X_0 + X_m \sin(2\pi F t), \tag{8}$$

where F is the frequency of $x(t)$, X_m is the amplitude of $x(t)$, and X_0 is the constant component of $x(t)$, the absolute total error is given by the formula:

$$\Delta_\Sigma = \frac{S T_g}{T_{xi}^2} + S X_m \left(1 - \frac{\sin \pi F T_{xi}}{\pi F T_{xi}} \right). \tag{9}$$

On the other hand, the total relative error of information processing related to the current value of the frequency f_{xi} is given by the formula:

$$\delta_\Sigma = \left(\frac{T_g}{T_{xi}} + \frac{f_m \left(1 - \frac{\sin \pi F T_{xi}}{\pi F T_{xi}} \right)}{f_{xi}} \right) \cdot 100\%. \tag{10}$$

where f_m is the amplitude of the frequency change calculated as the product of the amplitude of $x(t)$ and the sensitivity of the X/f converter.

The graphs of the total error of information processing by the f/N converter are shown in Figure 6 for different selected values of the converter counter clock frequency and the assumed frequency of the waveform $x(t)$ equal to 1 Hz.

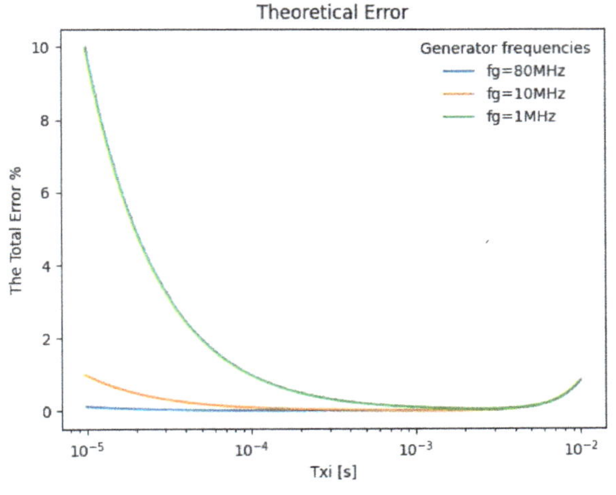

Figure 6. Examples of total error graphs for $F = 1$ Hz.

3. The f/N Converter Model

This article is intended to present a practical implementation of the method presented in [31] with the use of a selected microcontroller system. The experiments carried out during past studies [33] have revealed that there is a discrepancy between the theoretically achievable measurement range for a counter with a specific pulse summing capacity at a selected clock frequency f_g and the range achievable in real conditions. The performed experiments have shown that the speed of implementation of many transducer functionalities is very important, including the execution of numerical operations by the processor, counter service, communication with the memory, and the interface used for communication with the master computer. The size of the operational memory needed for calculations and data storage is also important. It turned out that the performance of the popular 8-bit microcontrollers of the AVR family [33] allowed theoretically to process signal frequencies in a wide range. After assuming a quantization error of 1% and using a clock frequency of 16 MHz and a 16-bit counter, it should theoretically be possible to measure frequencies up to 160 kHz. Unfortunately, popular microcontrollers usually have a small amount of RAM, which significantly limits their ability to collect data. An attempt to transmit the measurement data directly resulted in a dramatic limitation of the measurement range down to about 10 kHz. Increasing the frequency of the tested signal meant that the microprocessor did not correctly detect limit pulses for subsequent T_{xi} periods (Figure 5), and, as a result, the frequency was twice as high as that actually measured. The number of undetected limit pulses and the number of incorrect measurements increased with the increase in frequency measured [33].

Taking into account the above considerations, it was assumed that a more efficient 32-bit microcontroller with a memory size sufficient to collect enough data to evaluate the operation of the method would be used.

To create the f/N converter model, the NUCLEO-L476RG runtime board was chosen to contain a 32-bit STM32L476RG microcontroller equipped with an ARM Cortex-M4 core.

In addition, the Nucleo-L476RG set was equipped with an ST-LINK programmer that allows the microcontroller to be programmed directly into the operating system via the USB interface and a dedicated application. The programmer also allows for debugging the processor's operation, i.e., for current analysis of the program stored in the flash memory of the microcontroller. Various programming languages and environments can be used to create the code for the microcontroller, e.g., Keil, IAR, and the environment developed by STM Microelectronics. A set of programming tools included in the STM32Cube was used in the work on the project of the f/N converter discussed in the article.

A 16-bit general-purpose counter, T3, was selected to implement the f/N converter. According to the technical documentation of the module, this counter can be controlled at a maximum frequency of 80 MHz. The microcontroller clock signal frequency can be produced in various ways. After purchase, the development kit is configured to work with an internal RC resonant system producing a frequency of 16 MHz. Other useful alternatives for the implementation of the f/N converter, which will give a more stable clock signal, are the external resonator and the resonator of the ST-LINK programmer. Measurement experiments were carried out to check the stability of the f/N converter by measuring selected frequencies in the range of 1221 Hz to 200 kHz, each time collecting a set of data representing 24,000T_{xi} periods. Due to the stability of the RC generator specified by the manufacturer at a level of ±1%, it was concluded that carrying out the test of the RC system was unnecessary. The operation of the checked f/N converter was controlled sequentially from three clock signal sources: two external quartz resonators produced by different manufacturers with a resonant frequency of 16 MHz and the ST-LINK programmer generating a clock frequency of 8 MHz. For each set input frequency of the f/N converter, the average value from a series of 24,000 results was calculated, and then the absolute error was calculated as the difference between the calculated average value and the value that should be measured if the set frequency was measured without error. The results of the calculations in the form of the absolute error vs. frequency plot are shown in Figure 7. The waveforms f_{g1} and f_{g2} describe the graphs of errors obtained for clock signals from external resonators, while the graph f_{g3} represents the measurement error when using the clock signal from the ST-LINK programmer.

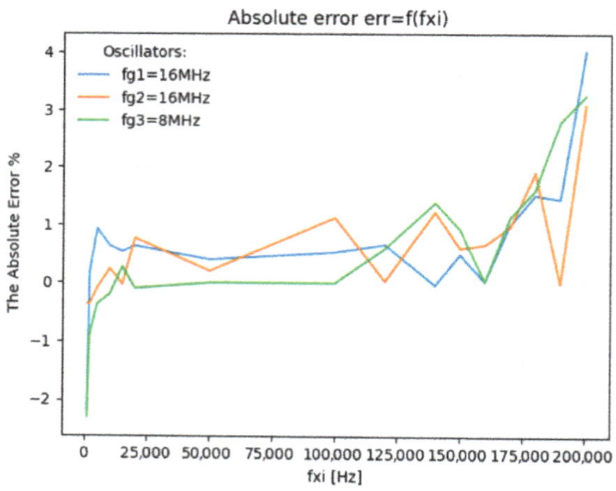

Figure 7. Absolute error of constant frequency measurement for individual resonators.

The average values of the absolute deviation of the N_{xi} value from the set value show that the work with the clock signal sent from the programmer is the most stable, where the error for frequencies up to 100 kHz is practically negligible. The clock signals from the 16 MHz crystal oscillators are basically comparable to each other. In these two cases, some instability can be observed, but its scale is not high.

The next comparison criterion used to assess the operation of the f/N converter was the number of large measurement errors generated, i.e., the frequency readings differing by more than 5% from the set value. The results are presented in Figure 8. The markings of the waveforms on the graph are analogous to those in Figure 7. Due to the need to present the stability of the f/N converter, the maximum of the y-axis representing the number of registered deviations is limited to 3. It can be seen that the occurrence of measured values deviating from the N_{xi} value by more than 5% is a great problem in the field of measurement errors. In this case, the transducer worked most stably using the programmer resonator.

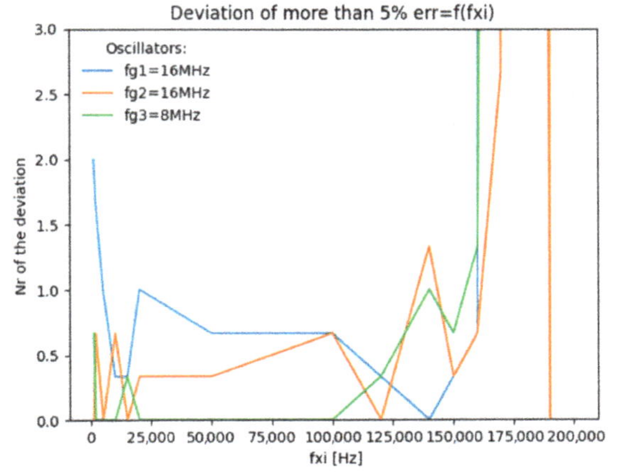

Figure 8. The average number of recorded deviations above 5% for individual resonators.

As a result of the conducted experiments, it was found that the clock signal generated in the programmer has the highest accuracy and stability and ensures the best working conditions for the f/N converter. Consequently, it was assumed that the clock signal generated by the programmer of the development kit would be used for further work. The configuration of the microcontroller enabling the generation of the clock frequency f_g = 80 MHz for the counter is shown in Figure 9.

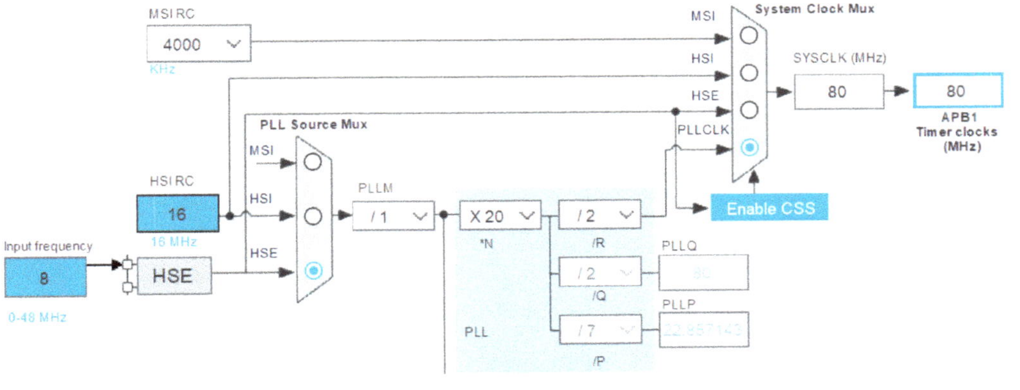

Figure 9. Clock frequency configuration in the STM32CubeIDE environment.

The converter has been developed according to the structure shown in Figure 1. A 16-bit T3 counter was selected to work as the working counter of the f/N converter. The counter was connected to the processor's internal APB1 bus. The maximum settable clock frequency of 80 MHz was used (Figure 9).

The T3 counter allows you to select and change the frequency division of the counter's clock signal. It also has a register that captures the current value stored in the summing register. A global interrupt for the T3 counter with the number 1 has been set to handle the operation of the counter.

Instead of the X/f converter, a programmable Agilent 33220A generator was used as the reference signal source [35]. The manual for the previous model of the generator marked HP 33120A [36], contains the information that the frequency signal is generated with an accuracy of 0.05% for each pulse for a modulation frequency F less than 600 Hz. Taking into account that the frequency F expected to be set in the tests is in the range $0 < F \leq 50$ Hz [31], this accuracy was found satisfactory. The use of a programmable generator instead of the X/f converter additionally made it possible to resign from the analysis of the X/f converter error, which was considered unnecessary from the point of view of the present work.

The generated frequency signal was passed to the PA6 line of the microcontroller. A total of 90 kB of RAM was used for data collection, which allowed for the collection of 45,000 two-byte numbers representing successive T_{xi} periods. The USART2 interface was used for data transmission to the master computer.

4. The Working Algorithm of the f/N Converter Model in the Basic Mode

Figure 10 shows the working algorithm that performs the processing of successive periods of the frequency signal in the basic mode. Turning on the microcontroller power supply starts the initialization of internal microcontroller modules and sets the default values of the system registers. The next step is to set the required configuration of the microcontroller pins and configure the USART2 interface. To obtain an acceptable transfer time, the data transfer rate was set to 1 Mb/s. A frame of 8 data bits, 1 stop bit, and no parity check was used. After the configuration, communication with the PC is started. The last step is setting the parameters of the T3 counter. It can be seen that saving the counter state during the processing of successive T_{xi} values will be forced at the rising edge of the frequency signal.

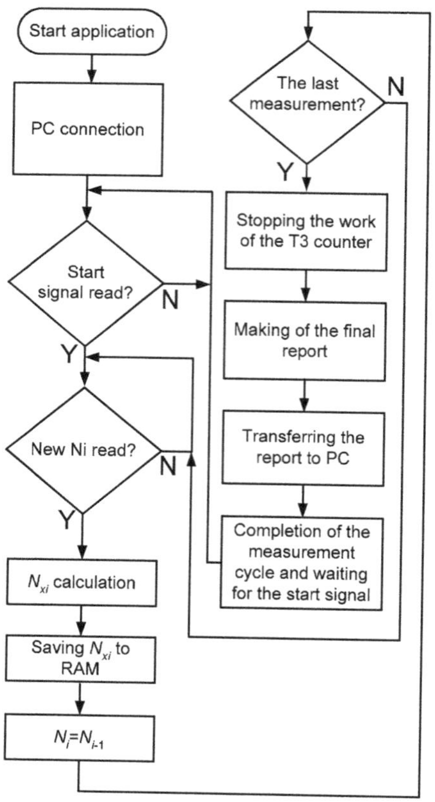

Figure 10. Algorithm of the f/N converter implementing the basic operating mode.

After completing the configuration, the f/N converter program sends its readiness for measurements to the PC.

Sending the command to start measurements by the PC starts the procedure of processing subsequent T_{xi} periods. It is carried out by the main program in cooperation with the interrupt handling procedure generated when the T3 counter detects the rising edge of the frequency signal. The working algorithm of the procedure handling the interrupt generated by the counter is shown in Figure 11. Due to the need to obtain the maximum speed of the f/N converter, the procedure contains only the commands necessary to carry out the measurements. Its task is to read the number N_i from the counter, assign it to a variable in the program, and set a flag informing the main program about the readout.

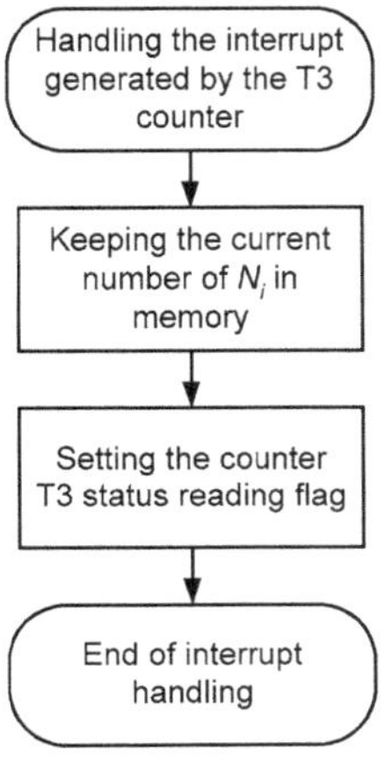

Figure 11. Algorithm of the procedure reading the current state of the T3 counter summing reg.

Further processing of the read information is performed in the main program. The stored value Ni is used to calculate the number N_{xi}, which represents the measured period T_{xi}. The current value of N_{xi} is calculated from the formula:

$$N_{xi} = N_{i+1} + N_{\max} * O - N_i. \tag{11}$$

The counter overflow is detected by analyzing the sign of the difference between N_i and N_{i-1}. A negative difference sign indicates that an overflow has occurred. Then, the maximum number of states that can be recorded by the counter is added to the difference; for a 16-bit counter, it is 2^{16}. The calculated N_{xi} is stored in the RAM. Controlling the sign of the difference between the numbers N_i and N_{i-1} is sufficient under the assumption that T_{xi} will not be greater than the product of $T_g{*}N_{max}$. Otherwise, it is necessary to use the counter overflow count control.

The value of N_i, after calculating N_{i-1}, is stored as N_{i-1} for calculation in the next iteration of the program. After reaching the complete set of data, the program stops the work of the T3 counter and generates the final report containing the saved data set and additional information about the measurement performed. The report is passed to the PC. Finally, the f/N converter program waits for the next measurement start command.

Before testing the f/N converter model, it was necessary to define the upper and lower measurement range limits. As already mentioned, the lower limit of the measurement range is the product of the period T_g of the clock signal used and the capacity of the meter used. For the assumed clock frequency f_g = 80 MHz and the counter capacity of 16 bits, the lower measurement limit is approximately equal to 1221 Hz.

Next, the upper limit of the measuring range was determined. It was assumed that a satisfactory method of determining the upper limit of the measurement range would be to analyze 1 million consecutive constant frequency measurements and verify that the measurement results are consistent with theoretical considerations and do not exceed the expected total error range (10). Measuring a frequency twice as low as expected would mean that the f/N converter model failed to detect the boundary between the T_{xi} data and summed up their lengths, thus producing an erroneous result. The maximum frequency for which erroneous measurements are not recorded was considered the upper limit of the measurement range. Due to the need to expose the exact maximum and minimum values of all measurements, it was decided that instead of a histogram, a graph of the measured frequency in time would be presented.

The results of the measurements performed are shown in Figures 12–14. Figure 12 shows that the measured frequency of 111 kHz is processed correctly, and all results can be considered correct.

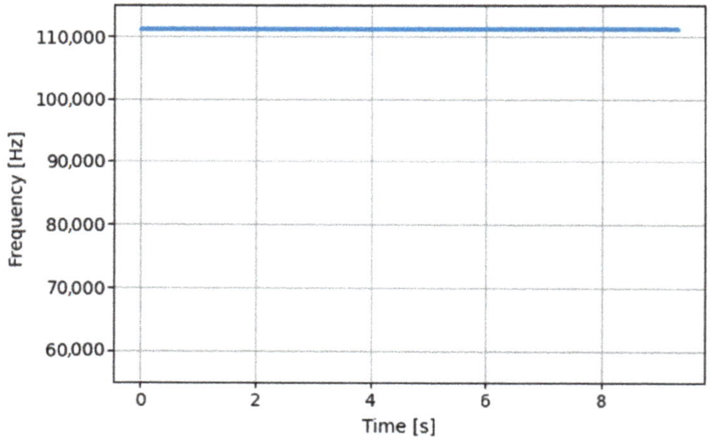

Figure 12. Measurement of f_{xi} = 111 kHz.

Figure 13 shows the effect of processing the 112 kHz frequency. It can be seen that there is a single frequency measurement with the value at half the set point range. In this case, there was also a problem with detecting the boundary between successive periods of the frequency signal.

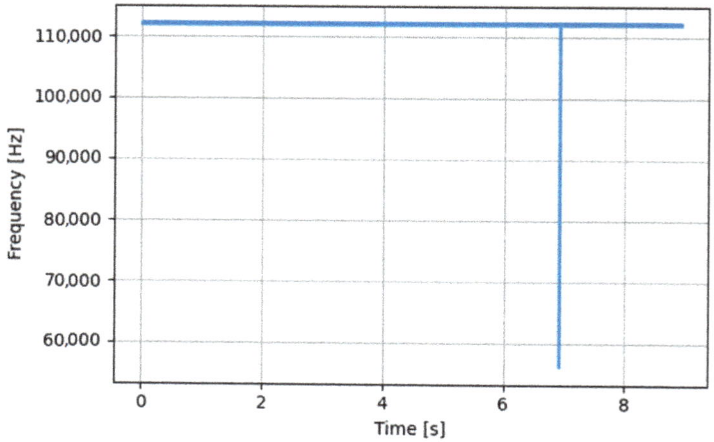

Figure 13. Measurement of f_{xi} = 112 kHz.

Figure 14 shows the effect of processing the frequency of 113 kHz. It can be seen that the measurement problem occurred for ten T_{xi} periods.

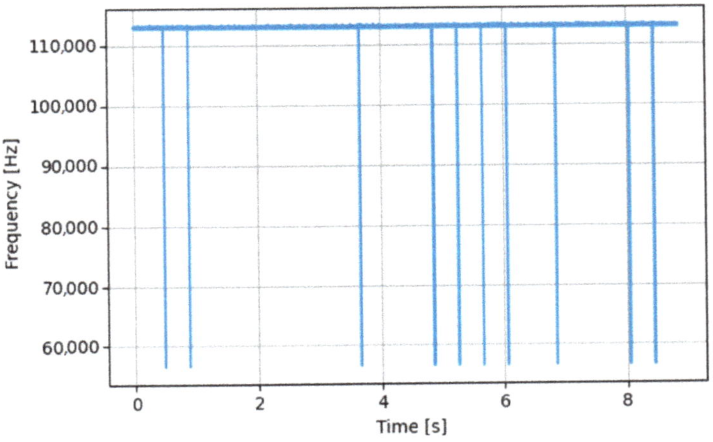

Figure 14. Measurement of f_{xi} = 113 kHz.

When converting the frequencies into numerical values, each subsequent measurement of the period is treated as a single measurement, and each incorrect reading of T_{xi} will have a significant impact on the presentation of the final result. To minimize incorrect reproduction of the graph of frequency changes and, consequently, the processed physical quantity, a decision was made to consider the frequency of 111 kHz as the upper limit of the measurement range.

The total error graphs of the developed model of the f/N converter are shown in Figure 15. The graph of the theoretical error is marked in red (10), while that of the error obtained from the experiment is marked in blue.

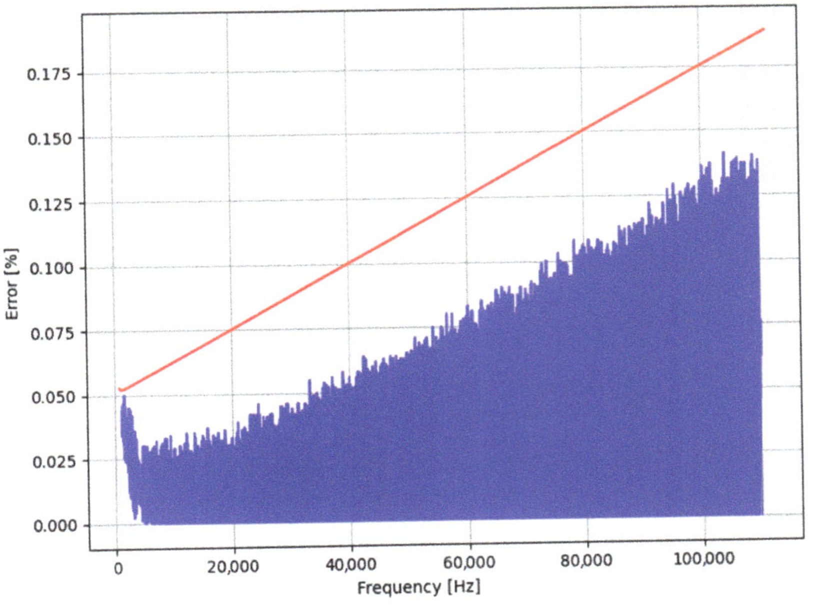

Figure 15. f/N converter error graph for f_g = 80 MHz.

It can be seen that the error recorded in the experiment does not exceed the theoretical value. There is also a significant influence of error averaging in the lower part of the measuring range. In order to determine the switching limit of the f_g frequency division for adaptive operation, the zoom of the obtained error graph for lower frequencies is shown in Figure 16. It can be seen that between 4 kHz and 5 kHz, there is an irregularity in the graph, after which the error value obtained from the experiment begins to increase noticeably. It was decided that the frequency of 5 kHz, located in the stable part of the graph, would be convenient and allow for stable operation of the f/N converter.

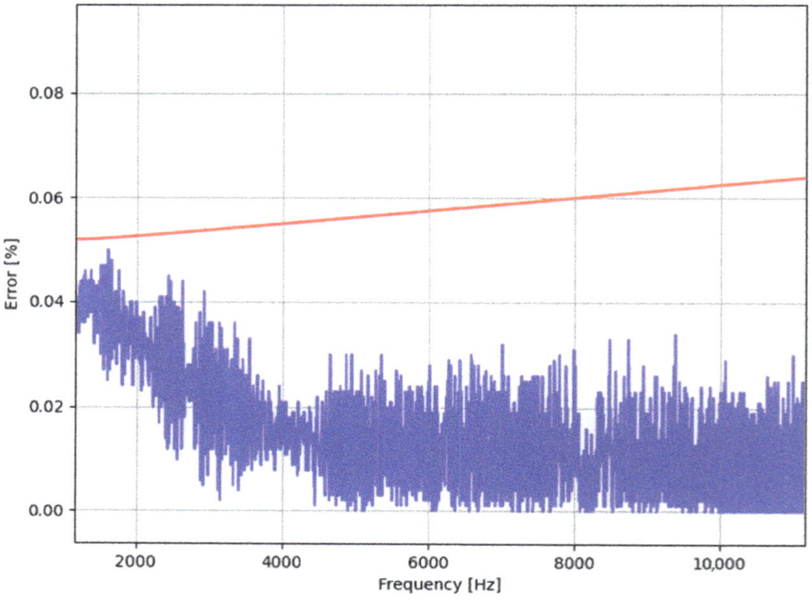

Figure 16. Zoom of the error graph for the lower part of the range.

5. Error Graph of the f/N Converter Model

The next stage of the research was creating an algorithm for the adaptive work of the f/N converter and introducing relevant modifications to the program code.

The f/N converter application should continuously analyze the current length of the T_{xi} period and change the degree of frequency division according to the established requirements if the threshold values are reached.

The details of the adaptive work of the method to change the degree of division of the clock frequency used by the counter are described in [31]. The results of the simulation analysis of this method for various operating conditions are also included there.

It was assumed that in the adaptive mode when reducing the measured frequency f_{xi} below 5 kHz, the frequency f_g would be divided by 8 in the prescaler. As a result, the T3 counter will add up the pulses occurring at a frequency of 10 MHz. The lower limit of the measurement range is 160 Hz, which is the product of the T_g period and the maximum capacity of the meter, rounded off to full tens.

The Increase of f_{xi} above 5 kHz will force the return of the division degree to 1 and set the clock signal for the counter T3 to the frequency of 80 MHz.

The adaptive operating mode changes the shape of the theoretical error graph. For the period T_{xi} with the value 20^{-4} s, corresponding to f_{xi} = 5 kHz, there is a step change in the total error value forced by the change in the quantization error. Figure 17 shows the error graph for the adaptive mode, plotted as a bold red line. For comparison, the waveforms of the theoretical total error graphs for f_g = 10 MHz and f_g = 80 MHz are added. Certainly, the change in the counter clock frequency division does not have to take place only for values close to f_{xi} = 5 kHz. The graph shows that the degree of frequency division can be switched for longer T_{xi} periods, which will further reduce the error. The presented threshold of change in the f_g division degree should be treated as an example, illustrating the effectiveness of the presented method.

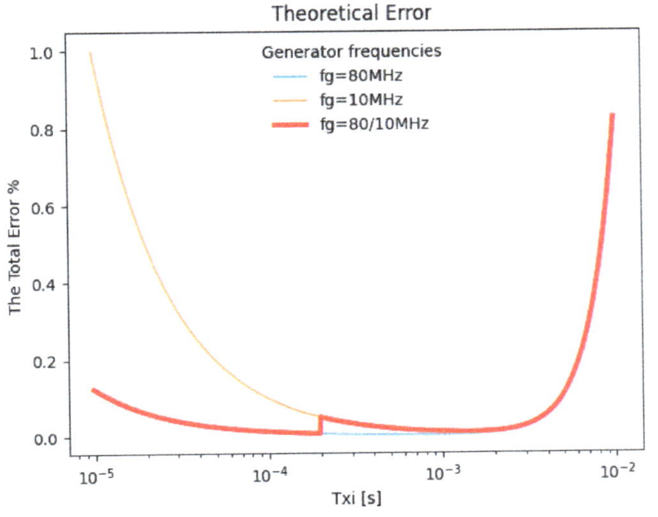

Figure 17. Graph of theoretical error for the f/N converter working in adaptive mode.

The algorithm for the adaptive work of the f/N converter is shown in Figure 18. Part of the functionality is identical to the algorithm presenting the converter's work in the basic mode (Figure 10). The blocks with unchanged functionality have a white filling, while new blocks and blocks with changed functionality are marked with a green filling.

The first modification of the algorithm consists of adding the start of f_g frequency division in the part initializing the counter T3 setting. Initially, the frequency division degree is set to 1, and the divider does not work.

The second modification is the inclusion of blocks that analyze the calculated current value of N_{xi} and, according to the obtained result, change the frequency division degree. When N_{xi} reaches a value of 16,000 or more, the division degree is set to 8. The program iteration number for the time of changing the division degree is also stored in a dedicated table. This allows the actual clock frequency to be decoded for subsequent N_{xi} when analyzing the data at a later time. In addition, in order to maintain subsequent division by 8, the program sets a dedicated memory location—a flag signaling that the division degree is currently set to 8. Thanks to this flag, the division coefficient is not changed until the next N_{xi} decreases below 2000, which corresponds to the frequency of 5 kHz for f_g = 10 MHz. Reducing the value of N_{xi} below 2000 forces the start of the procedure of setting the division degree to 1. Like in the previous case, the number of program iterations at which the division degree has been changed is stored for later use, but this time in the table storing the data about setting the f_g division by 1. In order to block duplicate settings in the f_g division, the division by 8 flags is cleared at the same time.

Depending on the calculated values of N_{xi}, the program changes the division degree in the prescaler on an ongoing basis, adjusting the counter clock frequency to the values set in the algorithm.

The last change introduced in the adaptive work algorithm refers to a different way of generating the final report. In order to be able to calculate T_{xi} from the transmitted N_{xi}, the N_{xi} strings for a given division degree are preceded by a division degree marker. Thanks to this, the software of the master computer, when reading subsequent data from the report, is able to calculate the duration of subsequent T_{xi} periods on an ongoing basis.

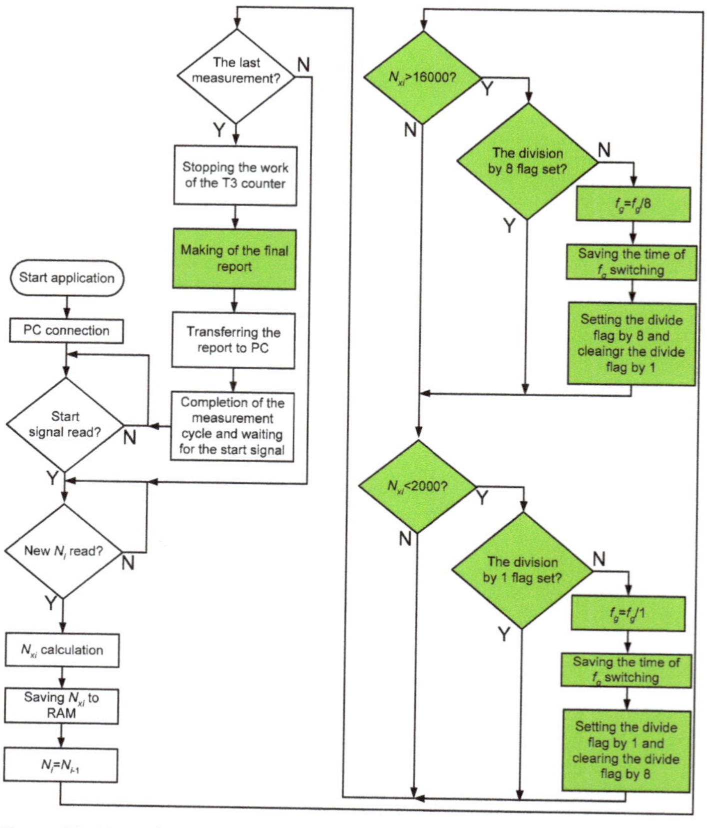

Figure 18. Algorithm of adaptive work of the f/N converter.

6. Verification of the Adaptive Work of the f/N Converter

The tests of adaptive operation of the converter consisted of checking how the converter works for frequencies in the lower part of the measurement range, the switching range in particular, by analyzing the increase in total error caused by the averaging error. It was assumed that the applied test signal is calculated based on the following relationship:

$$f_t = f_0 + f_m \sin(2\pi F t). \tag{12}$$

The following parameters were set: DC component f_0 = 5.160 kHz, amplitude f_m = 5 kHz, and frequency F = 1 Hz. For these conditions, it was possible to determine the error graph in the range of 160 Hz ÷ 10,160 Hz. The recorded variable frequency waveform is shown in Figure 19.

The analysis of the error graph for the entire recorded waveform would result in multiple plottings of the error curve on one graph, thus making the graphical presentation illegible. For this reason, it was assumed that the graph would show a data sequence representing one frequency change from the minimum to the maximum of the given test waveform f_t. The magnification of this waveform is shown in Figure 20. It can be seen that the division degree change generates a jump in the calculated f_{xi}. An incorrect value of f_{xi} is read when the procedure sets the new division degree value in the counter and reinitializes the counter after changing the setting.

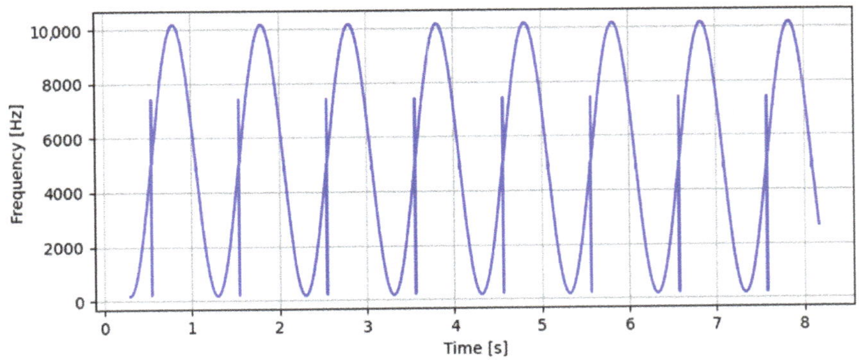

Figure 19. Result of recording the test signal after starting the adaptive mode.

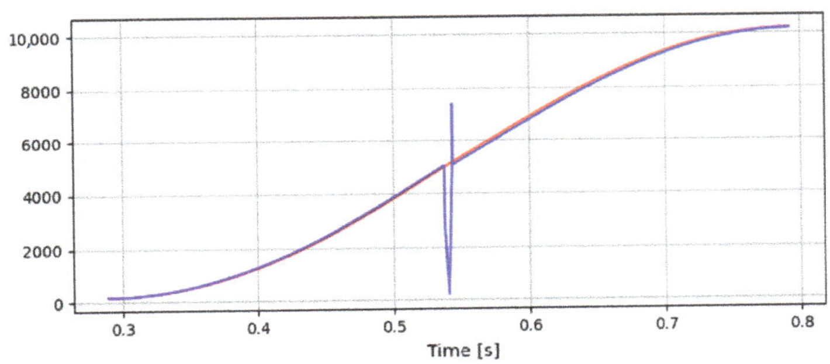

Figure 20. Graph of measurement results in which the process of adding clock pulses to the counter has been disturbed.

The presented magnification shows that for a threshold frequency close to 5 kHz, the reproduced waveform (blue) deviates significantly from the test waveform (red), with the difference between these waveforms reaching a value close to the amplitude f_m. Taking into account that the change of f_{xi} differs significantly from the set test waveform, it was concluded that an approximation should be used to reconstruct the waveform in the graph part covering the switching time of the counter's clock frequency division degree.

The first approximation used was extrapolation with a zero-degree polynomial replacing the incorrectly determined values with the last correct one. Figure 21 shows the error graph reconstructed in the above way. It can be seen that the theoretical error graph is consistent with that obtained from the experiment. The effect of reducing the processing error to a value of about 0.27% has been obtained.

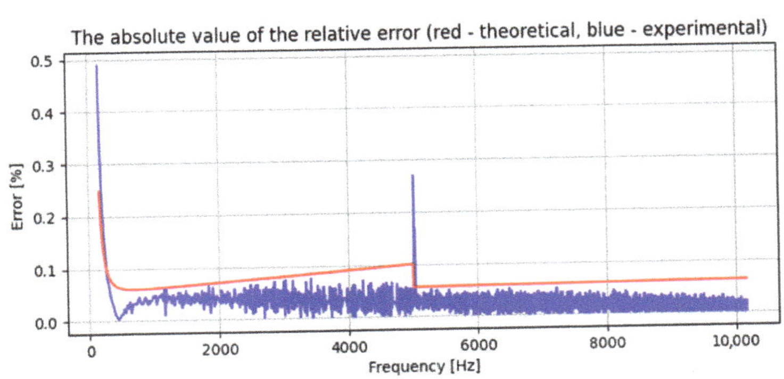

Figure 21. The error graph after zero-degree polynomial extrapolation.

Due to the rather large error value of the approximation used, it was decided to additionally test the extrapolation with a first-degree polynomial and the interpolation with zero and first-degree polynomials. It is obvious that the tested approximations caused differences in the experimentally obtained error graphs only in the switching range.

A decision was made that, in order to make the approximation effects comparable, a summary of the error graphs for all the approximations used will be presented in an enlarged form showing only the frequencies near 5 kHz, i.e., in the range in which the T3 clock frequency division degree was switched. This summary of graphs is shown in Figure 22. It can be seen that the best approximation of the test signal change trend is provided by linear interpolation.

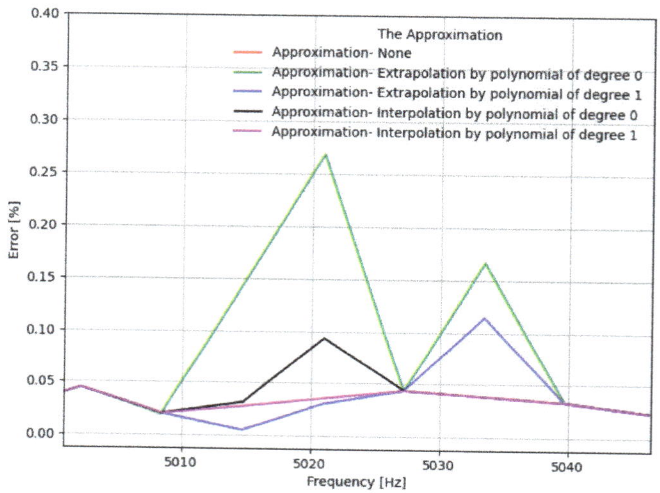

Figure 22. Enlarged error graphs for the approximations used in the range of T3 counter clock frequency switching.

Figure 23 shows the examined waveform after applying the first-degree polynomial interpolation. The applied approximation method caused the waveform to be undistorted in the regions of frequency division degree in the prescaler and to well correspond to the given test signal.

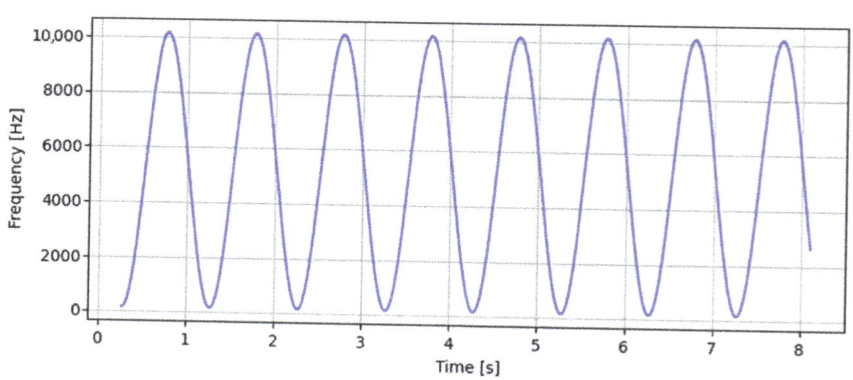

Figure 23. Signal recorded after first-degree polynomial interpolation.

7. Conclusions

A number of challenges had to be met when implementing an adaptive method for changing the frequency division of the counter clock signal in a frequency-to-code converter. The first was the selection of a microcontroller to implement the frequency-to-code converter. Due to the limitations of the factory-applied RC clock signal in the microcontroller, it was necessary to select and test a more stable signal source. The next challenge was to determine a method to check the upper limit of the measurement range. The main and most time-consuming challenge was to implement the basic and adaptive algorithms in the microcontroller structure. An additional challenge, not discussed in the article, was the development of software in Python to analyze the data and create final reports on each measurement. The measurement experiments conducted allow us to conclude that it is possible to implement the f/N converter structure in the STM32L476RG microcontroller. The obtained results confirm the compliance of the metrological analysis of the f/N converter model with the experimentally obtained error graph.

The efficiency of the microcontroller used allowed us to modify the operation of the f/N converter and improve its operating parameters. The design of the frequency-to-code converter developed in an earlier work [33] using an 8-bit microcontroller allowed error-free processing of successive periods of a waveform with a maximum frequency of 10.25 kHz. As shown in the paper, the device using a 32-bit microcontroller allowed the upper range of the measured frequencies to reach 111kHz. The above result proves the desirability of using microcontrollers with higher computing power in cases where processing of successive periods of the signal over a wider range is needed or when the computing power is used for additional data processing in the microcontroller. The implementation of an adaptive algorithm that allows for the extension of the lower limit of the f/N converter measurement range without requiring the use of a counter with increased capacity is presented.

This limit can be easily extended by selecting the clock frequency division degree of the meter to the value required in the measurements. The microcontroller used in the research allowed for the effective selection of the division degree of the meter's clock frequency to the required extent. If frequencies in other ranges need to be processed, the f/N converter system allows the user to divide the frequency within a very wide range.

When using a software change of the counter clock frequency division degree, one should take into account that during the division degree change, the meter's operation is disturbed and, as a result, a small number of incorrect measurement results are generated. In this case, good interference filtering effects are obtained using linear interpolation.

Author Contributions: Conceptualization, P.W.; methodology, D.Ś. and P.W.; software, D.Ś. and P.W.; validation, D.Ś. and P.W.; formal analysis, D.Ś. and P.W.; investigation, D.Ś. and P.W.; resources, D.Ś. and P.W.; data curation, D.Ś. and P.W.; writing—original draft preparation, D.Ś. and P.W.; writing—review and editing, D.Ś. and P.W.; visualization, D.Ś. and P.W.; supervision, D.Ś. and P.W.; project administration, D.Ś. and P.W.; funding acquisition, P.W. All authors have read and agreed to the published version of the manuscript.

Funding: This research was supported by the Lublin University of Technology, Lublin, Science Fund No. FD-20/EE-2/106.

Data Availability Statement: The data presented in this study is available on request from the corresponding author. The data is stored in a custom format implemented in the software. Some of the data was stored only temporarily in computer memory while Python was processing it. If necessary, it is possible to repeat the calculations presented and generate a summary containing the necessary information.

Conflicts of Interest: The authors declare no conflict of interest.

References

1. Szyduczyński, J.; Kościelnik, D.; Miśkowicz, M. Time-to-digital conversion techniques: A survey of recent developments. *Measurement* **2023**, *214*, 112762. [CrossRef]
2. Vaegae, N.K.; Komanapalli, V.L.N.; Annepu, B.R. Design and modeling of an intelligent temperature to frequency converter. *Measurement* **2016**, *85*, 54–64. [CrossRef]
3. Murmu, A.; Bhattacharyya, B.; Munshi, S. A synergy of voltage-to-frequency converter and continued-fraction algorithm for processing thermocouple signals. *Measurement* **2018**, *116*, 514–522. [CrossRef]

4. Martínez Ciro, R.A.; López Giraldo, F.E.; Betancur Perez, A.F.; Luna Rivera, M. Characterization of light-to-frequency converter for visible light communication systems. *Electronics* **2018**, *7*, 165. [CrossRef]
5. Ghassoul, M. A dual solar tracking system based on a light to frequency converter using a microcontroller. *Fuel Commun.* **2021**, *6*, 100007. [CrossRef]
6. Kirianaki, N.V.; Yurish, S.Y.; Shpak, N.O.; Deynega, V.P. *Data Acquisition and Signal Processing for Smart Sensors*; John Wiley & Sons, Ltd.: Chichester, UK, 2001.
7. Kurniawati, A.; Hamzah, T.; Indrato, T. A Low-Cost Transcutaneous Electrical Nerve Stimulation Measuring Device Using Frequency-to-Voltage and Current-to-Voltage. *J. Electron. Electromed. Med. Inform.* **2020**, *2*, 65–70. [CrossRef]
8. Rajan, D.; Ramachandran, E.T. Design and Development of Sun Tracking Solar Panel. *Int. J. Adv. Eng. Manag.* **2021**, *3*, 1343–1357.
9. Li, P.G.; Hu, W.; Hu, R.; Huang, Q.; Yao, J.; Chen, Z. Strategy for wind power plant contribution to frequency control under variable wind speed. *Renew. Energy* **2019**, *130*, 1226–1236. [CrossRef]
10. Yurish, S.Y.; Kirianaki, N.V. Interface circuit design for frequency-time domain MEMS sensors. *J. Phys. Conf. Ser.* **2006**, *34*, 17. [CrossRef]
11. *CTM-PER Conntinous-Period Counter*; User's Guide; Keithley MetraByte Co.: Taunton, MA, USA, 1990.
12. Helal, E.; Alvarez-Fontecilla, E.; Eissa, A.I.; Galton, I. A time amplifier assisted frequency-to-digital converter based digital fractional-N PLL. *IEEE J. Solid-State Circuits* **2021**, *56*, 2711–2723. [CrossRef]
13. Tancock, S.; Arabul, E.; Dahnoun, N. A review of new time-to-digital conversion techniques. *IEEE Trans. Instrum. Meas.* **2019**, *68*, 3406–3417. [CrossRef]
14. Garzetti, F.; Corna, N.; Lusardi, N.; Geraci, A. Time-to-digital converter IP-core for FPGA at state of the art. *IEEE Access* **2021**, *9*, 85515–85528. [CrossRef]
15. Portaluppi, D.; Pasquinelli, K.; Cusini, I.; Zappa, F. Multi-channel FPGA time-to-digital converter with 10 ps bin and 40 ps FWHM. *IEEE Trans. Instrum. Meas.* **2022**, *71*, 2002109. [CrossRef]
16. Mao, X.; Yang, F.; Wei, F.; Shi, J.; Cai, J.; Cai, H. A Low Temperature Coefficient Time-to-Digital Converter with 1.3 ps Resolution Implemented in a 28 nm FPGA. *Sensors* **2022**, *22*, 2306. [CrossRef] [PubMed]
17. Dorozhovets, M.; Pawlowski, E.; Swisulski, D. Frequency measurement research with weight averaging of pulse output signal of voltage-to-frequency converter. *Measurement* **2023**, *216*, 112912. [CrossRef]
18. Vlachogiannakis, G.S.; Ximenes, A.R.; Staszewski, R.B. Fractional-N Frequency Synthesizer Incorporating Cyclic Digital-To-Time And Time-To-Digital Circuit Pair. U.S. Patent No. 9,722,537; U.S. Patent and Trademark Office: Washington, DC, USA, 1 August 2017.
19. Mahajan, T.; Shetty, D.; Muthukaruppan, R. Time To_Digital Converter. U.S. Patent No. 10,175,655; U.S. Patent and Trademark Office: Washington, DC, USA, 8 January 2019.
20. Dutton, N.; Henderson, R.K.; Gnecchi, S. Time To Digital Converter And Applications Thereof. U.S. Patent No. 9,639,063; U.S. Patent and Trademark Office: Washington, DC, USA, 2 May 2017.
21. Song, H.W.; Abu-Rahma, M.H. Time To_Digital Converter. U.S. Patent No. 9,092,013; U.S. Patent and Trademark Office: Washington, DC, USA, 28 July 2015.
22. Kumar, P.; Staszewski, R.; Charbon, E. Time To Digital Converter And Method Therefor. U.S. Patent No. 10,079,608; U.S. Patent and Trademark Office: Washington, DC, USA, 18 September 2018.
23. Cho, Y.H.; Staszewski, R.B. Time To Digital System And Associated Frequency Synthesizer. U.S. Patent No. 9,037,886; U.S. Patent and Trademark Office: Washington, DC, USA, 19 May 2015.
24. Kumar, P.; Staszewski, R.; Charbon, E. Time To Digital Converter And Method Therefor. U.S. Patent No. 9,419,635; U.S. Patent and Trademark Office: Washington, DC, USA, 16 August 2016.
25. Pereira, J.M.; Postolache, O.A.; Girao, P.S. Analog to Digital Conversion Methods for Smart Sensing Systems. In *Advances in Measurement Systems*; Sharma, M.K., Ed.; InTech Open: Rijeka, Croatia, 2010.
26. Yurish, S.Y. Sensors and transducers: Frequency output versus voltage output. *Sens. Transducers* **2004**, *49*, 302–305.
27. Duoshan, L.; Hefei Lianxin Power Supply Co., Ltd. Inverter Direct Voltage Sampling Circuit for Concentration Storage Battery Emergency Power Supply. Chinese Patent CN107144728A, 28 May 2017.
28. Chu, L.; Wang, S.; Chen, P. Direct Current Bus Voltage Detecting Circuit of Power Unit. Chinese Patent CN202256484U, 30 May 2012.
29. Murillo, C.A.; López, B.C.; Celma, S. *Voltage-to-Frequency Converters: CMOS Design and Implementation*; Springer: New York, NY, USA, 2013.
30. Pawłowski, E. A simulation investigation into the signal reconstruction accuracy of the transducer with pulse frequency output. In Proceedings of the 2016 IEEE 21st International Conference on Methods and Models in Automation and Robotics (MMAR), Miedzyzdroje, Poland, 29 August–1 September 2016; pp. 762–766.
31. Warda, P. Simulation of an Adaptive Method of Improving the Accuracy and Extending the Range of Frequency Signal Processing in a Frequency-to-Code Converter. *Appl. Sci.* **2021**, *11*, 341. [CrossRef]
32. Goswami, J.C.; Hoefel, A.E. Algorithms for estimating instantaeous frequency. *Signal Process.* **2004**, *84*, 1423–1427. [CrossRef]
33. Warda, P. Frequency-to-code converter with direct data transmission. *Inform. Autom. Pomiary W Gospod. I Ochr. Sr.* **2022**, *12*, 74–77.

34. Weltin-Wu, C.; Zhao, G.; Galton, I. A 3.5 GHz Digital Fractional-PLL Frequency Synthesizer Based on Ring Oscillator Frequency-to-Digital Conversion. *IEEE J. Solid-State Circuits* **2015**, *50*, 2988–3002. [CrossRef]
35. *Agilent 33220A20MHz Function/Arbitrary Waveform Generator, User's Guide*; Agilent Technologies, Inc.: Santa Clara, CA, USA, 2007.
36. *HP 33120AFunction/Arbitrary Waveform Generator, User's Guide*, 5th ed.; Manual Part Number: 33120-90005; Hewlett-Packard Company: Loveland, CO, USA, 1997.

energies

MDPI

Article

Boost Converter with Main Switch Possessing ZVT and ZCT and Auxiliary Switch Possessing ZCS

Kuo-Ing Hwu [1,*], Zhi-Fan Lin [2] and Pei-Ching Tseng [1]

1 Department of Electrical Engineering, National Taipei University of Technique, 1, Sec. 3, Zhongxiao E. Rd., Taipei 10608, Taiwan; t111319007@ntut.org.tw
2 FSP Technique Inc., No. 22, Jianguo E. Rd., Taoyuan Dist., Taoyuan City 33068, Taiwan; fenlin@fsp-group.com.tw
* Correspondence: eaglehwu@ntut.edu.tw; Tel.: +886-2-27712171 (ext. 2159)

Abstract: In this paper, a zero-voltage and zero-current transition (ZVZCT) boost converter is presented with a small number of auxiliary components, such as a resonant capacitor, a resonant inductor and an auxiliary power switch, to produce a main power switch with both zero-voltage switching (ZVS) and zero-current switching (ZCS). Furthermore, the auxiliary power switch also has zero current switching. In addition, a look-up table is employed to implement an auto-tuning technique to regulate the trigger position and turn-on time of the auxiliary power switch, to further improve efficiency, especially at light load, thereby making the overall efficiency of the converter present a horizontal curve. Moreover, in terms of the system control, the digital controller is implemented directly from the z-domain, and the field programmable gate array (FPGA) is utilized as the system control kernel to achieve a fully digitalized control system. The simulated results are used to demonstrate the feasibility of the proposed converter, whereas the experimental results are used to verify its effectiveness.

Keywords: auto-tuning technique; boost converter; FPGA; fully-digitalized control; z-domain; controller design; ZVZCT; ZVT; ZCT; ZVS; ZCS

Citation: Hwu, K.-I.; Lin, Z.-F.; Tseng, P.-C. Boost Converter with Main Switch Possessing ZVT and ZCT and Auxiliary Switch Possessing ZCS. *Energies* **2023**, *16*, 5504. https://doi.org/10.3390/en16145504

Academic Editors: Abdelali El Aroudi and Salvatore Musumeci

Received: 29 May 2023
Revised: 29 June 2023
Accepted: 18 July 2023
Published: 20 July 2023

1. Introduction

As the techniques for using power converters have become more and more developed, high efficiency, high switching frequency, small size and good stability have become the basic requirements. However, as the switching frequency of the converter increases, the switching loss of the power switch also increases, thus reducing the efficiency of the power converter. In order to overcome the impact of traditional hard-switching power converters, the soft switching technique has been developed, which can effectively reduce the switching loss.

The methods of resonant power converters have been presented in [1,2], in which resonant inductors and resonant capacitors are added to converters to form resonant circuits, which are used to change the voltages on power switches or the currents flowing through the power switches to achieve zero-voltage or zero-current switching. However, the disadvantage is that the high voltage or high current caused by the resonance of the auxiliary circuit increases the component stress and, in turn, increases the conduction loss. In addition, since the resonance time of the auxiliary circuit is fixed, it is necessary to add variable frequency control in order to keep the output voltage stable at the prescribed value, which makes the design of the output filter difficult.

The literature [3,4] proposes adding auxiliary power switches and resonant elements to converters to generate resonant circuits to achieve zero-voltage or zero-current switching, and hence the fixed resonant time of the semi-resonant converter when it is on or off is improved, so that the converter does not need variable frequency control, and therefore fixed frequency control can be achieved.

The literature [5,6] also suggests that before the main power switch is turned on, the auxiliary power switch is turned on first or afterwards to form a resonant circuit, and by generating transient resonance, the voltage across the main power switch or the current flowing through the main power switch resonates to zero. This transient resonance occurs only during the instant of power switch switching, and does not resonate during the rest of the time, thus avoiding the problem of high voltage or high current stress, therefore reducing the conduction loss of the converter.

The literature [7,8] has proposed that the auxiliary power switch could generate two transient resonances in each switching cycle without affecting the circuit behavior. The main power switch can achieve zero voltage switching (ZVS) during the first transient resonance and zero current switching (ZCS) during the second transient resonance. Therefore, the main power switch has both ZVS and ZCS, but the disadvantage is that the auxiliary power switch cannot achieve ZVS or ZCS, and it is mostly floating, which not only increases the complexity of the gate driver circuit, but also requires multiple auxiliary components to make the main power switch have both ZVS and ZCS.

The main structure in [9] is an interleaved boost converter, which features only one auxiliary switch with common ground and enables two main switches to realize ZVS and ZCS; the drawback is that the auxiliary switch needs four gate driving signals over one switching cycle, and the resonance condition is strict, which makes the mathematical mode analysis extremely complicated and difficult.

The topology shown in [10] is a two-phase interleaved boost ZVT converter, whose characteristics are different from those shown in [a]. The circuit shown in [b] uses two auxiliary switches to realize the main switch with ZVS turn-on and ZCS turn-off and the used auxiliary switches have ZCS turn-off with only one gate driving signal over one switching cycle, but the drawback is that the two auxiliary switches are floating, thereby making the gate driving more difficult and greatly reducing the practicality of this converter.

The circuit shown in [11] is a simple structure of a boost ZVT converter, which features a zero-current quasi-resonant circuit and an auxiliary switch with MOSFETs in series with diodes combined, and the auxiliary switch possesses common ground. In addition, the advantage is that the current waveform of the main switch has the characteristics of zero-current turn-on and zero-current turn-off, the switching control mode of the converter is fixed frequency, and the structure is simple, easy to analyze and low in cost.

The circuits shown in [12,13] both are boost converters using ZVT and ZCT technologies, but the feature of the circuit shown in [12] is that it is easier to achieve ZVS and ZCS because the main switch and the auxiliary switch are common-grounded, but the drawback is that too many components are used, resulting in higher cost. The circuit shown in [13] uses fewer components for the resonant circuit of the auxiliary switch and is easier to analyze, but the disadvantage is that the condition of the main switch to realize soft switching is more severe and difficult and the soft switching feature cannot be implemented all over the whole load range.

The circuit shown in [14] is a two-phase interleaved four-switch boost ZVT converter, using only one coupling inductor and one clamping capacitor, which is characterized by the ability to give the four main switches ZVS turn-on in the current conduction mode (CCM) without any auxiliary switches. This circuit is characterized by few auxiliary circuit components, easy analysis and ZVS capability for most of the load range. However, the disadvantage is that the auxiliary switch is on the secondary side and floating, which makes the feedback control difficult and the transient response poor.

This circuit shown in [15] is characterized by a small number of components in the auxiliary circuit, easy analysis and the capability for ZVS for most of the load range, but the disadvantage is that the auxiliary switch is on the secondary side and is floating, resulting in a circuit with poor feedback control and poor transient response.

The circuit shown in [16] is a full-bridge DC to AC ZVT converter with an auxiliary circuit using only three components (i.e., one inductor and two MOSFETs) to achieve ZVS turn-on for all four main switches and ZCS turn-off for the auxiliary switch. The advantage is that a small number of components is used to achieve soft switching, but the disadvantage is that the auxiliary switches are floating.

Reference [17] presents a ZCZVT step-up converter. The advantage of this is that this circuit is the integration of three coupled inductors into a single magnetic element and the auxiliary switch has common ground and a soft switching feature. The disadvantage is that too many components are used to achieve soft switching, thereby making the auxiliary circuit too complex to analyze its operating principle.

Reference [18] proposes a step-up and step-down ZVT converter with a demagnetized auxiliary circuit. The advantage is that the demagnetized winding is used to reduce the voltage stress on the auxiliary circuit, but the disadvantage is that the auxiliary switch is hard-switched.

In this paper, a boost converter with soft switching is developed, which is implemented only by one auxiliary switch, one resonant inductor and one resonant capacitor as compared to the traditional boost converter. In this circuit, the main switch has zero-voltage and zero-current transition (ZVZCT) whereas the auxiliary switch has zero-current switching (ZCS). Aside from this, the PWM auto-tuning technique based on a given lookup table is added to adjust the turn-on instant and turn-on time of the auxiliary switch, so that the efficiency is further upgraded, particularly at light load. Therefore, the curve of efficiency versus load current is made nearly flat all over the load range. Regarding the system control, the digital controller is designed directly from the z-domain.

2. Operating Principle

Figure 1 shows the proposed step-up converter with soft switching, in which the dashed box is the auxiliary circuit, which consists of an auxiliary power switch S_a, a resonant inductor L_r and a resonant capacitor C_r. Figure 2 displays the illustrated waveforms of the proposed converter operating.

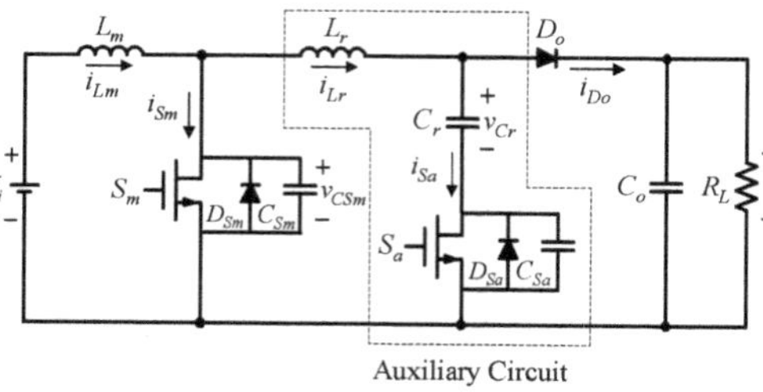

Figure 1. Boost converter with S_a having ZVT and ZVT and S_b having ZCS.

Prior to the analysis, the following assumptions are made:

(1) The power switches and diodes are regarded as ideal components;
(2) The parasitic resistances of the inductance and capacitance are negligible;
(3) The input inductance is extremely large and can be viewed as an ideal constant current source;
(4) The output capacitance is very large and can be considered as an ideal constant voltage source.

According to the above assumptions, the converter operating can be divided into twelve states over one switching cycle.

State 1 [$t_0 \leq t \leq t_1$]: Before the start of the switching cycle, both the main power switch S_m and the auxiliary power switch S_a are in the off-state, and the output diode D_o is in the on-state. As displayed in Figure 3a at the time t_0, the auxiliary power switch S_a turns on first, and the auxiliary power switch current i_{Sa} rises rapidly to equal the input current I_{Lm}, causing the output diode D_o to turn off and then state 1 to come to the end.

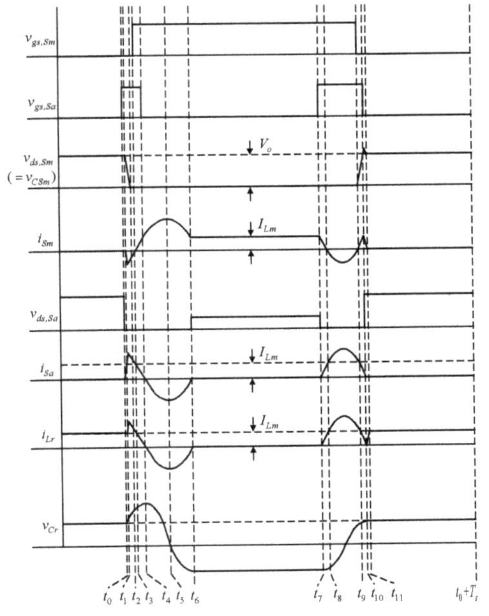

Figure 2. Illustrated waveforms of the proposed converter operating.

The initial current of the resonant inductor is the input current I_{Lm} and the initial voltage of the resonant capacitor is V_{Cr10}.

The corresponding equations in this state are

$$\begin{cases} v_{Cr}(t) = \frac{I_{Lm}-I_o}{C_r}(t-t_0) + V_{Cr10} \\ i_{Lr}(t) = I_{Lm} \end{cases} \quad (1)$$

By substituting the boundary condition $v_{Cr}(t_1) = V_{Cr1}$ into (1), we can find that the corresponding time experienced by this state is

$$T_1 = t_1 - t_0 = \frac{C_r(V_{Cr1} - V_{Cr10})}{(I_{Lm} - I_o)} \quad (2)$$

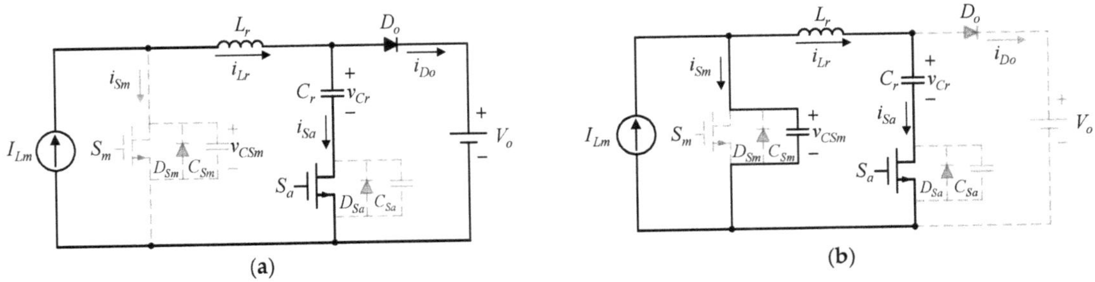

Figure 3. *Cont.*

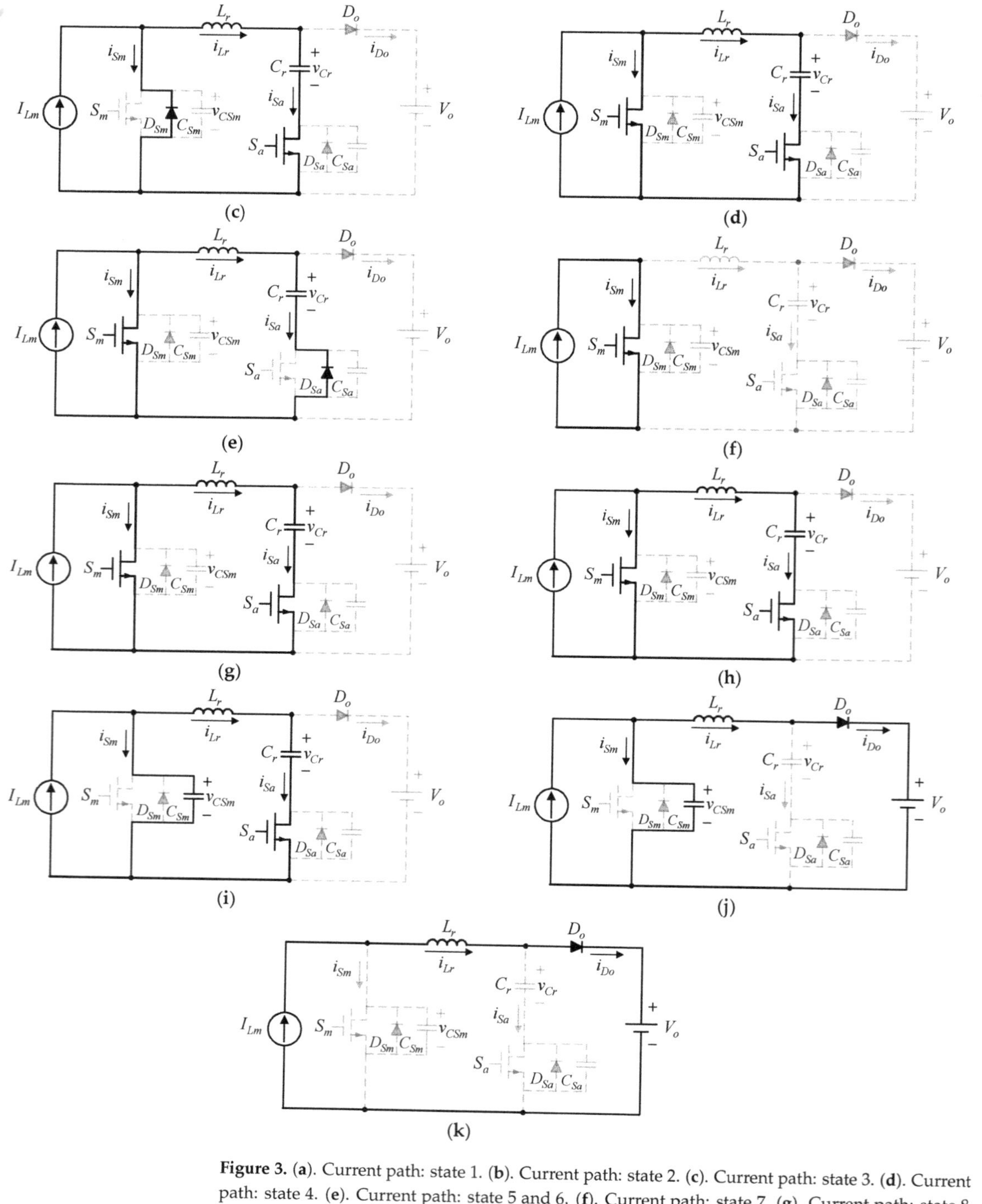

Figure 3. (**a**). Current path: state 1. (**b**). Current path: state 2. (**c**). Current path: state 3. (**d**). Current path: state 4. (**e**). Current path: state 5 and 6. (**f**). Current path: state 7. (**g**). Current path: state 8. (**h**). Current path: state 9. (**i**). Current path: state 10. (**j**). Current path: state 11. (**k**). Current path: state 12.

State 2 [$t_1 \le t \le t_2$]: As displayed in Figure 3b, when the auxiliary power switch current i_{Sa} rises to the input current I_{Lm}, the parasitic capacitor C_{Sm} of the main power switch discharges. When the parasitic capacitor C_{Sm} of the main power switch discharges to zero, state 2 ends.

The initial current of the resonant inductor is the input current I_{Lm}, the initial voltage of the resonant capacitor is V_{Cr1}, and the initial voltage of the parasitic capacitor across the main power switch is V_o.

The corresponding equations in this state are

$$\begin{cases} i_{Lr}(t) = \frac{(V_o - V_{Cr1})}{Z_1} \sin \omega_1 (t - t_1) + \frac{C}{C_r} I_{Lm} \cos \omega_1 (t - t_1) + \frac{C}{C_{Sm}} I_{Lm} \\ v_{Cr}(t) = \frac{C}{C_r} \left[\frac{I_{Lm}}{C_r \omega_1} \sin \omega_1 (t - t_1) - (V_o - V_{Cr1}) \cos \omega_1 (t - t_1) + (V_o - V_{Cr1}) \right] + \frac{I_{Lm}}{C_r + C_{Sm}} (t - t_1) + V_{Cr1} \\ v_{CSm}(t) = -\frac{C}{C_{Sm}} \left[\frac{I_{Lm}}{C_r \omega_1} \sin \omega_1 (t - t_1) - (V_o - V_{Cr1}) \cos \omega_1 (t - t_1) + (V_o - V_{Cr1}) \right] + \frac{I_{Lm}}{C_r + C_{Sm}} (t - t_1) + V_o \end{cases} \tag{3}$$

where

$$\omega_1 = \frac{1}{\sqrt{L_r C}},\ Z_1 = \sqrt{\frac{L_r}{C}},\ C = \frac{C_r C_{Sm}}{C_r + C_{Sm}} \tag{4}$$

By substituting the boundary conditions into (3), we can find the corresponding time taken by this state is

$$T_2 = t_2 - t_1 = \frac{1}{\omega_1} \left[\frac{Z_1 (I_{Lr2} - I_{Lm})}{V_o - V_{Cr1}} \right] \tag{5}$$

State 3 [$t_2 \le t \le t_3$]: As displayed in Figure 3c, when the parasitic capacitance C_{Sm} of the main power switch is discharged to zero, the body diode D_{Sm} of the main power switch S_m is turned on. The initial current of resonant inductor is I_{Lr2} and the initial voltage of the resonant capacitor is V_{Cr2}.

The corresponding equations in this state are

$$\begin{cases} i_{Lr}(t) = I_{Lr2} \cos \omega_2 (t - t_2) - \frac{V_{Cr2}}{Z_2} \sin \omega_2 (t - t_2) \\ v_{Cr}(t) = V_{Cr2} \cos \omega_2 (t - t_2) + Z_2 I_{Lr2} \sin \omega_2 (t - t_2) \end{cases} \tag{6}$$

where

$$\omega_2 = \frac{1}{\sqrt{L_r C_r}},\ Z_2 = \sqrt{\frac{L_r}{C_r}} \tag{7}$$

By substituting the boundary conditions into (6), we can find the corresponding time experienced by this state is

$$T_3 = t_3 - t_2 = \frac{1}{\omega_2} \left[\frac{Z_2 (I_{Lr2} - I_{Lm})}{V_{Cr2}} \right] \tag{8}$$

State 4 [$t_3 \le t \le t_4$]: As shown in Figure 3d, when the main power switch S_m is turned on with ZVS, the main power switch current i_{Sm} starts to rise from zero to the input current I_{Lm}. At this time, the auxiliary power switch S_a is turned off with ZCS.

The initial current of the resonant inductor is I_{Lm}, and the initial voltage of the resonant capacitor is V_{Cr3}.

The corresponding equations in this state are

$$\begin{cases} i_{Lr}(t) = I_{Lm} \cos \omega_2 (t - t_3) - \frac{V_{Cr3}}{Z_2} \sin \omega_2 (t - t_3) \\ v_{Cr}(t) = V_{Cr3} \cos \omega_2 (t - t_3) + Z_2 I_{Lm} \sin \omega_2 (t - t_3) \end{cases} \tag{9}$$

By substituting the boundary conditions into (9), we can find the time experienced by this state is

$$T_4 = t_4 - t_3 = \frac{1}{\omega_2} \left[\frac{Z_2 I_{Lm}}{V_{Cr3}} \right] \tag{10}$$

State 5 [$t_4 \leq t \leq t_5$]: As displayed in Figure 3e, when the main power switching current i_{Sm} is greater than the input current I_{Lm}, the resonant inductor current i_{Lr} starts to flow in the opposite direction. When the resonant capacitor C_r is discharged to zero, state 5 ends.

The initial current of the resonant inductor is zero, and the initial voltage of the resonant capacitor is V_{Cr4}.

The corresponding equations in this state are

$$\begin{cases} i_{Lr}(t) = -\frac{V_{Cr4}}{Z_2}\sin\omega_2(t-t_4) \\ v_{Cr}(t) = V_{Cr4}\cos\omega_2(t-t_4) \end{cases} \tag{11}$$

By substituting the boundary conditions into (11), we can find the time taken by this state is

$$T_5 = t_5 - t_4 = \frac{1}{\omega_2}\left[\frac{Z_2 I_{Lr5}}{V_{Cr4}}\right] \tag{12}$$

Sate 6 [$t_5 \leq t \leq t_6$]: As displayed in Figure 3e, this state continues to resonate. When the resonant inductor current i_{Lr} resonates to zero, this state comes to end.

The initial current of the resonant inductor is I_{Lr5}, and the initial voltage of the resonant capacitor is zero.

The corresponding equations in this state are

$$\begin{cases} i_{Lr}(t) = -I_{Lr5}\cos\omega_2(t-t_5) \\ v_{Cr}(t) = -Z_2 I_{Lr5}\sin\omega_2(t-t_5) \end{cases} \tag{13}$$

By substituting the boundary conditions into (13), we can find the time experienced by this state is

$$T_6 = t_6 - t_5 = \frac{1}{\omega_2}\left[\frac{V_{Cr6}}{Z_2 I_{Lr5}}\right] \tag{14}$$

State 7 [$t_6 \leq t \leq t_7$]: As displayed in Figure 3f, this state is like the magnetization state of the traditional boost converter, where the output capacitor C_o provides energy to the load. When the auxiliary power switch S_a is turned on again, this state ends.

State 8 [$t_7 \leq t \leq t_8$]: As displayed in Figure 3g, when the auxiliary power switch S_a is turned on, the resonant inductor current i_{Lr} rises to the input current I_{Lm}, and then state 8 comes to an end.

The initial current of the resonant inductor is zero, and the initial voltage of the resonant capacitor is $-V_{Cr7}$.

The corresponding equations in this state are

$$\begin{cases} i_{Lr}(t) = \frac{V_{Cr7}}{Z_2}\sin\omega_2(t-t_7) \\ v_{Cr}(t) = -V_{Cr7}\cos\omega_2(t-t_7) \end{cases} \tag{15}$$

By substituting the boundary conditions into (15), we can find the time experienced by this state is

$$T_8 = t_8 - t_7 = \frac{1}{\omega_2}\left[\frac{Z_2 I_{Lm}}{V_{Cr7}}\right] \tag{16}$$

State 9 [$t_8 \leq t \leq t_9$]: As shown in Figure 3h, when the resonant inductor current i_{Lr} is greater than the input current I_{Lm}, the main power switch current i_{Sm} starts to flow in the opposite direction. When the inductor current resonates to the input current I_{Lm} again, the main power switch is turned off at this time, so that the main power switch has ZCS turn-on.

The initial current of the resonant inductor is I_{Lm}, and the initial voltage of the resonant capacitor is $-V_{Cr8}$.

The corresponding equations in this state is

$$\begin{cases} i_{Lr}(t) = I_{Lm}\cos\omega_2(t-t_8) + \frac{V_{Cr8}}{Z_2}\sin\omega_2(t-t_8) \\ v_{Cr}(t) = -V_{Cr8}\cos\omega_2(t-t_8) + Z_2 I_{Lm}\sin\omega_2(t-t_8) \end{cases} \quad (17)$$

By substituting the boundary conditions into (17), we can find the time taken by this state is

$$T_9 = t_9 - t_8 = \frac{1}{\omega_2}\left[\frac{V_{Cr9}+V_{Cr8}}{Z_2 I_{Lm}}\right] \quad (18)$$

State 10 [$t_9 \le t \le t_{10}$]: As shown in Figure 3i, when the resonant inductor current i_{Lr} drops to zero, the auxiliary power switch S_a is turned off with ZCS. Once the auxiliary power switch is turned off, this state ends.

The initial current of the resonant inductor is I_{Lm}, the initial voltage of the resonant capacitor is V_{Cr9}, and the initial voltage of the parasitic capacitor of the main power switch is zero.

The corresponding equations of this state is

$$\begin{cases} i_{Lr}(t) = \frac{-V_{Cr9}}{Z_1}\sin\omega_1(t-t_9) + \frac{C}{C_r}I_{Lm}\cos\omega_1(t-t_9) + \frac{C}{C_{Sm}}I_{Lm} \\ v_{Cr}(t) = \frac{C}{C_r}\left[\frac{I_{Lm}}{C_r\omega_1}\sin\omega_1(t-t_9) + V_{Cr9}\cos\omega_1(t-t_9) - V_{Cr9}\right] + \frac{I_{Lm}}{C_r+C_{Sm}}(t-t_9) + V_{Cr9} \\ v_{CSm}(t) = -\frac{C}{C_{Sm}}\left[\frac{I_{Lm}}{C_r\omega_1}\sin\omega_1(t-t_9) + V_{Cr9}\cos\omega_1(t-t_9) - V_{Cr9}\right] + \frac{I_{Lm}}{C_r+C_{Sm}}(t-t_9) \end{cases} \quad (19)$$

By substituting the boundary conditions into (19), we can find the time experienced by this state is

$$T_{10} = t_{10} - t_9 = \frac{1}{\omega_1}\left[\frac{Z_1 I_{Lm}}{V_{Cr9}}\right] \quad (20)$$

State 11 [$t_{10} \le t \le t_{11}$]: As shown in Figure 3j, when the voltage v_{CSm} on the parasitic capacitance of the main power switch rises above the output voltage V_o and the resonant inductor current rises linearly from zero to the input current I_{Lm}, this state 11 ends.

The initial current of the resonant inductor is zero, and the initial voltage of the parasitic capacitor of the main power switch is V_{CSm10}.

The corresponding equations of state in this state are

$$\begin{cases} i_{Lr}(t) = I_{Lm} - I_{Lm}\cos\omega_{10}(t-t_{10}) + \frac{(V_{CSm10}-V_o)}{Z_{10}}\sin\omega_{10}(t-t_{10}) \\ v_{CSm}(t) = V_o + (V_{CSm10}-V_o)\cos\omega_{10}(t-t_{10}) + Z_{10}I_{Lm}\sin\omega_{10}(t-t_{10}) \end{cases} \quad (21)$$

where

$$\omega_{10} = \frac{1}{\sqrt{L_r C_{Sm}}},\ Z_{10} = \sqrt{\frac{L_r}{C_{Sm}}} \quad (22)$$

By substituting the boundary conditions into (21), we can find the time experienced in this state is

$$T_{11} = t_{11} - t_{10} = \frac{1}{\omega_{10}}\left[\frac{Z_{10}I_{Lm}}{V_{CSm10}-V_o}\right] \quad (23)$$

State 12 [$t_{11} \le t \le t_0 + T_s$]: As shown in Figure 3k, this state is just like the input inductor demagnetization state of the traditional boost converter, sending energy to the output. When the auxiliary power switch S_a is turned on again, this state ends and returns to state 1.

After deducing from the above-mentioned states, the soft switching status of the power switch can be found from Table 1.

Table 1. Soft switching status of the power switches.

State	Power Switch	Soft Switching Status
3	S_m	ZVT turn-on
4	S_a	ZCS turn-off
9	S_m	ZCT turn-off
10	S_a	ZCS turn-off

Table 1 display soft switching status of the power switches.

3. Modeling Based on Dual Time Scale Averaging Method

In this paper, the averaging method for dual time scale [19] is used to derive the small-signal mathematical model of the proposed circuit. This averaging method can be used to classify the system into slow state variables and fast state variables. The slow state variables are input inductance current i_{Lm} and output capacitance voltage v_{Co}, and the fast state variables are resonant capacitance voltage v_{Cr} and resonant inductor current i_{Lr}. Figure 4 shows the equivalent circuit of the converter used after averaging. The dashed line shows the averaging mode of the fast state variables relative to the slow state variables.

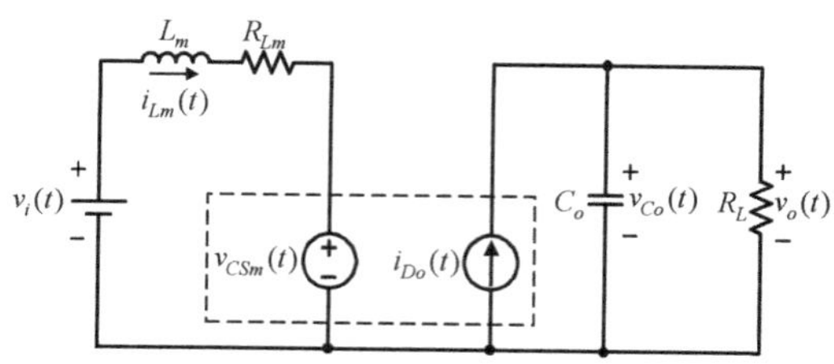

Figure 4. Equivalent circuit after averaging.

The equations for the slow state variables can be listed from Figure 4 as

$$\begin{cases} \frac{di_{Lm}(t)}{dt} = \frac{v_i(t)}{L_m} - \frac{R_{Lm}}{L_m} i_{Lm}(t) - \frac{v_{CSm}(t)}{L_m} \\ \frac{dv_{Co}(t)}{dt} = -\frac{v_{Co}(t)}{C_o R_L} + \frac{i_{Do}(t)}{C_o} \end{cases} \quad (24)$$

$$v_o(t) = v_{Co}(t) \quad (25)$$

where $v_{CSm}(t)$ is the average function of the main power switch over one switching period T_s, i.e., $v_{CSm}(t) = \langle v_{CSm}(t) \rangle_{Ts}$; $i_{Do}(t)$ is the average function of the output diode current over one switching period T_s, i.e., $i_{Do}(t) = \langle i_{Do}(t) \rangle_{Ts}$. Therefore, the averaging mode of the slow state variables can be obtained by finding the average function of $v_{CSm}(t)$ and $i_{Do}(t)$. The symbols are defined herein first to reduce the complexity of the analysis, as follows:

$$\begin{cases} i_{Lm}(t) = i_{Lm} \\ v_o(t) = v_o \\ v_{Co}(t) = v_{Co} \\ v_i(t) = v_i \\ d(t) = d \\ d'(t) = d' = 1 - d \\ v_{CSm}(t) = v_{CSm} \\ i_{Do}(t) = i_{Do} \end{cases} \quad (26)$$

where each time function includes the DC component and the AC small-signal component.

The definition of the average function of v_{CSm}, i.e., $\langle v_{CSm}\rangle_{Ts}=\langle v_{CSm}(t)\rangle_{Ts} = \frac{1}{T_s}\sum_{i=0}^{11}\int_{t_i}^{t_{i+1}} v_{CSm}(\tau)\, d\tau$ and the average function of i_{Do}, i.e., $\langle i_{Do}\rangle_{Ts} = \langle i_{Do}(t)\rangle_{Ts}= \frac{1}{T_s}\sum_{i=0}^{11}\int_{t_i}^{t_{i+1}} i_{Do}(\tau)\, d\tau$ and the solution v_{CSm} and i_{Do} at each state can be obtained from the derivation in Section 2, as follows:

$$\begin{aligned}\langle v_{CSm}\rangle_{T_s} &= \frac{1}{T_s}\sum_{i=0}^{11}\int_{t_i}^{t_{i+1}} v_{CSm}(\tau)\, d\tau \\ &= \frac{1}{T_s}\Big\{v_oT_1 + \Big[\frac{C}{C_{Sm}}\frac{1}{\omega_1}(v_o - V_{Cr1})\sin\omega_1 T_2 - \frac{C}{C_{Sm}}(v_o - V_{Cr1})T_2 + V_{Cr1}T_2\Big] \\ &\quad + \Big[-\frac{C}{C_{Sm}}\frac{1}{\omega_1}V_{Cr9}\sin\omega_1 T_{10} + \frac{C}{C_{Sm}}V_{Cr9}T_{10}\Big] \\ &\quad + \Big[v_oT_{11} + \frac{1}{\omega_{10}}(V_{CSm10} - v_o)\sin\omega_{10}T_{11} + \frac{1}{2}Z_{10}i_{Lm}\omega_{10}T_{11}^2\Big] \\ &\quad + v_o[d'T_s - (T_{10} + T_{11})]\Big\}\end{aligned} \tag{27}$$

$$\begin{aligned}\langle i_{Do}\rangle_{T_s} &= \frac{1}{T_s}\sum_{i=0}^{11}\int_{t_i}^{t_{i+1}} i_{Do}(\tau)\, d\tau \\ &= \frac{1}{T_s}\Big[\int_{t_0}^{t_1}\frac{v_o}{R_L}\, d\tau + \int_{t_{10}}^{t_{11}}\frac{v_o}{R_L}\, d\tau + \int_{t_{11}}^{t_0} i_{Lm}\, d\tau\Big] \\ &= \frac{1}{T_s}\Big\{\frac{v_o}{R_L}T_1 + \frac{v_o}{R_L}T_{11} + i_{Lm}[d'T_s - (T_{10} + T_{11})]\Big\}\end{aligned} \tag{28}$$

Equations (27) and (28) can be expressed as

$$\begin{cases}\langle v_{CSm}\rangle_{T_s} = f_{vCSm}(v_o, i_{Lm}, d') = v_od' + f_{vCSm}(v_o, i_{Lm}) \\ \langle i_{Do}\rangle_{T_s} = f_{iDo}(v_o, i_{Lm}, d') = i_{Lm}d' + f_{iDo}(v_o, i_{Lm})\end{cases} \tag{29}$$

where

$$\begin{cases}\langle v_{CSm}\rangle_{T_s} = V_{CSm} + \tilde{v}_{CSm},\ |\tilde{v}_{CSm}| << V_{CSm} \\ \langle i_{Do}\rangle_{T_s} = I_{Do} + \tilde{i}_{Do},\ \left|\tilde{i}_{Do}\right| << I_{Do}\end{cases} \tag{30}$$

Equations (24) and (25) can be rewritten according to Equation (30) as

$$\begin{bmatrix}\frac{di_{Lm}}{dt} \\ \frac{dv_{Co}}{dt}\end{bmatrix} = \begin{bmatrix}-\frac{R_{Lm}}{L_m} & 0 \\ 0 & -\frac{1}{C_oR_L}\end{bmatrix}\begin{bmatrix}i_{Lm} \\ v_{Co}\end{bmatrix} + \begin{bmatrix}\frac{1}{L_m} \\ 0\end{bmatrix}v_i + \begin{bmatrix}-\frac{1}{L_m} & 0 \\ 0 & \frac{1}{C_o}\end{bmatrix}\begin{bmatrix}V_{CSm} + \tilde{v}_{CSm} \\ I_{Do} + \tilde{i}_{Do}\end{bmatrix} \tag{31}$$

$$v_o(t) = \begin{bmatrix}0 & 1\end{bmatrix}\begin{bmatrix}i_{Lm} \\ v_{Co}\end{bmatrix} \tag{32}$$

Since there are nonlinear terms in (29), differential equations based on the averaging mode are nonlinear. To obtain linear equations, Taylor series expansions at the DC operating point of the converter must be performed to remove nonlinear terms, and so that

$$\begin{cases}\langle v_{CSm}\rangle_{T_s} = (V_o + \tilde{v}_o)(D' - \tilde{d}) + \frac{\partial f_{vCSm}(v_o, i_{Lm})}{\partial v_o}\tilde{v}_o + \frac{\partial f_{vCSm}(v_o, i_{Lm})}{\partial i_{Lm}}\tilde{i}_{Lm} \\ \langle i_{Do}\rangle_{T_s} = (I_{Lm} + \tilde{i}_{Lm})(D' - \tilde{d}) + \frac{\partial f_{iDo}(v_o, i_{Lm})}{\partial v_o}\tilde{v}_o + \frac{\partial f_{iDo}(v_o, i_{Lm})}{\partial i_{Lm}}\tilde{i}_{Lm}\end{cases} \tag{33}$$

where the small-signal AC equations are

$$\begin{cases}\tilde{v}_{CSm} = \Big[D' + \frac{\partial f_{vCSm}(v_o, i_{Lm})}{\partial v_o}\Big]\tilde{v}_o + \frac{\partial f_{vCSm}(v_o, i_{Lm})}{\partial i_{Lm}}\tilde{i}_{Lm} - V_o\tilde{d} \\ \tilde{i}_{Do} = \frac{\partial f_{iDo}(v_o, i_{Lm})}{\partial v_o}\tilde{v}_o + \Big[D' + \frac{\partial f_{iDo}(v_o, i_{Lm})}{\partial i_{Lm}}\Big]\tilde{i}_{Lm} - I_{Lm}\tilde{d}\end{cases} \tag{34}$$

At the quiescent DC operating point, applying (34) and small-signal AC disturbances shown in (35) to (31) and (32) can obtain (36) and (37), as follows:

$$\begin{cases} v_i = V_i + \tilde{v}_i \\ v_o = V_o + \tilde{v}_o \\ i_{Lm} = I_{Lm} + \tilde{i}_{Lm} \\ d = D + \tilde{d} \\ d' = D' - \tilde{d}' \end{cases} \tag{35}$$

$$\begin{bmatrix} \frac{d\tilde{i}_{Lm}}{dt} \\ \frac{d\tilde{v}_{Co}}{dt} \end{bmatrix} = \begin{bmatrix} \frac{-R_{Lm}}{L_m} & \frac{-D'}{L_m} \\ \frac{D'}{C_o} & \frac{-1}{R_L C_o} \end{bmatrix} \begin{bmatrix} \tilde{i}_{Lm} \\ \tilde{v}_o \end{bmatrix} + \begin{bmatrix} \frac{1}{L_m} & \frac{V_o}{L_m} \\ 0 & \frac{-I_{Lm}}{C_o} \end{bmatrix} \begin{bmatrix} \tilde{v}_i \\ \tilde{d} \end{bmatrix} + \begin{bmatrix} \frac{\frac{-\partial f_{vCSm}(v_o,i_{Lm})}{\partial i_{Lm}}}{L_m} & \frac{\frac{-\partial f_{vCSm}(v_o,i_{Lm})}{\partial v_o}}{L_m} \\ \frac{\frac{\partial f_{iDo}(v_o,i_{Lm})}{\partial i_{Lm}}}{C_o} & \frac{\frac{\partial f_{iDo}(v_o,i_{Lm})}{\partial v_o}}{C_o} \end{bmatrix} \begin{bmatrix} \tilde{i}_{Lm} \\ \tilde{v}_o \end{bmatrix} \tag{36}$$

$$\tilde{v}_o = \begin{bmatrix} 0 & 1 \end{bmatrix} \begin{bmatrix} \tilde{i}_{Lm} \\ \tilde{v}_{Co} \end{bmatrix} \tag{37}$$

After obtaining (36), the transfer function $G_{vg}(s)$ of input voltage $\tilde{v}_i$ to output voltage $\tilde{v}_o$ and the transfer function $G_{vd}(s)$ of duty cycle $\tilde{d}$ to output voltage $\tilde{v}_o$ can be obtained as follows:

$$G_{vg}(s) = \frac{\frac{D'}{L_m C_o}}{den(s)} + \frac{\frac{\frac{\partial f_{iDo}(v_o,i_{Lm})}{\partial i_{Lm}}}{L_m C_o}}{den(s)} \tag{38}$$

$$G_{vd}(s) = \frac{-\frac{I_{Lm}}{C_o}s + \frac{1}{L_m C_o}(-R_{Lm} I_{Lm} + D' V_o)}{den(s)} + \frac{\frac{1}{L_m C_o}\left[-\frac{\partial f_{vCSm}(v_o,i_{Lm})}{\partial i_{Lm}} I_{Lm} + \frac{\partial f_{iDo}(v_o,i_{Lm})}{\partial i_{Lm}} V_o\right]}{den(s)} \tag{39}$$

where the denominator is defined as

$$\begin{aligned} den(s) = \; & s^2 + \left(\frac{1}{R_L C_o} + \frac{R_{Lm}}{L_m}\right)s + \frac{1}{R_L L_m C_o}(R_{Lm} + D'^2 R_L) + \left[\frac{\frac{\partial f_{vCSm}(v_o,i_{Lm})}{\partial i_{Lm}}}{L_m} - \frac{\frac{\partial f_{iDo}(v_o,i_{Lm})}{\partial v_o}}{C_o}\right]s \\ & + \frac{1}{R_L L_m C_o}\left[\frac{\partial f_{vCSm}(v_o,i_{Lm})}{\partial i_{Lm}} - \left(R_{Lm} + \frac{\partial f_{vCSm}(v_o,i_{Lm})}{\partial i_{Lm}}\right)\frac{\partial f_{iDo}(v_o,i_{Lm})}{\partial v_o} R_L \right. \\ & \left. + R_L D'\left(\frac{\partial f_{vCSm}(v_o,i_{Lm})}{\partial v_o} + \frac{\partial f_{iDo}(v_o,i_{Lm})}{\partial i_{Lm}}\right) + R_L \frac{\partial f_{vCSm}(v_o,i_{Lm})}{\partial v_o}\frac{\partial f_{iDo}(v_o,i_{Lm})}{\partial i_{Lm}}\right] \end{aligned} \tag{40}$$

Table 2 shows the system and component specifications.

Table 2. System and component specifications.

Name	Specification
System Operation Mode	Continuous Conduction Mode (CCM)
Rated Input Voltage V_i	156 V
Rated Output Voltage V_o	200 V
System Switching Frequency f_s/Period T_s	100 kHz/10 µs
Rated Output Current $I_{o,rated}$/Power $P_{o,rated}$	2 A/400 W
Minimum Output Current $I_{o,min}$/Power $P_{o,min}$	0.2 A/40 W
Input Inductance L_m/Output Capacitance C_o	1 m H/470 µF
Resonance Inductance L_r/Resonant Capacitor C_r	1 µH/22 nF
Power Switch S_m and S_a	STW20NM60FD
Output Diode D_o	BYV29X-600

By substituting the system and component specifications shown in Table 2 into Equations (38) and (39), the input-to-output transfer function $G_{vg}(s)$ can be found as

$$G_{vg}(s) = \frac{1.05 \times 10^6}{s^2 + 561.92s + 816.33 \times 10^3} \tag{41}$$

Furthermore, the duty-to- output transfer function is

$$G_{vd}(s) = \frac{-4.085 \times 10^3 s + 208.1 \times 10^6}{s^2 + 561.92s + 816.33 \times 10^3} \tag{42}$$

After finding the transfer function of the proposed structure with resonant small signals by the dual time scale averaging method, the corresponding transfer function is compared with the small-signal transfer function of the traditional boost converter, which can be expressed as

$$\widetilde{G}_{vd}(s) = \frac{-4.085 \times 10^3 s + 215 \times 10^6}{s^2 + 151s + 843 \times 10^3} \tag{43}$$

From (42) and (43), the difference in zero value between them is not significant. However, the pole values in (42) are $-280.96 \pm j858.72$, and the pole values in (43) are $-75.5 \pm j915.04$. From these pole values of both equations, we can see that the proposed structure has a larger bandwidth because the resonance parameters are taken into considerations. That is to say, if the resonance parameters are ignored in the controller design, the designed controller is not suitable for the proposed structure.

4. Controller Design in Z-Domain

As shown in Figure 5, the design of the digital controller proposed in this paper is designed directly in the z – domain by the pole-zero configuration and the pole-zero cancellation design method. Figure 5 shows the digital control loop with loop gain $L(z)$:

$$L(z) = K \cdot C(z) \cdot G_{vd}(z) \cdot H(z) \cdot z^{-1} \tag{44}$$

where $G_{vd}(z)$ is the discrete transfer function of the boost converter, $H(z)$ is the divider gain, K is the PWM gain, AD gain, z^{-1} is the delay factor, and Equation (45) is the discrete transfer function of this controller, where z_{p1} and z_{p2} are poles, z_{o1} and z_{o2} are zeroes, and K_{c2} is the gain.

$$C(z) = K_{c2} \frac{(z - z_{o1})(z - z_{o2})}{(z - z_{p1})(z - z_{p2})} \tag{45}$$

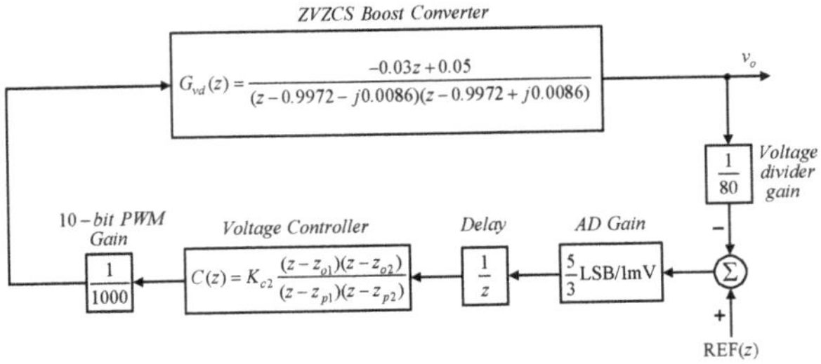

Figure 5. z-domain digital control loop.

When the controller is not added, the phase margin of the system known from the Bode plot of the loop gain is 14.7°. After designing the controller to make the system meet the prescribed specifications, the following steps will briefly describe the controller design.

Step 1: The gain margin is greater than 6 dB above and the crossover frequency is equal to one-tenth of the switching frequency.

Step 2: Configure poles z_{p1} and z_{p2} to match the gain margin and crossover frequency set by the system. Assuming a gain of $K_{c2} = 1$ and using the Matlab software assistant tool, named SISO, to configure and observe the two poles z_{p1} and z_{p2} several times, $z_{p1} = 0.178$ and $z_{p2} = 0.7$ are finally selected to meet the prescribed gain margin and switching frequency.

Step 3: After step 2, the crossover frequency is fixed and the system phase margin is adjusted. Finally, the phase margin is set to 60 degrees. The gained K_{c2} value, which is the value required to adjust the phase margin to 60 degrees, can be found by the Matlab syntax to obtain a gain K_{c2} value of 0.81.

$$\begin{aligned} [mag, phase, \omega] &= bode(Lz) \\ K_{c2} &= margin(mag, phase - 60\ ,\omega) \end{aligned} \tag{46}$$

After the above steps, the discrete transfer function of the controller can be obtained as follows along with $z_{o1} = 0.9972 + j0086$ and $z_{p2} = 0.9972 - j0086$:

$$C(z) = 0.81\frac{(z - 0.9972 - j0.0086)(z - 0.9972 + j0.0086)}{(z - 0.178)(z - 0.7)} \tag{47}$$

The discrete transfer function of the above equation is converted into a difference equation, so that the difference equation can be written into programming language for digital control of the system.

$$\begin{aligned} u(n) &= a_2u(n-2) + a_1u(n-1) + b_2e(n-2) + b_1e(n-1) + b_0e(n) \\ &= -0.12u(n-2) + 0.87u(n-1) + 0.8e(n-2) - 1.6e(n-1) + 0.8e(n) \end{aligned} \tag{48}$$

Figure 6 shows the Bode plot of the system loop gain to verify the correctness of the designed controller.

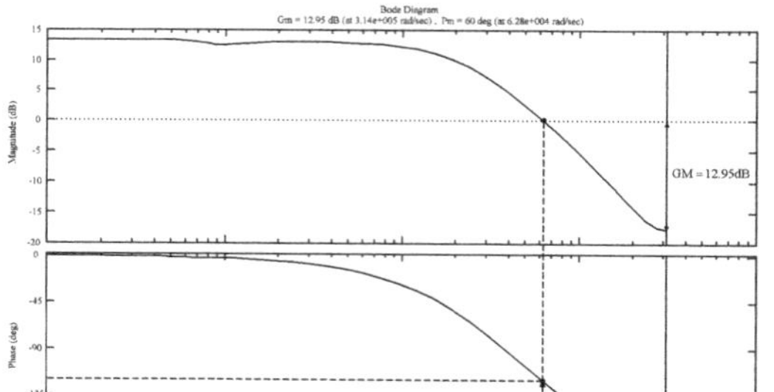

Figure 6. Bode plot of the system loop gain.

5. Auto-Adjustment Technique

In this paper, the auto-adjustment technique is implemented by using the lookup table to regulate the on-time and triggering instant of the auxiliary power switch to further improve the efficiency, especially at light loads, and to make the overall efficiency of the converter present a flat curve. In addition, when the load increases or decreases, since the efficiency does not vary regularly with the on-time of the preceding and following transients of the auxiliary power switch and the triggering instant relative to the main power switch, a look-up table is used instead of a complex calculation to determine the required on-time and triggering instant of the auxiliary power switch.

Figure 7a,b show the auto-adjustment technique for the auxiliary power switch operating at light load and rated load, respectively. This auto-adjustment technique is based on the following two conditions: first, the auxiliary power switch must reach zero current cutoff, as shown in Figure 7 at points A and B; second, when the auxiliary power switch current i_{Sa} resonates to equal to the input current I_{Lm}, the main power switch is turned on, as shown in Figure 7 at points C and D. Therefore, by using the above-mentioned technique and using the input inductance current as the self-variable of the lookup table, the design steps are as follows.

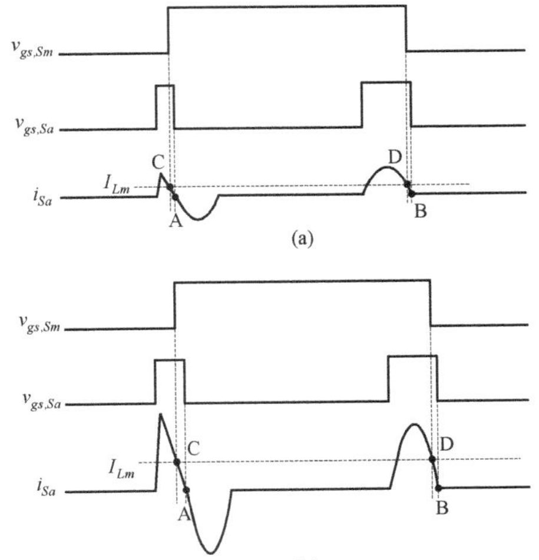

Figure 7. Auto-adjustment of turn-on time and triggering time of the auxiliary switch at (**a**) light load; (**b**) rated load.

Step 1: The light load to full load currents are divided into ten intervals and recorded separately based on no ADC sampling [20]. The sampled count value of the input inductance current is recorded.

Step 2: Within the set ten current intervals, the on-time before and after transients of the auxiliary power switch and the triggering instant relative to the main power switch are adjusted and recorded, respectively, from full load to light load.

Step 3: According to the above two steps, the turn-on time and triggering time of the auxiliary power switch can be determined by this prescribed look-up table.

It is worth mentioning that the addition of a hysteresis band is required to avoid the oscillation caused by the current interval change.

6. Design of Resonant Inductor and Resonant Capacitor

In this section, the resonant capacitor C_r and resonant inductor L_r are designed based on the results of states 1, 2 and 10 in Section 2. Without affecting the circuit operation behavior, it is determined that the overrun lead time of the auxiliary power switch must be less than or equal to one-tenth of the on time of the main power switch, i.e., less than or equal to 220 ns. Accordingly, the elapsed time T_1+T_2 shown in Figure 2 must be less than or equal to 220 ns, where the elapsed time T_1 is the time it takes for the current i_{Sa} of the auxiliary power switch S_a to rise to the input inductance current I_{Lm} after the auxiliary power switch is turned on, and the elapsed time T_2 is the time it takes for the parasitic capacitor C_{Sm} of the main power switch to discharge to zero. The rise time t_r and the fall time t_f in the power switch instruction manual correspond to the characteristics of the current flowing through the power switch, so we can know that the time $T_2 \cong 20$ ns. Therefore, the elapsed time T_1 is less than or equal to 200 ns [21], so the following equation can be found:

$$T_1 = \frac{C_r(V_{Cr1} - V_{Cr10})}{(I_{Lm} - I_o)} \leq 200 \text{ ns} \tag{49}$$

where V_{Cr1} can be obtained by state 2 as follows:

$$V_{Cr1} = V_o - \frac{L_r I_{Lm}}{T_2} \tag{50}$$

where I_{Lm} is the average value of the rating input inductance current, and V_{Cr10} can be obtained by state 10 as follows:

$$V_{Cr10} = \frac{I_{Lm} T_{10}}{C_r + C_{Sm}}(1 + \frac{C_{Sm}}{C_r}) + \frac{L_r I_{Lm}}{T_{10}} \tag{51}$$

where the elapsed time T_{10} can be known from the following equation:

$$\begin{aligned} \Delta Q &= C_r V_o = 0.5 I_{Lm} T_{10} \\ &= 550\text{p} \cdot 200 = 0.5(2.56) T_{10} \end{aligned} \tag{52}$$

Therefore, T_{10} can be found as 86 ns.

Sequentially, let the resonant frequency f_2 be greater than or equal to ten times the switching frequency, i.e., $\omega_2 \geq 2\pi \cdot 10^6$rad/sec, and substitute (47) and (48) into (49) to obtain

$$\begin{aligned} &\frac{C_r\left[V_o - \frac{L_r I_{Lm}}{T_2} - \frac{I_{Lm} T_2}{C_r + C_{Sm}}(1 + \frac{C_{Sm}}{C_r}) + \frac{L_r I_{Lm}}{T_2}\right]}{(I_{Lm} - I_o)} \leq 200\text{ ns} \\ \Rightarrow\ &\frac{C_r\left[200 - \frac{\frac{1}{4\pi^2\omega_2^2}(2.56)}{20\text{ns}} - \frac{(2.56)(85\text{ ns})}{C_r + 550\text{p}}(1 + \frac{550\text{p}}{C_r}) + \frac{\frac{1}{4\pi^2 f_2^2}(2.56)}{85\text{ns}}\right]}{(2.56 - 2)} \leq 200\text{ ns} \end{aligned} \tag{53}$$

The resonant capacitance C_r can be obtained as 21.45 nF from (53). The allowable error of the actual capacitance is considered, so the resonant capacitance C_r is chosen as 22 nF. After obtaining the resonant capacitance C_r, the resonant inductance L_r can be obtained as

$$L_r \leq \frac{1}{4\pi^2 f_2^2 C_r} = \frac{1}{4\pi^2 (10^6)^2 22\text{ n}} \tag{54}$$

Therefore, the value of resonance inductance can be obtained as $L_r \leq 1.15$ μH, so this paper selects the resonance inductance as 1 μH.

7. Simulated and Experimental Results

The system specifications are shown in Table 2 in Section 3. In this section, the proposed converter and control strategy will be verified by using Active-HDL combined with Matlab/Simulink as the simulation environment, and, finally, the effectiveness will be verified by using a real circuit. In this paper, the simulated controller parameters are simulated by the controller parameters designed in (50). The controller parameters are fine-tuned to $a_2 = -0.12$, $a_1 = 0.88$, $b_2 = 1.2$, $b_1 = 1$ and $b_0 = 1$. However, the parasitic components of the line must be considered in the implementation of this circuit. Consequently, the controller parameters are fine-tuned to $a_2 = -0.14$, $a_1 = 0.88$, $b_2 = 1$, $b_1 = 0.88$ and $b_0 = 1$.

7.1. Simulated Waveforms

The purpose of the software simulation is used to demonstrate the feasibility of the proposed converter. Figure 8 shows the system simulation block diagram, which includes the proposed structure, the digital control block generated by the Active-HDL software, the sampling circuit module, the peripheral auxiliary power supply, and the oscilloscope. Furthermore, the simulation environment is based on the 2021a MATLAB and Simulink software with the system specifications shown in Table 2. In addition, the power switches used herein are non-ideal, but the passive components are all ideal. Furthermore, the circuit is simulated at rated load. The value of C_{Sm} based on [21] is set at 550 pF, which is called equivalent capacitance affected by the value of v_{CSm}.

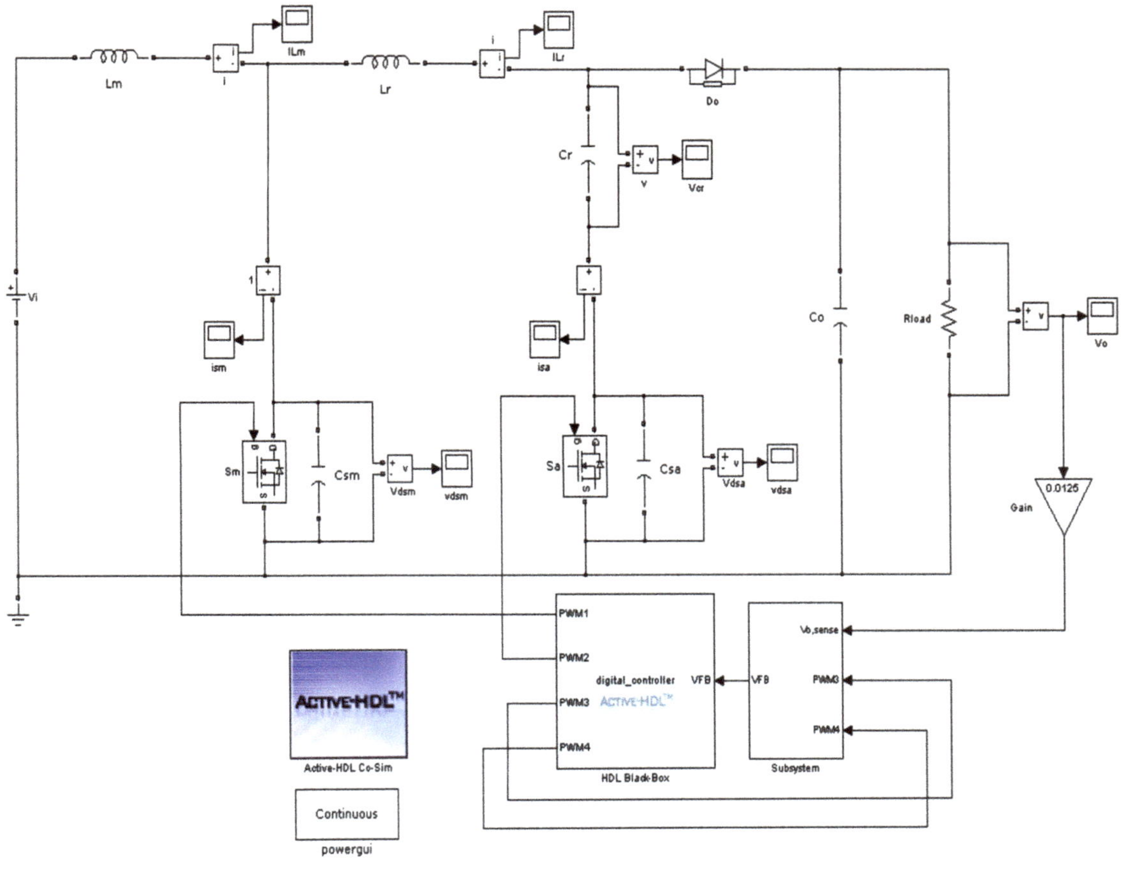

Figure 8. System simulation block diagram.

Figure 9 shows the gate driving signal $v_{gs,Sm}$ for the main power switch S_m and the gate driving signal $v_{gs,Sa}$ for the auxiliary power switch S_a, where the gate driving signal $v_{gs,Sa}$ possesses the pre- and post-duty cycle of the gate driving signal $v_{gs,Sm}$.

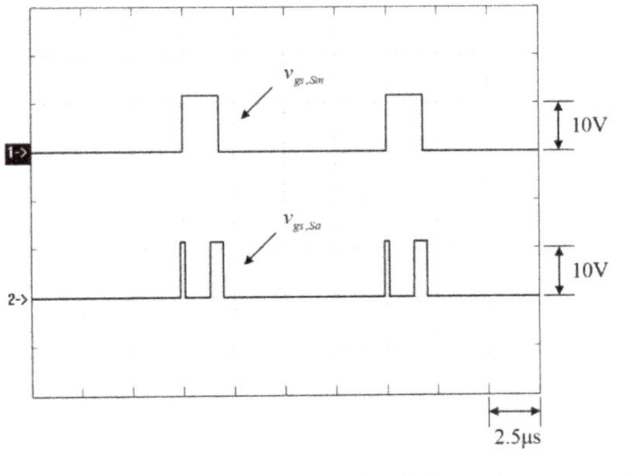

Figure 9. Simulated gate driving signals for main power switch S_m and auxiliary power switch S_a.

Figure 10 shows the waveforms of the main power switch, where $v_{gs,Sm}$ is the gate driving signal, $v_{ds,Sm}$ is the voltage across S_m, i_{Sm} is the current in the switch S_m. Figures 11 and 12 are zoomed-in waveforms of Figure 10.

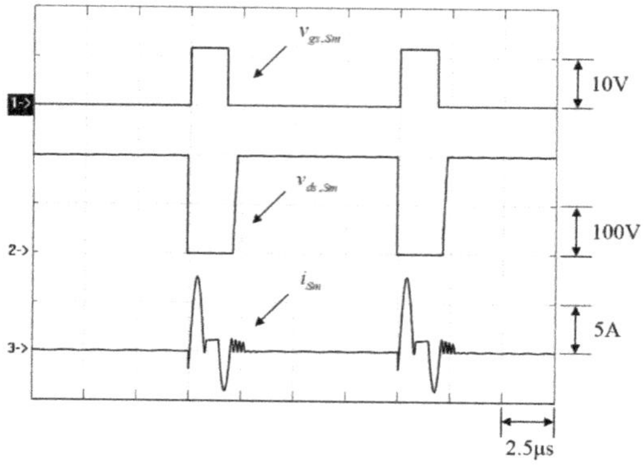

Figure 10. Simulated waveforms related to the main power switch S_m.

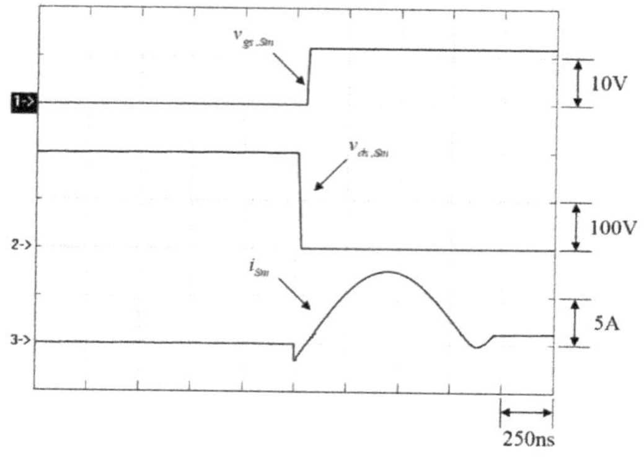

Figure 11. Simulated zoomed-in transient waveforms before main power switch S_m.

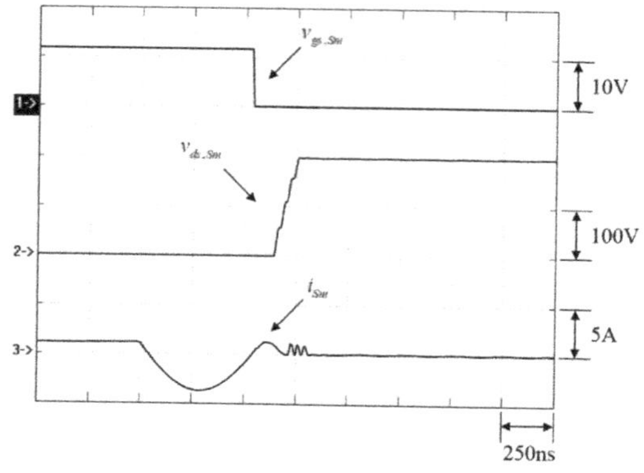

Figure 12. Simulated zoomed-in transient waveforms after the main power switch S_m.

Figure 13 shows the waveforms of the auxiliary power switch S_a, where $v_{gs,Sa}$ is the gate driving signal, $v_{ds,Sa}$ is the voltage on S_a and i_{Sa} is the current in S_a. Figures 14 and 15 are zoomed-in waveforms of Figure 13.

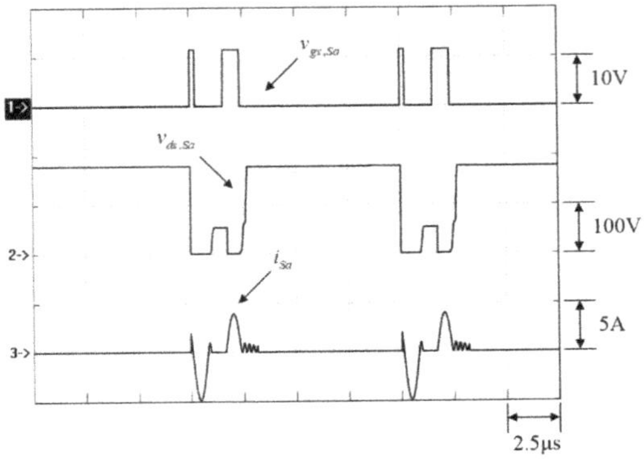

Figure 13. Simulated waveforms related to the auxiliary power switch S_a.

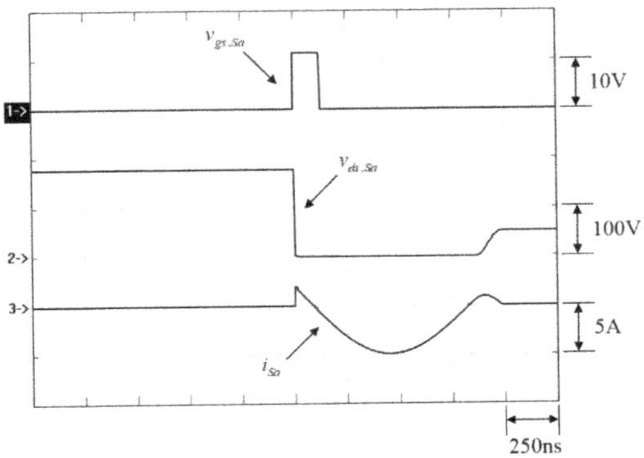

Figure 14. Simulated zoomed-in transient waveforms before the auxiliary power switch S_a.

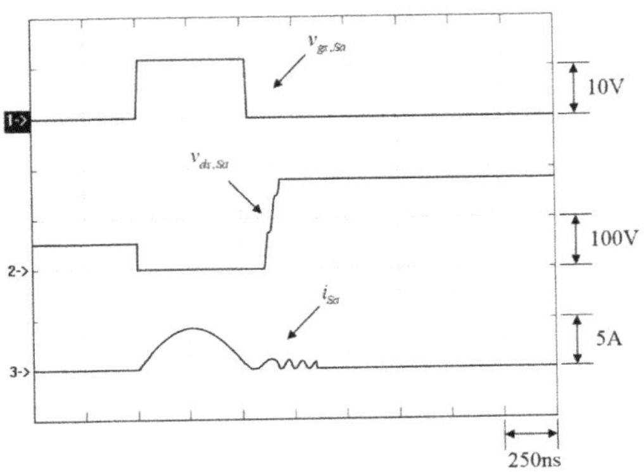

Figure 15. Simulated zoomed-in transient waveforms after the auxiliary power switch S_a.

Figure 16 shows the associated waveforms of the resonant elements, where i_{Lr} is the current in the resonant inductor L_r and v_{Cr} is the voltage across the resonant capacitor C_r.

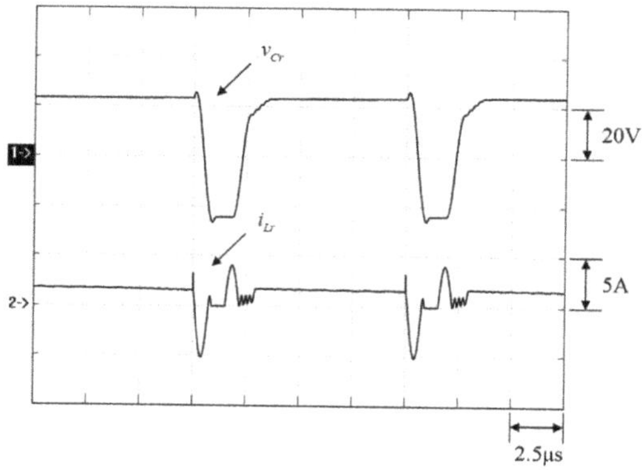

Figure 16. Simulated resonant waveforms.

From the above simulated results, it can be seen that when the auxiliary power switch S_a is turned on before the main power switch S_m, the auxiliary power switch current i_{Sa} rises rapidly to a level greater than the input inductance current I_{Lm}. According to Kirchhoff's current law, the parasitic capacitor C_{Sm} of the main power switch starts to discharge the resonant circuit until it discharges to zero, then the body diode of the main power switch D_{Sm} turns on, and the voltage $v_{ds,Sm}$ of the main power switch is clamped to zero, and after this, the main power switch S_m is turned on with ZVT, as shown in Figures 10 and 11.

The current waveforms in Figures 10, 12, 13, 15 and 16 show that when the auxiliary power switch S_a is cut off, the parasitic capacitance C_{Sa} of the auxiliary power switch, the resonant inductor L_r, the resonant capacitor C_r and the parasitic capacitor C_{Sm} of the main power switch form a resonance loop, resulting in a ringing phenomenon, so the corresponding currents will ring.

From Figures 11, 14 and 16, it can be seen that when the auxiliary power switch current i_{Sa} resonates to zero, the parasitic capacitor C_{Sa}, resonant inductor L_r and resonant capacitor C_r form a resonance loop, so a small resonant current is generated, and the corresponding small voltage across C_r will ring.

As can be seen from Figure 12, when the main power switch S_m is turned off, the ringing phenomenon on the main power switch current i_{Sm} causes the voltage $v_{ds,Sm}$ across the main power switch S_m to be charged slowly, so the voltage v_{CSm} rises slowly and the resulting overshoot current becomes small. In addition, the auxiliary power switch S_a is turned on after the main power switch S_m. When the auxiliary power switch current i_{Sa} resonates to the input inductance current I_{Lm}, according to Kirschhoff's current law, the main power switch S_m is turned off at this time, so that the main power switch S_m is turned off with ZCT.

As can be seen from Figures 13 and 15, when the auxiliary power switch current i_{Sa} drops to zero, the auxiliary power switch S_a is turned off with ZCS.

7.2. Experimental Waveforms

The purpose of the experimental measurement is used to verify the effectiveness of the proposed converter. In addition, the circuit is experimented at rated load.

Figure 17 shows the waveforms of the gate driving signal $v_{gs,Sm}$ of the main power switch S_m and the gate driving signal $v_{gs,Sa}$ of the auxiliary power switch S_a at rated load, where the gate driving signal $v_{gs,Sa}$ possesses the pre- and post-duty cycle of the gate driving signal $v_{gs,Sm}$.

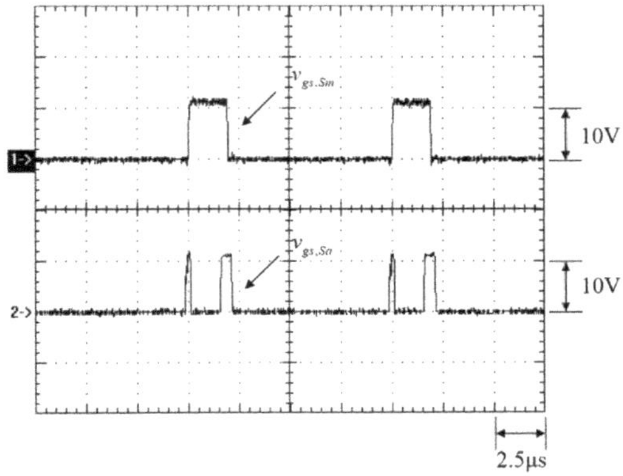

Figure 17. Measured gate driving signals of the main power switch S_m and auxiliary power switch S_a.

Figure 18 shows the waveforms of the main power switch S_m at rated load, where $v_{gs,Sm}$ is the gate driving signal, $v_{ds,Sm}$ is the voltage on S_m and i_{Sm} is the current in S_m. Figures 19 and 20 are the zoomed-in versions of Figure 18.

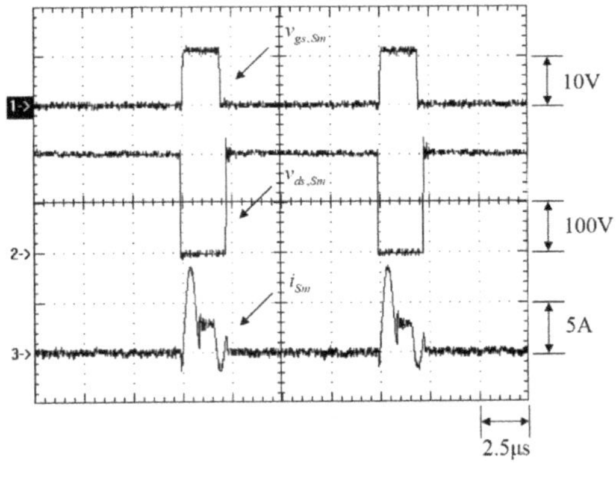

Figure 18. Measured waveforms related to the main power switch S_m.

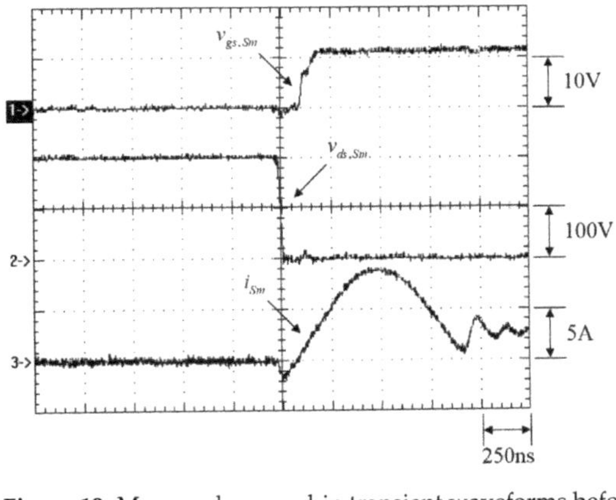

Figure 19. Measured zoomed-in transient waveforms before the auxiliary power switch S_m.

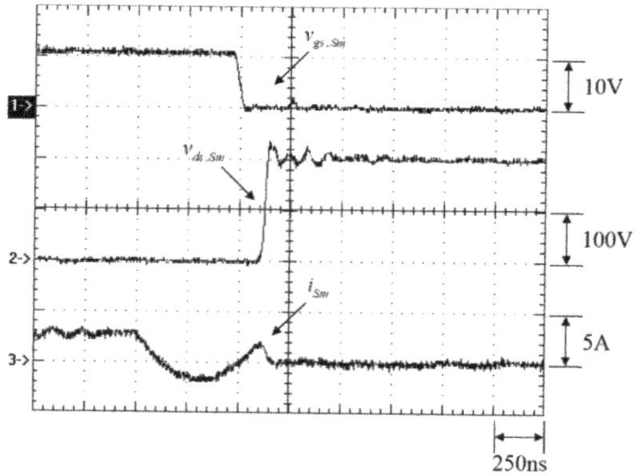

Figure 20. Measured zoomed-in transient waveforms before the auxiliary power switch S_m.

Figure 21 shows the waveforms of the auxiliary power switch S_a at rated load, where $v_{gs,Sa}$ is the gate driving signal, $v_{ds,Sa}$ is the voltage on S_a and i_{Sa} is the current in S_a. Figures 22 and 23 are the zoomed-in versions of Figure 21.

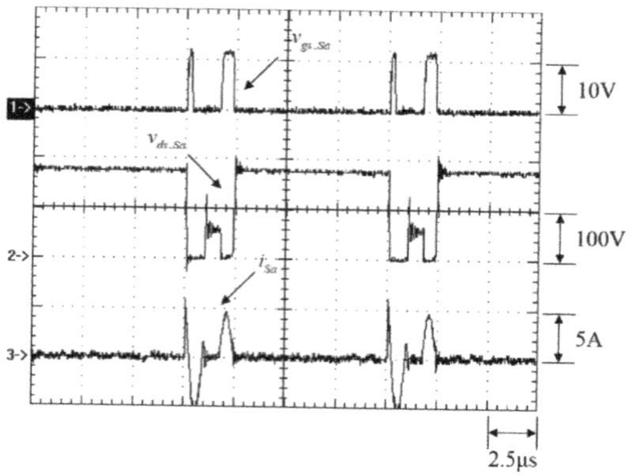

Figure 21. Measured waveforms related to the main power switch S_a.

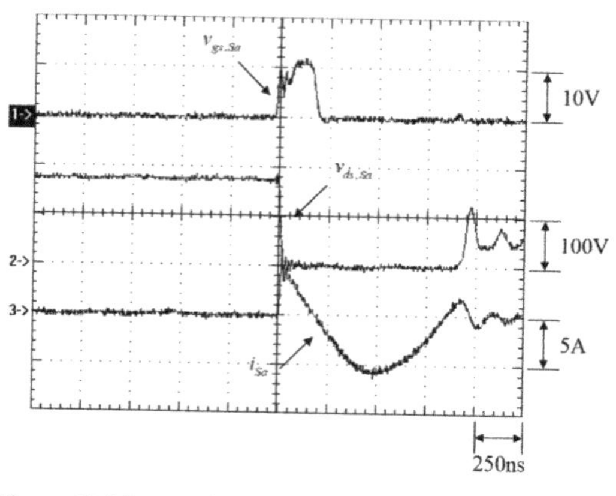

Figure 22. Measured zoomed-in transient waveforms before the auxiliary power switch S_a.

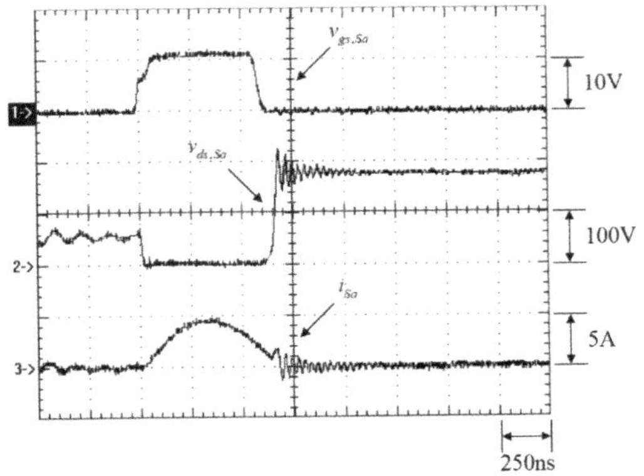

Figure 23. Measured zoomed-in transient waveforms after the auxiliary power switch S_a.

Figure 24 shows the waveforms of the resonant element under the rated load, where i_{Lr} is the current in the resonant inductor L_r and v_{Cr} is the voltage on the resonant capacitor C_r.

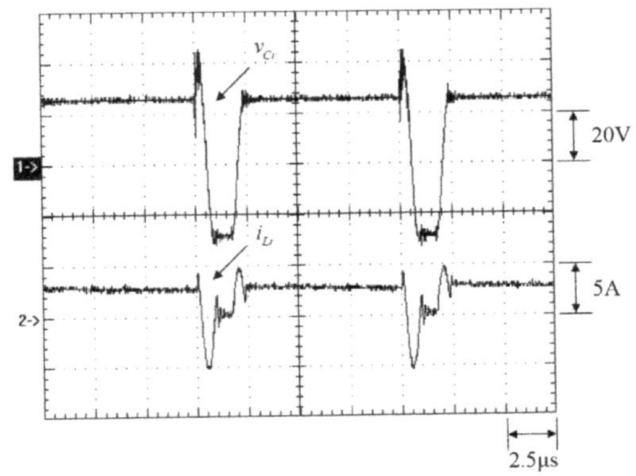

Figure 24. Measured resonant waveforms.

Figure 25 shows the efficiencies of the proposed structure under different output powers with soft switching and auto-adjustment technique, the proposed structure with soft switching and no auto-adjustment technique, and the traditional structure with hard switching.

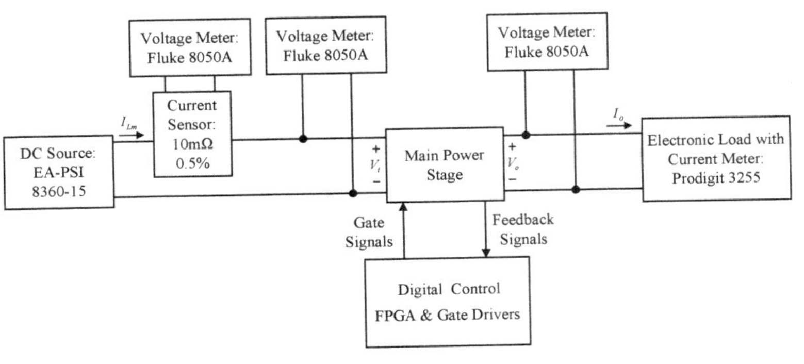

Figure 25. Efficiency measurement block diagram.

7.3. Efficiency Measurement

As shown in Figure 25, a current-sensing resistor is connected in series with the input current path, and a digital meter (Fluke 8050A, manufactured by FLUKE Co., Everett, Washington, USA) is used to measure the voltage across this resistor to obtain the input current value and the input voltage using a digital meter to obtain the input power. On the output side, an electronic load (Prodigit 3255, manufactured by PRODIGIT Co., Taipei, Taiwan) is used to provide the load current required by the converter, and the output voltage is measured using a digital meter to obtain the output power. Finally, the input power and output power are used to calculate the efficiency of the actual circuit operation. In Figure 26, curves of efficiency versus output power under soft switching with and without auto-adjustment and hard switching are displayed.

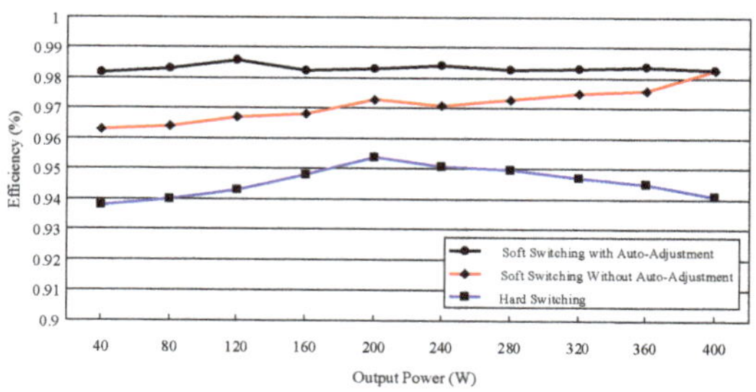

Figure 26. Curves of efficiency versus output power under soft switching with and without auto-adjustment, and hard switching.

7.4. Experimental Setup

As shown in Figure 27, the photo of the experimental setup is displayed.

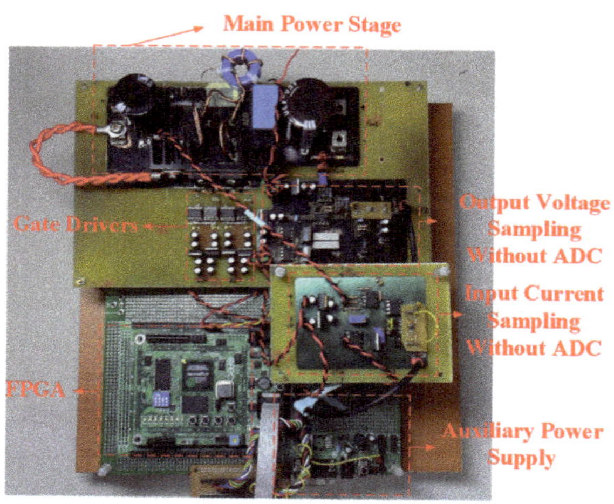

Figure 27. Photo of the experimental setup.

From the above experimental results, when the auxiliary power switch S_a precedes the main power switch S_m, the auxiliary power switch current i_{Sa} rises rapidly to greater than the input inductance current I_{Lm}; according to Kirchhoff's current law, the parasitic capacitor C_{Sm} of the main power switch starts to discharge the resonant circuit until the discharge reaches zero. Therefore, the body diode D_{Sm} of the main power switch turns on and the voltage $v_{ds,Sm}$ of the main power switch is clamped to zero, so the main power switch S_m is turned on with ZVT, as shown in Figures 17 and 19.

The current waveforms in Figures 19, 22 and 24 show that the resonant inductor current i_{Lr} flows in the opposite direction when operating in mode 6, and when the resonant inductor current i_{Lr} resonates to zero, the body diode D_{Sa} of the auxiliary power switch turns from on to off, generating the reverse recovery current, and this current flows through the parasitic capacitor C_{Sa} of the auxiliary power switch, the resonant inductor L_r, the resonant capacitor C_r, the parasitic capacitor C_{Sm} of the main power switch and the parasitic inductor of the line, causing the ringing phenomenon. Since all belong to the same resonant tank, so this resonance will be reflected in the currents as well as also being reflected in the voltage $v_{ds,Sa}$ of the auxiliary power switch.

As can be seen from Figure 20, there is a delay in the voltage $v_{ds,Sm}$ across the main power switch. The reason for this can be seen from state 9, when the main power switch S_m is turned off from on, because the power switch will delay the cut-off, so the main power switch current i_{Sm} continues to resonate up, and when it rises to the input current, the voltage $v_{ds,Sm}$ across the main power switch is larger than the output voltage, the resonant inductor L_r begins to be magnetized. According to the Kirchhoff's current law, the main power switch current i_{Sm} begins to fall to zero, so the main power switch S_m has ZCT turn-off.

From Figure 23, it can be seen that when the auxiliary power switch S_a is turned off, the voltage $v_{ds,Sa}$ across the auxiliary power switch and the current i_{Sa} have the ringing phenomenon. The reason is that the output diode D_o is delayed when changing from state 10 to state 11. Therefore, the resonant inductor current i_{Lr} will flow through the parasitic capacitor C_{Sa} of the auxiliary power switch, forming a resonant circuit and hence causing the auxiliary power switch to have ZCS turn-off.

From Figure 23, it can be seen that when the auxiliary power switch S_a is turned on, the rapid rise of the auxiliary power switch current i_{Sa} and the effect of the parasitic inductance of the line cause noise to be generated across the resonant capacitor C_r, but this noise does not affect the circuit operation behavior. In addition, from Figure 23, the voltage v_{Cr} across the resonant capacitor C_r also has a ringing at the negative voltage due to the ringing on the resonant inductor current i_{Lr}.

From Figure 26, it can be seen that under the input voltage of 156 V and the output voltage of 200 V as shown in Table 2, the proposed structure with/without auto-adjustment can effectively improve the overall efficiency of the converter in the output power range (40 W $\sim$ 400 W). The difference in efficiency between soft switching with auto-adjustment and hard switching is 4.8%, whereas the difference in efficiency between soft switching with auto-adjustment and soft switching without auto-adjustment is about 2.1%. Most importantly, the proposed structure with the addition of the auto-adjustment technique not only further improves the efficiency, but also makes the efficiency show an approximately horizontal curve.

8. Conclusions

From the experimental results, it can be seen that the main power switch has both ZVT and ZCT, and the auxiliary power switch also has ZCS. In addition, the lookup table is also added to realize the auto-adjustment technique to regulate the on-time and triggering position of the auxiliary power switch to further improve the efficiency, especially at light load, thus making the overall efficiency of the converter present a flat curve. From the experimental measurements, the maximum efficiency of the proposed converter is 98.5% and the light load efficiency is 98.1%, and the overall system efficiency is above 98%. Regarding the system control, the direct design of the digital controller under z-domain can effectively stabilize the output voltage of the converter and achieve the purpose of fully digitalized control.

Author Contributions: Conceptualization, K.-I.H., Z.-F.L. and P.-C.T.; methodology, K.-I.H. and P.-C.T.; software, Z.-F.L.; validation, Z.-F.L.; formal analysis, Z.-F.L.; investigation, Z.-F.L.; resources, K.-I.H.; data curation, P.-C.T.; writing—original draft preparation, K.-I.H.; writing—review and editing, K.-I.H.; visualization, Z.-F.L.; supervision, K.-I.H.; project administration, K.-I.H.; funding acquisition, K.-I.H. All authors have read and agreed to the published version of the manuscript.

Funding: This research was funded by the Ministry of Science and Technique, Taiwan, under the Grant Number: MOST 110-2221-E-027-045-MY2.

Data Availability Statement: No new data were created or analyzed in this study. Data sharing is not applicable to this article.

Conflicts of Interest: The authors declare no conflict of interest.

References

1. Tabisz, W.A.; Jovanic, M.M.; Lee, F.C. High-frequency multi-resonant converter technique and its applications. In Proceedings of the 1990 Fourth International Conference on Power Electronics and Variable-Speed Drives, London, UK, 17–19 July 1990; pp. 1–8.
2. Jabbari, M.; Farzanehfard, H. New resonant step-down/up converters. *IEEE Trans. Power Electron.* **2010**, *25*, 249–256. [CrossRef]
3. Wang, C.-M. New family of zero-current-switching PWM converters using a new zero-current-switching PWM auxiliary circuit. *IEEE Trans. Ind. Electron.* **2006**, *53*, 768–777. [CrossRef]
4. Shuai, P.; De Novaes, Y.R.; Canales, F.; Barbi, I. A non-insulated resonant boost converter. In Proceedings of the 2010 Twenty-Fifth Annual IEEE Applied Power Electronics Conference and Exposition (APEC), Palm Springs, CA, USA, 21–25 February 2010; pp. 550–556.
5. Huang, W.; Gao, X.; Bassan, S.; Moschopoulos, G. Novel dual auxiliary circuits for ZVT-PWM converters. *Can. J. Electr. Comput. Eng.* **2008**, *33*, 153–160. [CrossRef]
6. Adib, E.; Farzanehfard, H. Zero-voltage-transition PWM converters with synchronous rectifier. *IEEE Trans. Power Electron.* **2010**, *25*, 105–110. [CrossRef]
7. Yang, S.-P.; Lin, J.-L.; Chen, S.-J. A novel ZCZVT forward converter with synchronous rectification. *IEEE Trans. Power Electron.* **2006**, *21*, 912–922. [CrossRef]
8. Bodur, H.; Bakan, A.F. A new ZVT-ZCT-PWM DC-DC converter. *IEEE Trans. Power Electron.* **2010**, *19*, 676–684. [CrossRef]
9. Hwu, K.-I.; Shieh, J.-J.; Jiang, W.-Z. Interleaved boost converter with ZVT-ZCT for main switches and ZCS for auxiliary switch. *Appl. Sci.* **2020**, *10*, 2033. [CrossRef]
10. Yau, Y.-T.; Hwu, K.-I.; Jiang, W.-Z. Two-phase interleaved boost converter with ZVT turn-on for main switches and ZCS turn-off for auxiliary switches based on one resonant loop. *Appl. Sci.* **2020**, *10*, 3881. [CrossRef]
11. Yau, Y.-T.; Hwu, K.-I.; Shieh, J.-J. Simple structure of soft switching for boost converter. *Energies* **2020**, *13*, 5448. [CrossRef]
12. Akın, B. An Improved ZVT–ZCT PWM DC–DC boost converter with increased efficiency. *IEEE Trans. Power Electron.* **2014**, *29*, 1919–1926. [CrossRef]
13. Altintaş, N.; Bakan, A.F.; Aksoy, I. A novel ZVT-ZCT-PWM boost converter. *IEEE Trans. Power Electron.* **2014**, *29*, 256–265. [CrossRef]
14. Yi, J.H.; Choi, W.; Cho, B.H. Zero-voltage-transition interleaved boost converter with an auxiliary coupled inductor. *IEEE Trans. Power Electron.* **2017**, *32*, 5917–5930. [CrossRef]
15. Khorasani, R.R.; Adib, E.; Farzanehfard, H. ZVT resonant core reset forward converter with a simple auxiliary circuit. *IEEE Trans. Ind. Electron.* **2018**, *65*, 242–250. [CrossRef]
16. Xia, Y.; Ayyanar, R. Naturally adaptive, low-loss zero-voltage-transition circuit for high-frequency full-bridge inverters with hybrid PWM. *IEEE Trans. Power Electron.* **2018**, *33*, 4916–4933. [CrossRef]
17. Zhang, X.; Qian, W.; Li, Z. Design and analysis of a novel ZVZCT boost converter with coupling effect. *IEEE Trans. Power Electron.* **2017**, *32*, 8992–9000. [CrossRef]
18. Rezvanyvardom, M.; Mirzaei, A. Zero-voltage transition nonisolated bidirectional buck–boost DC–DC converter with coupled inductors. *IEEE J. Emerg. Sel. Top. Power Electron.* **2021**, *9*, 3266–3275. [CrossRef]
19. Sun, J.; Grotstollen, H. Averaged modelling of switching power converters: Reformulation and theoretical basis. In Proceedings of the 23rd Annual IEEE Power Electronics Specialists Conference, Toledo, Spain, 29 June–3 July 1992; pp. 1165–1172.
20. Hwu, K.I.; Yau, Y.T. Applying an FPGA-counter-based control strategy to multiple SR outputs of a forward converter. In Proceedings of the 2006 37th IEEE Power Electronics Specialists Conference, Jeju, Republic of Korea, 18–22 June 2006; pp. 1–4.
21. Available online: https://pdf1.alldatasheet.com/datasheet-pdf/view/158402/STMICROELECTRONICS/STB11NM60FD.html (accessed on 12 April 2021).

energies

MDPI

Article

Dynamic Analysis of a Supercapacitor DC-Link in Photovoltaic Conversion Applications

Fabio Corti [1], Antonino Laudani [2,*], Gabriele Maria Lozito [1,3], Martina Palermo [2], Michele Quercio [2], Francesco Pattini [3] and Stefano Rampino [3]

[1] Dipartimento di Ingegneria dell'Informazione, Università degli Studi di Firenze, 50139 Florence, Italy; fabio.corti@unifi.it (F.C.); gabrielemaria.lozito@unifi.it (G.M.L.)
[2] Dipartimento di Ingegneria Industriale, Elettronica e Meccanica, Università degli Studi Roma Tre, 00146 Rome, Italy; martina.palermo@uniroma3.it (M.P.); michele.quercio@uniroma3.it (M.Q.)
[3] Institute of Materials for Electronic and Magnetism-National Research Council (IMEM-CNR), delle Scienze 37/A, 43124 Parma, Italy; francesco.pattini@cnr.it (F.P.); stefano.rampino@cnr.it (S.R.)
* Correspondence: antonino.laudani@uniroma3.it

Abstract: In this work, a dynamic analysis describing the charge and discharge process of a supercapacitor for the DC-link between a photovoltaic source and a constant power load is presented. The analysis results in a complete nonlinear and dynamic model that can be used for simulation and control for DC–DC converters, achieving fast recharge times and accurate steady-state voltages in the DC link to avoid overcharging the supercapacitor during low power absorption scenarios. The proposed approach includes parasitic elements for the supercapacitor and efficiency effects on the conversion stage, proposing equations useful for design and control. Stability is also discussed for the charge process of the supercapacitor. Validation of the analytical model is performed by comparison with LTSpice simulation, confirming a good agreement between theory and simulation.

Keywords: photovoltaics; Supercapacitors; DC–DC; dynamic systems; circuit simulation

Citation: Corti, F.; Laudani, A.; Lozito, G.M.; Palermo, M.; Quercio, M.; Pattini, F.; Rampino, S. Dynamic Analysis of a Supercapacitor DC-Link in Photovoltaic Conversion Applications. *Energies* **2023**, *16*, 5864. https://doi.org/10.3390/en16165864

Academic Editor: Miguel Castilla

Received: 15 July 2023
Revised: 30 July 2023
Accepted: 5 August 2023
Published: 8 August 2023

1. Introduction

The power conversion of PV sources for DC and AC applications is almost always required to ensure that their strongly variable voltage and current are regulated in accordance with the electrical and electronic components for which the energy is generated [1–3]. This conversion makes use of DC–DC converters that, in general, adapt the voltage for the next stage of conversion (or the final load). This approach is also valid in the case of discrete PV + Storage hybrid systems, where the DC–DC converters are used for matching the output voltage of the solar cell with the charging voltage of the integrated storage device [4]. Since the PV source is a strongly non-linear device based on the photogeneration process [5], it is not only necessary to adapt the voltage to the load, but also to the source, to ensure that the operating point of the PV source is the one where the maximum power is delivered [6–9]. This process, commonly referred to as maximum power-point tracking (MPPT), introduces the necessity for a dedicated DC–DC converter, since a single converter does not have the degrees of freedom to adapt both input (for MPPT) and output (for correct load functioning) voltage. The two converters, usually connected in cascade, are interleaved with a storage element that has the purpose of stabilizing the voltage and decoupling the converter on the PV side from the converter on the load side. This stage is called the DC-link, and according to the applications, it can be implemented with batteries or capacitors. The former allows the DC-link stage to act as an energy-storage stage, which is particularly useful due to the intermittent nature of the PV source. However, a capacitor-based DC-link is much simpler, leaves less overall footprint, and is more durable [10–14].

The development of new technologies for high-capacity capacitors, known as Supercapacitors (SC), introduced the possibility of achieving a degree of energy-storage capabilities without resorting to a battery-based DC-link. Novel technologies of SC also include interesting results in terms of materials [15], including recycling materials [16] and natural and organic sources [17,18]. SC offers acceptable energy density (albeit still lower than Li-based batteries) with very high-power density. SCs are also less expensive and age much less quickly due to their charge and discharge patterns.

Unfortunately, the electrical behavior of a SC is different from a battery's, also due to the variable voltage across its terminal as dependent on the state of charge [19–21]. A battery exhibits an almost constant voltage across its terminals for the majority of its discharge (and charge) process. SC behavior is similar to a that of a classic capacitor, and thus, in the best of cases, discharges exponentially. Although the decoupling capabilities of the SC are ensured, the voltage range that it exhibits during operative conditions is large, and this requires special care in the control of the DC–DC converters, especially the one connecting the SC to the non-linear PV source.

The purpose of this work is to investigate the dynamics of the SC in an energy-conversion chain, starting from a PV source and ending with a constant-power load, which is a very common DC load able to support, for example, USB powered devices. To investigate the dynamic of the system, a circuit model of the chain, representing the conversion at frequencies below the ones of the dynamic response for the converter but including the ones regulating the charge and discharge processes of the SC, is proposed [22,23]. The model includes the non-linear nature of the PV source, the efficiency of the converter distributed among voltage and current drops, and both ideal and parasitic models for the SC.

With respect to current research, the novelty of this work contribution is twofold. First, an analytical complete dynamic model including parasitic components is presented for a very actual scenario of low–mid power generation from renewable sources. The investigation of this model highlights novel results such as the feasible operating ranges for the charge operation of the DC-link supercapacitor, determining stable and unstable equilibrium points of the dynamic system. Second, a numerical approach to estimate the duty cycle for the final-voltage charge is presented, considering the full non-linear nature of the photovoltaic source. Moreover, both results are validated against LTSpice time-domain simulations, considering both the low-frequency DC model of the power conversion stage and its fully dynamic high-frequency implementation. All LTSpice models are reported in their completeness for repeatability of the proposed study. The analytical and numerical analysis presented can be used as a foundation for advanced control algorithms and energy-storage management systems.

The manuscript is structured as follows. First, the energy conversion chain will be described, and individual circuit models will be proposed for the block constituting it. Then, according to the parasitic model considered in the SC, three different state–space models for three different chains are derived. Then, feasibility regions for the final voltage charge are discussed, and the numerical problem to derive the duty-cycle is formalized and solved. Following, the validation procedure in the LTSpice environment is presented along with the simulation results. Final remarks and a conclusion close the manuscript.

2. Materials and Methods

2.1. Modelling for the Blocks of the Conversion Chain

The energy-conversion chain considered in this work is composed of five blocks. The first block is the primary source of energy, represented by a generic PV device under variable conditions of irradiance and temperature. The second block is the first DC–DC converter, operating to regulate the energy conversion between the PV source and the following DC-link supercapacitor. The third block is the supercapacitor itself, which features a capacitance and operating voltage range according to the design of the remainder of the system. The fourth block is the second DC–DC converter, operating to regulate the energy conversion between the DC link and the final load. The fifth block is the load itself, which is assumed as a CV&CP load (constant voltage and constant power). Each individual block of the conversion chain can be represented by a suitable circuit model that can later be coupled to achieve a full dynamic representation of the system.

For the first block, the equivalent circuit model known as "single diode" or "one diode" is used to represent the behavior of the photovoltaic source. The model is lightweight and can be used to represent a large variety of silicon-based PV sources and other technologies [24], and the process for the identification from experimental data or the produced datasheet values is well understood in the literature [25]. The model-circuit representation is given in Figure 1.

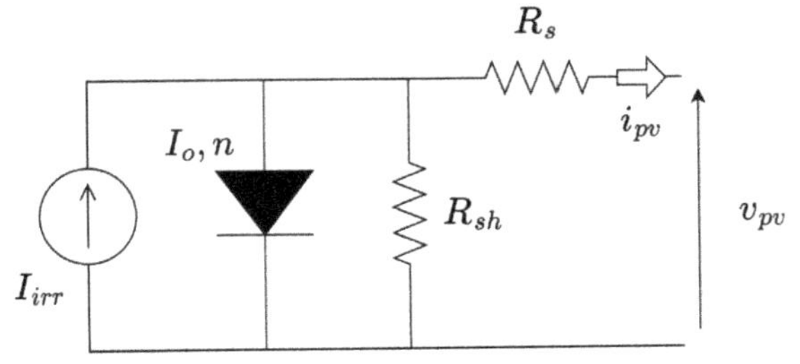

Figure 1. Single diode model for a PV device.

The model drawback is the implicit nature of its current-voltage relationship, which is given by (1):

$$i_{pv} = I_{irr} - I_o\left(e^{\frac{v_{pv}+R_s i_{pv}}{nV_T}} - 1\right) - \frac{v_{pv} + R_s i_{pv}}{R_{sh}} \tag{1}$$

where I_{irr} is the photogenerated current, I_o is the reverse saturation current of the diode, n is the modified quality factor of the diode, R_{sh} is the shunt resistance and R_s is the series resistance. The V_T term represent the thermal voltage ($\approx$26 mV at 293.15 K). The five parameters $\{I_{irr}, I_o, R_s, R_{sh}, n\}$ can be identified in specific irradiance G and temperature T conditions addressed in literature as Standard Test Conditions (STC). The conditions are $G = 1000\ \mathrm{W/m^2}$ and $T = 293.15$ K. Individual equations are reported in the literature to calculate the parameters in conditions different from STC.

The current-voltage relationship shown in (1) can be used in practical applications through numerical methods that calculate the current i_{pv} given a known voltage v_{pv}, or vice-versa. However, the numerical solution of (1) is not practical if, for example, the PV device should be included in Kirchhoff Voltage Laws (KVL) or Kirchhoff Current Laws (KCL) for formulating state-space equations. In this case, an alternative formulation can be used, which exploits the Lambert-W function to give an explicit expression of v_{pv} as a function of i_{pv} (2) and i_{pv} as a function of v_{pv} (3).

$$v_{pv} = R_{sh}(I_{irr} + I_o) - (R_s + R_{sh})i_{pv} - nV_T W\left(\frac{I_o R_{sh}}{nV_T} e^{\frac{R_{sh}(I_{irr}+I_o-i_{pv})}{nV_t}}\right) \tag{2}$$

$$i_{pv} = -\frac{nV_T}{R_s} W\left(\frac{R_s}{nV_T}\cdot\frac{I_o R_{sh}}{R_{sh}+R_s}\cdot e^{\frac{R_{sh}}{R_s+R_{sh}} * \frac{v_{pv}+R_s(I_{irr}+I_o)}{nV_t}}\right) + \frac{(I_{irr}+I_o)R_{sh} - v_{pv}}{R_{sh}+R_s} \tag{3}$$

The formulation with (2) and (3) are explicit and can be implemented easily in KVL and KCL. Indeed, the formulation is still numeric, but the computational burden is moved from the generic root-finding that is present in (1) to the solution of the Lambert-W, which can be optimized and is in general an easier task. Equations (2) and (3) provide a complete electrical characterization of the first block and can be completed with update equations for the parameters to account for variable irradiance and temperature conditions [26–28].

The second block is represented by the first DC–DC converter. The role of this converter is to set the operating point of the PV device to ensure proper current flowing inside the DC-link. In general, if the DC-link is implemented by means of a SC, its voltage should be low. For this reason, a step-down converter could be used to convert the voltage from the

PV source to the SC. However, this could limit the operating conditions of the PV device in the case of smaller systems. On the other hand, step-up-down topologies such as the SEPIC or Zeta converters can be implemented with very little effort and offer large voltage gains with high efficiency. The Zeta converter topology is the one considered for this block and is shown in its ideal lossless representation in Figure 2. The Zeta converter is a topology particularly suitable for photovoltaic (PV) applications due to several advantages. It can achieve a high voltage gain, making it suitable for boosting the low voltage output of PV panels to the desired level. This is especially useful when the PV array operates at low voltage levels, as it allows efficient energy conversion without the need for additional voltage-boosting stages. Unlike the Boost converter, which typically requires multiple stages for voltage conversion, the Zeta converter can perform the voltage step-up or step-down conversion in a single stage. This simplifies the converter topology, reduces the component count, and improves overall efficiency. Finally, the Zeta converter is capable of accommodating a wide input-voltage range, allowing it to handle the fluctuations in the PV panel output more effectively [29–32].

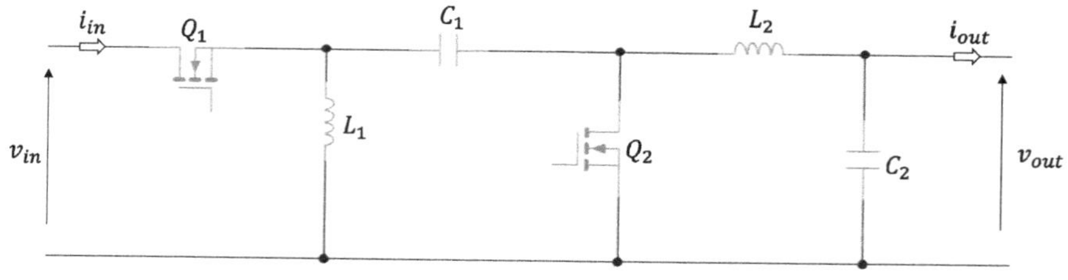

Figure 2. Lossless model for a step-up-down synchronous converter with Zeta topology.

The complete derivation and sizing for the converter can be found in the literature. Under the assumption of continuous-conduction mode (CCM), the converter-voltage gain and current gain in steady state can be approximated as:

$$M_v = \frac{v_{out}}{v_{in}} = \frac{D}{(1-D)} \tag{4}$$

$$M_i = \frac{1}{M_v} \tag{5}$$

where M_v is the lossless voltage gain, M_i is the lossless current gain, and D is the duty cycle for which the Q_1 transistor is in conduction and the Q_2 transistor is in interdiction. In case lossy elements are present in the converter, the efficiency is in general distributed between the voltage and current gain, since:

$$\eta = \frac{v_{out} \cdot i_{out}}{v_{in} \cdot i_{in}} = M_{v,l} \cdot M_{i,l} = \eta_v M_v \cdot \eta_i M_i \tag{6}$$

where η is the efficiency of the converter and $M_{v,l}$ and $M_{i,l}$ are the lossy voltage and current gain. The distribution of the efficiency η as the scaling terms η_v and η_i depends on the specific values of the parasitic components. Equations (4)–(6) yield a simple steady-state representation of the converter through two mutually-coupled controlled generators, as shown in Figure 3a.

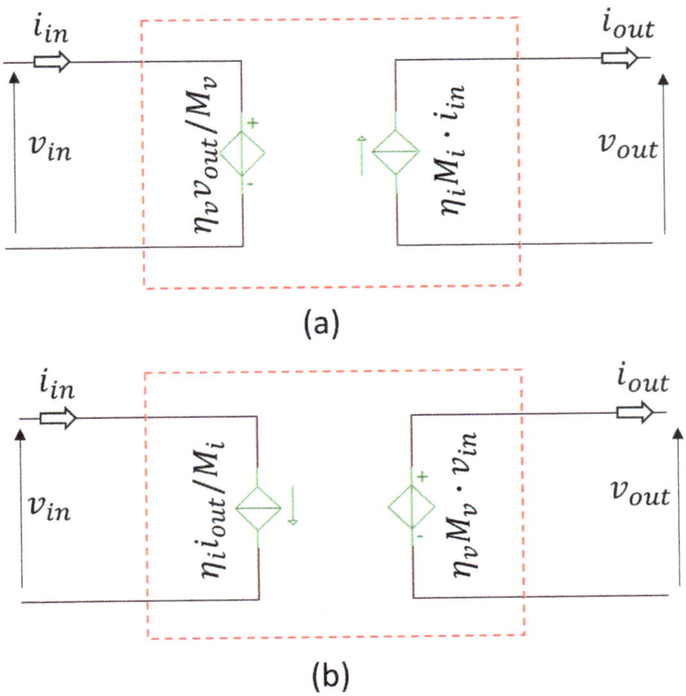

Figure 3. Two possible steady-state models for a lossy DC–DC converter, with imposed voltage at the primary side (**a**) and at the secondary side (**b**).

Indeed, a reciprocal representation with the current generator on the primary side and the voltage generator on the secondary side is possible as well, as shown in Figure 3b. Both representations are correct and allow the formulation of KVL/KCL. The choice of one over the other for formulation resides in the representation of the third block.

Under the reasonable assumption that the DC–DC converter operates to achieve Maximum Power Point Tracking (MPPT) on the PV device, the voltage gain will be set as:

$$\eta_v M_v = \frac{v_{out}}{v_{mp}};\quad \eta_v\left(\frac{1}{M_i}\right) = \frac{v_{out}}{v_{mp}};\quad M_i = \frac{v_{mp}}{v_{out}}\eta_v \tag{7}$$

where v_{mp} is the voltage where maximum power is exhibited by the PV source. Indeed, a reciprocal approach could be taken in case of imposing the current gain, albeit much less common. The third block is the DC-link, which is constituted by a SC. The SC can be represented by a capacitor with a capacitance C and a variable degree of parasitic elements. The simpler representation shown in Figure 4a is a simple, lossless capacitor. Inclusion of a parallel self-discharge resistor, r_{sh}, is shown in Figure 4b. Inclusion of an additional equivalent series resistance (ESR), r_s, is shown in Figure 4c, which constitutes the complete, lossy representation of the SC.

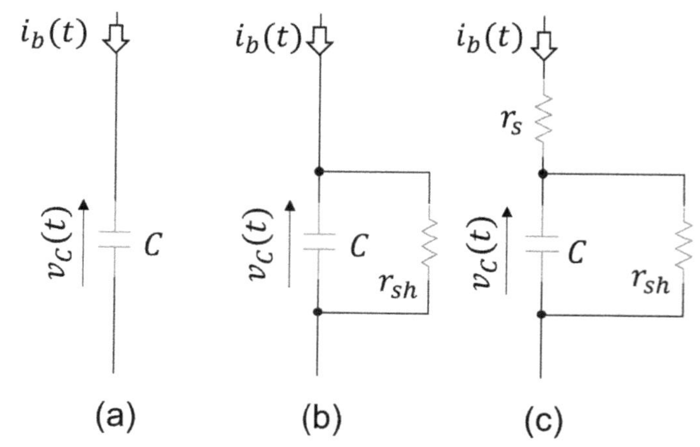

Figure 4. Three possible representations of a SC: (**a**) ideal; (**b**) ideal with self-discharge r_{sh} resistor; (**c**) with both self-discharge resistance and series-current limitation.

Where $i_b(t)$ is the branch current, the behaviour of the SC in the three represented equivalent circuits can be determined by the following state equations:

$$\dot{v}_c(t) = \frac{1}{C}(i_b(t)) \tag{8}$$

$$\dot{v}_c(t) = \frac{1}{C}\left(i_b(t) - \frac{v_c(t)}{r_{sh}}\right) \tag{9}$$

where (8) can be used for Figure 4a and (9) can be used for Figure 4b,c. The presence of the r_s element does not directly alter the state equation but influences the branch current according to how the element is connected to the rest of the circuit.

The fourth and fifth blocks operate in a coupled mode: since the load is expected to operate at a constant voltage v_{load} for variable absorbed power p_{load}, the secondary DC–DC converter operating the conversion between the DC-link and the load must be controlled to ensure that the output voltage v_{out} is always equal to v_{load}.

To ensure $v_{out} = v_{load}$, the voltage gain of the converter is:

$$M_v = \frac{v_{load}}{v_{in}} \tag{10}$$

If the load must operate with p_{load} and v_{load}, regardless of its inner nature, the absorbed current will be $i_{load} = i_{out} = p_{load}/v_{load}$. Assuming the converter operates, for simplicity, with unitary efficiency $\eta = M_v M_i = 1$, the output current seen from the primary side of the load is:

$$i_{in} = \frac{i_{load}}{M_i} = i_{load} \cdot M_v = \frac{i_{load} \cdot v_{load}}{v_{in}} = \frac{p_{load}}{v_{in}} \tag{11}$$

$$R = \frac{p_{load}}{v_{in}^2} \tag{12}$$

Thus from (9) it is possible to derive a representation as a controlled generator, and from (10) as a voltage-controlled resistor. Clearly, both representations are non-linear elements. The three representations are shown in Figure 5.

2.2. Full Chain Dynamic Model

Full modeling of the dynamic system should take into account both the interaction between the DC–DC converter and the storage elements [33–35] and between the DC–DC converter and the photovoltaic source [1,36–39]. Also, the model must implement a current sink for a constant power load operating at a specific desired voltage, such as the case for a USB powered device [40,41]. The proposed model taken individually exhibits a different dynamic behavior from the one shown in the full-chain dynamic model presented hereafter. The most notable difference is that the concurrence of a power-limited nonlinear source such as a PV device with a constant power load creates a non-feasible area where, even if the theoretical maximum delivered power by the PV device is higher than the power required by the load, the SC will not charge unless an appropriate MPPT strategy adapting the operating point is implemented.

Since the only dynamic element present is the capacitance in the SC block, the system behaves as a non-linear first-order system. Thus, its dynamics should be described from the complete state-equation describing the voltage across the capacitor. According to the representation of the SC block, different equations or sets of equations must be considered.

2.2.1. Ideal Supercapacitor Dynamic Model (IDSC)

In this case, the SC is represented without any parasitic component. A simple explicit state equation can be formulated as following:

$$\dot{v}_c = \frac{1}{C}\left(\eta_i M_i i_{pv}\left[\frac{v_c}{\eta_v M_v}\right] - \frac{p_{load}}{v_c}\right) \tag{13}$$

where the term $i_{pv}\left[v_{pv}\right]$ is given by (3). Under the assumption that the converter operates in MPPT according to Equation (7):

$$\dot{v}_c = \frac{1}{C}\left(\eta_i M_i i_{pv}\left[v_{mp}\right] - \frac{p_{load}}{v_c}\right) \tag{14}$$

$$\dot{v}_c = \frac{1}{C}\left(\eta_i\left(\frac{v_{mp}}{v_{out}}\eta_v\right)i_{mp} - \frac{p_{load}}{v_c}\right) \tag{15}$$

$$\dot{v}_c = \frac{1}{C}\left(\frac{\eta p_{mp} - p_{load}}{v_c}\right) \tag{16}$$

Interestingly, the state equation can be solved analytically leading to the time expression of the capacitor voltage. From a known initial voltage $v_c(0)$:

$$v_c(t) = \pm\sqrt{\frac{2\cdot t\cdot[\eta p_{MP} - p_{USB}]}{C} + v_c(0)^2} \tag{17}$$

2.2.2. Self-Discharge Supercapacitor Dynamic Model (SDSC)

The derivation in this case is very similar to the IDSC one, since the only difference is the presence of an additional term in the state equation describing the branch current.

$$\dot{v}_c = \frac{1}{C}\left(\eta_i M_i i_{pv}\left[\frac{v_c}{\eta_v M_v}\right] - \frac{p_{load}}{v_c} - \frac{v_c}{r_{sh}}\right) \tag{18}$$

$$\dot{v}_c = \frac{1}{C}\left(\frac{\eta p_{mp} - p_{load}}{v_c} - \frac{v_c}{r_{sh}}\right) \tag{19}$$

The main difference resides in the presence of an additional term relative to the current flowing on the shunt resistance. The solution in this case is a superposition of the previously found solution with a discharge exponential.

2.2.3. Full Supercapacitor Dynamic Model (FSC)

In this case, the voltage across the supercapacitor is different from the voltage across the output of the first DC–DC converter, and this difference depends on the current drawn from the PV source. This results in an implicit-state equation that must be solved in conjunction with a KCL to derive the branch current on the supercapacitor.

$$\begin{cases} \dot{v}_c = \frac{1}{C}\left(\eta_i M_i i_{pv} - \frac{p_{load}}{v_c}\right) \\ \eta_i M_i i_{pv} - \left(\eta_v M_v v_{pv} - \frac{v_c}{r_s}\right) - \frac{p_{load}}{v_c} \\ v_{pv}\left[i_{pv}\right] - v_{pv} = 0 \end{cases} \quad (20)$$

The system can only be integrated numerically, deriving at each time-step the v_{pv} voltage and i_{pv} current from the last two equations, and substituting the i_{pv} in the first-state equation to compute (e.g., thorough Euler or Runge–Kutta integration) the next timestep value of the capacitor voltage.

2.3. Equilibrium Points for Final Voltage Charging

In practical applications, the produced power from the PV source will be larger than the power drawn from the load. Assuming perfect MPPT, this leads to an unstable system without equilibrium points, where the voltage across the capacitor rises indefinitely. In particular, the square-root-of-time expression in (17) represents the fastest possible charging velocity that can be achieved in this system.

In real scenarios, however, this charge curve must be abandoned when in proximity of the SC maximum voltage to avoid damaging the component. In this case, a voltage gain (and thus, a duty-cycle) for the primary DC–DC converter must be determined with the aim of creating an equilibrium point for the SC voltage equal to the SC nominal maximum voltage $v_c = v_{MAX}$.

Let us first consider the IDSC chain with $\eta = 1$. The equilibrium point to be found is the one where the current drawn from the load p_{load}/v_{MAX} is equal to the current coming from the converter $\eta p_{mp}/v_{MAX}$. This is an operating condition where this equilibrium for a possible M_v, M_i can be better explained graphically. In Figure 6, several I–V characteristics of a 54 W PV source are represented, as seen from the secondary side of the first PV panel, for different M_v, M_i gains. The thick, black line is the IV characteristic at $M_v = M_i = 1$ (thus equal to one of the PV devices). The red and blue crosses represent four combinations of loads (imposing the p_{load}) and SC (imposing the v_{MAX}), as reported in Table 1. The dashed line represents the ensemble of the maximum power points.

Table 1. Load and supercapacitor combinations to determine the existence of a feasible equilibrium condition.

Symbol	p_{load}	v_{MAX}
Red Cross	60 W	10 V
Red Circle	60 W	20 V
Blue Cross	40 W	10 V
Blue Circle	40 W	20 V

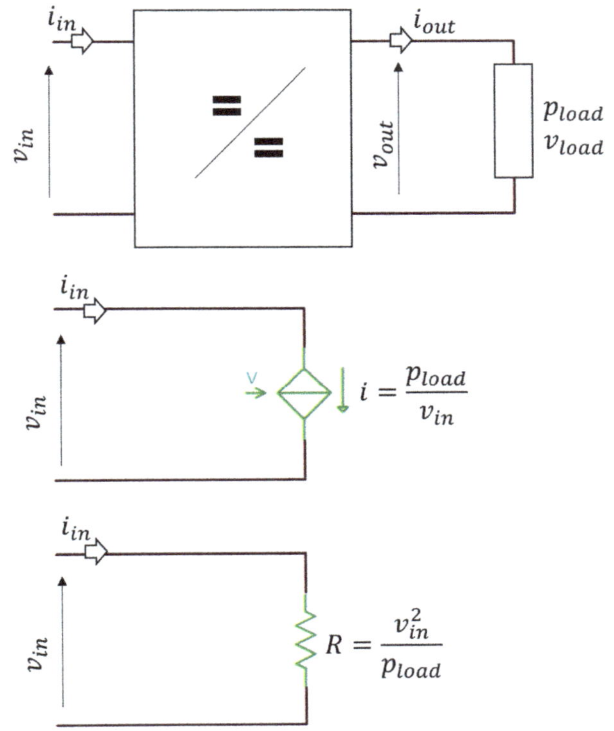

Figure 5. Fourth and fifth block: a constant voltage with a variable absorbed power load and a DC–DC converter. Below, the primary side representations of the load.

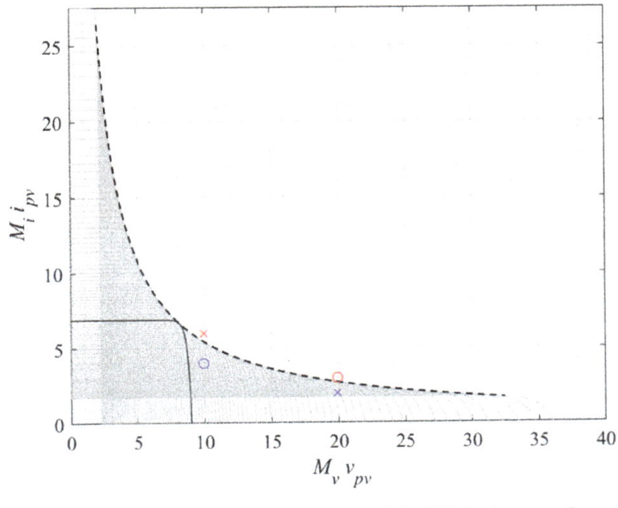

Figure 6. All possible I–V characteristics of the PV device seen from the SC side, for various duty-cycles. The full line is the PV characteristic, and the dashed line is the envelope of the maximum power points. Blue circle and ×, and red circle and × represent points from Table 1.

As can be seen from the figure, the two blue combinations fall within the possible operating points that can be determined on the secondary side of the first DC–DC converter, whereas the red combinations are not. In a lossless and completely ideal chain as the IDSC with $\eta = 1$, an equilibrium point is always found as long as $p_{load} < p_{mp}$.

Relaxing the condition on the efficiency, three possible scenarios can be studied: efficiency is distributed among voltage and current gain $\eta = \eta_v \eta_i$, efficiency is negligible on the current gain $\eta = \eta_v$ and efficiency is negligible on the voltage gain $\eta = \eta_i$. The result is shown in Figure 7a–c.

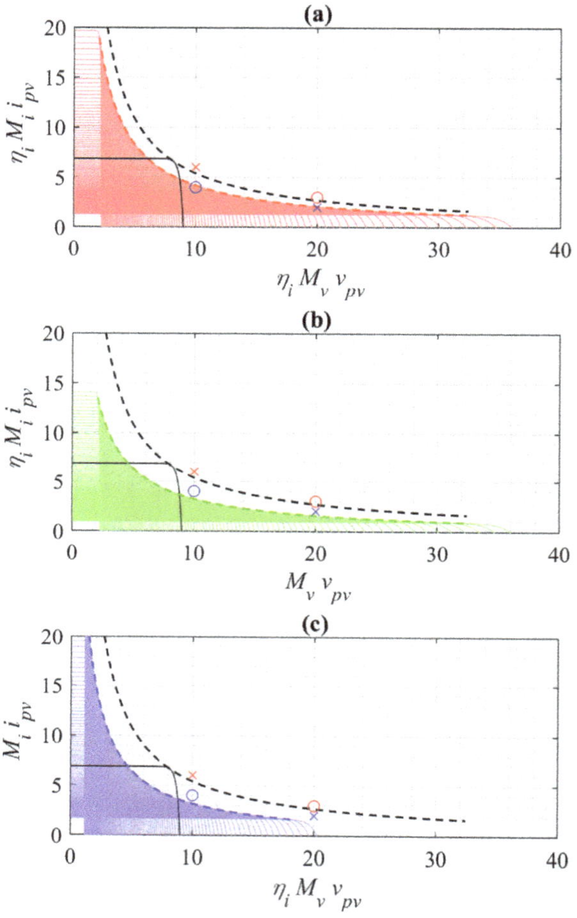

Figure 7. Effects of the efficiency, in terms of I–V PV characteristics seen from the SC side, on both voltage and current (**a**), only current (**b**) and only voltage (**c**). Blue circle and $\times$, and red circle and $\times$ represent points from Table 1.

The dashed black line represents the ensemble of the maximum power points for $\eta = 1$, kept as a reference. As can be seen, the voltage and current distribution of the efficiency strongly alters the lieu of feasible operating points that can be obtained at the secondary side of the first DC–DC converter. An even distribution of efficiency for voltage and current in Figure 7a shifts the ensemble towards the origin, slightly limiting both maximum voltage and current. A distribution of efficiency skewed towards voltage or current limits, respectively, the maximum voltage and maximum current. Due to the effect of efficiency, previously feasible combinations of p_{load}, v_{MAX} can fall outside the lieu of operating points.

The introduction of the non-ideality in the SC stage consists in using the SDSC or the FSC chain instead of the IDSC. The inclusion of SDSC results in an additional current to consider, the one absorbed by the load. This current is r_{sh}/v_{MAX} at the equilibrium. Thus, this shifts the operating points "upward" in the I–V plots shown before. A simple example is given for the $\eta = 1$ case, where an r_{sh} resistance with range $r_{sh} = \{10–1000\}$ Ω is considered (note that this range is highly unrealistic for an SC, but it is useful for visualization purposes). As could be expected, the effect of self-discharge is more present for higher v_{MAX} values (Figure 8).

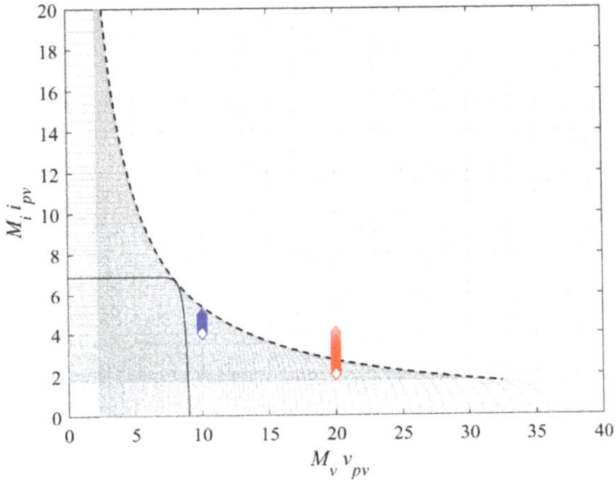

Figure 8. Effect of the shunt resistance in the SDSC case. Larger effects are seen at higher v_{MAX}.

However, it should be noted that the equivalent self-discharge resistance for series-connected SCs decreases for each added cell; thus, in practical cases, this effect is independent of the v_{MAX}. Finally, the FSC chain behaves in the same manner as the SDSC chain. This is due to the positioning of the r_s element, which does not alter the KCL used to study the SDSC.

2.4. *Duty Cycle for Final Voltage Charging*

In the previous, the IDSC, SDSC and FDC chains were analyzed to understand whether, given a couple of p_{load}, v_{MAX}, a combination of voltage and current gain could be found that would ensure that the SC would have $v_c = v_{MAX}$ at equilibrium. In this section, the strategy to determine the gains (and the relative duty-cycles) are discussed.

The desired gains are the ones that ensure:

$$\dot{v}_c = \begin{cases} \frac{1}{C}\left(\eta_i M_i i_{pv}\left[\frac{v_{MAX}}{\eta_v M_v}\right] - \frac{p_{load}}{v_{MAX}}\right) = 0 & \text{IDSC} \\ \frac{1}{C}\left(\eta_i M_i i_{pv}\left[\frac{v_{MAX}}{\eta_v M_v}\right] - \frac{p_{load}}{v_{MAX}} - \frac{v_{MAX}}{r_{sh}}\right) = 0 & \text{SDSC} \\ \frac{1}{C}\left(\eta_i M_i i_{pv}\left[\frac{v_{MAX}}{\eta_v M_v}\cdot\frac{r_s + r_{sh}}{r_{sh}}\right] - \frac{p_{load}}{v_{MAX}} - \frac{v_{MAX}}{r_{sh}}\right) = 0 & \text{FSC} \end{cases} \quad (21)$$

Albeit the current across the capacitor is null, in this case, the optimal gain value is different between SDSC and FSC due to the (in general negligible) voltage divider term.

Regardless of the specific chain, (21) expresses a numerical problem in the two unknowns M_v, M_i. Since those are the lossless gains, and $M_v = 1/M_i$, the problem can be reduced to a single unknown. Expressing the voltage gain as $M_v = \frac{D}{1-D}$, it is possible to rewrite (21) as a set of equations in D for which the zero corresponds to a duty-cycle leading the SC to charge at v_{MAX}.

$$f_{IDSC}(D, v_{MAX}, p_{load}) = \left(\eta_i \left(\frac{1-D}{D}\right) i_{pv}\left[\left(\frac{1-D}{D}\right)\frac{v_{MAX}}{\eta_v}\right] - \frac{p_{load}}{v_{MAX}}\right) \quad (22)$$

$$f_{SDSC}(D, v_{MAX}, p_{load}) = \left(\eta_i \left(\frac{1-D}{D}\right) i_{pv}\left[\left(\frac{1-D}{D}\right)\frac{v_{MAX}}{\eta_v}\right] - \frac{p_{load}}{v_{MAX}} - \frac{v_{MAX}}{r_{sh}}\right) \quad (23)$$

$$f_{FSC}(D, v_{MAX}, p_{load}) = \left(\eta_i \left(\frac{1-D}{D}\right) i_{pv}\left[\left(\frac{1-D}{D}\right)\frac{v_{MAX}}{\eta_v}\cdot\frac{r_s + r_{sh}}{r_{sh}}\right] - \frac{p_{load}}{v_{MAX}} - \frac{v_{MAX}}{r_{sh}}\right) \quad (24)$$

Any one of equations in (22)–(24) can be equaled to zero and solved numerically (using (3) for the current-voltage relationship) for different values of p_{load} and v_{MAX} to determine the desired duty cycle D. In general, for a given Equations (22)–(24), and a specific p_{load} and v_{MAX}, there are two numerical solutions to (22)–(24). This is because a given current, leading to equilibrium, can be achieved both in the near open-circuit area of the PV converter and near the short-circuit area of the PV converter, as seen in similar previous works [42]. Visualization of the solutions for (22)–(24) requires some constraints, since considering the variable D and the two parameters p_{load} and v_{MAX} results in a three-dimensional function for which the zero-crossing cannot be visualized. Thus, in the following plots, the p_{load} will be assumed as constant. In Figure 9a, the absolute values of the f_{ISDC} are shown for a $p_{load} < p_{mp}$. Then, to compare the magnitude of the variations among the different (22)–(24), in Figure 9b a cross-section of the surface for $v_{MAX} = 6$ V is shown also for f_{SDSC} and f_{FSC}. In this case, series $r_s = 1\ \Omega$ and shunt $r_{sh} = 50\ \Omega$ resistances were considered in the SDSC and FSC cases. Again, the quantities are elevated for visualization purposes, but it is relevant to understand that very little difference occurs at a steady state between the purely ideal IDSC case and the parasitic-affected SDSC and FSC cases. The two solutions are not both stable. In fact, the sign of the $\dot{v}_c$ can be studied to derive a phase portrait shown in Figure 10, where it can be seen that the equilibrium points found for higher D and lower v_{MAX} are a boundary towards an unstable region where the SC will simply discharge to zero instead of reaching the equilibrium point. In practical terms, since the low boundary of D is often limited by design, this creates a minimum voltage for the SC below which there is no way to recharge it anymore.

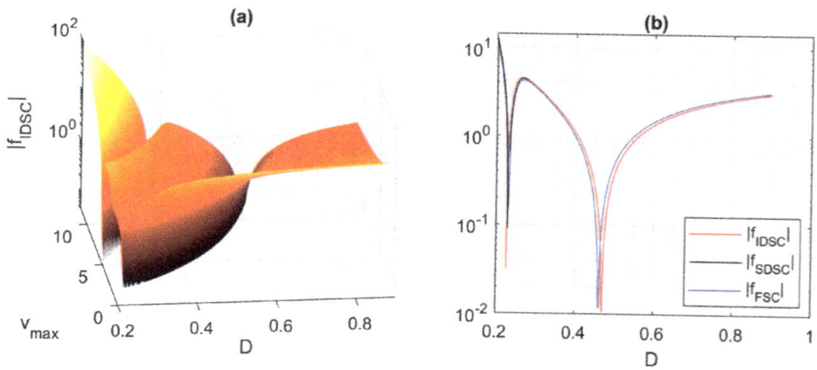

Figure 9. Absolute value of the time derivative of the SC voltage in logarithmic form (**a**), showing three regions separated by two curves where the function reaches zero for the IDSC. The IDSC, SDSC and FSC are shown for a fixed $v_{MAX} = 6$ V in (**b**), underlining the very little difference between the models at steady state.

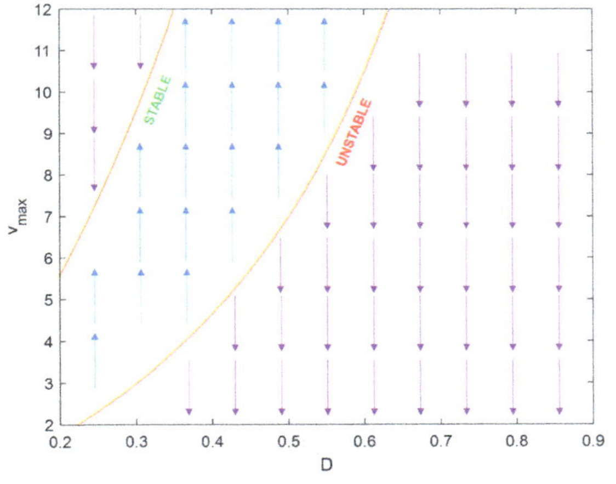

Figure 10. Phase portrait (normalized) of (21), showing the stability nature of the equilibrium points defined by the yellow curves.

3. Validation Results

The methodology and the equations proposed herein are validated through a time-domain simulation of the IDSC, SDSC and FSC chains in an LTSpice environment. The schematic is depicted in Figure 11. As can be seen, the PV source can be implemented using an ideal diode specifying the *N* and *Is* parameters. The PV device implemented in the figure delivers a power of $p_{mp} = 47$ W at $v_{mp} = 16.6$ V. The conversion stage is implemented via controlled generators, including the efficiency already divided in the current and voltage contribution. The SC link is represented in the FSC form with parametric resistances which can be set at very high (shunt) and low (series) values to restore the semi-ideal SDSC or the fully ideal IDSC chains. The load is represented by a current generator operating at constant power. It should be noted that this behavioral source rolls back to a pure resistor if the voltage across it falls below a certain threshold, to avoid divergence during simulation.

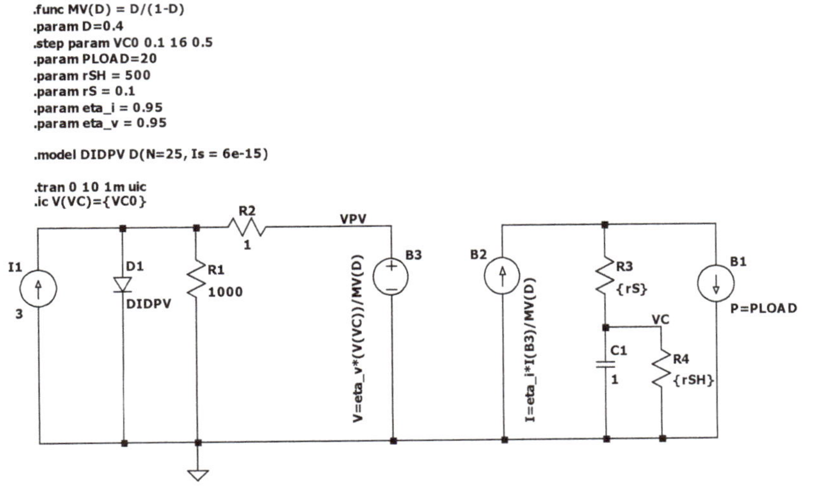

Figure 11. LTSpice schematic for the validation of the IDSC, SDSC and FSC chains.

3.1. Test A: Final Voltage Charge and Stability Region

In this test the SC is initially kept at a variable charge between 2 V and 16 V. The charging is performed at a fixed duty-cycle (D = 0.6), and the load absorbs a constant power of $p_{load} = 20$ W. The final voltage and the equilibrium points delimiting, for D = 0.6, the stability region, are computed with (24). The charging profiles are shown in Figure 12 (left) along with the vertical aligned $|f_{FSC}|$ in Figure 12 (right). Confirming the hypothesis, the SC charges (or discharges) towards the equilibrium point as long as the initial condition is within the stability region shown in green. If it falls below, in the region shown in red, it will simply discharge to zero. The slight difference in the equilibrium points (less than 0.5 V) can be attributed to the different model used by the diode in LTSpice.

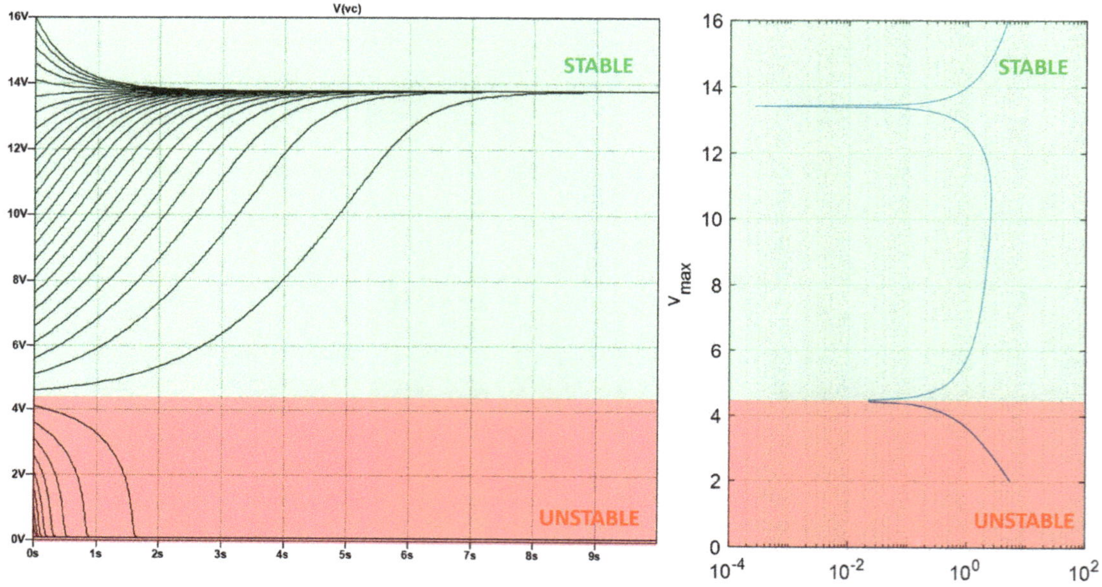

Figure 12. Time-domain voltage of the SC (**left**) and stability regions (**right**) according to (21). If the SC initial voltage is within the stable region, it charges towards the stable equilibrium voltage; otherwise, it discharges to zero.

3.2. Test B: Short and Heavy Load Perturbation

In this test, the chain initially works at a nominal load $p_{load} = 20$ W, and the SC is charged to the final value (initial value is 10 V). After it reaches the charged state, a heavy and short (5 s) load perturbation is applied. The load is incremented by 300% to $p_{load} = 50$ W, thus driving the system outside the feasible area. As can be seen, the voltage across the capacitor drops rapidly, and if the perturbation were to continue indefinitely, it would reach zero as expected. Time-domain simulations are shown in Figure 13.

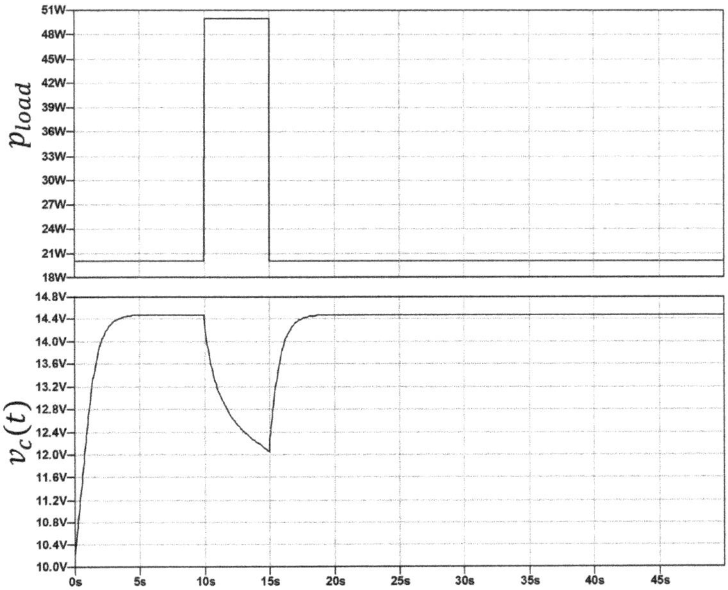

Figure 13. Load power time-domain profile (**top**) and SC voltage (**bottom**) for a short load perturbation.

3.3. *Test C: Long and Heavy Load Perturbation*

This test is similar to the previous one, but in this case, the perturbation from the nominal condition (p_{load} = 20 W) is going to be longer (15 s). The notable difference that can be appreciated is that the discharge induced in the capacitor by the load perturbation leads the SC in the unstable region; thus, even after the perturbation has ended, the SC is not able to recover and simply discharges to zero as expected. Time-domain simulations are shown in Figure 14.

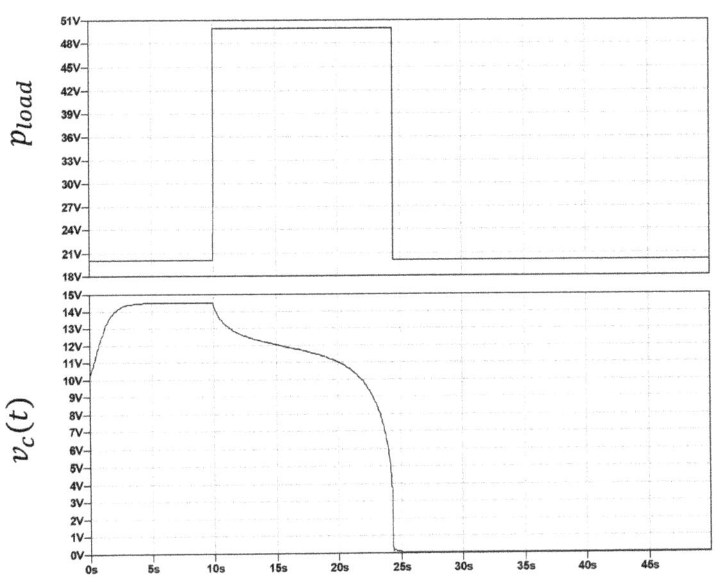

Figure 14. Load power time-domain profile (**top**) and SC voltage (**bottom**) for a long load perturbation.

3.4. Test D: Dynamic Response

This test compares the adynamic response of the chain, where the only dynamic element is composed by the SC, with the fully dynamic response of the circuit with a properly sized Zeta converter. The converter was sized to ensure a CCM for D = 0.5 and an operative switching frequency of 10 kHz. The LTSpice model of the dynamic converter is shown in the top part of Figure 15, and the comparison of the voltage waveforms between the adynamic previous results and the dynamic measured ones can be seen in the bottom part of Figure 14. As can be seen, the charging dynamics are slightly different in the dynamic case (as the transfer function of the dynamic converter influences the time-domain response), but the steady-state value is practically identical. Moreover, the same considerations concerning the stable and unstable regions are still valid in the dynamic time domain analysis.

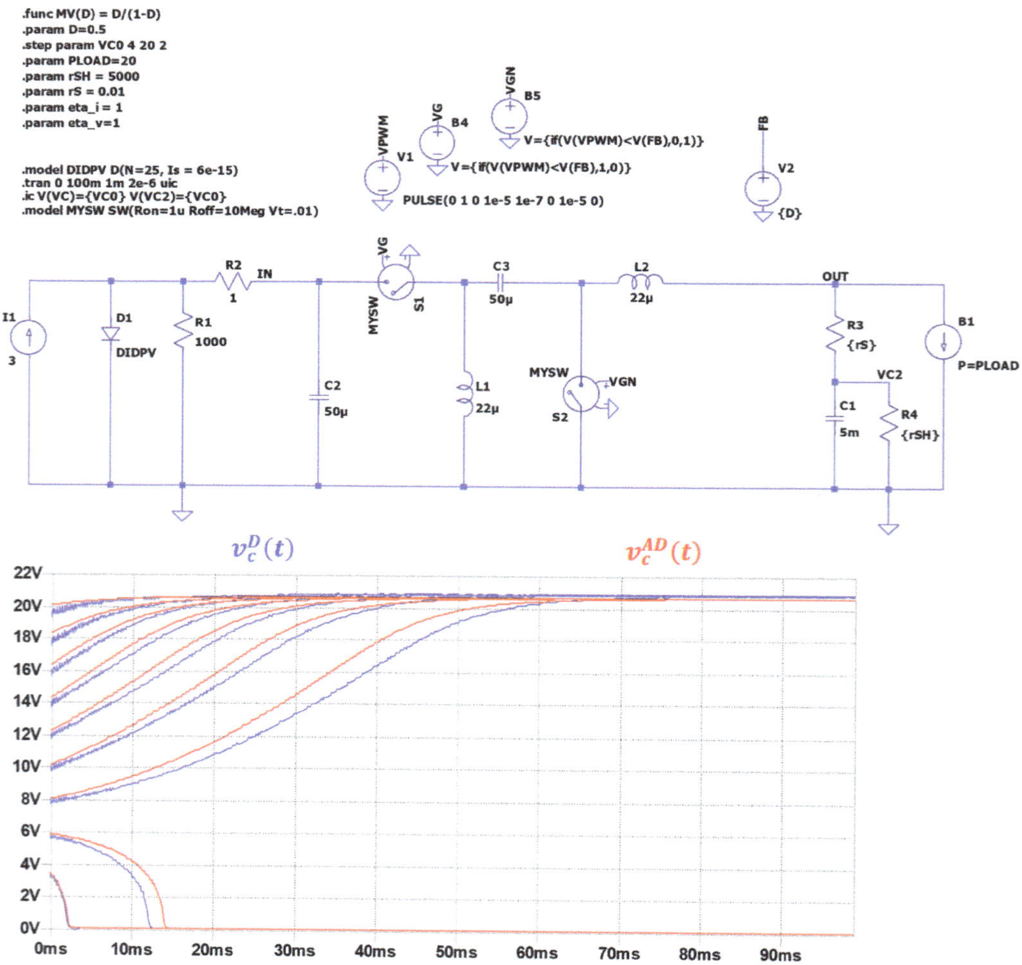

Figure 15. Dynamic LTSpice model of the FSC chain (**top**) and comparison (**bottom**) of the capacitor voltage curve in the dynamic case (blue) and adynamic case (red).

4. Discussion

The proposed model, the insight derived from its analysis, and the validation obtained from the simulations are useful tools to understand the possibilities that are available for a SC-based DC-link in the very common application of PV conversion for constant power loads. The dynamic-state equations derived in Section 2 are validated against time-domain simulations, offering an accurate and computationally lightweight alternative for time-domain simulation of these kinds of systems in environments that do not natively implement circuit simulation.

The estimation of the feasible region for the final-voltage charge with different distribution of the efficiency towards voltage and/or current highlights the necessity of deeper consideration when analyzing power converters. In fact, most of the analysis assumes a resistive load (for which the efficiency skewness is irrelevant), but for non-linear applications such as the one proposed in this manuscript, the effects of voltage and current drop are dramatically different. Since this contribution comes from different parasitic elements inside the converter, a possible evolution of this work is to consider the correlation between such parasitic elements and the shrinking of the feasible area.

The study of the stable and unstable regions opens a discussion on the possible operating conditions where the system can or cannot operate. In fact, operating in the unstable region leads to a discharge of the SC. The minimum possible duty-cycle supported by the converter creates a lower boundary for the SC voltage, thus limiting in fact the available energy that can be extracted from it. This is not necessarily a very limiting factor, since many modern SC technologies, such as the hybrid SC [14], are already lower-bound limited in voltage, yet the consideration of this boundary is at this point critical during the design stage of the system. It turns out from the simulations that the proposed voltage-conversion system equipped with an SC-based DC-link is very effective in the case of constant power loads (as in the case of resistive loads or battery storage devices). However, the effectiveness and stability of this system are highly dependent on the initial charging conditions of the SC, which must exhibit a starting voltage above a certain threshold to be charged under stability conditions. In order to avoid unstable working conditions, it would be more effective to use hybrid SCs with a minimum starting voltage [20] in the DC-link section, or to associate standard SC's with an undervoltage protection circuit.

5. Conclusions

In this work, a fully dynamic model for the conversion chain between a photovoltaic source and a constant power load with a supercapacitor-based DC-link is proposed. The system was studied considering the circuit model, including the non-linear nature of the PV source, the efficiency of the DC–DC converter, and the parasitic elements on the supercapacitor model. The study investigated the dynamic nature of the system, highlighting critical aspects such as a bounded region of feasibility for the charge of the SC element and the presence of a stable and unstable set of equilibrium points leading to a final-voltage charge. The proposed approach offers a valuable tool for implementation of control algorithms, further refinement of the analysis by considering the DC–DC converter individual parasitic components, and development of MPPT strategies incorporating both the non-linear nature of the PV source and the dynamic behavior of the DC-link stage. Validation in terms of hardware full-chain implementation or through hardware-in-the-loop approaches constitutes the future development of this work. Concerning the converter, a practical approach for the converter implementation can be found in [43]. Concerning the implementation of the storage, interesting results to be investigated can be found in [44], and, in a completely passive PV and battery configuration, in [45]. It should be noted, however, that as shown in this work, the converter efficiency plays a major role in the general performance of the system, and for this reason, efficient and isolated converters operating at high frequency could be the best choice [46].

Author Contributions: Conceptualization and Methodology, G.M.L., M.Q. and A.L.; resources and project administration, S.R. and F.P.; Software and Validation, G.M.L. and F.C.; investigation and formal analysis, M.P., A.L. and F.C.; supervision, A.L. and S.R. All authors have read and agreed to the published version of the manuscript.

Funding: This work was partially funded by the Italian Ministry of University and Research: "Ecosystem for Sustainable Transition in Emilia-Romagna" (Ecosister), funded under the National Recovery and Resilience Plan (NRRP), Mission 4 Component 2 Investment 1.5-Call for tender No. 3277 of 30 December 2021.

Data Availability Statement: Data and simulations are available at readers' requests.

Conflicts of Interest: The authors declare no conflict of interest.

References

1. Corti, F.; Laudani, A.; Lozito, G.M.; Reatti, A. Computationally efficient modeling of DC–DC converters for PV applications. *Energies* **2020**, *13*, 5100. [CrossRef]
2. Azuatalam, D.; Paridari, K.; Ma, Y.; Förstl, M.; Chapman, A.C.; Verbič, G. Energy management of small-scale PV-battery systems: A systematic review considering practical implementation, computational requirements, quality of input data and battery degradation. *Renew. Sustain. Energy Rev.* **2019**, *112*, 555–570. [CrossRef]
3. Sri Revathi, B.; Prabhakar, M. Non isolated high gain DC–DC converter topologies for PV applications—A comprehensive review. *Renew. Sustain. Energy Rev.* **2016**, *66*, 920–933. [CrossRef]
4. Fagiolari, L.; Sampò, M.; Lamberti, A.; Amici, J.; Francia, C.; Bodoardo, S.; Bella, F. Integrated energy conversion and storage devices: Interfacing solar cells, batteries and supercapacitors. *Energy Storage Mater.* **2022**, *51*, 400–434. [CrossRef]
5. Imran, A.; Zhu, Q.; Sulaman, M.; Bukhtiar, A.; Xu, M. Electric-Dipole Gated Two Terminal Phototransistor for Charge-Coupled Device. *Adv. Opt. Mater.* **2023**, 2300910. [CrossRef]
6. Laudani, A.; Lozito, G.M.; Lucaferri, V.; Radicioni, M.; Fulginei, F.R.; Salvini, A.; Coco, S. An analytical approach for maximum power point calculation for photovoltaic system. In Proceedings of the 2017 European Conference on Circuit Theory and Design, ECCTD 2017, Catania, Italy, 4–6 September 2017.
7. Kumar, N.; Hussain, I.; Singh, B.; Panigrahi, B.K. Self-Adaptive Incremental Conductance Algorithm for Swift and Ripple-Free Maximum Power Harvesting from PV Array. *IEEE Trans. Ind. Inform.* **2018**, *14*, 2031–2041. [CrossRef]
8. Chiang, S.J.; Shieh, H.J.; Chen, M.C. Modeling and control of PV charger system with SEPIC converter. *IEEE Trans. Ind. Electron.* **2009**, *56*, 4344–4353. [CrossRef]
9. Mohammadinodoushan, M.; Abbassi, R.; Jerbi, H.; Waly Ahmed, F.; Abdalqadir kh ahmed, H.; Rezvani, A. A new MPPT design using variable step size perturb and observe method for PV system under partially shaded conditions by modified shuffled frog leaping algorithm- SMC controller. *Sustain. Energy Technol. Assess.* **2021**, *45*, 101056. [CrossRef]
10. Cossutta, M.; Vretenar, V.; Centeno, T.A.; Kotrusz, P.; McKechnie, J.; Pickering, S.J. A comparative life cycle assessment of graphene and activated carbon in a supercapacitor application. *J. Clean. Prod.* **2020**, *242*, 118468. [CrossRef]
11. Diarra, B.; Zungeru, A.M.; Ravi, S.; Chuma, J.; Basutli, B.; Zibani, I. Design of a photovoltaic system with ultracapacitor energy buffer. *Procedia Manuf.* **2019**, *33*, 216–223. [CrossRef]
12. Pandey, S.; Dwivedi, B.; Tripathi, A. Performance analysis of super capacitor integrated PV fed multistage converter with SMC controlled VSI for varying nonlinear load conditions. *Int. J. Eng. Technol.* **2018**, *7*, 174–180. [CrossRef]
13. Melkia, C.; Ghoudlburk, S.; Soufi, Y.; Maamri, M.; Bayoud, M. Battery-Supercapacitor Hybrid Energy Storage Systems for Stand-Alone Photovoltaic. *Rev. Compos. Mater. Av.* **2022**, *24*, 265–271. [CrossRef]
14. Shchur, I.; Kulwas, D.; Wielgosz, R. Combination of distributed MPPT and distributed supercapacitor energy storage based on cascaded converter in photovoltaic installation. In Proceedings of the 2018 IEEE 3rd International Conference on Intelligent Energy and Power Systems, IEPS 2018, Kharkiv, Ukraine, 10–14 September 2018.
15. Al-Thabaiti, S.A.; Mostafa, M.M.M.; Ahmed, A.I.; Salama, R.S. Synthesis of copper/chromium metal organic frameworks—Derivatives as an advanced electrode material for high-performance supercapacitors. *Ceram. Int.* **2023**, *49*, 5119–5129. [CrossRef]
16. Mostafa, M.M.M.; Alshehri, A.A.; Salama, R.S. High performance of supercapacitor based on alumina nanoparticles derived from Coca-Cola cans. *J. Energy Storage* **2023**, *64*, 107168. [CrossRef]
17. Gouda, M.S.; Shehab, M.; Helmy, S.; Soliman, M.; Salama, R.S. Nickel and cobalt oxides supported on activated carbon derived from willow catkin for efficient supercapacitor electrode. *J. Energy Storage* **2023**, *61*, 106806. [CrossRef]
18. Gouda, M.S.; Shehab, M.; Soliman, M.M.; Helmy, S.; Salama, R. Preparation and characterization of supercapacitor electrodes utilizing catkin plant as an activated carbon source. *Delta Univ. Sci. J.* **2023**, *6*, 255–265.
19. Zhao, Y.; Xie, W.; Fang, Z.; Liu, S. A Parameters Identification Method of the Equivalent Circuit Model of the Supercapacitor Cell Module Based on Segmentation Optimization. *IEEE Access* **2020**, *19*, 92895–92906. [CrossRef]
20. Corti, F.; Gulino, M.S.; Laschi, M.; Lozito, G.M.; Pugi, L.; Reatti, A.; Vangi, D. Time-domain circuit modelling for hybrid supercapacitors. *Energies* **2021**, *14*, 6837. [CrossRef]
21. Yue, X.; Kiely, J.; Gibson, D.; Drakakis, E.M. Charge-Based Supercapacitor Storage Estimation for Indoor Sub-mW Photovoltaic Energy Harvesting Powered Wireless Sensor Nodes. *IEEE Trans. Ind. Electron.* **2020**, *14*, 6837. [CrossRef]
22. Reatti, A.; Corti, F.; Tesi, A.; Torlai, A.; Kazimierczuk, M.K. Effect of parasitic components on dynamic performance of power stages of DC–DC PWM buck and boost converters in CCM. In Proceedings of the IEEE International Symposium on Circuits and Systems, Sapporo, Japan, 26–29 May 2019.
23. Kazimierczuk, M.K. *Pulse-Width Modulated DC–DC Power Converters*; John Wiley & Sons: Hoboken, NJ, USA, 2012; ISBN 9780470773017.

24. Bronzoni, M.; Colace, L.; De Iacovo, A.; Laudani, A.; Lozito, G.M.; Lucaferri, V.; Radicioni, M.; Rampino, S. Equivalent circuit model for Cu(In,Ga)Se_2 solar cells operating at different temperatures and irradiance. *Electronics* **2018**, *7*, 324. [CrossRef]
25. De Soto, W.; Klein, S.A.; Beckman, W.A. Improvement and validation of a model for photovoltaic array performance. *Sol. Energy* **2006**, *80*, 78–88. [CrossRef]
26. Corless, R.M.; Gonnet, G.H.; Hare, D.E.G.; Jeffrey, D.J.; Knuth, D.E. On the Lambert W function. *Adv. Comput. Math.* **1996**, *5*, 329–359. [CrossRef]
27. Roberts, K.; Valluri, S.R. Solar Cells and the Lambert W Function 2016. In Proceedings of the Conference "Celebrating 20 Years of the Lambert W function", London, ON, Canada, 25–28 July 2016.
28. Blakesley, J.C.; Castro, F.A.; Koutsourakis, G.; Laudani, A.; Lozito, G.M.; Riganti Fulginei, F. Towards non-destructive individual cell I–V characteristic curve extraction from photovoltaic module measurements. *Sol. Energy* **2020**, *202*, 342–357. [CrossRef]
29. Oommen, S.; Ballaji, A.; Ankaiah, B.; Ananda, M.H. Zeta converter simulation for continuous current mode operation. *Int. J. Adv. Res. Eng. Technol.* **2019**, *10*, 2019. [CrossRef]
30. Chandran, I.R.; Ramasamy, S.; Nallaperumal, C. A high voltage gain multiport zeta-zeta converter for renewable energy systems. *Inf. MIDEM* **2020**, *50*, 215–230. [CrossRef]
31. Song, M.S.; Son, Y.D.; Lee, K.H. Non-isolated bidirectional soft-switching SEPIC/ZETA converter with reduced ripple currents. *J. Power Electron.* **2014**, *14*, 649–660. [CrossRef]
32. Banaei, M.R.; Bonab, H.A.F. A High Efficiency Nonisolated Buck-Boost Converter Based on ZETA Converter. *IEEE Trans. Ind. Electron.* **2020**, *67*, 1991–1998. [CrossRef]
33. Camara, M.B.; Gualous, H.; Gustin, F.; Berthon, A. Design and new control of DC/DC converters to share energy between supercapacitors and batteries in hybrid vehicles. *IEEE Trans. Veh. Technol.* **2008**, *57*, 2721–2735. [CrossRef]
34. Pena, R.A.S.; Hijazi, A.; Venet, P.; Errigo, F. Balancing Supercapacitor Voltages in Modular Bidirectional DC–DC Converter Circuits. *IEEE Trans. Power Electron.* **2022**, *37*, 137–149. [CrossRef]
35. Carpita, M.; De Vivo, M.; Gavin, S. Dynamic modeling of a bidirectional DC/DC interleaved converter working in discontinuous mode for stationary and traction supercapacitor applications. In Proceedings of the SPEEDAM 2012—21st International Symposium on Power Electronics, Electrical Drives, Automation and Motion, Sorrento, Italy, 20–22 June 2012.
36. Jagadeesh, I.; Indragandhi, V. Comparative Study of DC–DC Converters for Solar PV with Microgrid Applications. *Energies* **2022**, *15*, 7569. [CrossRef]
37. Hole, S.R.; Goswami, A. Das Quantitative analysis of DC–DC converter models: A statistical perspective based on solar photovoltaic power storage. *Energy Harvest. Syst.* **2022**, *9*, 113–121. [CrossRef]
38. Formisano, A.; Hernández, J.C.; Petrarca, C.; Sanchez-Sutil, F. Modeling of pv module and DC/DC converter assembly for the analysis of induced transient response due to nearby lightning strike. *Electronics* **2021**, *10*, 120. [CrossRef]
39. Van De Sande, W.; Ravyts, S.; Alavi, O.; Nivelle, P.; Driesen, J.; Daenen, M. The sensitivity of an electro-thermal photovoltaic DC–DC converter model to the temperature dependence of the electrical variables for reliability analyses. *Energies* **2020**, *13*, 2865. [CrossRef]
40. Karami, Z.; Shafiee, Q.; Sahoo, S.; Yaribeygi, M.; Bevrani, H.; Dragicevic, T. Hybrid Model Predictive Control of DC–DC Boost Converters with Constant Power Load. *IEEE Trans. Energy Convers.* **2021**, *36*, 1347–1356. [CrossRef]
41. Ali, M.S.; Soliman, M.; Hussein, A.M.; Hawash, S.A.F. Robust Controller of Buck Converter Feeding Constant Power Load. *J. Control. Autom. Electr. Syst.* **2021**, *32*, 153–164. [CrossRef]
42. Corti, F.; Laudani, A.; Lozito, G.M.; Reatti, A.; Bartolini, A.; Ciani, L. Model-Based Power Management for Smart Farming Wireless Sensor Networks. *IEEE Trans. Circuits Syst. I Regul. Pap.* **2022**, *69*, 2235–2245. [CrossRef]
43. Santosh Kumar Reddy, P.L.; Obulesu, Y.P.; Singirikonda, S.; Al Harthi, M.; Alzaidi, M.S.; Ghoneim, S.S.M. A Non-Isolated Hybrid Zeta Converter with a High Voltage Gain and Reduced Size of Components. *Electronics* **2022**, *11*, 483. [CrossRef]
44. Rana, M.M.; Uddin, M.; Sarkar, M.R.; Shafiullah, G.M.; Mo, H.; Atef, M. A review on hybrid photovoltaic—Battery energy storage system: Current status, challenges, and future directions. *J. Energy Storage* **2022**, *51*, 104597. [CrossRef]
45. Chibuko, U.; Merdzhanova, T.; Weigand, D.; Ezema, F.; Agbo, S.; Rau, U.; Astakhov, O. Module-level direct coupling in PV-battery power unit under realistic irradiance and load. *Sol. Energy* **2023**, *249*, 233–241. [CrossRef]
46. Lal, V.M.; Parihar, P.S.; Kanimozhi, G. Isolated SEPIC-Based DC–DC Converter for Solar Applications. *Smart Grids Smart Cities* **2023**, *1*, 309–322.

energies

MDPI

Article

LLC Resonant Converters as Isolated Power Factor Corrector Pre-Regulators—Analysis and Performance Evaluation

Matteo Sucameli and Claudio Adragna *

STMicroelectronics, 20864 Agrate Brianza, MB, Italy; matteo.sucameli@st.com
* Correspondence: claudio.adragna@st.com

Abstract: The power supply of many applications running off the power line is made up of an isolated dc-dc converter powered by a front-end power factor corrector (PFC) stage. The PFC stage ensures compliance with the electromagnetic compatibility regulations but does not usually provide safety isolation since it is typically implemented with a boost converter. Lately, the increase in multi-output power supplies, especially in lighting and USB Power Delivery applications, has raised the need for an isolated PFC at power levels where currently there is not an industry standard solution. This isolated PFC is intended to power one or more non-isolated post-regulators to enable a substantial simplification of the overall architecture and a cost reduction. The usage of an LLC resonant converter as an isolated PFC has been considered and demonstrated only quite recently, raising the industry's attention due to the favorable converter's characteristics that have led to its success as a dc-dc converter. This paper provides two significant contributions. Firstly, it provides a quantitative assessment of the difference in the results obtained by designing an LLC-based PFC converter based on the first harmonic approximation analysis or the time-domain analysis by applying them to the design of the same converter. Secondly, it demonstrates that designing an LLC-based PFC converter to work also in the above-resonance region optimizes its performance by reducing the (magnetizing) reactive current in the resonant tank and, therefore, the rms currents on both the input and the output side and the related power loss.

Keywords: power quality; THD; isolated PFC; LLC resonant converter; LLC PFC; first harmonic approximation; FHA; time-domain analysis; TDA

Citation: Sucameli, M.; Adragna, C. LLC Resonant Converters as Isolated Power Factor Corrector Pre-Regulators—Analysis and Performance Evaluation. *Energies* **2023**, *16*, 7114. https://doi.org/10.3390/en16207114

Academic Editor: Wadim Strielkowski

Received: 20 July 2023
Revised: 13 October 2023
Accepted: 15 October 2023
Published: 16 October 2023

1. Introduction

The industry standard solution to address the compliance of mains-operated power supplies with the IEC 61000-3-2 regulation [1], which sets limits to the harmonic content of their input current, is the addition of an electronic front-end circuit, the so-called power factor corrector (PFC). It is a switch-mode converter directly supplied by the rectified mains and controlled to draw a sinusoidal current in phase with the voltage. This results in a low total harmonic distortion (THD) and near-unity power factor (PF = 1), hence the name PFC, as if the electronic equipment was a resistive load.

Figure 1 shows the most common architecture of a power-factor-corrected switch-mode power supply (SMPS): a PFC pre-regulator front end providing a regulated output voltage and powering a cascaded dc-dc converter.

In most applications, the PFC pre-regulator is realized using a boost converter, a non-isolated topology. The cascaded dc-dc converter is therefore responsible for providing the isolation required in most power supplies operated from the mains to meet safety requirements and regulations.

The boost converter can cover a broad power range, whereas the cascaded dc-dc converter uses different topologies depending on the power level. Flyback converters are commonly used in the lower power range. LLC resonant converters [2] have become very popular at higher power levels because of their nice properties conferred by soft-switching operation (low power loss, high efficiency, high operating frequency, low electromagnetic emissions, high power density) without significant drawbacks.

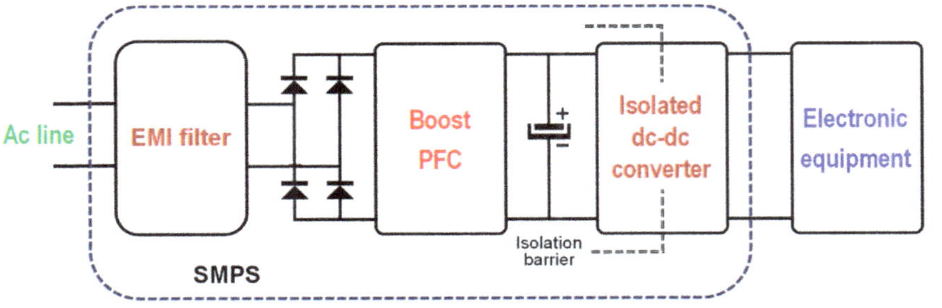

Figure 1. Typical two-stage architecture of a power-factor-corrected SMPS.

There are applications, however, where an isolated PFC cascaded by one or more non-isolated converters might be a better option: multi-output isolated converters involve a more complicated design and are more difficult to fine-tune, while non-isolated dc-dc converters are quite standard building blocks, can be very efficient and compact and eliminate cross-regulation issues among the various outputs. This is the case, for example, of multi-output SMPS or LED drivers, or chargers for mobile equipment, single and multiport [3], required to comply with the USB Power Delivery (USB-PD) protocol.

Figure 2 shows this type of architecture.

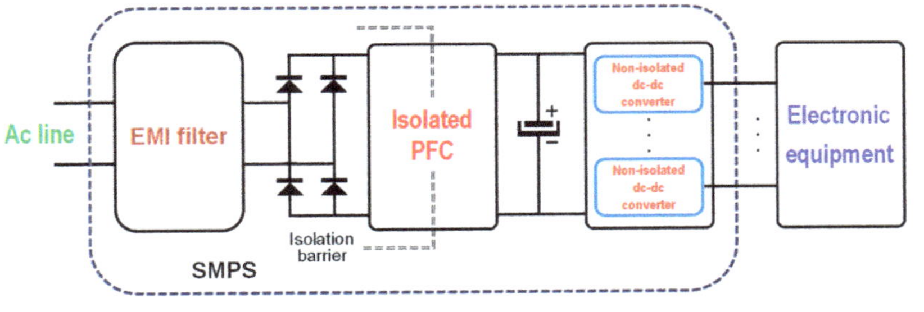

Figure 2. Alternative architecture of a power-factor-corrected, multioutput SMPS.

Additionally, there are other SMPS applications (e.g., battery chargers) where the load is tolerant to the low-frequency ripple of a PFC output. In this case, an isolated PFC might enable a single-stage architecture (see Figure 3) with a substantial cost saving.

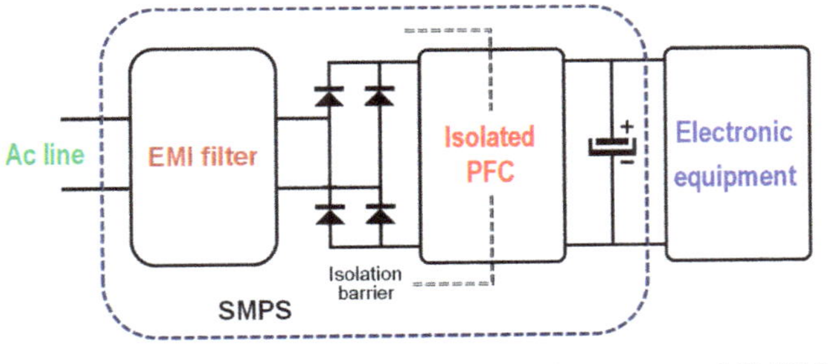

Figure 3. Architecture of a single-stage power-factor-corrected SMPS for low-frequency, ripple-tolerant loads.

Flyback-based isolated PFCs are a good choice up to 50–60 W: they are often used in lighting equipment [4–7]. For higher power levels, though the literature presents plenty of different solutions [8], none have become an industry standard like the flyback PFC.

Isolated boost converters [9] are not easy to handle, and the provisions needed to make them operate properly make them less cost-effective; other topologies, such as SEPIC [10], Ćuk [11] and Zeta [12] converters, can effectively work as an isolated PFC, but they are seldom used.

Likewise, a lot of new topologies combining a boost-type front end with an isolated converter that share the same control have been proposed over the years [13–20] but did not find broad industrial usage, with very few exceptions [20]. Many of them [15–19] considered the LLC converter to be the isolated converter.

Only recently, research has concentrated on the standalone LLC converter used as a PFC stage (an LLC-PFC in short), therefore supplied by a rectified sinusoidal voltage and not by a dc voltage like in two-stage SMPS architectures.

Ref. [21] uses the first harmonic approximation (FHA) analysis to demonstrate LLC's ability to work as a PFC stage, keeping its previously mentioned benign properties. It sets up a design procedure that is used in [22] to prove the feasibility of an LLC-PFC at the kW level and in [23] for the design of a modular PFC stage operated from the three-phase line. In these papers, it is assumed that the converter always works in the so-called "below-resonance" operating region, and the design constraints required to ensure soft switching in this region are not addressed.

As for the control of the output voltage/current of an LLC-PFC, the previously mentioned works consider average current mode control (ACMC), like in boost PFC operated in Continuous Conduction Mode [24], to achieve extremely low distortion of the input current. The implementation is microcontroller-based.

Other authors [25,26] consider modulating the switching frequency directly (they call it voltage-mode control) to simplify the control system at the price of a significantly higher THD of the input current, though still sufficient to comply with the IEC 61000-3-2 regulation.

In [25,27], the accuracy limits of the FHA approach are highlighted, and the time-domain (TD) analysis is proposed to overcome these limitations. Ref. [27] shows that the FHA approach results in a too-conservative design that does not fully utilize the operating region of the converter, whereas by using the TD approach, the operating region of the converter can be fully utilized. However, the degree of conservativeness of the FHA approach is shown only qualitatively, the TD-based design procedure is just sketched, and the details on how the design is carried out are few.

Based on these premises, the first objective of the present work is to assess how the usage of the TD approach can improve an FHA-based design. This goal is achieved by addressing the design of a specified converter with the two approaches and comparing the results. A more detailed TD-based design procedure is provided in this paper.

A second objective is to demonstrate that designing the LLC-PFC converter to work also in the above-resonance region optimizes its performance by reducing the reactive current in the resonant tank, which in turn reduces the rms currents on both the input and the output sides.

Thirdly, to assess the feasibility of implementing ACMC using low-cost analog components, this control is considered and verified by simulations and bench experiments.

Ultimately, the goal of the present work is to demonstrate that an LLC-PFC can be an attractive solution in many use cases, with all the credentials to become an industry standard for an isolated PFC at the power levels that a flyback PFC cannot support.

Therefore, the paper is organized as follows.

In Section 2, the FHA analysis is reviewed; compared to previously published procedures, here, operation above the upper resonance frequency is accounted for. The resulting step-by-step design procedure is given in Appendix A.

In Section 3, the TD analysis is reviewed. Due to its mathematical complexity, only the basic definitions and the results are provided in this section, the details of the analysis are given in Appendix B.

Section 4 provides the electrical specification of an LLC-PFC intended for a high-power LED lamp driver and uses two different design strategies with the FHA approach: the first one assumes that the converter operates at the upper resonance frequency on the peak of the maximum ac input voltage, whereas the second one assumes that the converter operates at the upper resonance frequency on the peak of the nominal ac input voltage, thus utilizing the above-resonance region to handle ac input voltages higher than the nominal one. The same is done with the TD approach.

Section 5 shows a prototype constructed so that the two designs based on the TD approach can be implemented in a single board and provides the results of their bench evaluation. The results are commented on to provide the basis for the conclusions drawn in Section 6.

2. First Harmonic Approximation (FHA) Analysis of the LLC-PFC

The simplified circuit diagram of the LLC-PFC converter is shown in Figure 4. The input capacitor *Cin* is not a bulk capacitor as it is usually placed after the input bridge in a non-power-factor-corrected converter; it is only a filter for the high-frequency switching noise, like a standard boost PFC.

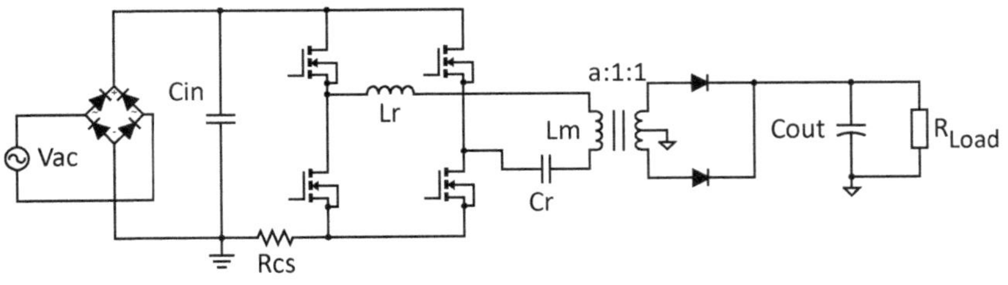

Figure 4. Simplified circuit diagram of an ac-dc LLC-PFC converter.

The FHA analysis, whose foundations were laid in a paper published in 1988 [28], is now widely used when designing an LLC resonant converter, especially because a handy design procedure can be found that starts from the electrical specification and leads to the definition of the resonant tank's parameter (a, Lr, Lm, Cr), e.g., like the 10-step procedure in [29].

Before extending this procedure to the design of an LLC-PFC, it is convenient to remember some basic definitions used in the FHA analysis:

Primary-to-secondary turn ratio	$a = \frac{Np}{Ns}$
Topology factor	$\alpha = \begin{cases} 2a & \text{half bridge} \\ a & \text{full bridge} \end{cases}$
Normalized voltage conversion ratio (voltage gain)	$\lvert M(f_n, \lambda, Q) \rvert = \alpha \frac{Vout}{Vin} = \frac{1}{\sqrt{\left[1+\lambda\left(1-\frac{1}{f_n^2}\right)\right]^2 + Q^2\left(f_n - \frac{1}{f_n}\right)^2}}$
Resonance frequencies	$f_{R1} = \frac{1}{2\pi\sqrt{Lr\,Cr}};\; f_{R2} = \frac{1}{2\pi\sqrt{(Lr+Lm)Cr}}$ (1)
Characteristic impedance	$Z_{R1} = \sqrt{\frac{Lr}{Cr}} = 2\pi\, f_{R1}\, Lr = \frac{1}{2\pi\, f_{R1}\, Cr}$
Series-to-magnetizing inductance ratio	$\lambda = \frac{Lr}{Lm} = \frac{f_{R2}^2}{f_{R1}^2 - f_{R2}^2}$
Normalized switching frequency	$f_n = \frac{f_{sw}}{f_{R1}}$
Output ac resistance	$Rout_{ac} = \frac{8}{\pi^2}\frac{Vout^2}{Pout} = \frac{8}{\pi^2} R_{Load}$
Quality factor	$Q = \frac{Z_{R1}}{R_{ac}} = \frac{Z_{R1}}{a^2 Rout_{ac}} = \frac{\pi^2}{8}\frac{Z_{R1}\,Pout}{a^2\,Vout^2}$

The FHA analysis developed for LLC converters supplied by a substantial dc input voltage can be extended to the PFC case, based on a quasi-static approximation. In fact, although the input voltage is a rectified sinusoid that goes all the way from zero to the peak, the line frequency f_{line} is such that the variations are much slower than the converter dynamics, making it possible to consider the system operating in steady-state conditions for all the instantaneous phase angles θ of the rectified sinusoid.

In order to act as a PFC stage and achieve a unity PF, the instantaneous input power along a line half cycle, $Pi(\theta)$, swings from zero at the zero crossing of the input voltage and current to twice the average power Pin (equal to the output power $Pout$ divided by the efficiency η) at the peak of the input voltage, as plotted in Figure 5:

$$Pi(\theta) = \left(Vin_{pk}sin\theta\right)\left(Iin_{pk}sin\theta\right) = 2Vin\ Iin\ \sin^2\theta = 2Pin\ \sin^2\theta = 2\frac{Pout}{\eta}\sin^2\theta. \quad (2)$$

where Vin and Iin are the rms values, and Vin_{pk} and Iin_{pk} are the peak values of the input voltage and current, respectively.

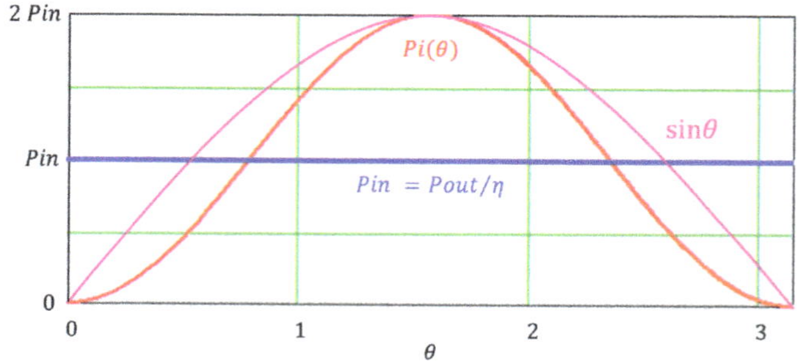

Figure 5. Instantaneous power and average (dc) power.

Evaluating (2) at $\theta = \pi/2$, we obtain

$$Pi(\pi/2) = 2\ Vin\ Iin = 2\ Pin = Vin_{pk}Iin_{pk}. \quad (3)$$

This means that at the peak of the sinusoidal line voltage, the converter is powered with an equivalent dc line voltage Vin_{pk} and absorbs an equivalent dc current equal to Iin_{pk}, resulting in the input power being twice the average value. Based on the concept of quasi-static approximation, the design based on the FHA approach can be carried out treating the peak values of the line voltage and current as if they were dc values.

To ensure proper operation as a PFC-LLC, some additional analysis is needed. To this purpose, it is fundamental to consider that not only the instantaneous input power but also the instantaneous output power varies along θ too:

$$Po(\theta) = \eta\ Pi(\theta) = 2\ \eta\ Pin\ \sin^2\theta = 2\ Pout\ \sin^2\theta, \quad (4)$$

and so do the ac resistance and the quality factor. Substituting (4) in (1), we get

$$Rout_{ac}(\theta) = \frac{4}{\pi^2}\frac{Vout^2}{Pout\ \sin^2\theta}Q(\theta) = \frac{\pi^2}{4}\frac{Z_{R1}}{a^2}\frac{Pout}{Vout^2}\sin^2\theta = Q_0\ \sin^2\theta. \quad (5)$$

Finally, also the voltage gain $|M|$ becomes a function of θ:

$$|M(f_n, \lambda, Q_0, \theta)| = \frac{1}{\sqrt{\left[1+\lambda\left(1-\frac{1}{f_n^2}\right)\right]^2 + Q_0^2\sin^4\theta\left(f_n - \frac{1}{f_n}\right)^2}}. \quad (6)$$

This voltage gain needs to be compared to the voltage gain of an LLC converter supplied by a rectified sinusoid $Vin_{pk}\sin\theta = \sqrt{2}\,Vin\sin\theta$ required to provide a regulated output voltage $Vout$:

$$M_{req}(Vin,\theta) = \alpha\frac{Vout}{\sqrt{2}\,Vin\sin\theta} = \alpha\frac{Vout}{Vin_{pk}\sin\theta}. \tag{7}$$

The plot of (7) vs. θ is shown in Figure 6: M_{req} is minimum on the peak of the sinusoid ($\theta = \pi/2$) and tends to infinity approaching the zero crossings.

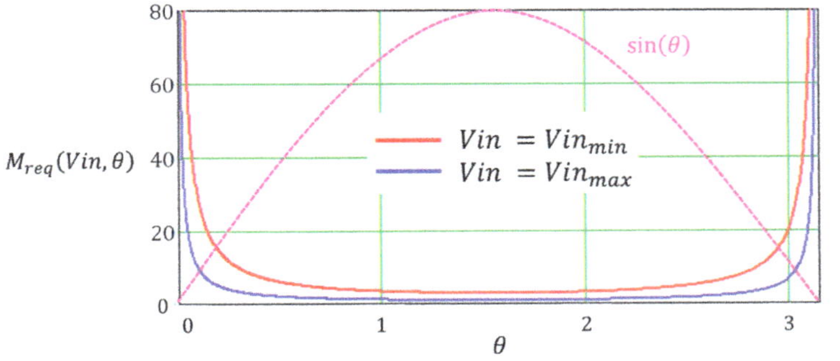

Figure 6. Minimum gain required for regulation in a single-stage PFC-LLC.

To ensure that the converter can regulate the output voltage, with fixed Q_0 and λ, it must be $|M| > M_{req}$ at some frequency for all θ values. Also, the operating points must be in the inductive region, where the converter works with zero-voltage switching (ZVS).

We can visualize this constraint considering a hypothetical LLC converter and plotting its voltage gain $|M|$ vs. the normalized frequency f_n at a fixed power level (i.e., at a fixed Q_0) with the phase angle θ as a parameter, as shown in Figure 7. In particular, the plot is drawn with θ equal to $\pi/2$, $\pi/3$ and $\pi/4$ (due to the symmetry of the sin function, these plots represent the supplementary angles too). The highest required gain M_{req} (calculated at the minimum input voltage $Vin_{pk_{min}}$ and, in case of variable output voltage, at $Vout = Vout_{max}$) for the same phase angles is plotted in dashed lines.

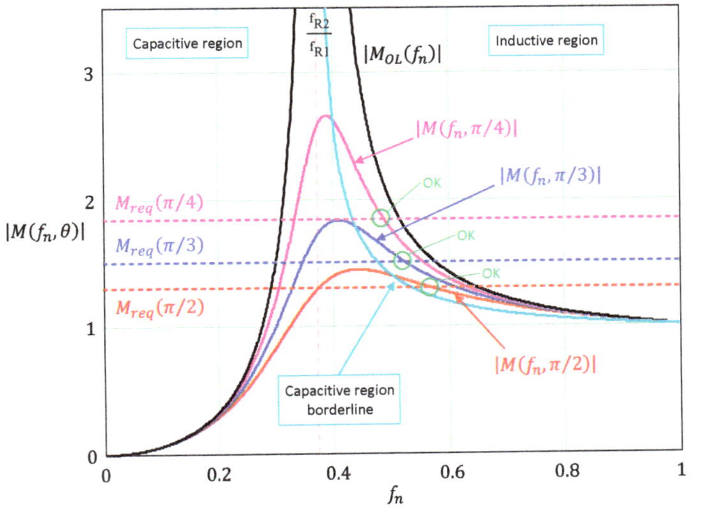

Figure 7. Voltage gain curves for different phase angles and comparison to the minimum required gain to achieve output voltage regulation.

For all of these three phase angles of the input voltage, there is an intersection between the required gain and the LLC voltage gain: these are the operating points for those phase angles. This means that the condition $|M| > M_{req}$ is met at some frequency. These points are in the inductive region, so the converter works with ZVS as required.

However, we need to ensure that this condition is verified for any phase angle θ included in $[0, \pi/2]$ (if it is so, for symmetry, it will be also in $[\pi/2, \pi]$).

The $|M|$ curves also show that the gain has a peak M_{pk} in the capacitive region and the maximum useful value M_{MAX} at the boundary between the capacitive and the inductive region. Notice that the difference between M_{MAX} and M_{req} gets smaller as θ tends to $\pi/2$ or, equivalently, when Q_0 increases. Also notice that M_{MAX} is larger than the gain at $f_{sw} = f_{R2}$ (or, equivalently, at $f_n = f_{R2}/f_{R1}$). The latter gain can be easily calculated by inserting $f_n = f_{R2}/f_{R1}$ in (6) and considering the definitions in (1):

$$|M(f_{R2}/f_{R1}, \lambda, Q_0, \theta)| = \frac{\sqrt{\lambda(1+\lambda)}}{Q_0 \sin^2\theta}. \tag{8}$$

Imposing that (8) be greater than (or at least equal to) the required voltage gain (7) evaluated at the minimum input voltage and maximum output voltage, where the needed gain is at a maximum, we will make sure that the gain will be always sufficient to achieve output voltage regulation:

$$M_{MAX}(\lambda, Q_0, \theta) \geq \frac{\sqrt{\lambda(1+\lambda)}}{Q_0 \sin^2\theta} \geq \alpha \frac{Vout_{max}}{\sqrt{2}\, Vin_{min} \sin\theta}. \tag{9}$$

Since the inequality $1/\sin^2\theta \geq 1/\sin\theta$ is always true for any phase angle between zero and π, the condition that must be fulfilled to ensure output voltage regulation is:

$$\frac{\sqrt{\lambda(1+\lambda)}}{Q_0} \geq \alpha \frac{Vout_{max}}{\sqrt{2}\, Vin_{min}}. \tag{10}$$

Likewise, the maximum value of the minimum voltage gain occurs when $f_{sw} > f_{R1}$ and $Q_0 = 0$ (i.e., the output load is zero):

$$|M(f_n, \lambda, 0, \theta)| = \frac{1}{1 + \lambda\left(1 - \frac{1}{f_n^2}\right)} = |M_{OL}(f_n, \lambda)|. \tag{11}$$

Its minimum value is called M_∞ and occurs when $f_{sw} \gg f_{R1}$, as shown in Figure 8.

Indeed, if the maximum switching frequency is fixed at $f_{max} > f_{R1}$, the maximum value of the minimum voltage gain can be evaluated:

$$|M_{OL}(f_{max}/f_{R1}, \lambda)| = \frac{1}{1 + \lambda\left[1 - \left(\frac{f_{R1}}{f_{max}}\right)^2\right]}. \tag{12}$$

Imposing that (12) be lower than the required voltage gain (7) evaluated at the peak of the maximum input voltage and the minimum output voltage, where the required gain is at a minimum, it is possible to find the minimum inductance ratio λ that fulfills the minimum gain requirement:

$$\frac{1}{1 + \lambda\left[1 - \left(\frac{f_{R1}}{f_{max}}\right)^2\right]} \leq \alpha \frac{Vout_{min}}{\sqrt{2}\, Vin_{max}} = M_{req_{min}} \rightarrow \lambda \geq \frac{\frac{1}{M_{req_{min}}} - 1}{1 - \left(\frac{f_{R1}}{f_{max}}\right)^2}. \tag{13}$$

The FHA-based design procedure used in [21–23] and described step-by-step in [30] sets the maximum operating frequency at the upper resonance frequency f_{R1}(or equivalently $f_n = 1$) where the gain is unity and independent of the load. Since this gain must be lower than or equal to the minimum required voltage gain $M_{req_{min}}$, it is possible to derive the following constraint on α:

$$1 \leq \alpha \frac{Vout_{min}}{\sqrt{2}\,Vin_{max}} \rightarrow \alpha \geq \frac{\sqrt{2}\,Vin_{max}}{Vout_{min}}. \quad (14)$$

If we extend the operating region at frequencies higher than f_{R1}(i.e., $f_n > 1$), the constraint on α can be derived from (13):

$$\alpha \geq \frac{\sqrt{2}\,Vin_{max}}{Vout_{min}} \left\{ 1 + \lambda \left[1 - \left(\frac{f_{R1}}{f_{max}} \right)^2 \right] \right\}. \quad (15)$$

From (14) or (15), along with (1), it is possible to derive the required turn ratio a.

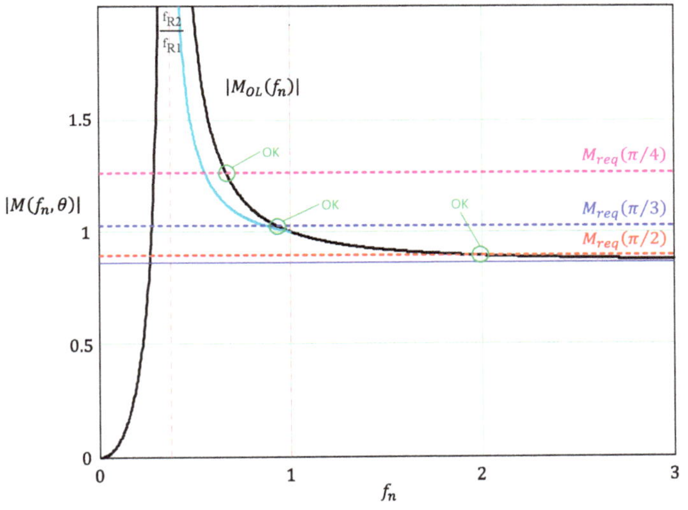

Figure 8. Voltage-gain curve at no load and comparison to the minimum required gain to achieve output voltage regulation.

Condition (10), along with (14) (or (15), depending on the design), is a fundamental design constraint because it determines the maximum and minimum gain of the LLC converter to fulfill the necessary gain to perform as a PFC.

By equating the required gain (7) to the voltage gain (6), it is possible to find how the normalized switching frequency varies along the phase angle θ of a line half cycle, as illustrated in the plot of Figure 9.

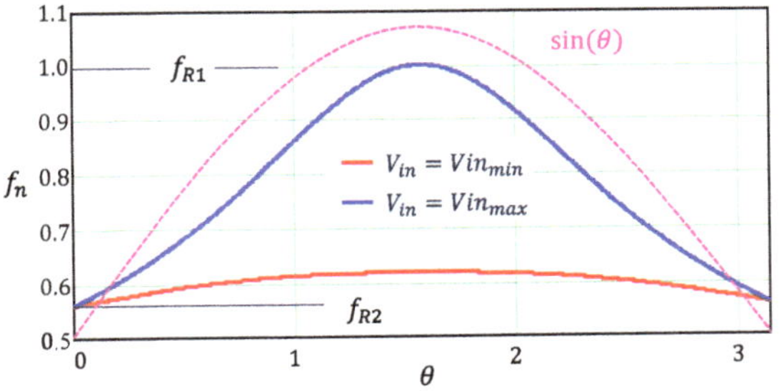

Figure 9. Normalized switching frequency vs. instantaneous phase angle (assumption: $f_{max} = f_{R1}$).

The switching frequency peaks at $\theta = \pi/2$ and decreases as the instantaneous line voltage moves toward the zero crossing, where it reaches the lower resonance frequency f_{R2} (the only possible equilibrium point when $Vin(\theta)$ and $Iin(\theta)$ are both zero). The relationship is strongly nonlinear and dependent on the rms input voltage, difficult to synthesize in an analog circuit or in a look-up table. This explains why controlling the switching frequency directly like in [25,26] results in a high THD of the input current.

Unfortunately, as already said, the accuracy of the FHA analysis is quite good when the system is working near the upper resonance frequency f_{R1} but is worse if we move away from this point. Since the LLC converter is designed to work in the entire range between the lower and the upper resonance frequencies and above when working as a PFC, a time-domain analysis is necessary to assess how the approximations inherent in the FHA analysis affect the results.

3. Time Domain (TD) Analysis of the LLC-PFC

The study of the differential equations of the current and voltages of the resonant tank allows us to better understand the behavior below the upper resonance frequency without the approximations inherent in the FHA approach.

As with the FHA, the quasi-static approximation allows us to solve the system of differential equations considering the input voltage fixed.

There are some takeaways from the FHA theory that can be used also in the TD analysis. One is the fact that the most stringent condition for the gain is at the peak of the minimum input voltage, where the difference between the available gain and the required gain is minimized, as shown in Figure 7. Also, the transformer turn ratio a can be derived considering (14) or (15), depending on whether the maximum switching frequency f_{max} is set at f_{R1} or above f_{R1}.

It is convenient to start the TD analysis by inspecting the key waveforms of voltage and current in the resonant tank. We will assume that at the peak of Vin_{min} (i.e., $Vin_{min_{pk}}$) and full output load, the resonant tank operating mode is DCMB2 as defined in [2] and characterized by two distinct time intervals in each switching half cycle.

Figure 10 shows the typical resonant tank currents and resonant capacitor voltage during the below resonance operation, in the first half switching cycle in DCMB2 mode. The waveforms in the second half switching cycle are mirror symmetric with respect to the horizontal axis (the initial conditions and the evolution in time are just opposite).

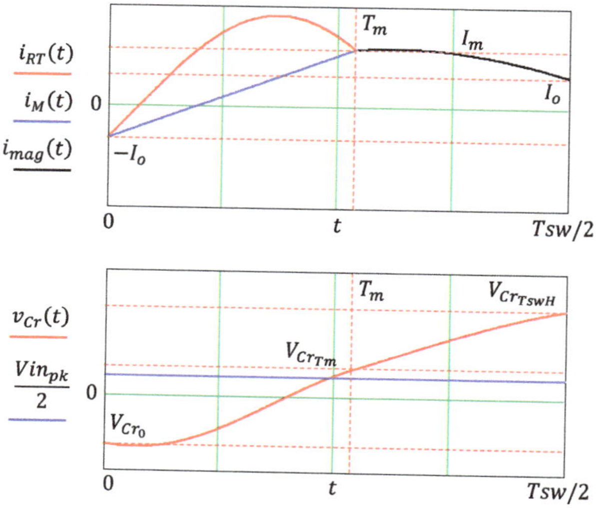

Figure 10. Resonant tank currents and resonant capacitor voltage at the peak of Vin_{min} in a half switching cycle.

The plots of the currents in Figure 10 show the following:

- The red portion is the resonant current i_{RT} during the time interval $(0 - Tm)$ during which current flows on the secondary side as well.
- The blue portion is the magnetizing current i_M flowing into Lm during the time interval $(0 - Tm)$; this current is subtracted from i_{RT} and does not contribute to the secondary current to form the output dc current.
- The black portion is the tank circuit current i_{mag} in the time interval $(Tm - Tsw/2)$; this is a magnetizing current too, flowing in both Lr and Lm; in this interval, the secondary current is zero.

The following analysis is performed supposing that the primary switch configuration is a half bridge (then, $\alpha = 2a$).

The circuit is described by two sets of differential equations in the two-time intervals $(0 - Tm)$ and $(Tm - Tsw/2)$. Referring to Figure 10 for the symbolism, in the time interval $(0 - Tm)$, it is possible to write:

$$\begin{cases} Vin_{pk} = Lr\frac{d}{dt}i_{RT}(t) + v_{Cr}(t) + a\ Vout \\ i_{RT}(t) = Cr\frac{d}{dt}v_{Cr}(t) \\ a\ Vout = Lm\frac{d}{dt}i_M(t) \end{cases} \tag{16}$$

with initial conditions:

$$\begin{cases} i_{RT}(0) = i_M(0) = -I_o \\ v_{Cr}(0) = V_{Cr_0} \end{cases}. \tag{17}$$

The solution of the system (16) is:

$$\begin{cases} v_{Cr}(t) = -Z_{R1}\ I_o\sin(2\pi f_{R1}t) - \left(Vin_{pk} - aVout - V_{Cr_0}\right)\cos(2\pi f_{R1}t) + \left(Vin_{pk} - aVout\right) \\ i_{RT}(t) = -I_o\cos(2\pi f_{R1}t) + \frac{1}{Z_{R1}}\left(Vin_{pk} - aVout - V_{Cr_0}\right)\sin(2\pi f_{R1}t) \\ i_M(t) = a\frac{Vout}{Lm}t - I_o \end{cases}, \tag{18}$$

where f_{R1} and Z_{R1} are those defined by (1).

During the same time interval, there is a current flowing on the secondary side of the converter:

$$i_{sec}(t) = a[i_{RT}(t) - i_M(t)]. \tag{19}$$

During the second time interval $(Tm - Tsw/2)$, there is no current flowing on the secondary side, and the inductances Lr and Lm are effectively in series. The second set of differential equations is:

$$\begin{cases} Vin_{pk} = (Lr + Lm)\frac{d}{dt}i_{RT}(t) + v_{Cr}(t) \\ i_{mag}(t) = Cr\frac{d}{dt}v_{Cr}(t) \end{cases} \tag{20}$$

with initial conditions:

$$\begin{cases} i_{mag}(Tm) = I_m \\ v_{Cr}(Tm) = V_{Cr_{Tm}} \end{cases}. \tag{21}$$

The solution of the system (20) is:

$$\begin{cases} v_{Cr}(t) = Z_{R2}\cdot I_m\sin[2\pi f_{R2}(t - Tm)] - \left(Vin_{pk} - V_{Cr_{Tm}}\right)\cos[2\pi f_{R2}(t - Tm)] + Vin_{pk} \\ i_{mag}(t) = I_m\cos[2\pi f_{R2}(t - Tm)] + \frac{1}{Z_{R2}}\left(Vin_{pk} - V_{Cr_{Tm}}\right)\sin[2\pi f_{R2}(t - Tm)] \end{cases}, \tag{22}$$

where f_{R2} has been already defined in (1), while it is:

$$Z_{R2} = \sqrt{\frac{Lr + Lm}{Cr}} = 2\pi f_{R2}(Lr + Lm) = \frac{1}{2\pi f_{R2}Cr}. \tag{23}$$

The complete TD analysis is detailed in Appendix B. Here, we show only the results, in particular, the following system of four nonlinear equations in the four unknowns I_o, I_m, Tm and Tsw that are highlighted in Figure 10:

$$\begin{cases} Iout_{pk} = a\frac{I_m - I_o}{Tsw}\left[\frac{\tan(\pi f_{R1} Tm)}{\pi f_{R1}} - Tm\right] \\ Iin_{pk} - \frac{Iout_{pk}}{2a} = \frac{1}{2Tsw}\left[Tm(I_m - I_o) + (I_m + I_o)\frac{\tan\left(\pi f_{R2}\left(\frac{Tsw}{2} - Tm\right)\right)}{\pi f_{R2}}\right] \\ \left[\frac{I_m + I_o}{2\pi f_{R1} Tm}\frac{K_V - 1}{\lambda} + \pi f_{R1} Tsw Iin_{pk}\right]^2 = \left[\frac{I_m + I_o\cos(2\pi f_{R1} Tm)}{\sin(2\pi f_{R1} Tm)}\right]^2 \\ \tan(\varphi_{min}) = \frac{I_o\sin(2\pi f_{R1} Tm)}{I_m + I_o\cos(2\pi f_{R1} Tm)} \end{cases}. \tag{24}$$

All the other parameters are known quantities:

- Iin_{pk} is the peak input current (averaged on a complete switching cycle) evaluated at the peak of the minimum input voltage (i.e., with phase angle $\theta = \pi/2$):

$$Iin_{pk} = 2\frac{Pin}{Vin_{pk}} \approx 2\frac{Pout}{\eta Vin_{pk}}.$$

- $Iout_{pk}$ is the peak output current (averaged on a complete switching cycle) evaluated at the peak of the minimum input voltage that is equal to twice the rated output current for a PFC circuit:

$$Iout_{pk} = 2\frac{Pout}{Vout}.$$

- K_V is the inverse of the voltage gain required of the converter that is the ratio between the minimum input voltage and the nominal input voltage at resonance:

$$K_V\frac{Vin_{min}}{Vin_{res}} = \frac{\sqrt{2}Vin_{min}}{\alpha Vout}.$$

- φ_{min} is the minimum phase angle between the resonant current and the half-bridge voltage, to ensure ZVS operation, strictly correlated to the dead time T_D that is purposely inserted between the turn-off of one switch of the half-bridge leg and the turn-on of the other one to allow ZVS and the turn-off delay T_{off} of the power switch:

$$\varphi_{min} = 2\pi f_{R1}\left(T_D - T_{off}\right).$$

The system (24) needs to be solved with a numerical method using a calculation tool. Once solutions (I_o, I_m, Tm and Tsw) are found, the parameters of the resonant tank circuit and the operating frequency can be calculated:

$$Lm = aVout\frac{Tm}{I_m + I_o}; Lr = \lambda Lm; Cr = \frac{1}{(2\pi f_{R1})^2 Lr}; fsw = \frac{1}{Tsw}.$$

4. Design of an LLC-PFC with Both FHA and TD Analysis

Table 1 shows the electrical specification of an LLC-PFC for a high-power LED lamp driver supplied by the European mains.

Based on this specification, two different design strategies are used.

The first one is such that the converter works at the upper resonance frequency f_{R1} at the peak of the maximum input voltage Vin_{max}. This automatically sets $f_{max} = f_{R1}$.

The second one is designed to work at the upper resonance frequency f_{R1} at the peak of the nominal input voltage Vin_{nom}, which is lower than Vin_{max}. As a result, the switching frequency will exceed f_{R1} in the voltage range included between Vin_{nom} and Vin_{max}. The maximum switching frequency f_{max} is specified, and the lower resonance frequency f_{R2} is determined by f_{max}.

Table 1. Electrical specification of the reference LLC-PFC for a high-power LED lamp driver.

Symbol	Name	Value	Unit
$Vin_{min} - Vin_{max}$	Input voltage range (rms)	176–305	Vac
Vin_{nom}	Nominal input voltage (rms)	230	Vac
$Vout$	Regulated output voltage (dc)	$60 \pm 5\%$	Vdc
V_{Rect}	Secondary rectifier voltage drop (dc)	0.1	Vdc
$Pout_{max}$	Maximum output power	240	W
η	Estimated efficiency @ $Pout_{max}$, Vin_{min}	94	%
f_{R1}	Upper resonance frequency	150	kHz
f_{R2}	Lower resonance frequency (*)	60	kHz
f_{max}	Maximum switching frequency (*)	300	kHz
C_{HB}	Half-bridge midpoint estimated capacitance	660	pF
T_D	Dead time	270	ns

(*) these two specifications are mutually exclusive: f_{R2} is specified if the design is to be carried out with $f_{max} = f_{R1}$; f_{max} is specified if the design is to be carried out with $f_{max} > f_{R1}$ (f_{R2} will be the result of computation).

The two design strategies are carried out using the FHA and the TD approach, for a total of four designs, which are labeled and summarized in Table 2. These design labels are used hereafter in this document.

Table 2. Summary of the four designs and their label.

Name	Description	Frequency Range
FHA1	Design with FHA-f_{R1} at $Vin_{max_{pk}}$	$f_{R2} \leq f_{sw} \leq f_{R1}$
FHA2	Design with FHA-f_{R1} at $Vin_{nom_{pk}}$	$f_{R2} \leq f_{sw} \leq f_{max}$
TD1	Design with TD-f_{R1} at $Vin_{max_{pk}}$	$f_{R2} \leq f_{sw} \leq f_{R1}$
TD2	Design with TD-f_{R1} at $Vin_{nom_{pk}}$	$f_{R2} \leq f_{sw} \leq f_{max}$

The resulting LLC resonant tanks are given in Table 3. Figure 11 shows the corresponding currents in the resonant tank circuit and the secondary side. The pictures with the same physical quantity have identical time bases and magnitudes, to easily compare the waveforms. Table 4 shows the corresponding calculated current stress.

Table 3. Calculation results for the four designs listed in Table 2.

Symbol	Name	FHA1	TD1	FHA2	TD2	Unit
a	Primary-to-secondary turn ratio	3.8	3.8	2.8	2.8	---
Vin_{res}	Input voltage at the resonance (rms)	323	323	238	238	Vac
Cr	Resonant capacitor	54	44	44	22	nF
Lr	Series resonant inductance	20.8	25.5	25.6	51	µH
Lm	Parallel resonant inductance	109.2	134	68.2	101	µH
f_{R1}	Upper resonance frequency	150.2	150.2	150	150.2	kHz
f_{R2}	Lower resonance frequency	60.1	60.1	78.3	87	kHz
I_o	Initial resonant current for ZVS	3.2	2.09	3.93	2.02	A

Table 4. Calculated rms currents for the four designs listed in Table 2. Rows #1–5: values averaged over a switching cycle at $Vin_{min_{pk}}$. Rows #6–7: values averaged over a line cycle at Vin_{min}.

#	Parameter	FHA1	TD1	FHA2	TD2	Unit
1	Resonant tank current	4.563	4.535	4.962	4.622	A
2	Magnetizing current	2.927	2.482	2.869	1.919	A
3	Initial current (instantaneous value)	3.203	2.096	3.929	2.022	A
4	Secondary current (x diode)	8.851	9.075	7.380	7.302	A
5	Secondary current (total)	12.517	12.834	10.437	10.327	A
6	Resonant tank current (over a line cycle)	3.843	3.524	4.026	3.318	A
7	Secondary current (total, over a line cycle)	7.37	7.615	6.343	6.293	A

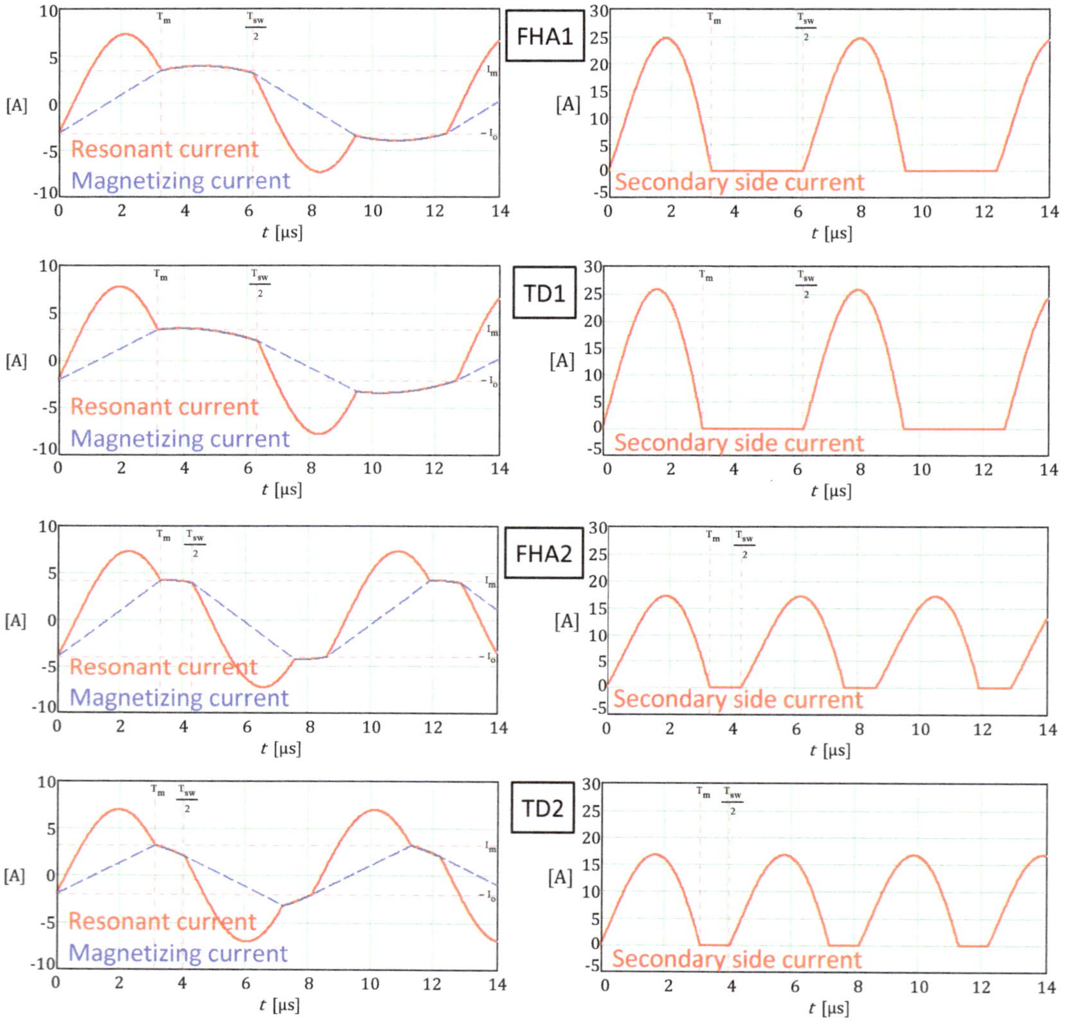

(a) (b)

Figure 11. Currents at full load and the peak of the minimum input voltage for the four different designs listed in Table 3: (**a**) resonant tank currents; (**b**) secondary side current.

With reference to Table 3, comparing the FHA designs with the TD designs (FHA1 vs. TD1, FHA2 vs. TD2), the FHA designs have a larger resonant capacitor *Cr* with both design strategies. Consequently, the resonant inductances obtained from the FHA designs are lower, and this leads to some key facts:

- The lower *Lm* increases its peak-to-peak current (magnetizing current) when the output current is flowing because the voltage across it is fixed and proportional to the output voltage times the transformer turn ratio, which is the same in both cases. This also increases the rms current and, therefore, the power dissipation. Analytical calculations (please refer to Table 4) show an increment of the rms value by more than 15% in the FHA1 vs. TD1 comparison and by more than 33% in the FHA2 vs. TD2 comparison. Of course, this positively affects the total rms current as well.

- The higher resonant capacitor Cr reduces the impedance of the resonant tank and, therefore, the quality factory Q_0. According to the FHA theory, the lower the quality factor, the higher the maximum gain that the converter can achieve, hence moving the converter's operation away from the capacitive region. This is confirmed by the initial resonant tank current I_o at the peak of the minimum input voltage, which is more than 34% greater in the FHA1 design with respect to the TD1 design, while it is over 48% larger in the FHA2 design compared to the TD2 design.

Comparing the two analytical results of the FHA analysis (FHA1 vs. FHA2) FHA2 has an initial current over 22% greater than FHA1, but there is a reduction in the rms secondary current by more than 16%. Vice versa, comparing the two results of the TD analysis (TD1 vs. TD2), the initial current is essentially the same since it mostly depends on the capacitance of the half-bridge node, while the rms currents in Lm and the output diodes are reduced by more than 22% and 19%, respectively, in the TD2 design compared to TD1. Table 4 summarizes all these results.

The same remarks can be made by observing the waveforms of Figure 11. Comparing FHA1 with TD1, we notice that the initial current I_o is lower in TD1, which means that the converter operates closer to the inductive-capacitive boundary and utilizes a larger portion of the available operating region. Also, it is confirmed that the peak-to-peak (and, consequently, the rms) magnetizing current, which is purely reactive, is smaller in TD1.

The advantage of designing with $f_{max} > f_{R1}$ is even more conspicuous. Comparing FHA1 with FHA2, the $(Tm - Tsw/2)$ interval is much shorter in FHA2, and, consequently, the conduction angle of the secondary current is larger, which significantly reduces their peak and rms value. Comparing FHA2 with TD2, we observe a reduction in the initial current I_o in TD2, with the same consequences.

It is interesting to visualize the difference between the results of the FHA approach with those of the TD approach shown in Table 3, using the key tool of the FHA approach.

The plots of the voltage gain evaluated with the FHA analysis at the peak of the input voltage for FHA1 and TD1 designs are shown in Figure 12.

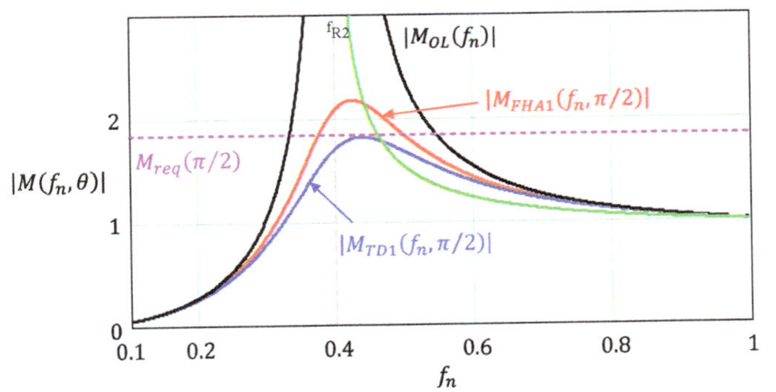

Figure 12. Voltage gain of FHA1 and TD1 designs, evaluated with FHA analysis.

The TD1 design (blue line) seems not to be working: the peak voltage gain gets very close to but does not reach the required gain, and this happens in the capacitive region. The conclusion would be that the converter cannot regulate the output voltage and works in capacitive mode, then without ZVS. In contrast, there is enough gain margin with the FHA design (red line). However, the time-domain analysis shows that the TD1 design works, and with ZVS. This depends, as already said, on the approximation of the FHA analysis that is less accurate when moving away from the upper resonance frequency f_{R1}.

Regarding the FHA2 and TD2 designs, they cannot be compared in the same plot because the lower resonance frequencies are quite different, and so they are the curves of the gain, but the same remarks apply if we look at their plots separately in Figures 13 and 14.

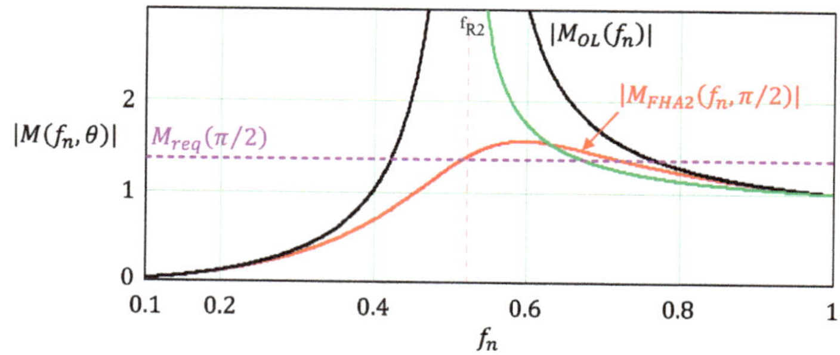

Figure 13. Voltage gain of the FHA2 design, evaluated with FHA analysis.

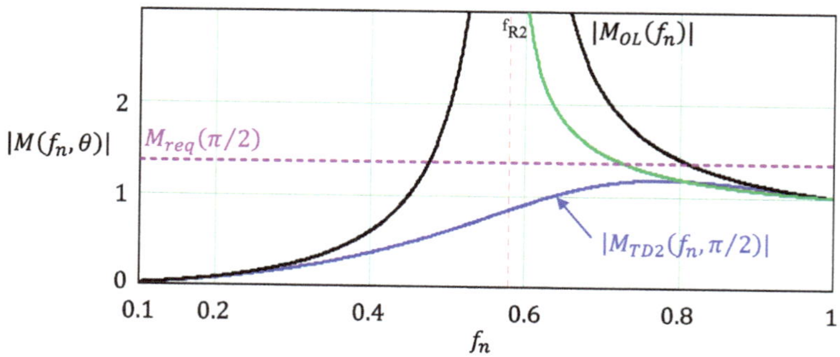

Figure 14. Voltage gain of the TD2 design, evaluated with FHA analysis.

The conclusion is that the FHA designs are more conservative with respect to the TD designs and, consequently, more lossy, and less efficient. Therefore, the FHA approach is useful for carrying out a preliminary design or a design where efficiency and power density are not the primary concerns. In fact, the real advantage of the FHA is the possibility to make a design easily and quickly without sophisticated calculation tools. For an optimized design, the TD analysis must be used, which allows for a reduction in the rms magnetizing current while still ensuring the ZVS of the primary switches.

The other key point worth highlighting is that designing the converter to work also in the above resonance region improves the shape of the magnetizing and secondary currents, thus reducing their peak and rms values. As a result, power dissipation is reduced too, and efficiency is increased. However, if the design is based on the FHA approach, this efficiency gain could be completely offset by the higher losses in the resonant tank, due to the excess of magnetizing current.

Before building the prototype, a series of simulations were run to check the paper designs and, if needed, refine them. The results, not shown here, do not differ significantly from those calculated and listed in Table 4. Also, the waveforms were very similar to those shown in Figure 11.

5. Prototype and Performance

A prototype was built able to accommodate both TD designs. Its picture, with the key components highlighted, is shown in Figure 15. Table 5 shows the electric characteristics of the resonant tank of both prototypes.

Table 6 shows the measured current stress at the line frequency time scale. The values are slightly larger than those predicted by the calculations (shown in Table 4, lines #6–7) which took power loss sources into account quite roughly.

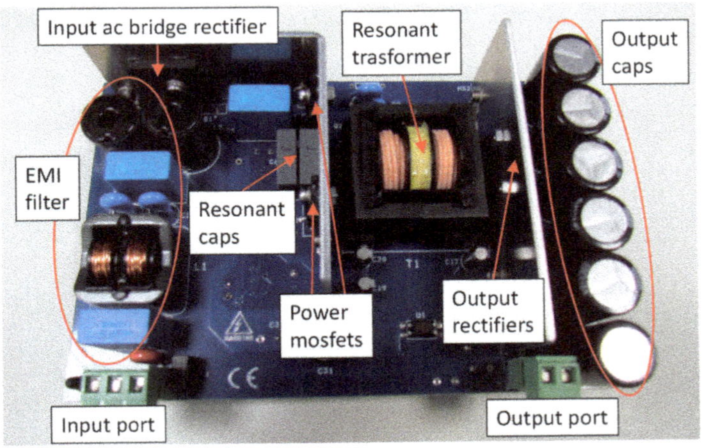

Figure 15. Realized prototype with the most important components highlighted.

Table 5. Electrical characteristics of the realized resonant tanks.

Symbol	Name	TD1	TD2	Unit
a	Primary-to-secondary turn ratio	4	2.8	---
Cr	Resonant capacitor	2 × 22	22	nF
Lr	Series resonant inductance	23.7	50	μH
Lm	Parallel resonant inductance	138	101	μH
f_{R1}	Upper resonance frequency	155.9	151.7	kHz
f_{R2}	Lower resonance frequency	59.7	87.3	kHz

Table 6. Measured current stress (rms values) over a line cycle at the minimum input voltage for the two TD designs listed in Table 5.

Parameter	TD1	TD2	Unit
Resonant tank current	3.79	3.53	A
Secondary current (total)	8.01	6.54	A

Figures 16 and 17 show the most important waveform of both TD designs over one line cycle at the minimum mains voltage. In particular, the half-bridge voltage (CH1, gold color) follows the ac input voltage (CH5, gray color), and the resonant current (CH4, green color) is modulated over the line cycle. Also, the output voltage (CH3, blue color) has a low-frequency ripple (twice the line frequency), typical of an active PFC circuit.

At the peak of the minimum input voltage and maximum load, the HB node swings inside the dead time, and both switches operate with ZVS, as shown in Figures 18 and 19 for the TD1 and TD2 designs, respectively.

As predicted by both the FHA and the TD analysis, when the instantaneous input voltage is below the peak, there is more margin for ZVS. Figures 20 and 21 show the resonant tank current and the HB voltage for the TD1 and TD2 designs, respectively, at a phase of the minimum input voltage $\theta = \pi/4$, while Figures 22 and 23 show the same waveforms at the valley of the minimum input voltage. Notice that the reactive (magnetizing) current is in quadrature with the applied voltage (inferable from the gate-drive signal), consistent with the instantaneous input power being zero at the zero crossing of the line voltage.

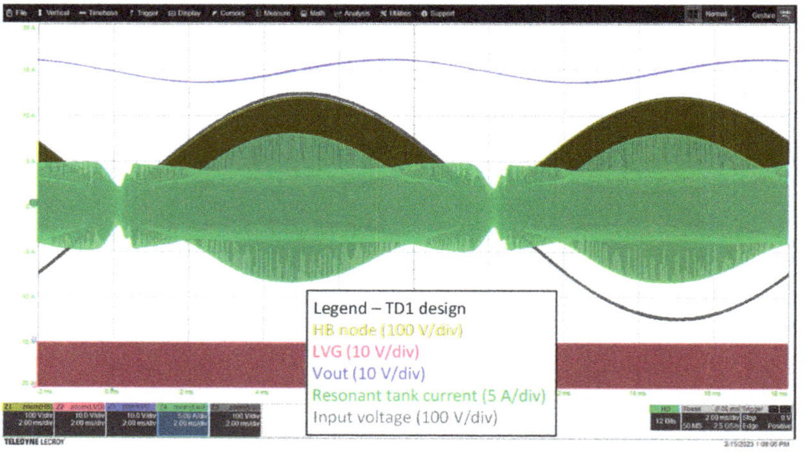

Figure 16. Prototype's main waveforms of the TD1 design of Table 5.

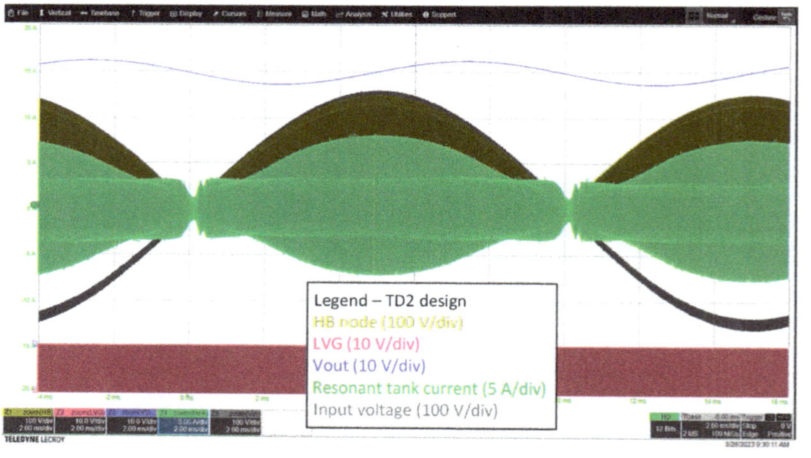

Figure 17. Prototype's main waveforms of the TD2 design of Table 5.

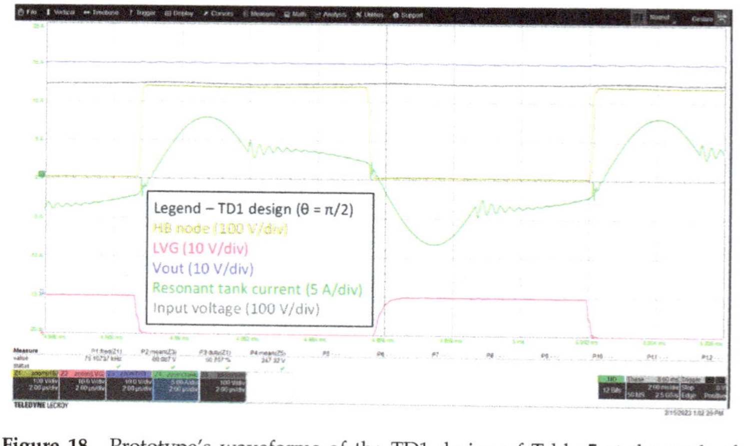

Figure 18. Prototype's waveforms of the TD1 design of Table 5 at the peak of the minimum input voltage.

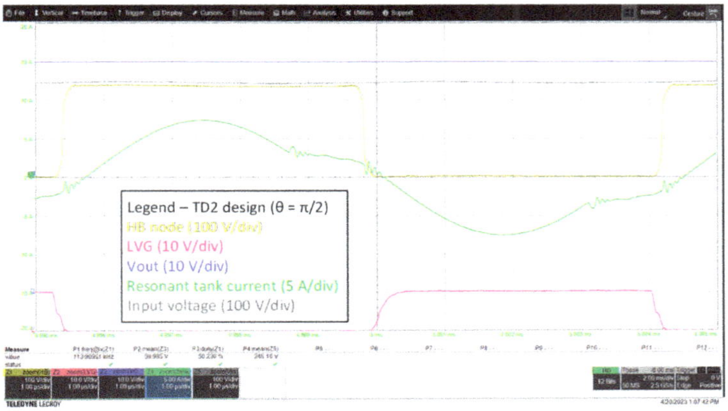

Figure 19. Prototype's waveforms of the TD2 design of Table 5 at the peak of the minimum input voltage.

Figure 20. Prototype's waveforms of the TD1 design of Table 5 at the minimum input voltage and phase $\theta = \pi/4$.

Figure 21. Prototype's waveforms of the TD2 design of Table 5 at the minimum input voltage and phase $\theta = \pi/4$.

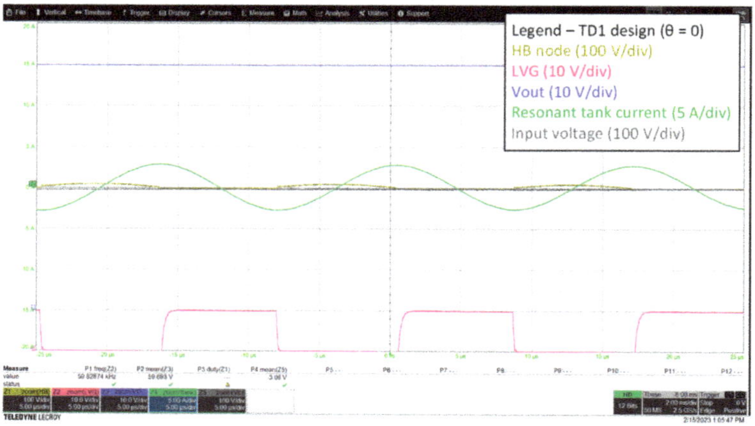

Figure 22. Prototype's waveforms of the TD1 design of Table 5 at the valley of the minimum input voltage (i.e., the zero crossing of the line voltage).

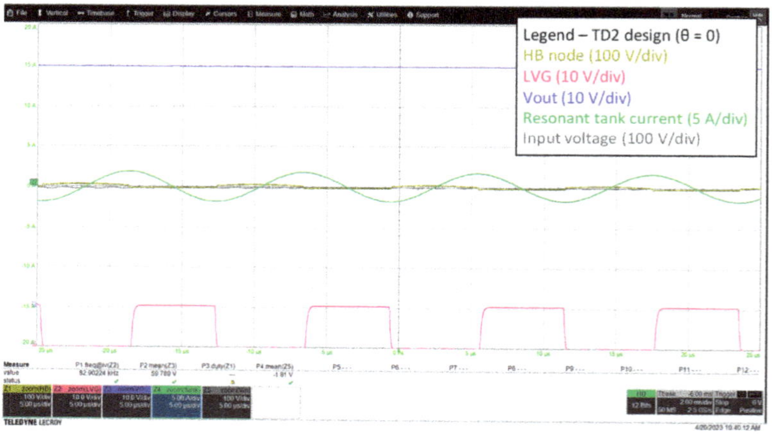

Figure 23. Prototype's waveforms of the TD2 design of Table 5 at the valley of the minimum input voltage (i.e., the zero crossing of the line voltage).

Finally, the input current and voltage at the minimum and nominal input voltage are shown in Figure 24: the shape of the current tracks the shape of the input voltage very closely and with very low phase shift, consistently with the low THD and high PF measured values, thus confirming the effectiveness of the topology.

The efficiency comparison between the TD designs is shown in Figure 25. The peak efficiency is 96% for both of them at 100% load, but the TD2 design outperforms the TD1 design at lower loads, where the dissipation due to the greater reactive current of the TD1 design is more impacting. At 10% load, TD2 is about 4% more efficient. The four-point average efficiency (mean value of the efficiency at 100%, 75%, 50% and 25% load) of TD2 is notably higher as well.

The total power loss at 100% load is about 10W and seems to be distributed quite evenly among the various parts that handle power (EMI filter, input bridge, primary MOSFETs, synchronous rectifier MOSFETs, resonant transformer). The transformer is the hottest spot, which suggests that its construction needs to be considered carefully.

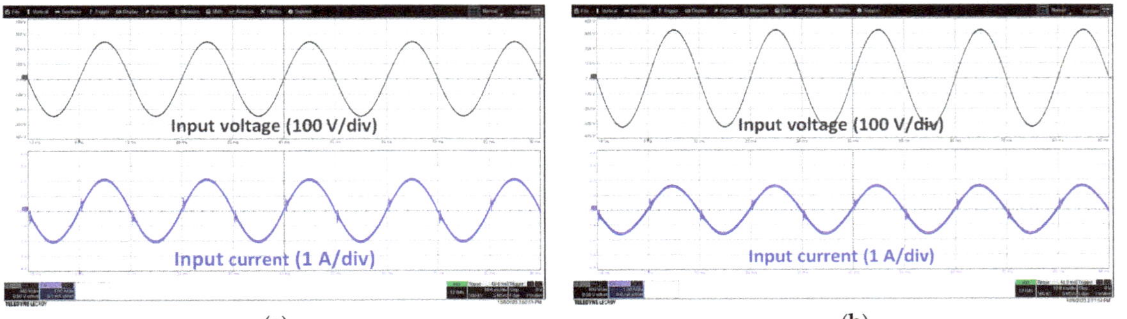

(**a**) (**b**)

Figure 24. Input voltage (in gray) and current (in blue) at full load in the TD2 design: (**a**) at 176 Vac; (**b**) at 230 Vac.

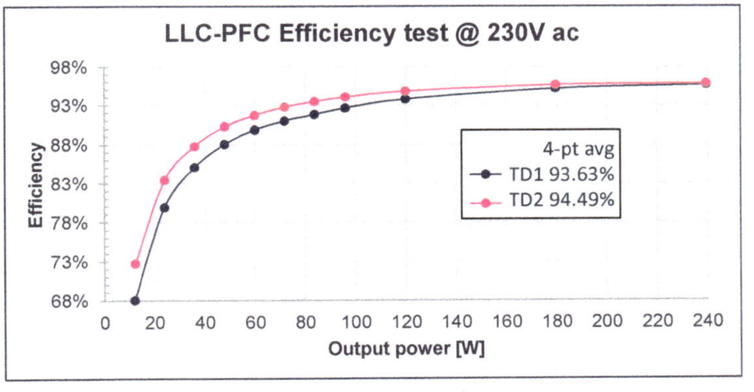

Figure 25. Efficiency comparison between the TD designs.

Figure 26 shows the THD and the PF of the TD2 design. The THD is below 10% down to 60W (25% of output load), while the PF peaks at 0.997 at full load and stays above 0.9 down to 25% load. These figures are comparable to those of boost PFC converters of similar power [31].

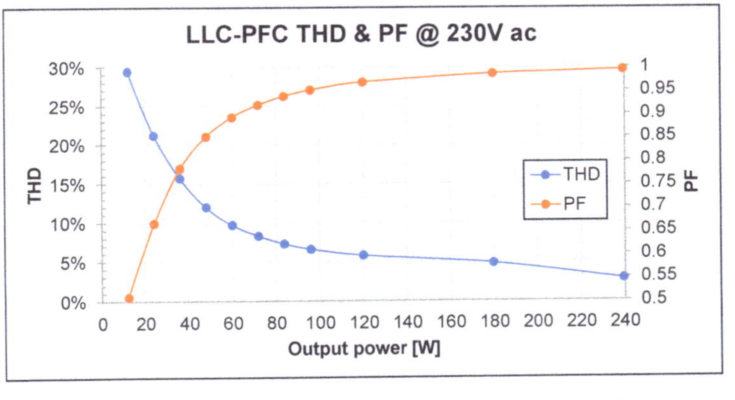

Figure 26. Prototype performance in terms of THD and PF of the TD2 design.

6. Conclusions

In this paper, the usage and the design of an LLC resonant converter as a power factor corrector, the LLC-PFC, were discussed. An LLC-PFC has many interesting features: safety isolation between the mains and the user, the ability to address any output voltage by adjusting the turn ratio of the transformer and soft-switching operation with high efficiency and low EMI emissions. From the system perspective, the usage of an isolated PFC enables the use of non-isolated downstream dc-dc converters (e.g., buck converters) to generate the final voltage for the equipment, which is particularly advantageous in multiple output systems.

The first harmonic approximation (FHA) analysis can prove that the LLC converter can perform as a PFC, but its accuracy as a design tool was questioned because of the approximations inherent in it. For this reason, a time domain (TD) analysis was performed, showing that the FHA analysis leads to a more conservative design of the resonant tank. If efficiency is a primary design target, a design based on the TD analysis is preferable despite its higher complexity because it results in a lower reactive current in the resonant tank, especially when the converter is designed to work partly above the upper resonance frequency.

Finally, two different designs (one working always below the upper resonance frequency, the other working also above this frequency) based on the TD analysis were carried out, and a prototype was realized, to compare and validate the results found. The experiments showed a significant reduction in the reactive current in the transformer and a higher efficiency, especially at low loads, when the converter is designed to work at the upper resonance frequency with the nominal input voltage.

Based on these results, the next steps will be the preparation of properly engineered hardware to evaluate the level of power density achievable with this converter. Another target of future work is the development of a PFC-LLC able to work with a wide range of mains, from 90 to 305 Vac. In both cases, the biggest expected challenge is in the transformer.

Author Contributions: Conceptualization, M.S. and C.A.; methodology, M.S.; validation, M.S.; formal analysis, M.S. and C.A.; investigation, M.S.; resources, M.S.; data curation, M.S.; writing—original draft preparation, M.S.; writing—review and editing, M.S. and C.A.; visualization, M.S.; supervision, C.A.; project administration, C.A. All authors have read and agreed to the published version of the manuscript.

Funding: This research received no external funding.

Data Availability Statement: Data are contained within the article.

Acknowledgments: The authors thank their retired colleague Silvio De Simone for his precious contribution to this work and ITACOIL s.r.l. for supplying the magnetic samples.

Conflicts of Interest: The authors declare no conflict of interest.

Appendix A

The 12-step procedure illustrated in [30] allows the design of an LLC-PFC converter with the FHA approach such that it works at the upper resonance frequency f_{R1} at the peak of the maximum input voltage. This procedure was followed for the FHA1 design.

This design procedure was slightly modified to design the converter such that it works at the upper resonance frequency f_{R1} at the peak of the nominal input voltage, thus allowing the frequency range to extend beyond f_{R1}. The FHA2 design was carried out with the following step-by-step design procedure based on the electrical spec in Table 1.

Step 1. Calculate the turn ratio a so that the converter works at resonance at the peak of the nominal input and output voltages:

$$a = \frac{Vin_{nom}}{\sqrt{2}(Vo + V_{Rect})} = 2.71 \rightarrow a = 2.8$$

Step 2. Calculate the output resistance $Rout_{ac}$:

$$Rout_{ac} = \frac{4}{\pi^2} a^2 \frac{(Vo + V_{Rect})^2}{Pout_{max}} = 47.82\ \Omega$$

Step 3. Calculate the maximum voltage gain M_{max} at the peak of Vin_{min} and the minimum voltage gain M_{min} at the peak of Vin_{max} and minimum output voltage:

$$M_{max} = 2a \frac{Vo + V_{Rect}}{\sqrt{2}\ Vin_{min}} = 1.352$$

$$M_{min} = 2a \frac{Vo_{min} + V_{Rect}}{\sqrt{2}\ Vin_{max}} = 0.78$$

Step 4. Calculate λ so the required minimum gain is fulfilled at f_{max}:

$$\lambda = \frac{\frac{1}{M_{min}} - 1}{1 - \left(\frac{f_{R1}}{f_{max}}\right)^2} = 0.375$$

Step 5. Calculate the maximum Q_0 value, Q_{max1}, necessary to stay in the inductive region at minimum Vin and maximum load:

$$Q_{max1} = \frac{\lambda}{M_{max}} \sqrt{\frac{{M_{max}}^2}{{M_{max}}^2 - 1} + \frac{1}{\lambda}} = 0.613$$

Step 6. Calculate the maximum Q_0 value, Q_{max2}, to ensure ZVS at zero load and maximum Vin:

$$Q_{max2} = \frac{2}{\pi} \lambda \frac{T_D}{Rout_{ac}\ C_{HB}} = 2.045$$

Step 7. Calculate the maximum Q_0 value, Q_{max3}, to ensure that the minimum value requirement of the maximum gain is fulfilled:

$$Q_{max3} = \frac{\sqrt{\lambda(1+\lambda)}}{M_{max}} = 0.531$$

Step 8. Choose a value of Q_0, Q_S, such that:

$$Q_S \leq min(Q_{max1}, Q_{max2}, Q_{max3}) = 0.531$$

Step 9. Calculate the normalized minimum operating frequency $f_{n_{min}}$ at Vin_{min} at $\theta = \pi/2$ and maximum output power:

$$f_{n_{min}} \cong \frac{1}{\sqrt{1 + \frac{1}{\lambda}\left(1 - \frac{1}{\frac{1+\left(\frac{Q_S}{Q_{max1}}\right)^5}{M_{max}}}\right)}} = 0.713$$

Step 10. Calculate the phase-shift φ_{min} of the resonant tank current at the peak of minimum input voltage and maximum output power and check if the ZVS condition is fulfilled. If so, proceed to step 11, otherwise choose a smaller value for Q_S and go back to step 9.

$$\varphi_{min} = tan^{-1} \frac{\left[\lambda^2 + \lambda + Q_S^2\left(f_{n_{min}}^2 - 1\right)\right] f_{n_{min}}^2 - \lambda^2}{Q_S\ f_{n_{min}}^3} = 0.26\text{rad}$$

$$\frac{\varphi_{min}}{2\pi f_{R1}} \frac{1}{f n_{min}} > T_D \rightarrow 386 \text{ ns} > 270 \text{ ns}$$

Step 11. Calculate the characteristic impedance of the resonant tank circuit and all components:

$$Z_0 = Re\ Q_S = 25.415\Omega$$

$$Cr = \frac{1}{2\pi f_{R1} Z_0} = 41.748 \text{ nF} \rightarrow Cr = 2{\cdot}22nF = 44 \text{ nF}$$

$$Ls = \frac{Z_0}{2\pi f_{R1}} = \frac{1}{(2\pi f_{R1})^2 Cr} = 25.586\ \mu\text{H} \rightarrow Ls = 25.6\ \mu\text{H}$$

$$Lp = \frac{Ls}{\lambda} = 68.2\ \mu\text{H}$$

Appendix B

Before manipulating the equations of the time-domain analysis, the following expressions and definitions are introduced:

$$\begin{gathered} \frac{Z_{R1}}{Z_{R2}} = \frac{f_{R1}}{f_{R2}} \frac{Lr}{Lr+Lm} = \frac{f_{R2}}{f_{R1}}, \\ Lr = Lm \frac{f_{R2}^2}{f_{R1}^2 - f_{R2}^2}, \\ Cr = \frac{1}{(2\pi f_{R1})^2 Lr}, \\ f_{n2} = \frac{f_{R2}}{f_{R1}} = \sqrt{\frac{Lr}{Lr+Lm}} = \sqrt{\frac{\lambda}{1+\lambda}}, \\ \lambda = \frac{Lr}{Lm} = \frac{f_{R2}^2}{f_{R1}^2 - f_{R2}^2} = \frac{f_{n2}^2}{1 - f_{n2}^2}. \end{gathered} \tag{A1}$$

For the solution (18) of the system (16), it is convenient to define also:

$$I_x = \frac{1}{Z_{R1}} \left(Vin_{pk} - a\ Vout - V_{Cr_0} \right) V_x = Z_{R1} \cdot I_x = \left(Vin_{pk} - a\ Vout - V_{Cr_0} \right). \tag{A2}$$

The solution (18) becomes:

$$\begin{cases} v_{Cr}(t) = -Z_{R1}\left[I_o \sin(2\pi f_{R1} t) + I_x \cos(2\pi f_{R1} t)\right] + \left(Vin_{pk} - a\ Vout\right) = \left(Vin_{pk} - a\ Vout\right) - Z_{R1}\ I_{pk} \cos(2\pi f_{R1} t - \varphi) \\ i_{RT}(t) = I_x \sin(2\pi f_{R1} t) - I_o \cos(2\pi f_{R1} t) = I_{pk} \sin(2\pi f_{R1} t - \varphi) \\ i_M(t) = a \frac{Vout}{Lm} t - I_o \end{cases}, \tag{A3}$$

where:

$$\begin{gathered} I_{pk} = \sqrt{I_x^2 + I_o^2} \\ \varphi = \arctan\left(\frac{I_o}{I_x}\right) = \arcsin\left(\frac{I_o}{I_{pk}}\right). \end{gathered} \tag{A4}$$

The first interval ends when the secondary current stops flowing in the Tm time instant, when the resonant current equals the magnetizing current:

$$i_{RT}(Tm) = I_{pk} \sin(2\pi f_{R1} Tm - \varphi) = I_M(Tm) = a \frac{Vout}{Lm} Tm - I_o = I_m \rightarrow Lm = a\ Vout \frac{Tm}{I_m + I_o}. \tag{A5}$$

The above expression allows us to express the magnetizing current as follows:

$$I_M(t) = \frac{I_m + I_o}{Tm} t - I_o.$$

The expression of the resonant capacitor voltage in $t = 0$ and $t = Tm$ is:

$$v_{Cr}(0) = V_{Cr_0} = \left(Vin_{pk} - a\ Vout\right) - Z_{R1}\ I_{pk}\cos(\varphi)$$
$$v_{Cr}(Tm) = V_{Cr_{Tm}} = \left(Vin_{pk} - a\ Vout\right) - Z_{R1}\ I_{pk}\cos(2\pi f_{R1} Tm - \varphi).$$

For the solution (22) of the system (20), it is convenient to define also:

$$I_y = \frac{1}{Z_{R2}}\left(Vin_{pk} - V_{Cr_{Tm}}\right) V_y = Z_{R2} \cdot I_y = \left(Vin_{pk} - V_{Cr_{Tm}}\right). \tag{A6}$$

The solution (22) becomes:

$$\begin{cases} v_{Cr}(t) = Z_{R2}\{I_m \sin[2\pi f_{R2}(t - Tm)] - I_y \cos[2\pi f_{R2}(t - Tm)]\} + Vin_{pk} = Vin_{pk} - Z_{R2}\ I_{pkM}\cos[2\pi f_{R2}(t - Tm) + \phi] \\ i_{mag}(t) = I_m \cos[2\pi f_{R2}(t - Tm)] + I_y \sin[2\pi f_{R2}(t - Tm)] = \\ = I_{pkM}\sin[2\pi f_{R2}(t - Tm) + \phi] \end{cases}, \tag{A7}$$

where:

$$I_{pkM} = \sqrt{I_y^2 + I_m^2}$$
$$\phi = \arctan\left(\frac{I_m}{I_y}\right) = \arcsin\left(\frac{I_m}{I_{pkM}}\right). \tag{A8}$$

The expressions of the resonant capacitor voltage at $t = Tm$ and $t = Tsw/2$ are the following:

$$v_{Cr}(Tm) = V_{Cr_{Tm}} = Vin_{pk} - Z_{R2}\ I_{pkM}\cos(\phi)$$
$$v_{Cr}\left(\frac{Tsw}{2}\right) = V_{Cr_{TswH}} = Vin_{pk} - Z_{R2}\ I_{pkM}\cos\left[2\pi f_{R2}\left(\frac{Tsw}{2} - Tm\right) + \phi\right].$$

Table A1 summarizes the time-domain equations in a half switching cycle.

Table A1. Time-domain equations in $(0 - Tsw/2)$.

Description	Equation	
Series inductor current	$i_{Lr}(t) = \begin{cases} i_{RT}(t) = I_{pk}\sin(2\pi f_{R1} t - \varphi) \forall t \in [0, Tm] \\ i_{mag}(t) = I_{pkM}\sin[2\pi f_{R2}(t - Tm) + \phi] \forall t \in \left[Tm, \left(\frac{Tsw}{2}\right)\right] \end{cases}$	(A9)
Parallel inductor current	$i_{Lm}(t) = \begin{cases} i_M(t) = \left(\frac{I_m + I_o}{Tm}\right)t - I_o \forall t \in [0, Tm] \\ i_{mag}(t) = I_{pkM}\sin[2\pi f_{R2}(t - Tm) + \phi] \forall t \in \left[Tm, \left(\frac{Tsw}{2}\right)\right] \end{cases}$	(A10)
Resonant capacitor voltage	$v_{Cr}(t) = \begin{cases} Vin_{pk} - aVout - Z_{R1} I_{pk}\cos(2\pi f_{R1} t - \varphi) \forall t \in [0, Tm] \\ Vin_{pk} - Z_{R2} I_{pkM}\cos[2\pi f_{R2}(t - Tm) + \phi] \forall t \in \left[Tm, \left(\frac{Tsw}{2}\right)\right] \end{cases}$	(A11)
Secondary side current	$i_{sec}(t) = a[i_{Lr}(t) - i_{Lm}(t)] = \begin{cases} a\left[I_{pk}\sin(2\pi f_{R1} t - \varphi) - \left(\left(\frac{I_m + I_o}{Tm}\right)t - I_o\right)\right] \forall t \in [0, Tm] \\ 0 \forall t \in \left[Tm, \left(\frac{Tsw}{2}\right)\right] \end{cases}$	(A12)

Since the resonant capacitor is supposed to be referred to ground (i.e., it is not split), the operation in the first switching half cycle $(0 - Tsw/2)$ is the one where the resonant tank current circulates in the high side switch of the half-bridge converter, and it is the only current drawn from the input supply voltage. In the subsequent half cycle $(Tsw/2 - Tsw)$, the circulating current in the resonant tank is the one flowing through the low side switch, and, thanks to the symmetry, all the currents and voltage expressions are known also in this second time interval.

The following continuity relationships for the first-time interval $(0 - Tm)$ need to be considered:

$$i_{RT}(0) = I_{pk}sin(-\varphi) = -I_o \rightarrow I_{pk} = \frac{I_o}{\sin(\varphi)}$$

$$i_{RT}(Tm) = I_{pk}\sin(2\pi f_{R1}Tm - \varphi) = I_m \rightarrow \frac{I_m}{I_{pk}} = \sin(2\pi f_{R1}Tm)\cos(-\varphi) + \sin(-\varphi)\cos(2\pi f_{R1}Tm).$$

Using the previous expression of I_{pk}, we get

$$\frac{I_m}{I_o} = \frac{1}{\tan(\varphi)}\sin(2\pi f_{R1}Tm) - \cos(2\pi f_{R1}Tm) \rightarrow \tan(\varphi) = \frac{I_o\sin(2\pi f_{R1}Tm)}{I_m + I_o\cos(2\pi f_{R1}Tm)}. \quad \text{(A13)}$$

The same must be done for the second time interval $(Tm - Tsw/2)$:

$$i_{mag}(Tm) = I_{pkM}\sin(\phi) = I_m \rightarrow I_{pkM} = \frac{I_m}{\sin(\phi)}$$

$$i_{mag}\left(\frac{Tsw}{2}\right) = I_{pkM}\sin\left[2\pi f_{R2}\left(\frac{Tsw}{2} - Tm\right) + \phi\right] = I_o \rightarrow \frac{I_o}{I_{pkM}} = \sin\left[2\pi f_{R2}\left(\frac{Tsw}{2} - Tm\right)\right]\cos(\phi) + \sin(\phi)\cos\left[2\pi f_{R2}\left(\frac{Tsw}{2} - Tm\right)\right].$$

Using the previous expression of I_{pkM}, we get

$$\frac{I_o}{I_m} = \frac{1}{\tan(\phi)}\sin[\pi f_{R2}(Tsw - 2Tm)] + cos[\pi f_{R2}(Tsw - 2Tm)] \rightarrow \tan(\phi) = \frac{I_m\sin[\pi f_{R2}(Tsw - 2Tm)]}{I_o - I_m\cos[\pi f_{R2}(Tsw - 2Tm)]}. \quad \text{(A14)}$$

The following trigonometric relationships hold true:

$$\begin{array}{lll} \sin(\varphi) = \frac{I_o}{I_{pk}}; & \cos(\varphi) = \sqrt{\frac{I_{pk}^2 - I_o^2}{I_{pk}^2}}; & \tan(\varphi) = \sqrt{\frac{I_o^2}{I_{pk}^2 - I_o^2}}; \\ \sin(\phi) = \frac{I_m}{I_{pkM}}; & \cos(\phi) = \sqrt{\frac{I_{pkM}^2 - I_m^2}{I_{pkM}^2}}; & \tan(\phi) = \sqrt{\frac{I_m^2}{I_{pkM}^2 - I_m^2}}. \end{array} \quad \text{(A15)}$$

Considering the above trigonometric relationships, Equations (A13) and (A14) become:

$$\frac{I_m}{I_o} = \sqrt{\frac{I_{pk}^2 - I_o^2}{I_o^2}}\sin(2\pi f_{R1}Tm) - \cos(2\pi f_{R1}Tm)$$

$$\frac{I_o}{I_m} = \sqrt{\frac{I_{pkM}^2 - I_m^2}{I_m^2}}\sin[\pi f_{R2}(Tsw - 2Tm)] + \cos[\pi f_{R2}(Tsw - 2Tm)].$$

From the above expressions, these two peak current expressions are found:

$$I_{pk} = \frac{\sqrt{I_m^2 + I_o^2 + 2 \cdot I_o \cdot I_m\cos(2\pi f_{R1}Tm)}}{\sin(2\pi f_{R1}Tm)} \quad \text{(A16)}$$

$$I_{pkM} = \frac{\sqrt{I_m^2 + I_o^2 - 2 \cdot I_o \cdot I_m\cos[\pi f_{R2}(Tsw - 2Tm)]}}{\sin[\pi f_{R2}(Tsw - 2Tm)]}. \quad \text{(A17)}$$

Keeping in mind that we are considering the operation at the peak input voltage $(\theta = \pi/2)$, where the input and output peak power in a PFC circuit equals twice the rated dc output power, we have that the peak output current equals twice the average value of the secondary current expression in one switching half cycle:

$$Iout_{pk} = 2Iout = 2\frac{Pout}{Vout} = \frac{2}{Tsw}\int_0^{\frac{Tsw}{2}} i_{sec}(t)dt = \frac{2a}{Tsw}\int_0^{Tm} [i_{RT}(t) - i_M(t)]dt \quad \text{(A18)}$$

using the expression of $i_{sec}(t)$ defined in (19).

In the same way, on the primary side, the peak input current at the peak input voltage is equal to the average value (in a complete switching cycle) of the current flowing through the HS switch of the half-bridge converter:

$$Iin_{pk}=\sqrt{2}\,Iin=2\frac{Pin}{Vin_{pk}}=\frac{1}{Tsw}\int_0^{Tsw} i_{Lr}(t)dt=\frac{1}{Tsw}\left[\int_0^{Tm} i_{RT}(t)dt+\int_{Tm}^{\frac{Tsw}{2}} i_{mag}(t)dt\right]. \tag{A19}$$

Subtracting (A18) from (A19) and with some algebra manipulations, we get

$$Iin_{pk}-\frac{Iout_{pk}}{2a}=\frac{1}{Tsw}\left[\int_0^{Tm} i_M(t)dt+\int_{Tm}^{\frac{Tsw}{2}} i_{mag}(t)dt\right]. \tag{A20}$$

Substituting in (A18) and (A20) in the expressions of the currents of solutions (A3) and (A7), using the continuity relationships and after some manipulations, we obtain:

$$i_{RT}(t)=\tfrac{I_o}{\sin(\varphi)}\sin(2\pi f_{R1}t-\varphi)$$

$$i_M(t)=\tfrac{I_m+I_o}{Tm}t-I_o$$

$$i_{mag}(t)=\tfrac{I_m}{\sin(\phi)}\sin[2\pi f_{R2}(t-Tm)+\phi]$$

$$Iout_{pk}=\tfrac{a\,I_o}{Tsw}\left\{\tfrac{1}{\pi f_{R1}}\left[\tfrac{1-\cos(2\pi f_{R1}Tm)}{\tan(\varphi)}-\sin(2\pi f_{R1}Tm)\right]-Tm\left(\tfrac{I_m}{I_o}-1\right)\right\}$$

$$Iin_{pk}-\tfrac{Iout_{pk}}{2a}=\tfrac{I_m}{2Tsw}\left\{Tm\left(1-\tfrac{I_o}{I_m}\right)+\tfrac{1}{\pi f_{R2}}\left[\tfrac{1-\cos(\pi f_{R2}(Tsw-2Tm))}{\tan(\phi)}+\sin(\pi f_{R2}(Tsw-2Tm))\right]\right\}.$$

Further manipulating the above expressions by using relationships (A13) and (A14), we finally obtain the following equations:

$$Iout_{pk}=a\frac{I_m-I_o}{Tsw}\left[\frac{\tan(\pi f_{R1}Tm)}{\pi f_{R1}}-T_m\right] \tag{A21}$$

$$Iin_{pk}-\frac{Iout_{pk}}{2a}=\frac{1}{2Tsw}\left[T_m(I_m-I_o)+(I_m+I_o)\frac{\tan\left[\pi f_{R2}\left(\frac{Tsw}{2}-Tm\right)\right]}{\pi f_{R2}}\right]. \tag{A22}$$

The resonant capacitor voltage is symmetric with respect to its dc component ($Vin/2$ in half-bridge converters) for any phase angle of the input voltage: considering the case at the peak of the input voltage (i.e., with $\theta=\pi/2$), the following relationship holds true:

$$\frac{Vin_{pk}}{2}-v_{Cr}(0)=v_{Cr}\left(\frac{Tsw}{2}\right)-\frac{Vin_{pk}}{2}\leftrightarrow v_{Cr}(0)+v_{Cr}\left(\frac{Tsw}{2}\right)=Vin_{pk}. \tag{A23}$$

Furthermore, the resonant capacitor voltage must be consistent with the resonant current. Therefore, this voltage can be expressed as follows:

$$v_{Cr}(t)=\left|\begin{pmatrix} v_{Cr}(0)+(\frac{1}{Cr})\int_0^t i_{RT}(\tau)d\tau+\forall\, t\in[0,Tm] \\ v_{Cr}(Tm)+(\frac{1}{Cr})\int_{Tm}^t i_{mag}(\tau)d\tau\forall\, t\in\left[Tm,(\frac{Tsw}{2})\right]\end{pmatrix}\right.. \tag{A24}$$

From the above relationship, and using the expression in (A19), we obtain:

$$v_{Cr}\left(\frac{Tsw}{2}\right)=v_{Cr}(Tm)+\frac{1}{Cr}\int_{Tm}^{\frac{Tsw}{2}} i_{mag}(\tau)d\tau=v_{Cr}(0)+\frac{1}{Cr}\left(\int_0^{Tm} i_{RT}(\tau)d\tau+\int_{Tm}^{\frac{Tsw}{2}} i_{mag}(\tau)d\tau\right)=v_{Cr}(0)+\frac{Tsw}{Cr}Iin_{pk}. \tag{A25}$$

From (A23) and (A25), the following relationships are then obtained:

$$v_{Cr}(0)=V_{Cr_0}=\frac{1}{2}\left(Vin_{pk}-Iin_{pk}\frac{Tsw}{Cr}\right) \tag{A26}$$

$$v_{Cr}\left(\frac{Tsw}{2}\right) = V_{Cr_{TswH}} = \frac{1}{2}\left(Vin_{pk} + Iin_{pk}\frac{Tsw}{Cr}\right). \quad (A27)$$

Calculating the value of the capacitor voltage in the Tm time instant from (A24) and using (A27) and the latter of (A13), we get:

$$\begin{aligned} v_{Cr}(Tm) = V_{Cr_{Tm}} &= v_{Cr}(0) + \tfrac{1}{Cr}\int_0^{Tm} i_{RT}(\tau)d\tau = V_{Cr_0} + \tfrac{1}{Cr}\int_0^{Tm} \tfrac{I_o}{\sin(\varphi)}\sin(2\pi f_{R1}\tau - \varphi)d\tau \\ &= V_{Cr_0} + \tfrac{I_o}{2\pi f_{R1}Cr}\left[\tfrac{1-\cos(2\pi f_{R1}Tm)}{\tan(\varphi)} - \sin(2\pi f_{R1}Tm)\right] \\ &= \tfrac{1}{2}\left(Vin_{pk} - Iin_{pk}\tfrac{Tsw}{Cr}\right) + \tfrac{I_m - I_o}{2\pi f_{R1}\,Cr}\tan(\pi f_{R1}Tm) \end{aligned} \quad (A28)$$

By using the relationships in (A1) and (A5), the expression of Cr can be obtained:

$$Cr = \frac{1}{(2\pi f_{R1})^2 Lr} = \frac{f_{R1}^2 - f_{R2}^2}{(2\pi f_{R1}\, f_{R2})^2 Lm} = \frac{f_{R1}^2 - f_{R2}^2}{(2\pi f_{R1}\, f_{R2})^2}\frac{I_m + I_o}{a\ Vout\ Tm}. \quad (A29)$$

From (A2), using expressions (A26) and (A29), we get:

$$V_x = \left(Vin_{pk} - a\ Vout - V_{Cr_0}\right) = \frac{Vin_{pk}}{2} - a\ Vout + \frac{Iin_{pk}}{2}\frac{Tsw}{Cr}$$

$$I_x = 2\pi f_{R1} Cr V_x = \frac{I_m + I_o}{2\pi f_{R1}Tm}\frac{f_{R1}^2 - f_{R2}^2}{f_{R2}^2}\left(\frac{Vin_{pk}}{2aVout} - 1\right) + \pi f_{R1} Tsw Iin_{pk}.$$

Therefore, the first expression in (A4) can be written as:

$$I_{pk} = \sqrt{I_x^2 + I_o^2} = \sqrt{\left[\frac{I_m + I_o}{2\pi f_{R1}Tm}\frac{f_{R1}^2 - f_{R2}^2}{f_{R2}^2}\left(\frac{Vin_{pk}}{2a\ Vout} - 1\right) + \pi f_{R1} Tsw\ Iin_{pk}\right]^2 + I_o^2}.$$

Finally, equating the above expression to the one in (A16), we obtain:

$$\left[\frac{I_m + I_o}{2\pi f_{R1}Tm}\frac{f_{R1}^2 - f_{R2}^2}{f_{R2}^2}\left(\frac{Vin_{pk}}{2a\ Vout} - 1\right) + \pi f_{R1} Tsw\ Iin_{pk}\right]^2 + I_o^2 = \frac{I_m^2 + I_o^2 + 2\ I_o\ I_m cos(2\pi f_{R1}Tm)}{sin^2(2\pi f_{R1}Tm)}. \quad (A30)$$

From the first relationship in (A6), by using the expressions (A28) and (A29), we obtain:

$$\begin{aligned} I_y = 2\pi f_{R2}\ Cr\left(Vin_{pk} - V_{Cr_{Tm}}\right) &= 2\pi f_{R2}\ Cr\left[\tfrac{Vin_{pk}}{2} + \tfrac{Iin_{pk}}{2}\tfrac{Tsw}{Cr} - \tfrac{I_m - I_o}{2\pi f_{R1}Cr}\tan(\pi f_{R1}Tm)\right] \\ &= \tfrac{I_m + I_o}{2\pi f_{R2}Tm}\tfrac{f_{R1}^2 - f_{R2}^2}{f_{R1}^2}\left(\tfrac{Vin_{pk}}{2a\ Vout}\right) + \pi f_{R2} Tsw\ Iin_{pk} - \tfrac{f_{R2}}{f_{R1}}(I_m - I_o)\tan(\pi f_{R1}Tm). \end{aligned}$$

Therefore, the first expression in (A8) can be written as:

$$I_{pkM} = \sqrt{I_y^2 + I_m^2} = \sqrt{\left[\frac{I_m + I_o}{2\pi f_{R2}Tm}\frac{f_{R1}^2 - f_{R2}^2}{f_{R1}^2}\left(\frac{Vin_{pk}}{2a\ Vout}\right) + \pi f_{R2} Tsw\ Iin_{pk} - \frac{f_{R2}}{f_{R1}}(I_m - I_o)tan(\pi f_{R1}Tm)\right]^2 + I_m^2}.$$

Finally, equating the above expression to the one in (A17), we obtain:

$$\begin{aligned} &\left[\tfrac{I_m + I_o}{2\pi f_{R2}Tm}\tfrac{f_{R1}^2 - f_{R2}^2}{f_{R1}^2}\left(\tfrac{Vin_{pk}}{2a\ Vout}\right) + \pi f_{R2} Tsw\ Iin_{pk} - \tfrac{f_{R2}}{f_{R1}}(I_m - I_o)tan(\pi f_{R1}Tm)\right]^2 + I_m^2 \\ &= \tfrac{I_m^2 + I_o^2 - 2\ I_o\ I_m cos[\pi f_{R2}(Tsw - 2Tm)]}{sin^2[\pi f_{R2}(Tsw - 2Tm)]}. \end{aligned} \quad (A31)$$

Now, we must introduce the conditions to achieve ZVS for the switches of the half-bridge. According to the FHA analysis, to achieve ZVS, the impedance of the resonant tank must be inductive: this means that the resonant current must lag the half-bridge voltage as shown in Figure A1.

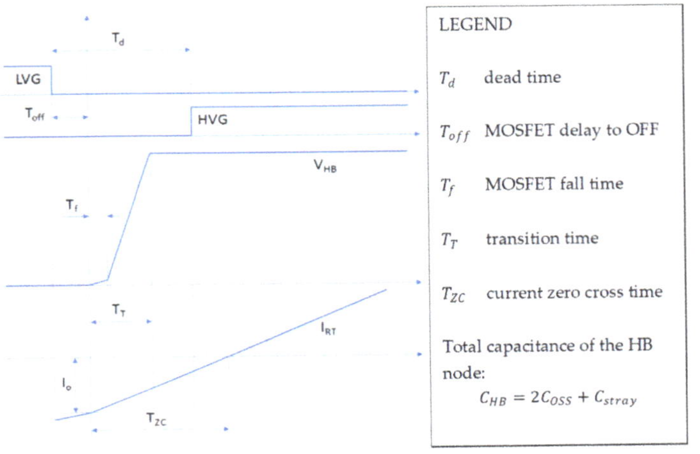

Figure A1. Half-bridge rising transition at LVG turn-off with times.

With reference to Figure A1, which refers to the turn-off of the LS switch and the turn-on of the HS switch, the following conditions are required for ZVS operation:

1. The initial current I_o must be large enough to swing the HB node rail-to-rail within the dead time T_d;
2. The circulating current (which lags the HB voltage by the angle φ) must not change sign during T_d

where T_d is the dead time between the turn-off of one switch of the HB converter and the turn-on of the other one.

After the LS gate signal (i.e., LVG) goes low, the HB node voltage remains unchanged due to the turn-off delay T_{off} of the switch. Then, the HB node swings rail-to-rail during the transition time T_T.

We make the following simplifying assumptions:

- The secondary side rectified current starts flowing at the beginning of the transition time T_T;
- The resonant tank current at the beginning of the transition time T_T equals the initial current I_o;
- The MOSFET fall time T_f is negligible with respect to the transition time T_T.

Therefore, the charging current of the HB node capacitance must be equal to the opposite of the resonant tank current:

$$i_{chg}(t) = -i_{RT}(t) = -I_{pk}\sin(2\pi f_{R1}t - \varphi) = -\frac{I_o}{\sin(\varphi)}\sin(2\pi f_{R1}t - \varphi).$$

By integrating the relationship

$$i_{chg}(t) = C_{HB}\frac{d}{dt}V_{HB}(t).$$

the expression of the half-bridge node voltage can be obtained:

$$V_{HB}(t) = \frac{1}{C_{HB}}\int_0^t i_{chg}(\tau)d\tau = \frac{1}{C_{HB}}\int_0^t -\frac{I_o\sin(2\pi f_{R1}t - \varphi)}{\sin(\varphi)}d\tau = \frac{I_o}{\sin(\varphi)}\frac{\cos(2\pi f_{R1}t - \varphi) - \cos(\varphi)}{2\pi f_{R1}\,C_{HB}}.$$

By imposing that the derivative of the HB node voltage is zero, the expression of the time instant where this voltage gets the peak value, and the expression of the peak value are found:

$$\frac{d}{dt}V_{HB}(t) = -\frac{I_o \sin(2\pi f_{R1} t - \varphi)}{C_{HB} \sin(\varphi)} = 0 \rightarrow \begin{cases} T_{HB_{pk}} = \frac{\varphi}{2\pi f_{R1}} = T_{ZC} \\ V_{HB_{pk}} = V_{HB}\left(T_{HB_{pk}}\right) = \frac{I_o}{\sin(\varphi)} \cdot \frac{1-\cos(\varphi)}{2\pi f_{R1} \cdot C_{HB}} \end{cases}. \quad \text{(A32)}$$

Mind that, referring to the first of Equation (A32), the time instant $T_{HB_{pk}}$ where the HB node voltage gets its peak $V_{HB_{pk}}$ corresponds to the time instant where the current in the resonant tank reaches zero; that is, it equals T_{ZC}.

The transition time T_T is calculated by imposing that the HB node voltage equals the peak input voltage Vin_{pk}; that is,

$$V_{HB}(t) = \frac{I_o}{sin(\varphi)} \frac{\cos(2\pi f_{R1} t - \varphi) - \cos(\varphi)}{2\pi f_{R1}\, C_{HB}} = Vin_{pk} \rightarrow T_T = \frac{\varphi - \arccos\left[cos(\varphi) + 2\pi f_{R1}\, C_{HB}\, Vin_{pk} \frac{\sin(\varphi)}{I_o}\right]}{2\pi f_{R1}} = T_{ZC} - T_{RES}, \quad \text{(A33)}$$

where T_{RES} is the residual time where the resonant current is still negative, and the HB voltage has reached the peak input voltage.

It is possible to define the minimum phase angle considering the maximum time between the switch turn-off and the turn-on of the other one:

$$\varphi_{min} = 2\pi f_{R1}\left(T_{d_{max}} - T_{off}\right). \quad \text{(A34)}$$

The ZVS condition can be imposed by setting the phase angle of the current in the resonant tank in the latter of Equation (A13) equal to φ_{min}; that is,

$$\tan(\varphi_{min}) = \frac{I_o \sin(2\pi f_{R1} Tm)}{I_m + I_o \cos(2\pi f_{R1} Tm)}. \quad \text{(A35)}$$

From the first of expressions (A33), and considering that $\varphi = \varphi_{min}$, we can get a guess value for the initial current I_o:

$$V_{HB_{pk}} = \frac{I_o}{\sin(\varphi_{min})} \frac{1 - \cos(\varphi_{min})}{2\pi f_{R1}\, C_{HB}} \geq Vin_{pk} \rightarrow I_o \geq 2\pi f_{R1}\, C_{HB}\, Vin_{pk} \frac{1 - \cos(\varphi_{min})}{\sin(\varphi_{min})} = I_{o_{min}}. \quad \text{(A36)}$$

Based on the time-domain analysis of the circuit, we found four equations (refer to (A21), (A22), (A30) and (A31)) and four unknowns (I_o, I_m, Tm and Tsw).

However, Equations (A30) and (A31) are not independent: this means that the system presents a degree of freedom that we can use to ensure ZVS.

By introducing the known quantities here (λ and K_V):

$$\lambda = \frac{L_r}{L_m} = \frac{f_{R2}^2}{f_{R1}^2 - f_{R2}^2}; K_V = \frac{\sqrt{2} Vin_{min}}{\alpha\, Vout} = \frac{Vin_{min\,pk}}{2a\, Vout}$$

and considering that the maximum required gain occurs when the input voltage is minimum, the two Equations (A30) and (A31) can be further simplified after some algebra, leading to the following equations:

$$\left[\frac{I_m + I_o}{2\pi f_{R1} Tm} \frac{K_V - 1}{\lambda} + \pi f_{R1} Tsw\, Iin_{pk}\right]^2 = \left[\frac{I_m + I_o cos(2\pi f_{R1} Tm)}{sin(2\pi f_{R1} Tm)}\right]^2 \quad \text{(A37)}$$

$$\left\{\frac{f_{R2}}{f_{R1}}\left[\frac{I_m + I_o}{2\pi f_{R2} Tm} \frac{K_V}{\lambda} + \pi f_{R1} Tsw\, Iin_{pk} - (I_m - I_o) tan(\pi f_{R1} Tm)\right]\right\}^2 = \left\{\frac{I_o - I_m cos[\pi f_{R2}(Tsw - 2Tm)]}{sin[\pi f_{R2}(Tsw - 2Tm)]}\right\}^2. \quad \text{(A38)}$$

Therefore, we can use Equations (A21), (A22) and (A37) (or (A38) alternatively) as the first three system equations, then add Equation (A35) to impose the ZVS condition, getting the following system of four equations with four unknowns (I_o, I_m, Tm and Tsw):

$$\begin{cases} Iout_{pk} = a\frac{I_m - I_o}{Tsw}\left[\frac{\tan(\pi f_{R1} Tm)}{\pi f_{R1}} - Tm\right] \\ Iin_{pk} - \frac{Iout_{pk}}{2a} = \frac{1}{2Tsw}\left[Tm(I_m - I_o) + (I_m + I_o)\frac{\tan\left(\pi f_{R2}\left(\frac{Tsw}{2} - Tm\right)\right)}{\pi f_{R2}}\right] \\ \left[\frac{I_m + I_o}{2\pi f_{R1} Tm}\frac{K_V - 1}{\lambda} + \pi f_{R1} Tsw Iin_{pk}\right]^2 = \left[\frac{I_m + I_o \cos(2\pi f_{R1} Tm)}{\sin(2\pi f_{R1} Tm)}\right]^2 \\ \tan(\varphi_{min}) = \frac{I_o \sin(2\pi f_{R1} Tm)}{I_m + I_o \cos(2\pi f_{R1} Tm)} \end{cases}.$$

Once the system solutions (I_o, I_m, Tm and Tsw) are found, the parameters of the LLC resonant tank circuit can be calculated:

$$Lm = aVout\frac{Tm}{I_m + I_o}; Lr = \lambda Lm; Cr = \frac{1}{(2\pi f_{R1})^2 Lr}; fsw = \frac{1}{Tsw}.$$

References

1. *IEC 61000-3-2:2018*; Electromagnetic Compatibility (EMC)—Part 3-2: Limits—Limits for Harmonic Current Emissions (Equipment Input Current ≤16 A per Phase). International Electrotechnical Commission: Geneva, Switzerland, 2020. Available online: https://webstore.iec.ch/publication/63492 (accessed on 13 October 2023).
2. STMicroelectronics. An Introduction to LLC Resonant Half-Bridge Converter, AN2644. 2008. Available online: https://www.st.com/resource/en/application_note/an2644-an-introduction-to-llc-resonant-halfbridge-converter-stmicroelectronics.pdf (accessed on 13 October 2023).
3. Ahmad, I.; Fernandes, B.G. Concept of Universal USB Charger. In Proceedings of the 2020 IEEE Industry Applications Society Annual Meeting, Detroit, MI, USA, 10–16 October 2020; pp. 1–5.
4. Deo, R.N.; Shrivastava, A.; Chatterjee, K. A Highly Efficient PFC Flyback Converter for Residential Lighting with Universal Input. *IETE J. Res.* **2023**. [CrossRef]
5. Qi, W.; Jie, W.; Lang, B. Design of Single-Stage Flyback PFC Converter for LED Driver. *TELKOMNIKA (Telecommunication, Comput. Electron. Control.* **2016**, *14*, 1263–1268. [CrossRef]
6. Lee, S.-W.; Do, H.-L. Single-Stage Single-Switch AC-DC LED Driver with High Power Factor. *Int. Rev. Electr. Eng. (IREE)* **2017**, *12*, 1–8. [CrossRef]
7. Chern, T.-L.; Liu, L.-H.; Huang, C.-N.; Chern, Y.-L.; Kuang, J.-H. High power factor Flyback converter for LED driver with Boundary Conduction Mode control. In Proceedings of the 2010 5th IEEE Conference on Industrial Electronics and Applications, Taichung, Taiwan, 15–17 June 2010. [CrossRef]
8. Mishra, A.K.; Pathak, M.K.; Das, S. Isolated converter topologies for power factor correction—A comparison. In Proceedings of the 2011 International Conference on Energy, Automation and Signal, Bhubaneswar, India, 28–30 December 2011; IEEE: New York, NY, USA, 2012; pp. 1–6.
9. Jang, Y.; Jovanovic, M.M. Isolated boost converters. In Proceedings of the Twenty-First Annual IEEE Applied Power Electronics Conference and Exposition, APEC '06, Dallas, TX, USA, 19–23 March 2006; p. 7.
10. Ali, M.; Orabi, M.; Ahmed, M.E.; El-Aroudi, A. A single stage SEPIC PFC converter for LED street lighting applications. In Proceedings of the 2010 IEEE International Conference on Power and Energy, Kuala Lumpur, Malaysia, 29 November–1 December 2010; pp. 501–506.
11. Jha, A.; Singh, B. Cuk PFC converter for high brightness LED driver with brightness control. In Proceedings of the 2016 IEEE 7th Power India International Conference (PIICON), Bikaner, India, 25–27 November 2016; pp. 1–6.
12. Singh, S.; Singh, B.; Bhuvaneswari, G.; Bist, V. Power Factor Corrected Zeta Converter Based Improved Power Quality Switched Mode Power Supply. *IEEE Trans. Ind. Electron.* **2015**, *62*, 5422–5433. [CrossRef]
13. Huber, L.; Jovanovic, M.M. Single-stage, single-switch input-current-shaping technique with reduced switching loss. In Proceedings of the APEC '99: Fourteenth Annual Applied Power Electronics Conference and Exposition, Dallas, TX, USA, 14–18 March 1999; Volume 1, pp. 98–104.
14. Qiao, C.M.; Smedley, K.M. A topology survey of single-stage power factor corrector with a boost type input-current-shaper. In Proceedings of the APEC 2000: Fifteenth Annual IEEE Applied Power Electronics Conference and Exposition, New Orleans, LA, USA, 6–10 February 2000; Volume 1, pp. 460–467.
15. Lai, C.M.; Lee, R.C. A Single-Stage AC/DC LLC Resonant Converter. In Proceedings of the 2006 IEEE International Conference on Industrial Technology, Mumbai, India, 15–17 December 2006; pp. 1386–1390. [CrossRef]
16. Lai, C.-M.; Lee, R.-C.; Wang, T.-W.; Shyu, K.-K. Design and Implementation of a Single-Stage LLC Resonant Converter with High Power Factor. In Proceedings of the 2007 IEEE International Symposium on Industrial Electronics, Vigo, Spain, 4–7 June 2007; pp. 455–460. [CrossRef]

17. Saasaa, R.; Eberle, W.; Agamy, M. A Single-Stage Interleaved LLC PFC Converter. In Proceedings of the 2016 IEEE Energy Conversion Congress and Exposition (ECCE), Milwaukee, MI, USA, 18–22 September 2016. [CrossRef]
18. Zhang, G.; Zeng, J.; Xiao, W.; Yu, S.S.; Zhang, B.; Zhang, Y. A Self-Protected Single-Stage LLC Resonant Rectifier. *IEEE J. Emerg. Sel. Top. Power Electron.* **2020**, *9*, 3361–3372. [CrossRef]
19. Peter, A.K.; Mathew, J. A Single Phase, Single Stage AC-DC Multilevel LLC Resonant Converter with Power Factor Correction. *IEEE Access* **2021**, *9*, 70884–70895. [CrossRef]
20. Huber, L.; Zhang, J.; Jovanovic, M.M.; Lee, F.C. Generalized topologies of single-stage input-current-shaping circuits. *IEEE Trans. Power Electron.* **2001**, *16*, 508–513. [CrossRef]
21. Qiu, Y.; Liu, W.; Fang, P.; Liu, Y.-F.; Sen, P.C. A mathematical guideline for designing an AC-DC LLC converter with PFC. In Proceedings of the 2018 IEEE Applied Power Electronics Conference and Exposition (APEC), San Antonio, TX, USA, 4–8 March 2018; pp. 2001–2008. [CrossRef]
22. Liu, W.; Yurek, A.; Sheng, B.; Chen, Y.; Liu, Y.-F.; Sen, P.C. A Single Stage 1.65kW AC-DC LLC Converter with Power Factor Correction (PFC) for On-Board Charger (OBC) Application. In Proceedings of the 2020 IEEE Energy Conversion Congress and Exposition (ECCE), Detroit, MI, USA, 11–15 October 2020; pp. 4594–4601. [CrossRef]
23. Forouzesh, M.; Liu, Y.F.; Sen, P.C. Implementation of an Isolated Phase-Modular-Designed Three-Phase PFC Rectifier Based on Single-Stage LLC Converter. In Proceedings of the 2021 IEEE Energy Conversion Congress and Exposition (ECCE), Vancouver, BC, Canada, 10–14 October 2021; pp. 2266–2273. [CrossRef]
24. de Souza, A.F.; Ribeiro, E.R.; Vicente, E.M.; Tofoli, F.L. Experimental evaluation of active power factor correction techniques in a single-phase AC-DC boost converter. *Int. J. Circuit Theory Appl.* **2019**, *47*, 1529–1553. [CrossRef]
25. Luo, H.; Zhou, B.; Liu, G.; Zang, T.; Chen, S.; Zhou, X.; Xu, J. College of Electrical Engineering, Southwest Jiaotong University, Chengdu, China., Voltage-Mode Variable-Frequency Controlled LLC Resonant Power Factor Correction Converter and Its Accurate Numerical Calculation Analysis. *IEEE J. Emerg. Sel. Top. Power Electron.* **2023**, *11*, 1979–1994. [CrossRef]
26. Wattenberg, M.; Schwalbey, U.; Pfostz, M. Single-Stage LLC Charger with PFC Functionality and Wide Input Voltage Range. In Proceedings of the 2019 IEEE Applied Power Electronics Conference and Exposition (APEC), Anaheim, CA, USA, 17–21 March 2019; pp. 750–756.
27. Chen, Y.; Yang, D.; Shao, S.; Zhang, J. Optimal Design of LLC-based Isolated Single-Stage PFC Converter. In Proceedings of the 2022 IEEE International Power Electronics and Application Conference and Exposition (PEAC), Guangzhou, China, 4–7 November 2022; pp. 1649–1654.
28. Steigerwald, R. A comparison of half-bridge resonant converter topologies. *IEEE Trans. Power Electron.* **1988**, *3*, 174–182. [CrossRef]
29. De Simone, S. LLC resonant half-bridge converter design guideline, AN2450, 2014. Available online: https://www.st.com/resource/en/application_note/an2450-llc-resonant-halfbridge-converter-design-guideline-stmicroelectronics.pdf (accessed on 13 October 2023).
30. Adragna, C. LLC Resonant Converters: An Overview of Modeling, Control and Design Methods and Challenges. *Found. Trends®Electr. Energy Syst.* **2022**, *5*, 75–491. [CrossRef]
31. STMicroelectronics. Low-THD 350W CCM PFC pre-regulator based on L4985, AN5734. 2021. Available online: https://www.st.com/content/ccc/resource/technical/document/application_note/group2/50/b1/4f/b8/6b/9b/43/40/DM00827238/files/DM00827238.pdf/jcr:content/translations/en.DM00827238.pdf (accessed on 13 October 2023).

energies

MDPI

Article

A Dynamic Transition Algorithm Integrated with Hybrid Modulation for CLLC Resonant Converters

Jiaxin Qian, Huipin Lin, Hongtao Tong, Mingyu Gao and Zhiwei He *

School of Electronic Information, Hangzhou Dianzi University, Hangzhou 310018, China; qianjiaxin@hdu.edu.cn (J.Q.); linhuipin@hdu.edu.cn (H.L.); tonghongtao@hdu.edu.cn (H.T.); mackgao@hdu.edu.cn (M.G.)

* Correspondence: zwhe@hdu.edu.cn; Tel.: +86-0571-86873861

Abstract: The CLLC resonant converter is a widely used bidirectional power converter known for its high energy transfer efficiency. To extend its operating range, the converter often employs a hybrid modulation strategy, which is valued for its simplicity and efficiency. However, at the critical transition point between pulse frequency modulation and phase-shift modulation, instability in the output can be observed due to repetitive and unnecessary mode changes caused by noise. In order to address this issue, this paper introduces a dynamic transition algorithm integrated with hybrid modulation. This approach adaptively updates the controller parameters to mitigate oscillations resulting from improper initial parameter settings. Additionally, it incorporates error analysis and hysteresis comparison to prevent false triggers caused by noise, enabling intelligent mode adjustments. Finally, a 1 kW prototype is designed to conduct experiments, demonstrating an approximately 50% improvement in dynamic performance when the converter operates with the dynamic transition algorithm.

Keywords: dynamic transition algorithm; CLLC resonant converters; hybrid modulation; fundamental harmonic approximation method; control loop deign

Citation: Qian, J.; Lin, H.; Tong, H.; Gao, M.; He, Z. A Dynamic Transition Algorithm Integrated with Hybrid Modulation for CLLC Resonant Converters. *Energies* **2023**, *16*, 7173. https://doi.org/10.3390/en16207173

Academic Editor: Salvatore Musumeci

Received: 17 September 2023
Revised: 17 October 2023
Accepted: 19 October 2023
Published: 20 October 2023

1. Introduction

Benefiting from the rapid development of energy management technologies, battery energy storage systems (BESS), as one of the vital components, have gained extensive applications. They are typically constructed by connecting low-voltage, high-capacity cells in series [1–4]. Due to factors such as manufacturing processes and operating temperatures, discrepancies in the state of charge (SOC) of individual cells among battery packs within a series-connected structure are common, imposing limitations on the performance and functionality of the BESS [5–7]. CLLC resonant converters, as one of the high-performance and wide-range bidirectional DC/DC topologies [8], are well suited for applications in battery balancing to enhance the performance of BESS.

Figure 1 illustrates the typical topology of a CLLC resonant converter, which can be functionally simplified into three main units: the chopper unit, the resonator unit, and the rectifier unit, denoted by green, gray, and blue color blocks, respectively. Influenced by the voltage fluctuations in individual cells under various operational conditions, the converter has to achieve both high energy transfer efficiency with heavy load conditions and cell balancing among packs with light load conditions. With a heavy load, the converter applies pulse frequency modulation (PFM) to adjust the switching frequency to be lower than the resonant frequency, which causes the converter to operate in the under-resonant mode to achieve a high voltage gain. In this mode, the CLLC resonant converter can attain zero voltage switching (ZVS) for the switch devices in the chopping unit, ensuring efficient operation [9,10]. Similarly, under light load conditions, the converter applies PFM to adjust the operating frequency to be higher than the resonant frequency, which enables the converter to function in the over-resonant mode to achieve a low voltage gain [11,12]. This mode ensures ZVS for the switch devices in the chopping unit and zero current switching (ZCS) for the switch devices in the rectification unit of the converter. However, as the load decreases even further, reducing the voltage gain with PFM becomes challenging due to the extremely high switching frequencies, requiring a high-performance digital controller. To address this issue, researchers have proposed a hybrid modulation mode that seamlessly combines PFM and phase-shift modulation (PSM) [13–15]. In most cases, the CLLC resonant converter applies PFM to either increase or decrease the switching frequency of the chopping unit to achieve output gain adjustment. Once it is unable to reach the desired output voltage requirement even at the highest switching frequency, the hybrid modulation algorithm helps the converter to transition from PFM to PSM as soon as possible. With PSM, the converter regulates the phase-shift angle to control the RMS output voltage of the chopping unit, which is applied as the input of the resonant tank. Moreover, J.Y. et al. proposed a synchronous rectification method for the case in which the converter is in PSM mode [16–18]. Moreover, T. et al. discussed the losses that may occur during PFM and PSM in CLLC resonant converters [19,20]. Although these analyses are informative, the detailed design of modulation mode transition has not been thoroughly explored and revealed.

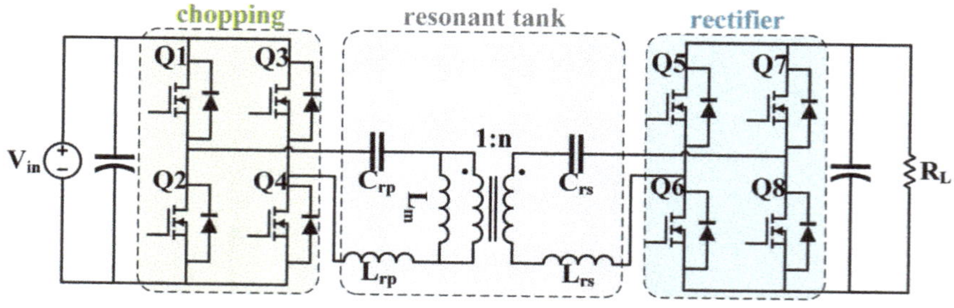

Figure 1. The typical CLLC resonant converter. V_{in} is the input bus voltage, R_L is the output load, C_{rp} and C_{rs} are the resonant capacitors, L_{rp} and L_{rs} are the resonant inductors, L_m is the excitation inductor, n is the transformer turn ratio, and $Q1 - Q8$ are switch devices.

To optimize the hybrid modulation strategy of CLLC resonant converters, this paper introduces a dynamic transition algorithm aimed at reducing the likelihood of repetitive and unnecessary modulation mode changes caused by noise at critical states. Compared to traditional hybrid modulation, the proposed algorithm manages the transition process more precisely due to adaptive updates of the controller parameters. Furthermore, building upon the traditional approach, it employs error analysis and hysteresis control under the current switch frequency and phase-shift angle to prevent undesired modulation mode changes. Finally, a 1 kW prototype CLLC circuit is utilized in experiments to verify the performance and feasibility of the proposed algorithm.

2. Analysis of Hybrid Modulation

In the buck mode of a full-bridge CLLC resonant converter, the hybrid modulation strategy encompasses two fundamental modulation techniques: PFM and PSM.

In this section, a comparative analysis of the two fundamental modulation methods is conducted concerning system gain analysis and control loop design. Concerning system gain analysis, the focus will be on considering the components contributing to the gain of the CLLC resonant converter and deriving quantitative expressions for each modulation method using an efficient simplified analysis approach. As for control loop design, the approach will involve utilizing computer simulation techniques to obtain the open-loop dynamic characteristics of CLLC resonant converters, thereby providing valuable guidance for the design of the PI compensator. Based on the two points mentioned above, a comprehensive hybrid modulation strategy with the traditional mode selector algorithm is constructed.

2.1. Gain Analysis of Modulations

The system gain of CLLC resonant converters is the cumulative result of three gains, each determined by the chopping unit, the resonant tank unit, and the rectifier unit, respectively.

As the input part of the CLLC resonant converter, the chopping unit converts the DC voltage into AC voltage with distinct characteristics, which are influenced by the various modulation techniques. As illustrated by the red curve in Figure 2a, when the CLLC resonant converter is modulated with PFM, this unit generates a square wave signal with a 50% duty cycle and the same amplitude but varying periods. However, when the converter is modulated with PSM, this unit generates a square wave signal with the same amplitude and the same periods but varying duty cycles, which is depicted by the red curve in Figure 2b.

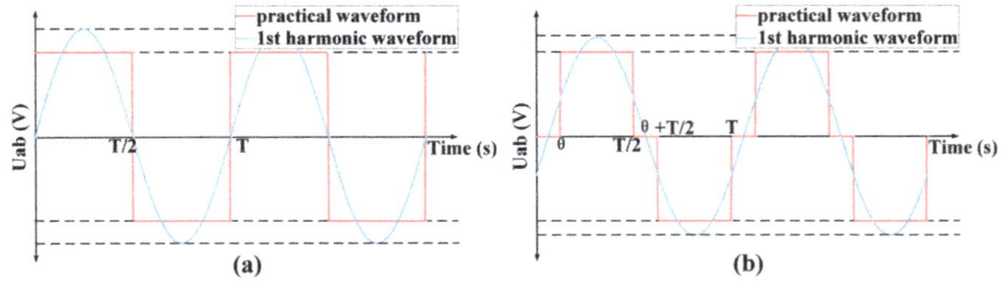

Figure 2. The practical output waveform and the corresponding fundamental harmonic for (**a**) PFM and (**b**) PSM. T represents the switch period, and θ represents the phase-shift angle.

The resonant tank unit is one of the most critical parts in the converter. Within the resonant tank unit, the combined interaction of five resonant components results in entirely distinct output voltages for varying duty cycles and frequencies of input signals, which are determined by the chopping unit. In general, there are two classical methods for the analysis of the effects of the input signals: the time domain analysis method (TDA) and the fundamental harmonic approximation method (FHA) [19,21,22]. Compared to TDA, the most significant advantage of FHA is its lower computational complexity and higher accuracy. Thanks to these advantages, FHA has become the preferred choice for the analysis of resonant tanks in engineering applications. It focuses solely on the impact of the fundamental harmonic on the energy transfer process and simplifies the transfer function using impedance analysis techniques [23–25].

The foundation of FHA is Fourier expansion, which can be expressed as

$$f(x) = \frac{a_0}{2} + \sum_{n=1}^{\infty}\left(a_n \times cos\left(\frac{2\pi}{T}nx\right) + b_n \times sin\left(\frac{2\pi}{T}nx\right)\right), \tag{1}$$

where $x \in [0, T]$, $a_n = \frac{2}{T}\int_0^T f(x) \times cos(\frac{2\pi}{T}nx)dx$, and $b_n = \frac{2}{T}\int_0^T f(x) \times sin(\frac{2\pi}{T}nx)dx$.

The blue sinusoidal curve in Figure 2 illustrates the first harmonic waveform corresponding to different modulation modes. In the PFM modulation mode, an equivalent input waveform can be obtained as in Equation (2)

$$V_{in.PFM} = b_{1.PFM} \times sin\left(\frac{2\pi}{T}t\right), \tag{2}$$

where

$$b_{1.PFM} = \frac{2}{T}\int_0^T V_{in}(t) \times sin\left(\frac{2\pi}{T}t\right)dt = \frac{4}{\pi}V_{in}, \tag{3}$$

and V_{in} is the input bus voltage.

Similarly, in the PSM modulation mode, an equivalent input waveform can be obtained as in Equation (4)

$$\begin{aligned} V_{in.PSM} &= a_{1.PSM} \times cos\left(\frac{2\pi}{T}t\right) + b_{1.PSM} \times sin\left(\frac{2\pi}{T}t\right) \\ &= c_{1.PSM} \times sin\left(\frac{2\pi}{T}(t+t_0)\right) \end{aligned}, \tag{4}$$

where

$$a_{1.PSM} = \frac{2}{T}\int_0^T V_{in}(t) \times cos\left(\frac{2\pi}{T}t\right)dt = -\frac{2}{\pi}sin(\theta)V_{in}, \tag{5}$$

$$b_{1.PSM} = \frac{2}{T}\int_0^T V_{in}(t) \times sin\left(\frac{2\pi}{T}t\right)dt = \frac{2}{\pi}(cos(\theta)+1)V_{in}, \tag{6}$$

$$c_{1.PSM} = \sqrt{a_{1.PSM}^2 + b_{1.PSM}^2} = \frac{4}{\pi}cos\left(\frac{\theta}{2}\right)V_{in}, \tag{7}$$

$$t_0 = \frac{\theta}{4\pi}T. \tag{8}$$

The amplitude and phase angle of the fundamental harmonics modulated by PFM and PSM can be obtained by referring to Equations (2) and (4), respectively. It is important to note that the phase angle does not affect the gain variation of the resonant tank. Thus, the only factor that affects the operation of the resonant tank is the amplitude of the fundamental harmonic, which can also be considered as the effective value of the fundamental harmonic. When PFM and PSM operate at the same switching frequency, the normalized expression for the input of the resonant tank can be simplified as in :

$$G_{c.normal} = \frac{rms(V_{in.PSM})}{rms(V_{in.PFM})} = cos\left(\frac{\theta}{2}\right). \tag{9}$$

Figure 3 illustrates the simplified model of a symmetric CLLC resonant tank unit. The analysis of this simplified model in the frequency domain leads to the derivation of the transfer function of the resonant tank unit under fundamental excitation, which can be expressed as in :

$$H(jw) = \frac{V_{out.FHA}(jw)}{U_{ab.FHA}(jw)} = \frac{Z_m//(Z_r+R_{ac})}{Z_m//(Z_r+R_{ac})+Z_r} \times \frac{R_{ac}}{Z_r+R_{ac}}, \tag{10}$$

where $Z_r = jwL_r + 1/(jwC_r)$, and $Z_m = jwL_m$.

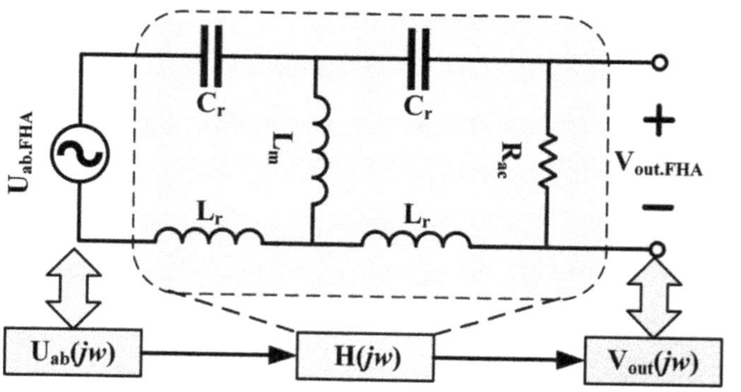

Figure 3. The simplified model of the symmetric CLLC resonant tank. $C_r = C_{rp} = C_{rs} \times n^2$ is the equivalent resonant capacitor, $L_r = L_{rp} = L_{rs}/n^2$ are the equivalent resonant inductors, and $R_{ac} = 8R_L/(2\pi n^2)$ is the equivalent AC load.

Taking the magnitude of Equation (10) and simplifying it yields the normalized expression for the resonant tank gain, which can be obtained as in

$$G_{rt.normal} = |H(jw)| = \frac{1}{\sqrt{\left(\frac{Q}{k}\right)^2\left((2k+1)f_n - \frac{2k+2}{f_n} + \frac{1}{f_n^3}\right)^2 + \left(1 + \frac{1}{k} - \frac{1}{kf_n^2}\right)^2}} \tag{11}$$

where $Q = \sqrt{L_r/C_r}/R_{ac}$ is the quality factor, $k = L_m/L_r$ is the inductor ratio, $f_n = f_s/f_r$ is the normalized frequency ratio, f_s is the switch frequency, and $f_r = 1/(2\pi\sqrt{L_rC_r})$ is the quasi-resonant frequency.

The rectification unit, being a fixed structure (full-bridge rectification) after the resonant tank unit, is not influenced by the modulation strategy of the chopping unit. Consequently, its normalized gain can be considered as $G_{r.normal} = 1$ [26].

By integrating the analysis of Equations (9) and (11), the normalized system gain of CLLC resonant converters can be expressed as in

$$G_{sys.normal} = \frac{cos\left(\frac{\theta}{2}\right)}{\sqrt{\left(\frac{Q}{k}\right)^2\left((2k+1)f_n - \frac{2k+2}{f_n} + \frac{1}{f_n^3}\right)^2 + \left(1 + \frac{1}{k} - \frac{1}{kf_n^2}\right)^2}}. \tag{12}$$

Based on Equation (12), the normalized gain curves of the CLLC resonant converter can be visualized under different load conditions, as illustrated in Figure 4. The left half of Figure 4 exhibits the converter operating in PFM modulation mode: when $f_n < 1$, the gain curve exhibits pronounced changes; meanwhile, when $f_n > 1$, the gain curve tends to change more smoothly, and it almost loses its regulation capability. The right part of Figure 4 depicts the converter operating in the PSM modulation mode with the maximum switch frequency: as the phase-shift angle increases, compared to the PFM modulation mode with the same switch frequency, the converter's gain significantly decreases. Therefore, the hybrid modulation mode combining PFM and PSM can effectively broaden the operating range of the converter and improve its efficiency.

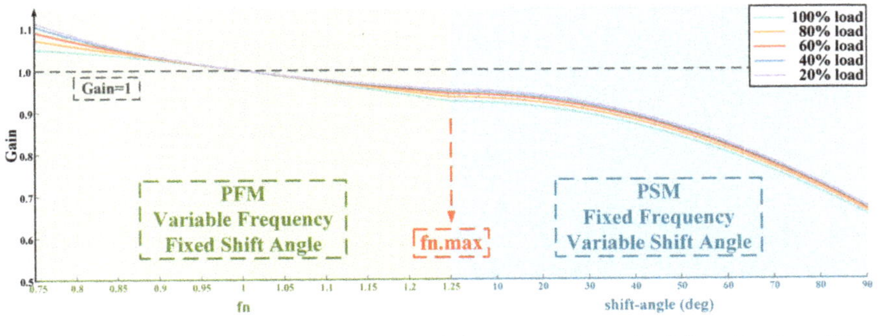

Figure 4. Visualized gain curves with varying modulation strategies under different load conditions. Test conditions: $L_r = 12.8$ µH, $C_r = 54.4$ nF, $L_m = 96.2$ µH.

2.2. Design of the Control Loop

In practical power converter systems, noise caused by the parasitic parameters of devices occurs frequently, and the consequences can range from significant power supply output ripples to more severe effects such as output voltage oscillations and damage to power components. To address the mentioned issues, it is necessary to introduce an appropriate compensator in the system feedback loop to regulate the power converter's performance in the presence of disturbances. A well-designed compensator can ensure the stability of the converter and improve its dynamic response characteristics.

The stability of a control loop is specifically depicted on a Bode plot through the parameters known as the phase margin (PM) and gain margin (GM). PM is defined as the difference between the phase angle at the system's crossover frequency and $-180°$. Typically, engineers set PM to be greater than 45° based on their experience, striking a balance between system damping and transient response capability. GM, on the other hand, is defined as the difference between the gain at the phase angle of $-180°$ and 0 dB. It is generally required to be greater than 6 dB for system stability. When the cumulative impact of the compensator and power converter displays adequate PM and GM on the Bode plot, it indicates that the compensator can enhance the performance of the converter [27,28].

The proportional-integral (PI) controller is one of the most commonly used compensators [29,30], and its transfer function is typically represented as in Equation (13)

$$PI(s) = \frac{K_p \cdot s + K_i}{s}, \tag{13}$$

where K_p is the proportional coefficient, and K_i is the integral coefficient.

Based on Equation (13), one can deduce the gain expression,

$$Gain.pi(f) = |PI(j \cdot 2\pi f)| = \sqrt{K_p^2 + \frac{K_i^2}{4\pi^2 f^2}}, \tag{14}$$

and phase expression,

$$Phase.pi(f) = arg(PI(j \cdot 2\pi f)) = arctan\left(\frac{2\pi f \times K_p}{K_i}\right) - 90° \tag{15}$$

for the compensator.

The gain and phase characteristics of the PI compensator are both depicted in Figure 5. The trend of the gain curve is primarily decided by the ZERO. When the frequency is lower than $K_i/(2\pi K_p)$ Hz, the curve follows an approximate -20 dB/decade slope; however, when the frequency exceeds $K_i/(2\pi K_p)$ Hz, the interaction between the pole and zero gradually brings the gain curve closer to 0 dB. Similarly, the trend of the phase curve is influenced by the ZERO. As the frequency approaches 0 Hz, the phase lags by 90°; when the frequency equals $K_i/(2\pi K_p)$ Hz, the phase exhibits a lag of 45°; and as the frequency tends towards infinity, the interaction between the ZERO and POLE causes the phase to approach 0°. To avoid complex nonlinear computations, the design of the PI is simplified by approximating the curves with linearization. Therefore, once the crossover frequency is selected, the calculation of constraint equations can be carried out based on the gain and phase of the power converter. With the adjustment of the PI compensator, when $f = f_c$, the open-loop gain is compensated to reach 1. Similarly, compared to the original phase without the compensator, the open-loop phase is increased by 45° when $f = f_c$. Combining the discussion above and the simplified analysis in Figure 5, the constraint equations can be expressed as in

$$\begin{cases} Gain.pi(f_c) \times Gain.openloop(f_c) = 1 \\ \dfrac{K_i}{2\pi K_p} = 0.1 f_c \end{cases} \tag{16}$$

where f_c is the selected crossover frequency, and *Gain.openloop* represents the gain of the open-loop system analyzed using the small signal model and computer simulations.

By utilizing computer simulation techniques, the loop characteristics of the CLLC resonant converter can be easily obtained. The Bode plot in Figure 6 displays the open-loop characteristic curve of the converter when it operates in PFM mode, which shows a significantly low gain at low frequencies. Additionally, due to the complex structure of the resonant tank, there are two resonance peaks located near $f_s = 2.0$ kHz and $f_s = 10.0$ kHz, making compensator bandwidth design challenging. Fortunately, the phase curve presents a favorable trend, indicating that the system's PM will not be affected by the phase lag of the compensator. To meet the dynamic performance requirements, a relatively conservative crossover frequency of 3.5 kHz has been set, which provides a gain of −65.8 dB and a phase angle of 5°, making it ideal for compensation.

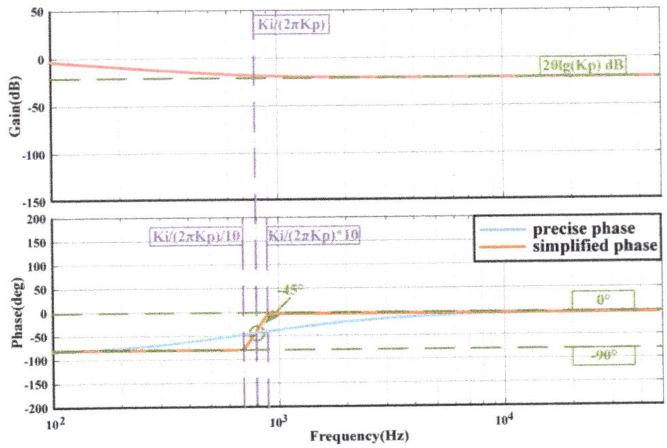

Figure 5. The analysis of a typical PI compensator. It's important to note that the red curve represents the gain variation, and the blue curve represents the phase variation. Demo's test conditions: $K_p = 0.093$, $K_i = 436.215$.

On the other hand, when the converter operates using the PSM modulation method, its open-loop characteristic curve has a high bandwidth gain curve of approximately 40 kHz, as shown in Figure 6b. However, the phase curve's trend is non-ideal, which poses a significant constraint on compensator design. To ensure the stable operation of the converter, the PM with compensation should be at least 45°, indicating that the original PM should be selected in the range of 85° to 95°. To meet the dynamic performance requirements and considering the above analysis, a conservative crossover frequency of 7.5 kHz has been set. At this frequency, the original open-loop characteristic curve has a gain of 20 dB and a phase of −95°, providing ample room for compensation.

By substituting the chosen crossover frequency into Equation (16), the corresponding compensator parameters can be obtained by solving the equation set. The solutions of the equation set for PFM mode are $K_p = 1.4 \times 10^3$ and $K_i = 3.1 \times 10^6$, while the solutions for PSM mode are $K_p = 9.3 \times 10^{-2}$ and $K_i = 4.3 \times 10^2$.

Figure 7 shows the loop characteristics of the CLLC resonant converter under the influence of the PI compensator. When the converter operates using the PFM modulation method, the GM is 58.8 dB, and the PM is 185°, ensuring the stable operation of the CLLC resonant converter under PFM mode based on engineering experience. Similarly, when the converter operates using the PSM modulation method, the GM is 6.2 dB, and the PM is 45°, demonstrating its outstanding system stability and exceptional system dynamic characteristics.

2.3. Introduction of the Traditional Mode Selector Method

Figure 8 illustrates the elements of the hybrid modulation for CLLC resonant converters. The error, which is the difference between the user-set voltage reference and the feedback voltage, is input into the controller unit. Taking into account the current operating frequency and phase-shift angle, the controller unit selects the appropriate modulation technique to generate the driver signal, thereby achieving closed-loop control of the CLLC resonant converter. The blue section in Figure 8 represents the basic structure of the controller unit, comprising the mode selector, the PFM modulator with its PI compensator, and the PSM modulator with its PI compensator. It is worth noting that the compensators are associated with the modulation methods individually, meaning that, when switching between modulation methods, the corresponding PI compensator will also switch accordingly. Based on the analysis of the gain and the design of loop stability compensation for CLLC resonant converters under PFM and PSM modulation mode in the previous section, the converter is capable of stable operation in a single modulation mode, which satisfies the output requirements of the converter in most scenarios.

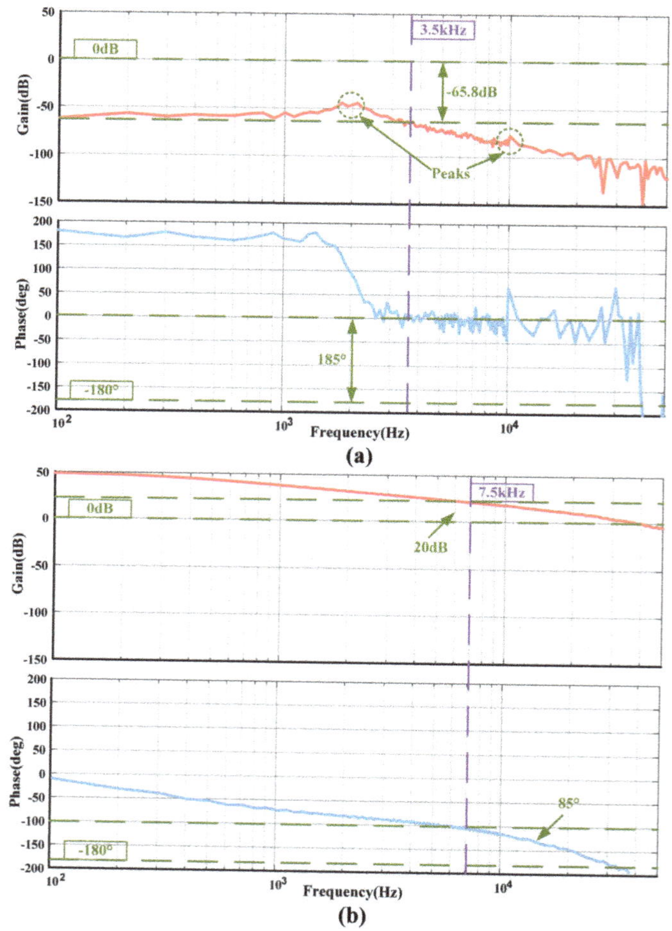

Figure 6. The loop analysis of a symmetric CLLC resonant converter modulated by (**a**) PFM mode and (**b**) PSM mode using computer simulation. It's important to note that the red curve represents the gain variation, and the blue curve represents the phase variation. Test conditions: $L_r = 12.8$ µH, $C_r = 54.4$ nF, and $L_m = 96.2$ µH.

However, when the converter operates at the highest switching frequency in PFM or the minimum phase-shift angle in PSM, tiny noise disturbances can trigger repetitive transitions between the two modulation modes, resulting in inevitable high-frequency oscillations in the output and consequently deteriorating the system's dynamic response capability.

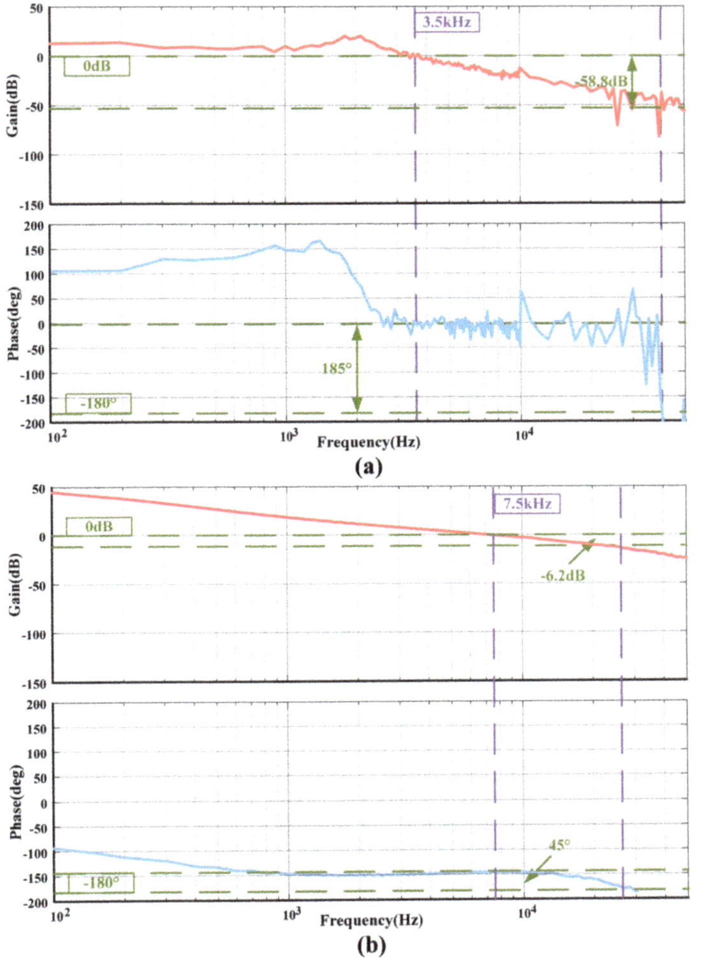

Figure 7. The loop analysis of a symmetric CLLC resonant converter with a PI compensator modulated by (**a**) PFM mode and (**b**) PSM mode. It's important to note that the red curve represents the gain variation, and the blue curve represents the phase variation. Test conditions: $L_r = 12.8$ μH, $C_r = 54.4$ nF, and $L_m = 96.2$ μH.

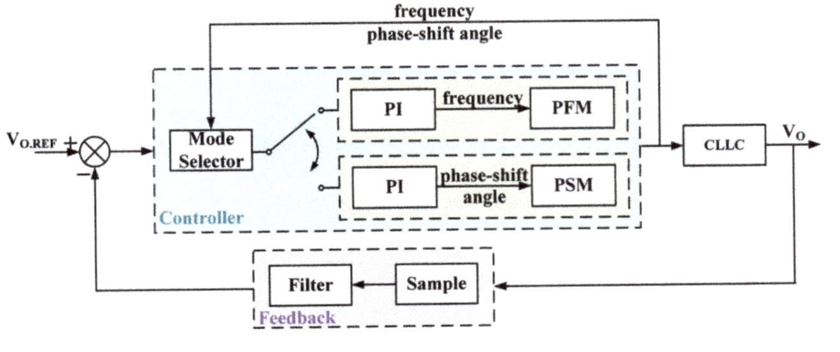

Figure 8. The description of loop control for CLLC resonant converters.

Figure 9 illustrates the detailed operational principle with hybrid modulation using the traditional mode selector method. After powering up the converter, the controller initially operates using the PFM modulation method. It first checks whether the initialization of the PI compensator is necessary. Once the initialization is completed or deemed unnecessary, the PI compensator calculates the switching frequency. When the calculated result matches the converter's maximum operating frequency, the flag will be set to 1. Otherwise, it will remain as 0. The main state machine then evaluates the flag value to determine the modulation method for the next cycle. Specifically, if the flag is equal to 1, the converter can operate using the PSM method. Otherwise, it will continue using the current modulation method, namely PFM. Similarly, the transition from PSM modulation to PFM modulation is entirely symmetrical. Benefiting from the advantages of digital control techniques, the state machine can be executed periodically. In brief, there are two primary tasks when a new control instruction is incoming. During the control cycle, two primary tasks are executed step by step. Firstly, the PI compensator is activated to calculate the appropriate switching frequency or phase-shift angle based on the calculation results from the previous control cycle. Secondly, the corresponding modulation unit is activated to output a pulse that adjusts the operational state of the chopping unit. It is worth noting that the initial switching frequency and initial phase used for compensator parameter initialization remain unchanged due to the absence of necessary updates. However, the initial error is updated at the end of each control cycle, benefiting from the error analysis performed during the PI calculation process.

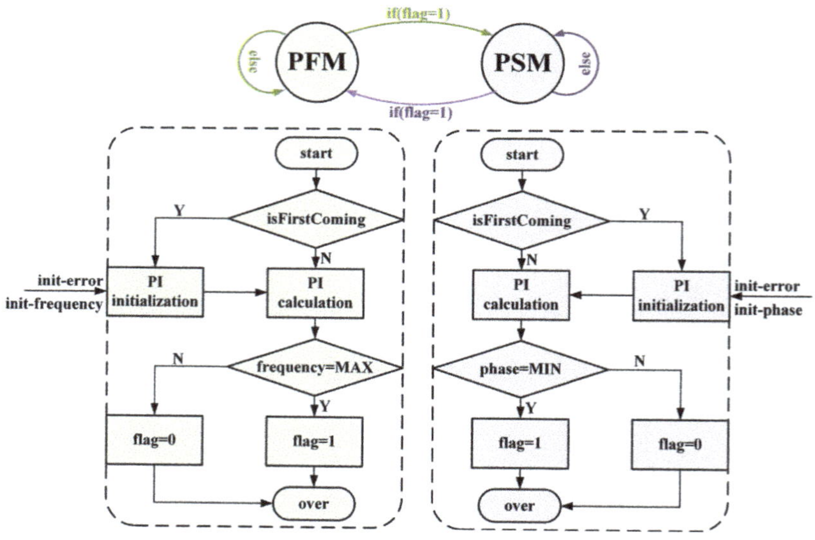

Figure 9. The description of a control unit using hybrid modulation with the traditional method.

In summary, the traditional modulation mode selector method primarily relies on assessing the output frequency with PFM modulation and the output phase-shift angle with PSM modulation. It is worth noting that PI initialization executes only once during the controller's startup, and the initialization parameters are not updated after this. This characteristic is also a contributing factor to the observed output voltage instability at critical states.

This approach is characterized by its simplicity, efficiency, and minimal computational overhead. However, when the converter operates in proximity to its critical state, the traditional modulation mode selector algorithm becomes vulnerable to noise interference, potentially leading to erroneous mode transitions. Additionally, due to the influence of the high-bandwidth PI compensator, these incorrect modulation mode switches can induce substantial output adjustments, resulting in avoidable voltage oscillations.

3. Proposed Dynamic Transition Algorithm

Based on the analysis above, to further enhance the performance of hybrid modulation, it is imperative to tackle the matter of compensator parameter updates, while simultaneously minimizing the influence of noise on mode switching.

This paper introduces an adaptive parameter updating method to mitigate the performance degradation resulting from improper parameter settings in the converter. Specifically, it compares the current modulation mode with the previous one to decide whether to update the compensator's previous error and previous output. If the current modulation mode matches the previous one, parameter updates are skipped; otherwise, the latest error is employed to update the controller. Additionally, it incorporates an error analysis method and hysteresis comparison to prevent unnecessary mode changes at critical states. This method ensures that modulation mode transitions only occur when the judgment condition persists for several consecutive cycles. Compared to other improvement approaches in the field, this method offers greater flexibility in mode switching control due to its unique error analysis and hysteresis comparison. As a result, the potential risk of noise interference is further reduced.

Figure 10 illustrates the workflow of the proposed mode selector. As with the traditional mode selector method, after the converter starts up, the controller initially operates using the PFM modulation method. Firstly, the controller evaluates the need for initialization based on the last modulation mode. Once the initialization is either completed or considered unnecessary, the PI compensator calculates the switching frequency. If the calculated result matches the user-defined maximum frequency, and the error between the reference voltage and the output voltage exceeds the acceptable range, the counter is enabled. When the count value exceeds the hysteresis threshold, the modulation mode transitions to PSM in the next cycle. When the converter is currently operating using the PSM modulation method, its operation process is also similar to the former. Similarly, the first step is evaluating the need for initialization based on the last modulation mode. Once the initialization is either completed or considered unnecessary, the PI compensator calculates the phase-shift angle. If the calculated result matches the user-defined minimum phase-shift angle, and the error between the reference voltage and the output voltage exceeds the acceptable range, the counter is enabled. When the count value exceeds the hysteresis threshold, the modulation mode transitions to PFM in the next cycle.

In comparison to the traditional control methods, the proposed approach offers two key optimizations. Firstly, it supplements the conditions for the initialization of the PI compensator, automatically updating the initial error when the modulation modes differ between consecutive cycles, which reduces overshoot during the transient process of modulation mode switching. Secondly, it introduces hysteresis in the transition process, requiring the satisfaction of multiple consecutive conditions before allowing a modulation mode switch. For instance, the transition from PFM to PSM necessitates two specific conditions: firstly, the output modulation frequency must match the maximum modulation frequency; secondly, the output voltage should fall outside the precision error range defined by the user for a consecutive number of cycles. This significantly mitigates the issue of noise interference with the mode selector.

With these improvements, the converter is able to effectively utilize the capabilities of the PI compensator during the switch to a new modulation mode, resulting in a significant increase in the dynamic performance of the output.

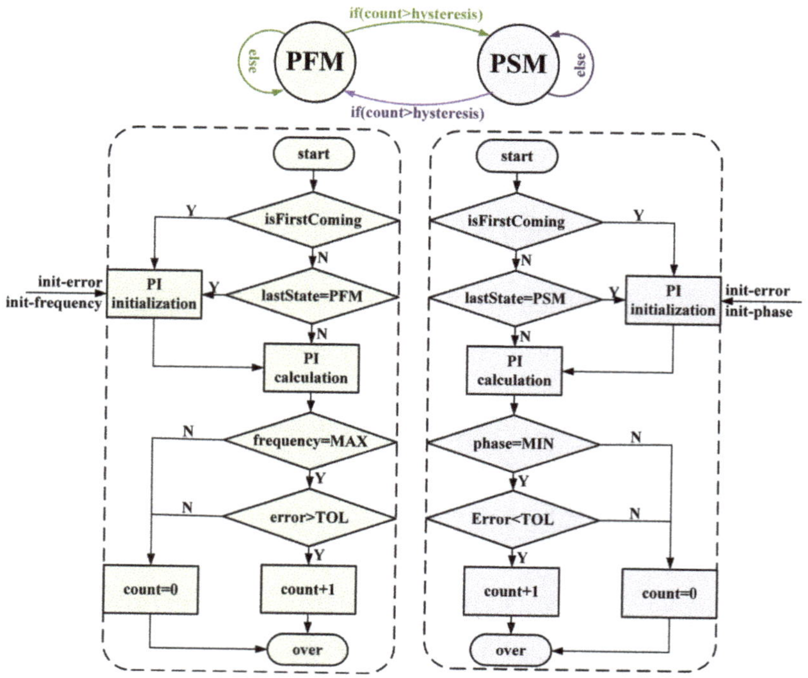

Figure 10. The description of a control unit using hybrid modulation with the proposed method.

4. Experimental Results and Discussion

To evaluate the robustness and feasibility of the proposed dynamic transition algorithm integrated with the hybrid modulation method, this section presents a 1 kW symmetric CLLC prototype with experimental results.

Due to manufacturing errors in the capacitors and inductors of the resonant tank, the operation performance of the CLLC resonant converter may differ from the theoretical analysis. To improve the accuracy, the resonant tank parameters are measured using the LCR meter after the prototype is assembled. All of the parameters of the prototype are shown in Table 1. It is worth noting that the resonant capacitance is composed of a parallel combination of 6.8 nF film capacitors, rather than a single capacitor. The purpose of this approach is to flexibly adjust the capacitance parameters while ensuring maximum current and voltage tolerance to avoid damage to the resonant tank.

Table 1. The practical parameters of the prototype.

Parameter	Value	Parameter	Value
Q1–Q8	C3M0032120K	f_r	200 kHz
inductor core	NPA158019	L_{rs}	9.9 µH
transformer core	PQ6562	L_{rp}	13.0 µH
n	24:19	C_{rp}	47.6 nF
V_{in}	80–200 V	C_{rs}	71.4 nF
V_{out}	50–180 V	L_m	92.6 µH

The test instruments and the symmetric CLLC prototype are depicted in Figure 11, and they include a four-channel oscilloscope with a compatible Hall current probe and isolated voltage probe, a 750 V 5 kW DC power source, and a 5 kW DC load.

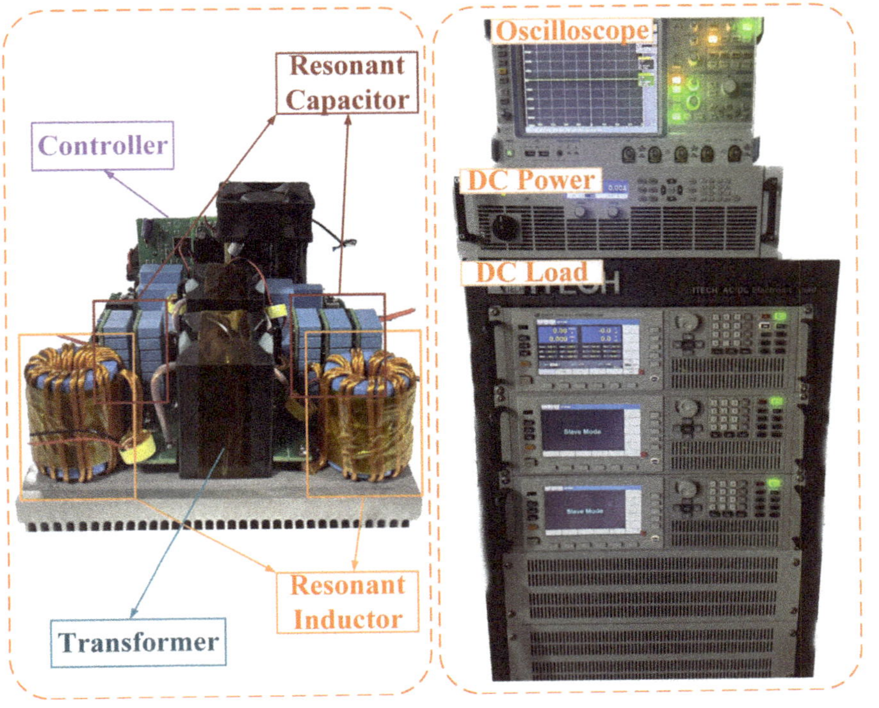

Figure 11. 1 kW CLLC prototype and instruments.

To facilitate the observation of experimental data, U_{ab} and I_{out} are both saved. U_{ab} indicates the voltage between the midpoints of the two bridges in the chopping unit, which reflects the actual excitation signal of the resonant tank. The density of U_{ab} can also demonstrate the operating frequency of the converter. On the other hand, I_{out} represents the output current of the converter.

Figure 12 illustrates the typical operation waveforms with different loads captured by an oscilloscope. Under heavy load conditions, the resonant tank input waveform of the prototype using PFM, as shown in Figure 12a, has a frequency of 155 kHz. Conversely, under light load conditions, the waveform of the prototype using PFM, as shown in Figure 12b, operates at a frequency of 163 kHz. Similarly, under heavy load conditions, the resonant tank input waveform of the prototype using PSM, as depicted in Figure 12c, exhibits a phase-shift angle of 7°, while, under light load conditions, the waveform of the prototype using PSM, as shown in Figure 12d, features a phase-shift angle of 63°. It is worth noting the current ripple observed in Figure 12, which represents common-mode noise caused by the switching of MOSFETs in the chopping unit. It is also worth noting that when the converter maintains a constant voltage output of 100 V, the current noise increases as the output power decreases.

Figure 13 demonstrates the closed-loop stability of the converter with a single modulation mode. When the converter's output load steps up, the PFM compensator quickly responds to increase the output current, as shown in Figure 13a. Conversely, when the converter's output load steps down, the PFM compensator promptly decreases the output current to stabilize the voltage output, as illustrated in Figure 13b. Similarly, when the converter operates in PSM mode, the compensator can quickly respond to adjust the output voltage, whether it is a step-up or step-down scenario, as depicted in Figure 13c,d. The calculation results analyzed in the previous section for the PI compensator parameter settings corresponding to the PFM method and the PSM method can be directly applied. These parameters are set to ensure the stability of the converter while maximizing its theoretical dynamic response capability.

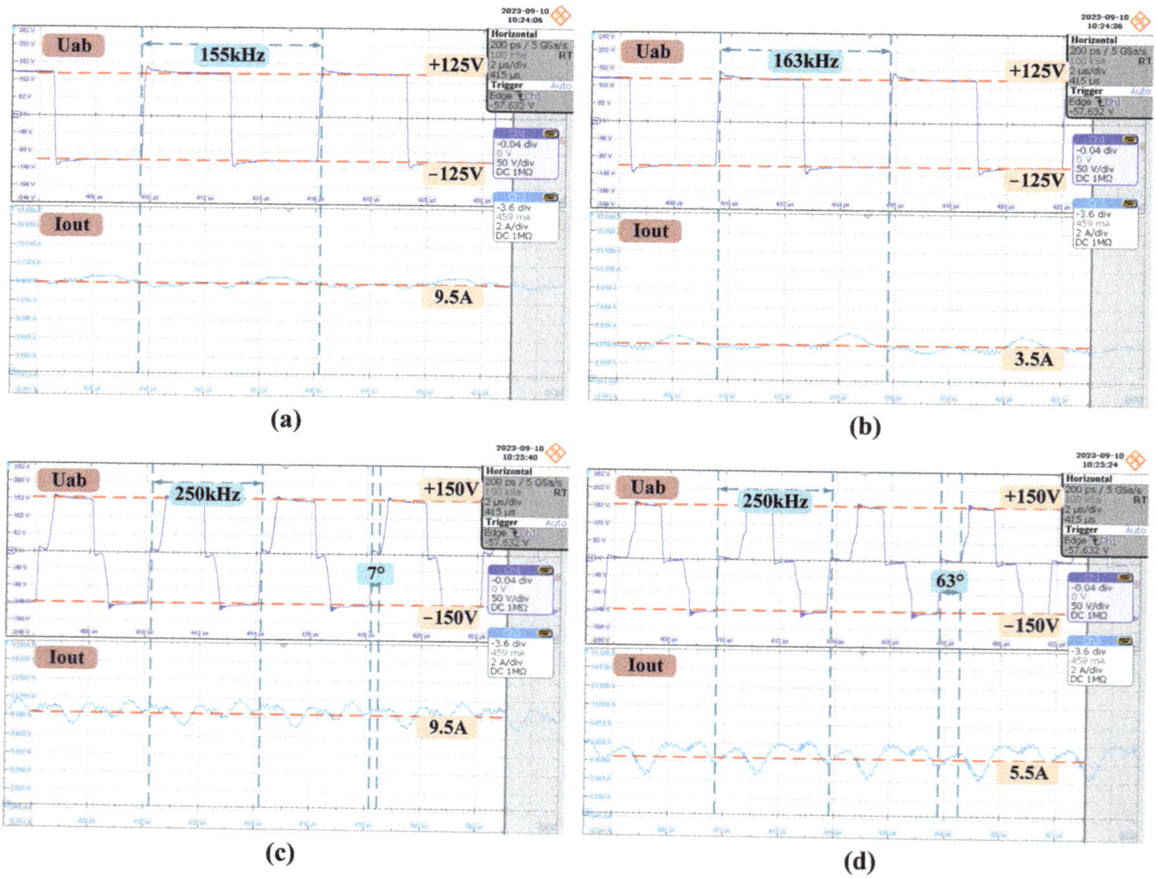

Figure 12. The typical waveforms of the CLLC resonant converters under different modulation modes. The waveforms of PFM under a heavy load and light load are presented in (**a**) and (**b**), respectively; and the waveforms of PSM under a heavy load and light load are presented in (**c**) and (**d**), respectively. The increment for U_{ab} is 50 V/div, while the increment for I_{out} is 2 A/div.

It is evident that when there is a load change, the converter with the appropriate and accurate parameter settings of the PI compensator can respond rapidly and adapt to a new steady state. This indicates that with a single modulation mode, the converter can achieve favorable dynamic and steady-state characteristics with the assistance of the corresponding compensator.

Based on trial-and-error, it is discovered that when the I_{out} is equal to 7.5 A, the converter operates in PSM mode with a relatively small phase-shift angle, and when the I_{out} is equal to 9.0 A, the converter operates in PFM mode with a relatively high switching frequency. These two load conditions not only ensure the stable operation of the converter in a single modulation mode before manually switching the load but also facilitate the observation of the converter's response speed and adjustment capability. Based on the above analysis, the current probe of the oscilloscope is set to 2 A/div to display waveform details and the overall trend of changes while avoiding signal observation being affected by noise interference during the mode transition.

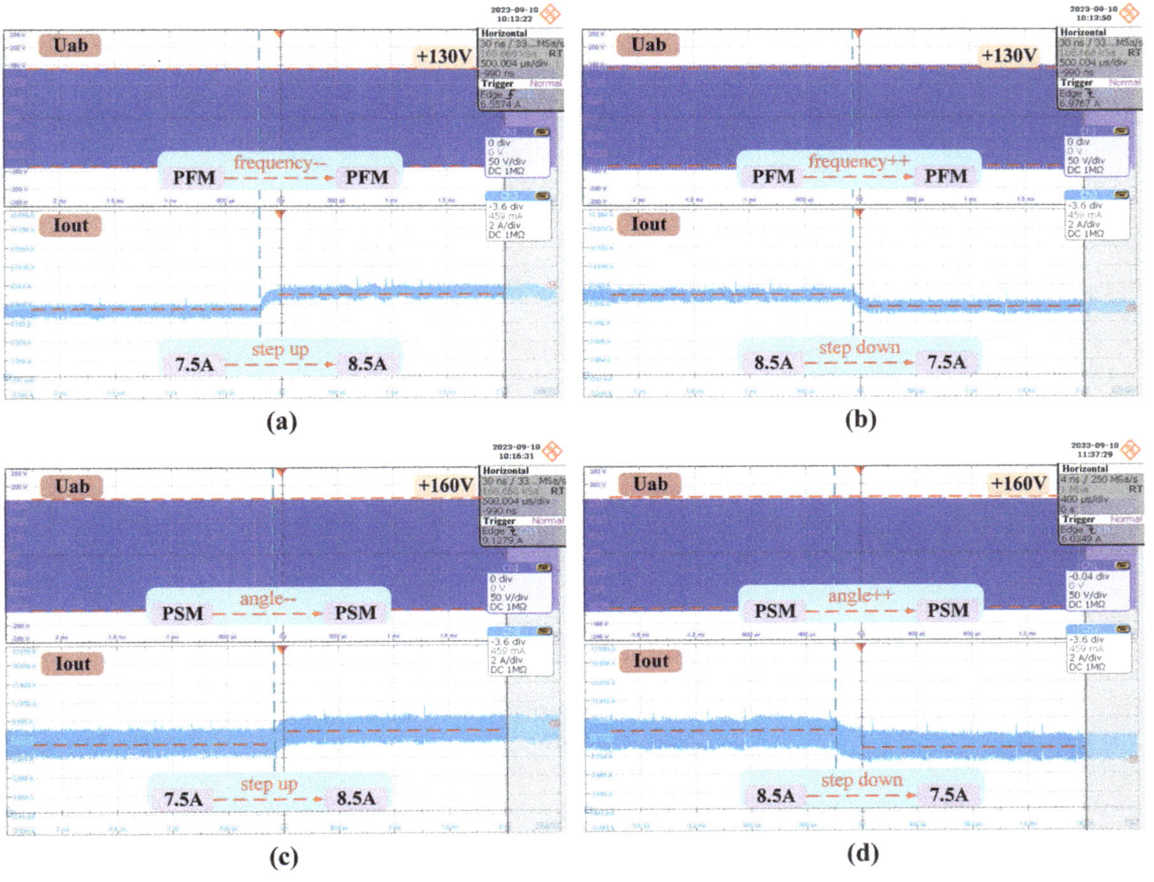

Figure 13. The typical dynamic response waveforms of the CLLC resonant converter. The waveforms of step up and step down with PFM are shown in (**a**) and (**b**); the waveforms of step up and step down with PSM are shown in (**c**) and (**d**). The increment for U_{ab} is 50 V/div, while the increment for I_{out} is 2 A/div.

Figure 14 illustrates the dynamic response of the converter when the output load steps up from 7.5 A to 9.0 A under the same compensator settings. Among them, the implementation results of the converter with the proposed dynamic transition algorithm are shown in Figure 14a, and it takes a total of 240 µs to transition from the load change to entering the new steady state. However, under the same conditions, the converter using the traditional method takes 440 µs, as shown in Figure 14b. While both execution strategies can achieve the mode transition from PSM to PFM, the time consumed by the two methods differs by nearly a factor of two. This highlights that the proposed dynamic algorithm enhances the performance of hybrid modulation when the load increases.

Correspondingly, Figure 15 illustrates the dynamic response performance of the converter when the output load steps down from 9.0 A to 7.5 A using the same compensator settings. The output current of the prototype is presented in Figure 15a, where it takes a total of 190 µs to transition from PFM to PSM. In contrast, when the converter employs the traditional method, as shown in Figure 15b, it takes a total of 320 µs to switch from PFM mode to PSM mode. Consistent with the previous comparative results, the proposed algorithm is approximately twice as fast as the traditional method in this scenario.

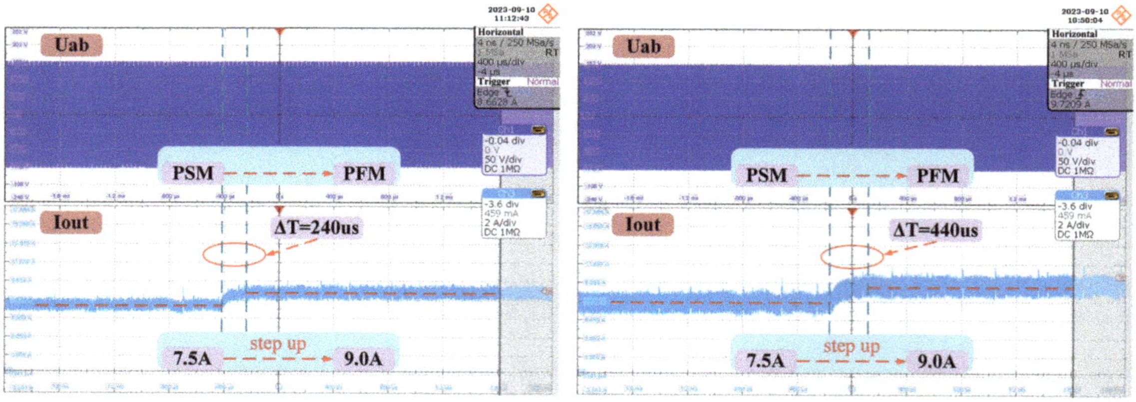

(a) (b)

Figure 14. Experimental waveforms of the load step up. (**a**) shows the operation state of the hybrid modulation with the proposed dynamic transition algorithm, and (**b**) shows the operation state of the traditional hybrid modulation. The increment for U_{ab} is 50 V/div, while the increment for I_{out} is 2 A/div.

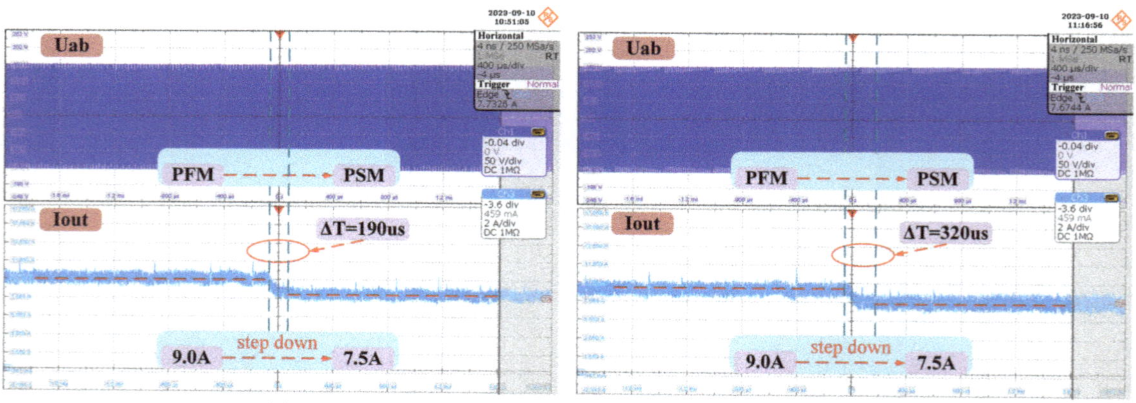

(a) (b)

Figure 15. Experimental waveforms of the load step down. (**a**) shows the operation state of the hybrid modulation with the proposed dynamic transition algorithm, and (**b**) shows the operation state of the traditional hybrid modulation. The increment for U_{ab} is 50 V/div, while the increment for I_{out} is 2 A/div.

Upon analyzing the experimental results stated above, it is found that the converter reaches a new steady state in only 240 μs when the output current rapidly increases from 7.5 A to 9.0 A, which is a typical step-up testing scenario. This marks a 45% improvement in dynamic performance compared to traditional methods. Notably, in a single modulation mode, the compensator parameters for both methods are identical, indicating that error analysis and hysteresis comparison play a crucial role in the mode switching process. Moreover, under the test conditions where the output current rapidly decreases from 9.0 A to 7.5 A, which is a typical step-down testing scenario, the converter reaches a new steady state in only 190 μs. This represents a 40% improvement in dynamic performance, demonstrating that the proposed method has a performance advantage under step-down conditions. Additionally, benefiting from the error analysis and the hysteresis comparison, the converter using the proposed method exhibits significantly reduced current ripple compared to the traditional method under the same conditions, which enhances its resistance to interference and increases the system's robustness.

To summarize, the dynamic transition algorithm, with its excellent anti-interference ability, can help the converter to achieve a nearly 50% improvement in dynamic performance, whether in step-up or step-down conditions. This is critical for the efficient operation of the system. Based on this method, the performance of the hybrid modulation approach has been significantly enhanced, making it better suited for applications with a larger load regulation range.

5. Conclusions

To further enhance the dynamic performance of the CLLC resonant converter using the hybrid modulation strategy, a dynamic transition algorithm is proposed. This algorithm aims to reduce the risk of repetitive and unnecessary modulation mode changes caused by noise at critical states by introducing more comprehensive and well-considered criteria for judgment. Building upon the traditional method, an adaptive parameter update scheme has been introduced to address the issue of unstable output due to improper parameter settings. Additionally, the proposed algorithm incorporates error analysis and hysteresis control to suppress the interference of high-frequency noise during mode transitions. Finally, a 1 kW symmetrical CLLC prototype is designed to validate the feasibility and robustness of hybrid modulation integrated with the dynamic transition algorithm. The experimental results show that when the converter operates under typical step-up conditions, the time taken to regulate the converter to reach a new state is reduced by 45%. Similarly, under typical step-down conditions, the proposed method reduces the adjustment time of the converter by around 40%. In the course of the research, some minor flaws in practical applications have been identified. For instance, the control time setting of hysteresis comparison relies heavily on trial-and-error methods. Additionally, the design of error analysis tolerance has not taken into account the actual sampling accuracy. In the future, these flaws will be addressed to ensure accurate and reliable results in practical applications.

Author Contributions: Conceptualization, J.Q. and H.L.; methodology, J.Q. and H.T.; software, J.Q. and H.T.; validation, J.Q., H.L. and H.T.; formal analysis, J.Q. and Z.H.; investigation, Z.H.; resources, J.Q.; data curation, J.Q. and H.T.; writing—original draft preparation, J.Q.; writing—review and editing, J.Q.; visualization, J.Q.; supervision, M.G.; project administration, Z.H.; funding acquisition, H.L. and M.G. All authors have read and agreed to the published version of the manuscript.

Funding: This work was supported by the National Major Scientific Research Instrument Development Project of China under Grant 62227802, the Central Guiding Local Science and Technology Development Fund Projects of China under Grant 2023ZY1008, the Provincial Public University Fundamental Research Fund under Grant GK229909299001-404, and the Zhejiang Provincial Key Lab of Equipment Electronics.

Institutional Review Board Statement: Not applicable.

Informed Consent Statement: Not applicable.

Data Availability Statement: No new data were created or analyzed in this study. Data sharing is not applicable to this article.

Conflicts of Interest: The authors declare no conflict of interest.

Abbreviations

The following abbreviations are used in this manuscript:

FHA	Fundamental harmonic approximation method
TDA	Time domain analysis method
PFM	Pulse frequency modulation method

PSM	Phase-shift modulation method
PM	Phase margin
GM	Gain margin
V_{in}	Input bus voltage
V_{out}	Output bus voltage
I_{out}	Output current
U_{ab}	Voltage between the midpoints of bridges in the chopping unit
R_L	Practical load
R_{ac}	Equivalent AC load
L_m	Excitation inductance
L_r	Equivalent resonant inductor
C_r	Equivalent resonant capacitor
Q	Quality factor ($\sqrt{L_r/C_r}/R_{ac}$)
k	Inductor ratio (L_m/L_r)
f_r	Resonant frequency
f_n	Normalized frequency
f_c	Crossover frequency

References

1. Lawder, M.T.; Suthar, B.; Northrop, P.W.C.; De, S.; Hoff, C.M.; Leitermann, O.; Crow, M.L.; Santhanagopalan, S.; Subramanian, V.R. Battery Energy Storage System (BESS) and Battery Management System (BMS) for Grid-Scale Applications. *Proc. IEEE* **2014**, *102*, 1014–1030. [CrossRef]
2. Yang, Y.; Bremner, S.; Menictas, C.; Kay, M. Battery Energy Storage System Size Determination in Renewable Energy Systems: A Review. *Renew. Sustain. Energy Rev.* **2018**, *91*, 109–125. [CrossRef]
3. Li, X.; Hui, D.; Lai, X. Battery Energy Storage Station (BESS)-Based Smoothing Control of Photovoltaic (PV) and Wind Power Generation Fluctuations. *IEEE Trans. Sustain. Energy* **2013**, *4*, 464–473. [CrossRef]
4. Oudalov, A.; Chartouni, D.; Ohler, C. Optimizing a Battery Energy Storage System for Primary Frequency Control. *IEEE Trans. Power Syst.* **2007**, *22*, 1259–1266. [CrossRef]
5. Cheng, K.W.E.; Divakar, B.P.; Wu, H.; Ding, K.; Ho, H.F. Battery-Management System (BMS) and SOC Development for Electrical Vehicles. *IEEE Trans. Veh. Technol.* **2011**, *60*, 76–88. [CrossRef]
6. Qian, J.; He, Z.; Gao, M.; Bao, Z.; Lin, H. State of Charge Estimation Based on Enchanced Sliding Mode Observer for Lithium-ion Battery. In Proceedings of the 2022 IEEE International Power Electronics and Application Conference and Exposition (PEAC), Guangzhou, China, 4–7 November 2022 ; pp. 539–544. [CrossRef]
7. Maharjan, L.; Inoue, S.; Akagi, H.; Asakura, J. State-of-Charge (SOC)-Balancing Control of a Battery Energy Storage System Based on a Cascade PWM Converter. *IEEE Trans. Power Electron.* **2009**, *24*, 1628–1636. [CrossRef]
8. Liu, Y.; Du, G.; Wang, X.; Lei, Y. Analysis and Design of High-Efficiency Bidirectional GaN-Based CLLC Resonant Converter. *Energies* **2019**, *12*, 3859. [CrossRef]
9. Cao, Y.; Ngo, M.; Burgos, R.; Ismail, A.; Dong, D. Switching Transition Analysis and Optimization for Bidirectional CLLC Resonant DC Transformer. *IEEE Trans. Power Electron.* **2022**, *37*, 3786–3800. [CrossRef]
10. Zhao, L.; Pei, Y.; Wang, L.; Pei, L.; Cao, W.; Gan, Y. Design Methodology of Bidirectional Resonant CLLC Charger for Wide Voltage Range Based on Parameter Equivalent and Time Domain Model. *IEEE Trans. Power Electron.* **2022**, *37*, 12041–12064. [CrossRef]
11. Jin, F.; Nabih, A.; Li, Q. Light load efficiency improvement of three phase CLLC resonant converter for on-board charger applications. In Proceedings of the 2022 IEEE Applied Power Electronics Conference and Exposition (APEC), Houston, TX, USA, 20–24 March 2022; pp. 1–7. [CrossRef]
12. Bu, H.; Guan, J.; Zhang, C.; Li, Z. Optimal Control Strategy of CLLC Bidirectional Resonant Converter under Light Load. In Proceedings of the 2022 9th International Forum on Electrical Engineering and Automation (IFEEA), Zhuhai, China, 4–6 November 2022; pp. 921–924. [CrossRef]
13. Tang, C.Y.; Wang, C.W.; Chien, H.C. A Dynamic Smooth Transition Control Integrated With Hybrid Modulation for Wide Output Voltage Range Bidirectional CLLC Resonant Converters. *IEEE Trans. Power Electron.* **2023**, *38*, 1–7. [CrossRef]
14. Li, L.; Cao, Y.; Hou, L.; Mao, T.; Ning, W. PFM-PS Hybrid Control Strategy CLLC Resonant Converter Research. In Proceedings of the 2023 8th Asia Conference on Power and Electrical Engineering (ACPEE), Tianjin, China, 14–16 April 2023; pp. 1278–1284. [CrossRef]
15. Liu, H.; He, L.; Sun, Q.; Men, C. Design and Simulation of a PFM-PWM Hybrid Controller for DCDC Converter with CLLC Topology. In Proceedings of the 2023 China Semiconductor Technology International Conference (CSTIC), Virtual Conference, 29 June–26 July 2023; pp. 1–3. [CrossRef]
16. Lin, J.Y.; Yueh, H.Y.; Lin, Y.F.; Liu, P.H. Variable-Frequency and Phase-Shift With Synchronous Rectification Advance On-Time Hybrid Control of LLC Resonant Converter for Electric Vehicles Charger. *IEEE J. Emerg. Sel. Top. Ind. Electron.* **2023**, *4*, 348–356. [CrossRef]

17. Sun, K.; Chen, H.; Lu, L.; Zhang, K.; Wang, L.; Qin, Y. A Simplified Sensorless Synchronous Rectification Method for Inner Phase Shift Control in CLLC Converters. *IEEE J. Emerg. Sel. Top. Power Electron.* **2023**, *11*, 3815–3826. [CrossRef]
18. Wang, Y.; Wang, F.; Zhuo, F.; Yu, K.; Tian, J.; Song, R. Analysis of High Performance Synchronous Rectified CLLC Resonant Converter with Reactive Power Optimization. *IEEE Trans. Power Electron.* **2023**, *3*, 1–18. [CrossRef]
19. Zhu, T.; Zhuo, F.; Zhao, F.; Wang, F.; Yi, H.; Zhao, T. Optimization of Extended Phase-Shift Control for Full-Bridge CLLC Resonant Converter With Improved Light-Load Efficiency. *IEEE Trans. Power Electron.* **2020**, *35*, 11129–11142. [CrossRef]
20. Gu, W.J.; Liu, R. A Study of Volume and Weight vs. Frequency for High-Frequency Transformers. In Proceedings of the Proceedings of IEEE Power Electronics Specialist Conference—PESC '93, Seattle, WA, USA, 20–24 June 1993; pp. 1123–1129. [CrossRef]
21. Li, B.; Chen, M.; Wang, X.; Chen, N.; Sun, X.; Zhang, D. An Optimized Digital Synchronous Rectification Scheme Based on Time-Domain Model of Resonant CLLC Circuit. *IEEE Trans. Power Electron.* **2021**, *36*, 10933–10948. [CrossRef]
22. Jin, T.; Wu, W.; Zhang, Z.; Xiao, X.; You, W. Optimal Design of Turn-Off Switching Loss of Full-Bridge CLLC Converter. *IEEE J. Emerg. Sel. Top. Power Electron.* **2023**, *11*, 3302–3313. [CrossRef]
23. Jung, J.H.; Kim, H.S.; Ryu, M.H.; Baek, J.W. Design Methodology of Bidirectional CLLC Resonant Converter for High-Frequency Isolation of DC Distribution Systems. *IEEE Trans. Power Electron.* **2013**, *28*, 1741–1755. [CrossRef]
24. Sankar, A.; Mallik, A.; Khaligh, A. Extended Harmonics Based Phase Tracking for Synchronous Rectification in CLLC Converters. *IEEE Trans. Ind. Electron.* **2019**, *66*, 6592–6603. [CrossRef]
25. Kim, H.S.; Ryu, M.H.; Baek, J.W.; Jung, J.H. High-Efficiency Isolated Bidirectional AC–DC Converter for a DC Distribution System. *IEEE Trans. Power Electron.* **2013**, *28*, 1642–1654. [CrossRef]
26. Malagon-Carvajal, G.; Duarte, C.; Ordoñez-Plata, G.; Almeida, C.F.M.; Kagan, N. A Harmonic Frequency Domain Modeling Method for Single-Phase Full Bridge Rectifiers. *J. Control. Autom. Electr. Syst.* **2020**, *31*, 1322–1333. [CrossRef]
27. Petric, I.Z.; Mattavelli, P.; Buso, S. Feedback Noise Propagation in Multisampled DC–DC Power Electronic Converters. *IEEE Trans. Power Electron.* **2022**, *37*, 150–161. [CrossRef]
28. Rocha, E.M.; Barra, W.; Lucas, K.E.; Medeiros, R.L.P.; Vaca-Benavides, D.A. Design and Experimental Assessment of a Robust Voltage Control for DC-DC Converters Considering Components Parametric Uncertainties. *IEEE Access* **2020**, *8*, 109217–109231. [CrossRef]
29. Åströo, K.J.; Panagopoulos, H.; Hägglund, T. Design of PI Controllers Based on Non-Convex Optimization. *Automatica* **1998**, *34*, 585–601. [CrossRef]
30. Wang, Y.G.; Shao, H.H. Optimal Tuning for PI Controller. *Automatica* **2000**, *36*, 147–152. [CrossRef]

energies

MDPI

Article

Zero-Voltage-Switching Analysis Model of the Triple-Active-Bridge Converter

Hyung-Jun Byun [1], Jung-Min Park [2], Junsin Yi [1] and Chung-Yuen Won [1,*]

[1] Department of Electrical and Computer Engineering, College of Information and Communication Engineering, Sungkyunkwan University, Suwon 16419, Republic of Korea; gudwns313@skku.edu (H.-J.B.); junsin@skku.edu (J.Y.)

[2] Industrial Standards Division Electrical Standards Center, Korea Testing Laboratory (KTL), Ansan-si 15588, Republic of Korea; stui86@ktl.re.kr

* Correspondence: woncy@skku.edu

Abstract: This study aims to analyze the zero-voltage-switching (ZVS) region of a Triple-Active-Bridge (TAB) converter with five degrees of freedom. A TAB converter is an isolated converter derived from a dual-active-bridge (DAB) converter and composed of three full bridges (FBs) coupled to three winding transformers. To reduce the switching loss of the 12 active switches that compose 3 FBs, the ZVS operation is essential. However, owing to the numerous operation modes derived by five-phase shift ratios, ZVS analysis is complicated, particularly in the time domain. Therefore, this study presents the ZVS analysis model of the TAB converter based on the generalized harmonic approximation (GHA). Through the GHA of a TAB converter, the proposed model consists of unified formulas applicable to all operating ranges of the converter. Unified formulas consider all parameters, such as series inductance, port voltage, parasitic capacitance, transformer voltage, and turn ratio. In the proposed model, the ZVS area is confirmed using five-phase ratios with voltage modulation ratios as variables and verified using MATLAB and experiments.

Keywords: triple-active-bridge converter (TAB); phase-shift modulation (PSM); zero-voltage-switching (ZVS)

Citation: Byun, H.-J.; Park, J.-M.; Yi, J.; Won, C.-Y. Zero-Voltage-Switching Analysis Model of the Triple-Active-Bridge Converter. *Energies* **2023**, *16*, 7763. https://doi.org/10.3390/en16237763

Academic Editor: Salvatore Musumeci

Received: 31 October 2023
Revised: 23 November 2023
Accepted: 23 November 2023
Published: 24 November 2023

1. Introduction

As a countermeasure to the greenhouse effect caused by fossil fuels, eco-friendly energy industries such as renewable energy sources, electric vehicles, ships, and direct-current (DC) grids are growing [1]. Energy storage systems (ESS) are essential in these renewable energy sources to compensate for intermittence and meet load demand [2,3]. The conventional solution for combining ESS with renewable energy is to use various converters independently. However, this approach increases the system volume and cost and reduces the life cycle of the ESS by generating circulating power between the converters. Moreover, limitations are observed in that an additional communication system is required to control the power transmission of each converter, making the entire system complex. To overcome these limitations, the demand for multiport converters (MPCs) has increased [4]. Compared to the approach using multiple two-port converters, MPCs exhibit a simple structure with fewer power switches and passive components, higher power density, and higher dynamics [5,6]. In addition, it can be applied to various sources, storage systems, and loads with various voltage and current ratings. Therefore, their applications are expanding from electric vehicles [7] and electric aircraft [8] to DC grids [9,10]. Because MPCs must support various voltage and current ratings for each port, galvanic isolation between different ports is required for safety [11]. Therefore, a multi-winding high-frequency transformer was used.

Among isolated MPCs, the triple-active-bridge (TAB) converter is a promising topology derived from a dual-active-bridge (DAB) converter. Three full bridges (FBs) are coupled through the three-winding transformer as shown in Figure 1 and unlike resonant-type converters, since additional passive elements are not required and power flow is formed through one transformer, power density can be increased compared to other MPCs [12–14]. Similar to a DAB converter, a TAB converter is operated through phase-shift modulation with five degrees of freedom consisting of two outer phases and three inner phases, which are the phase-shift ratios between the three FBs and between the legs of each FB.

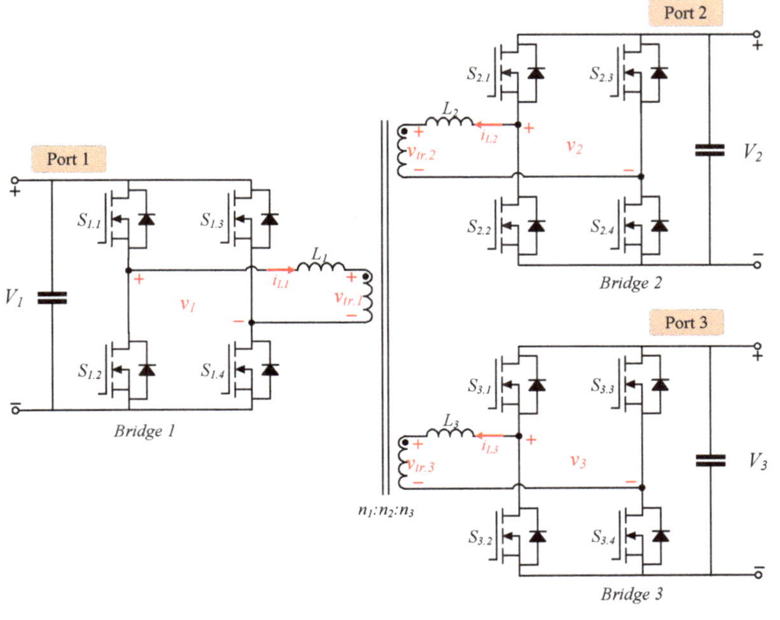

Figure 1. Schematic of a TAB converter.

Similar to other FB converters, a TAB converter is composed of several power switches, and hard switching in the commutation process causes high losses. Therefore, soft switching is essential, particularly when the switching frequency increases. Zero-voltage switching (ZVS) is the most widely used soft-switching method in PSM operation converters. Satisfaction with the ZVS operation depends on the phase shift ratios, and the ZVS region of the DAB converter, which is a family with a TAB converter that operates through PSM with three degrees of freedom, has been analyzed. In [15,16], different dual phase-shift modulations are proposed, and triple phase-shift modulations are proposed in [17–19] to improve the operation efficiency of the DAB converter. Hence, using the time domain analysis, the DAB operation was separated into 12 operating modes in [20]. Furthermore, computationally intensive numerical approaches that incorporate the effects of parasitic capacitances were used to derive the ZVS region [21].

However, unlike a DAB converter, which has relatively few operating modes with three degrees of freedom, the five degrees of freedom of a TAB converter are derived from numerous operating conditions. Moreover, because the three FBs are coupled, equalizing the series inductance by substituting one side is impossible. Therefore, port-voltage superposition in the three-winding transformer must be considered. Because of these issues, the ZVS analysis of the TAB converter is more complicated than that of the DAB converter.

A ZVS analysis of a TAB converter was attempted in [22]; however, it considered only an outer phase shift and did not consider an inner phase shift. In addition, the voltage superposition in the transformer and the effect of parasitic capacitance were neglected. To analyze the ZVS accurately, owing to the amount of charge in the parasitic capacitance of the metal-oxide-semiconductor-field-effect-transistor (MOSFET), not only the direction of the inductor current but also the minimum magnitude must be considered when discharging the capacitor [23,24]. In [25], an approach for analyzing the ZVS region of a TAB converter using a Thevenin-equivalent circuit was proposed. The series inductances of the three FBs were substituted through a Thevenin equivalent circuit, and the voltage superposition of the transformer was analyzed according to the switching state of the switches. In addition, the parasitic capacitor of the MOSFET was considered, and the amount of current required to establish ZVS was derived. However, because the TAB converter was analyzed in the discontinuous time domain, numerous conditions were derived based on the switching states of 12 switches. Therefore, various formulas were required for application in numerous cases; consequently, ZVS could not be analyzed using a unified formula. For example, to establish whether ZVS operates, the inductor current has to be derived. The slope of the inductor current is determined by the potential difference between the switching pole voltage and the transformer voltage. Therefore, in time domain analysis, numerous inductor current slopes must be derived according to the numerous operating regions of the TAB converter. Moreover, in the case of a multi-active bridge converter with a structure in which more FBs were added, the limitation of the time-domain analysis was even more pronounced. In addition, additional conversions of the Thevenin equivalent circuit were required. Consequently, the application of different equivalent circuits according to the viewpoint of each FB is required, resulting in difficulties in intuitive interpretation.

To address the limitations of time-domain analysis, this study presents a TAB converter ZVS region analysis model based on the generalization-harmonic approximation (GHA). The switching-pole voltage of a TAB converter is a periodic function of a square or quasi-square wave. Therefore, it can be expressed as a unified formula using the cumulative sum of harmonics through a Fourier series. This approach can directly derive the inductor current and voltage of a three-winding transformer and the switching pole voltage. Therefore, regardless of the conditions, a unified formula can be derived for the entire switching operation area. The effect of the parasitic capacitor was also considered, and the magnitude of the minimum inductor current for discharging was derived using the energy balance equation concept applied in the ZVS analysis of the DAB converter [26]. The proposed approach and model are simple to extend and apply, even to a multi-active-bridge converter that adds FBs and can be easily calculated through MATLAB. The remainder of this paper is organized as follows: a description of the TAB converter and derivation of the harmonic form; analysis of commutation processes in the FB; derivation of conditions to achieve ZVS considering parasitic capacitors; and derivation of the ZVS analysis model. The proposal is verified using MATLAB simulations and experiments.

2. Operation and the GHA Modeling of the TAB Converter

2.1. Operating Principles

The equivalent circuit model of a TAB converter is shown in Figure 2. Considering the turn ratio of the transformer, the components of the circuit are replaced as follows:

$$\begin{cases} v'_x &= v_x / n_x \\ i'_{L.x} &= n_x i_{L.x} \\ L'_x &= L_x / n_x^2 \end{cases} \tag{1}$$

In (1), the x means unspecific FB. For the unified formulas, the post equations were derived using x, y, and z. The actual TAB converter is in the form of the Y-connected structure, as shown in Figure 2a, but the Δ-connected model shown in Figure 2b is also used in circuit analysis. Through the Y-Δ conversion, the equivalent inductance L'_{xy} between unspecific two FBs x and y in the Δ-connected model is derived as follows:

$$L'_{xy} = L'_x + L'_y + \frac{L'_x L'_y}{L'_z}. \tag{2}$$

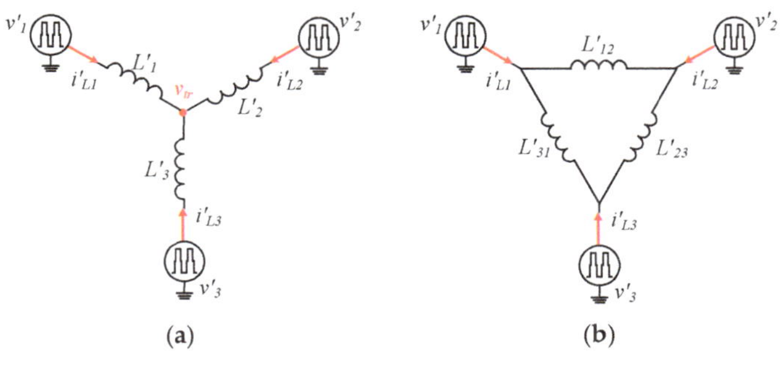

Figure 2. Equivalent circuit model of a TAB converter: (**a**) Y-connected model; (**b**) Δ-connected model.

Figure 3 shows the operating waveform of the TAB converter. In the waveforms, α_x is the inner phase shift ratio between the leading leg ($S_{x.1}$, $S_{x.2}$) and lagging leg ($S_{x.3}$, $S_{x.4}$), and ϕ_{xy} is the outer phase shift ratio between FB x and y ($\phi_{xy} = \phi_x - \phi_y$). The inner phase shift as α results in an equal switching state formed in the leading and lagging legs, and the voltage of the switching pole becomes a quasi-square wave. The voltages of each switching pole are out of phase as ϕ, and all voltages are superposed to the transformer and form the v_{tr}. Consequently, the inductor current i'_{Lx} and delivered power of each FB are determined by the potential difference between the switching pole and transformer.

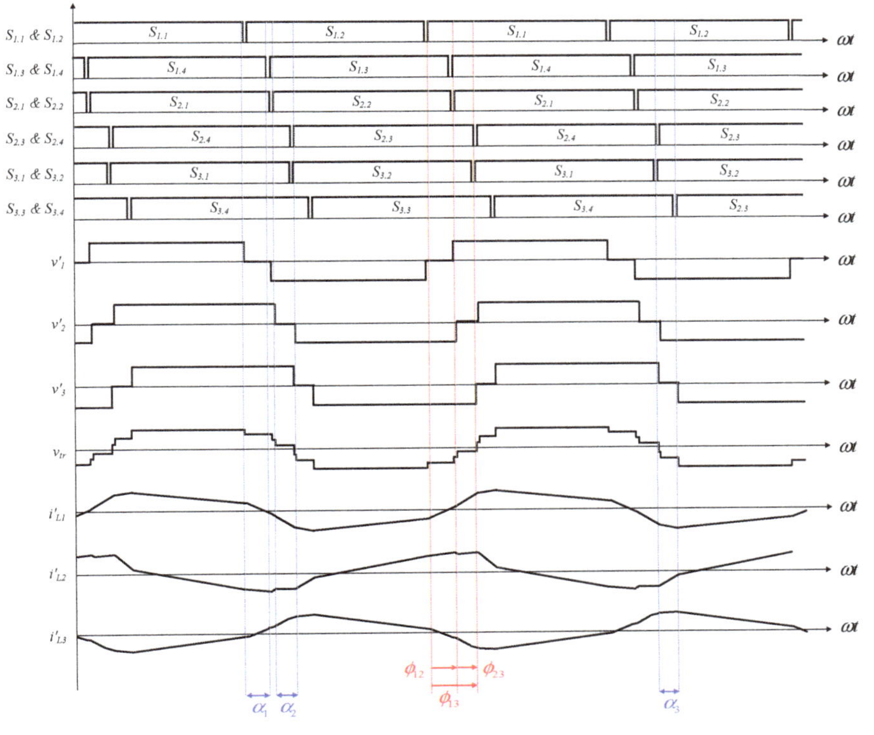

Figure 3. Operating waveforms of a TAB converter.

2.2. GHA Modeling

In the mathematical modeling of the TAB converter, the conventional method of deriving the state equation according to the on-off state of the power switch is too complex to derive the formula because of the 12 power switches with five degrees of freedom. As an alternative to settling this problem, GHA modeling was used. As the switching of the FB is periodic, a Fourier series can be used. Therefore, the switching pole voltage v'_x can be expressed as the cumulative sum of the odd harmonics, as follows:

$$v'_x(t) = \frac{4V_x}{n_x\pi}\sum_{k=0}^{\infty}\frac{1}{(2k+1)}\cos\left\{(2k+1)\frac{\alpha_x}{2}\right\}\sin\{(2k+1)(\omega t+\phi_x)\}. \tag{3}$$

The transformer neutral-point voltage v_{tr} is determined using three voltage sources and three series of inductances, as shown in Figure 2a. Through the principle of superposition between multiple sources, v_{tr} can be expressed as

$$v_{tr}(t) = F\left\{v'_x(t)+v'_y(t)+v'_z(t)\right\} = F\{v'_x(t)\}+F\left\{v'_y(t)\right\}+F\{v'_z(t)\}. \tag{4}$$

By ignoring the voltage sources of FB y and z as short circuits, the function $F\{v'_x(t)\}$ for v'_x of v_{tr} is as follows:

$$F\{v'_x(t)\} = \frac{L'_yL'_zv'_x(t)}{L'_xL'_y+L'_yL'_z+L'_zL'_x}. \tag{5}$$

Therefore, v_{tr} can be derived using (4) and (5), and the $v_{tr.x}$ the transformer voltage viewed from FB x can be expressed as:

$$\begin{aligned} v_{tr.x}(t) &= n_x\left[F\{v'_x(t)\}+F\left\{v'_y(t)\right\}+F\{v'_z(t)\}\right] \\ &= n_x\frac{L'_yL'_zv'_x(t)+L'_zL'_xv'_y(t)+L'_xL'_yv'_z(t)}{L'_xL'_y+L'_yL'_z+L'_zL'_x} \end{aligned}. \tag{6}$$

The inductor current i'_{Lx} can be directly calculated through the voltage sources on the switching pole and transformer side derived in (3) and (5). However, it is very complicated to derive owing to the variables (ϕ_x, ϕ_y, ϕ_z, α_x, α_y, α_z) of voltage sources. Alternatively, it can be derived indirectly through the Δ-connected model shown in Figure 2b. (2) and (3), the inductor current i'_{Lxy} between FB x and y is derived as follows:

$$i'_{L.xy}(0) = -i'_{L.xy}(\pi/\omega) \tag{7}$$

$$\begin{aligned} i'_{L.xy}(t) &= i'_{L.xy}(0)+\frac{1}{L'_{xy}}\int_0^t\left\{v'_x(\tau)-v'_y(\tau)\right\}d\tau \\ &= \frac{4}{\pi\omega L'_{xy}}\sum_{k=0}^{\infty}\frac{1}{(2k+1)^2}\sqrt{A_{xy}^2+B_{xy}^2}\sin\left\{(2k+1)\omega t+\tan^{-1}\frac{A_{xy}}{B_{xy}}\right\} \end{aligned}. \tag{8}$$

The A_{xy} and the B_{xy} are:

$$\begin{cases} A_{xy} = V'_y\cos\left\{(2k+1)\frac{\alpha_y}{2}\right\}\cos\{(2k+1)\phi_{xy}\}-V'_x\cos\left\{(2k+1)\frac{\alpha_x}{2}\right\} \\ B_{xy} = V'_y\cos\left\{(2k+1)\frac{\alpha_y}{2}\right\}\sin\{(2k+1)\phi_{xy}\} \end{cases}. \tag{9}$$

By applying the process of (7) to (9) between the FB x and z, i'_{Lx} can be finally derived as:

$$i'_{L.x}(t) = i'_{L.xy}(t)+i'_{L.xz}(t). \tag{10}$$

3. Zero Voltage Switching

An ideal MOSFET consists of only a channel and body diode, and in this case, only the inductor current direction in the commutation process is used to confirm the ZVS operation. However, capacitance exists in an actual MOSFET, and this parasitic capacitor stores charge during the turn-off period of the switch. Therefore, in the commutation process, before the switch turns on, the energy of the parasitic capacitor must be completely discharged to achieve zero voltage in the switch. Hence, the direction and magnitude of the inductor current should be considered to confirm ZVS operation.

Figure 4 shows the commutation process in dead time before the switch is turned on. Because the commutation process is instantaneous, the voltages on the switching pole and transformer do not change, and the voltage at the transformer port is expressed as the initial magnitude of the process as $V_{tr.x}$ the DC value. Figure 4a,b show the switching cases without inner phase shift ($\alpha_x = 0$), and the leading and lagging legs operate simultaneously. Further, Figure 4c–f shows the switching cases with an inner phase shift ($\alpha_x \neq 0$), and the leading and lagging legs operate differently. Before each switch is turned on, the current flows in the direction of discharging the parasitic capacitor $C_{oss.x}$, as indicated by the orange line, and the reference direction of each current is indicated by a red arrow. The upper and lower switches of one leg operate complementarily; therefore, if one switch of the leg satisfies the condition for the ZVS operation, the other switch can also perform the ZVS operation. Because of the nonlinear capacitance characteristics of MOSFETs, direct calculation of the energy stored in a parasitic capacitor is difficult. Therefore, the energy to be discharged during the commutation process and the magnitudes of the minimum inductor energy and current are derived by applying the energy-balance equation. During FB operation, two parasitic capacitors in one switching leg exchange energy with each other and do not transfer to the other components. Consequently, the energies of the two capacitors on one leg can be ignored, and the energy balance equation for the commutation process is as follows:

$$E_s - E_L - E_{tr.x} = 0. \tag{11}$$

E_s, E_L, and $E_{tr.x}$ represent the source, inductor, and transformer-side energies, respectively, during commutation.

$$\begin{cases} E_s = \int_{t_0}^{t_d} V_x i_{in.x} dt \\ E_L = \int_{t_0}^{t_d} v_{L.x} i_{L.x} dt = \frac{1}{2} L_x i_{L.x}^2(t_d) - \frac{1}{2} L_x i_{L.x}^2(t_0) \\ E_{tr} = \int_{t_0}^{t_d} V_{tr.x} i_{L.x} dt \end{cases} \tag{12}$$

Integration sections t_0–t_d are the commutation process sections before the subject switches on. Meanwhile, the minimum magnitude condition of the inductor current and energy for ZVS operation occurs when the current and energy become zero at the end of the commutation process, as shown in (13). Applying (12) to (11), we obtain the following inequality:

$$i_{L.x}(t_d) = 0, \ E_L(t_d) = 0 \tag{13}$$

$$E_{diff} = \pm(E_{tr} - E_s) \tag{14}$$

$$E_L(t_0) \geq E_{diff}. \tag{15}$$

E_{diff} is the energy difference between the source and transformer sides. Because the energy flow changes depending on the direction of the inductor current, the sign follows that of the inductor current. As shown in (15), when E_L (t_0) is higher than E_{diff}, the parasitic capacitor can be discharged entirely through inductive energy, and ZVS operation is established. If E_{diff} has a negative value, the energy supply is higher than the energy consumption between the two voltage sources. Therefore, unlike the minimum inductor current and energy conditions for ZVS in (13), the surplus energy is stored in the inductor, the parasitic capacitor can be sufficiently discharged regardless of the magnitude of the initial inductive energy, and ZVS operation is obtained.

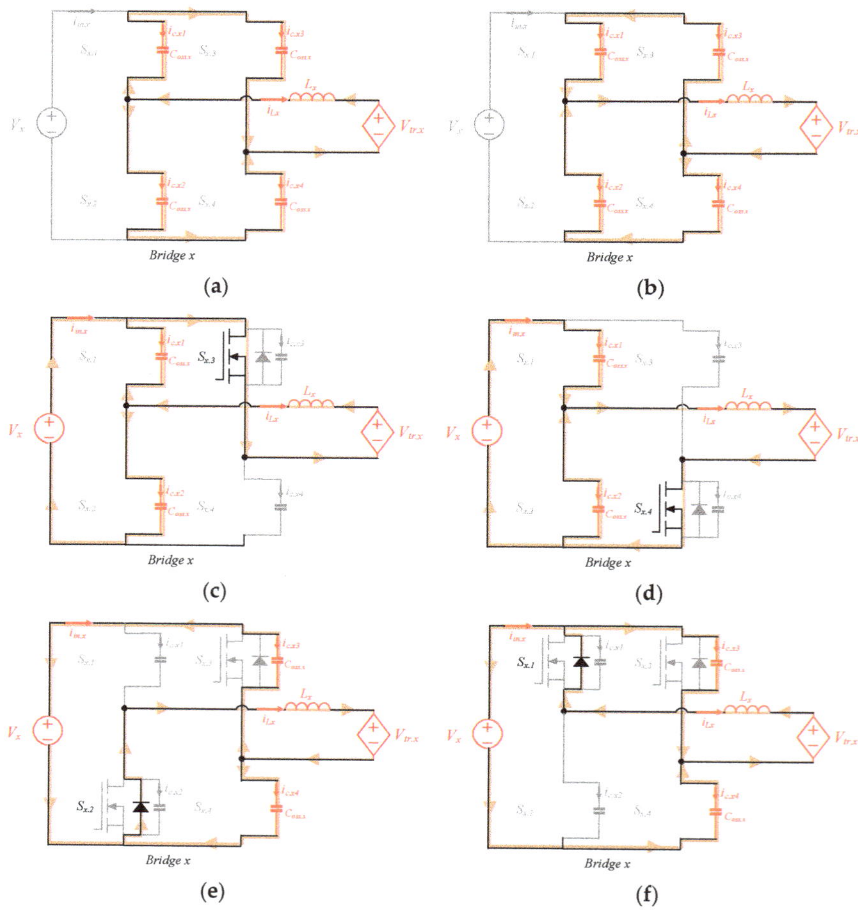

Figure 4. Commutation process. (**a**) Case A: before Sx.1 and Sx.4 turn on. (**b**) Case A′: before Sx.2 and Sx.3 turn on. (**c**) Case B: before Sx.1 turns on. (**d**) Case B′: before Sx.2 turns on. (**e**) Case C: before Sx.3 turns on. (**f**) Case C′: before Sx.4 turns on.

3.1. Without Inner Phase Shift ($\alpha_x = 0$)

In Case A in Figure 4a, a closed circuit is formed without an input voltage source, and the direction of the inductor current is negative. The circulating energy during the commutation process is supplied from the voltage source on the transformer side, and the input and inductor currents are derived using the node equation as follows:

$$i_{in.x} = 0 \tag{16}$$

$$\begin{aligned} i_{L.x} = i_{c.x1} - i_{c.x2} &= C_{oss.x}\frac{dv_{c.x1}}{dt} - C_{oss.x}\frac{dv_{c.x2}}{dt} \\ &= C_{oss.x}\frac{d(V_x - v_{c.x2})}{dt} - C_{oss.x}\frac{dv_{c.x2}}{dt} = -2C_{oss.x}\frac{dv_{c.x2}}{dt}. \end{aligned} \tag{17}$$

Applying (16) and (17) to (12) and (14), E_{diff} can be expressed as

$$E_{diff} = -2C_{oss.x}V_{tr.x}V_x \ (i_{Lx} < 0). \tag{18}$$

Conversely, in case A′ in Figure 4b, where the inductor current is in the positive direction, E_{diff} is obtained as

$$E_{diff} = 2C_{oss.x}V_{tr.x}V_x \ (i_{Lx} > 0). \tag{19}$$

If E_{diff} derived from (18) and (19), satisfies the energy inequality in (15), then the ZVS operation is established for cases A and A′.

3.2. With Inner Phase Shift ($\alpha_x \neq 0$)

In Case B in Figure 4c, the inductor current is negative, the leading and lagging legs do not operate simultaneously because of the inner phase shift, and a closed circuit is formed with the input voltage source. The input and inductor currents derived from the node equation are:

$$i_{in.x} = i_{c.x2} = C_{oss.x}\frac{dv_{c.x2}}{dt} \tag{20}$$

$$\begin{aligned} i_{L.x} &= i_{c.x1} - i_{c.x2} = C_{oss.x}\frac{dv_{c.x1}}{dt} - C_{oss.x}\frac{dv_{c.x2}}{dt} \\ &= C_{oss.x}\frac{d(V_x - v_{c.x2})}{dt} - C_{oss.x}\frac{dv_{c.x2}}{dt} = -2C_{oss.x}\frac{dv_{c.x2}}{dt} \end{aligned}. \tag{21}$$

Similarly, as in Section 3.1, E_{diff} can be derived as:

$$E_{diff} = -2C_{oss.x}V_{tr.x}V_x + C_{oss.x}V_x^2 \ (i_{Lx} < 0). \tag{22}$$

Conversely, in case B′ in Figure 4d, where the inductor current is in the positive direction, E_{diff} is:

$$E_{diff} = 2C_{oss.x}V_{tr.x}V_x - C_{oss.x}V_x^2 \ (i_{Lx} > 0). \tag{23}$$

Similarly, if E_{diff} derived from (22) and (23) satisfies the energy inequality in (15), a ZVS operation is established. Cases B′ and C, which have an inner phase shift and the same inductor current direction for the discharge parasitic capacitor, have the same E_{diff} formula, and cases B and C′ also have the same. Therefore, E_{diff} can be defined as

$$E_{diff(\alpha_x = 0)} = \pm\int_{t_0}^{t_d} (V_{tr.x}i_{L.x} - V_x i_{in.x})dt = \pm 2C_{oss.x}V_{tr.x}V_x \tag{24}$$

$$E_{diff(\alpha_x \neq 0)} = \pm\int_{t_0}^{t_d} (V_{tr.x}i_{L.x} - V_x i_{in.x})dt = \pm\left(2C_{oss.x}V_{tr.x}V_x - C_{oss.x}V_x^2\right). \tag{25}$$

As in (14), the sign of E_{diff} in (24) and (25) follows that of the inductor current, as shown in Figure 4. Finally, the minimum magnitude of the inductor current required to satisfy the ZVS operation in each case is listed in Table 1. Applying $V_{tr.x}$, V_x, and $i_{L.x}$ derived in Section 2.2 to the equations in Table 1, the ZVS operation can be confirmed. For example, when the turn-on point of the switch in full bridge x is t_0, the transformer side voltage and inductor current, which are variables to confirm whether ZVS is established, are as follows:

$$V_{tr.x} = v_{tr.x}(t_0) = n_x\left[F\{v_x'(t_0)\} + F\left\{v_y'(t_0)\right\} + F\{v_z'(t_0)\}\right] \tag{26}$$

$$i_{L.x}(t_0) = i_{L.xy}(t_0) + i_{L.xz}(t_0). \tag{27}$$

Table 1. ZVS operating conditions for each switching case.

Case	Switches	Energy Difference between the Two Voltage Sources	Conditions for ZVS Operation
A	*Sx.1* and *Sx.4*	$E_{diff} = -2C_{oss.x}V_{tr.x}V_x$	$i_{Lx}(t_0) \le \sqrt{\frac{-4C_{oss.x}V_{tr.x}V_x}{L_x}}$ or $E_{diff} \le 0$
A′	*Sx.2* and *Sx.3*	$E_{diff} = 2C_{oss.x}V_{tr.x}V_x$	$i_{Lx}(t_0) \ge \sqrt{\frac{4C_{oss.x}V_{tr.x}V_x}{L_x}}$ or $E_{diff} \le 0$
B	*Sx.1*	$E_{diff} = C_{oss.x}V_x^2 - 2C_{oss.x}V_{tr.x}V_x$	$i_{Lx}(t_0) \le \sqrt{\frac{2C_{oss.x}V_x^2 - 4C_{oss.x}V_{tr.x}V_x}{L_x}}$ or $E_{diff} \le 0$
B′	*Sx.2*	$E_{diff} = -C_{oss.x}V_x^2 + 2C_{oss.x}V_{tr.x}V_x$	$i_{Lx}(t_0) \ge \sqrt{\frac{-2C_{oss.x}V_x^2 + 4C_{oss.x}V_{tr.x}V_x}{L_x}}$ or $E_{diff} \le 0$
C	*Sx.3*	$E_{diff} = -C_{oss.x}V_x^2 + 2C_{oss.x}V_{tr.x}V_x$	$i_{Lx}(t_0) \ge \sqrt{\frac{-2C_{oss.x}V_x^2 + 4C_{oss.x}V_{tr.x}V_x}{L_x}}$ or $E_{diff} \le 0$
C′	*Sx.4*	$E_{diff} = C_{oss.x}V_x^2 - 2C_{oss.x}V_{tr.x}V_x$	$i_{Lx}(t_0) \le \sqrt{\frac{2C_{oss.x}V_x^2 - 4C_{oss.x}V_{tr.x}V_x}{L_x}}$ or $E_{diff} \le 0$

4. Simulation and Experimental Results

The operation of the TAB converter involves seven variables: five phase-shift ratios and two voltage-modulation ratios.

$$m_{xy} = \frac{V_y/n_y}{V_x/n_x},\ m_{xz} = \frac{V_z/n_z}{V_x/n_x} \tag{28}$$

However, simulations can only have two variables because they use one dimension as the output in three dimensions. Therefore, the ZVS analysis model was simulated in two parts using MATLAB, as listed in Table 1. A case in which use the outer phase shift ratio ϕ as variables; a case in which use voltage modulation ratio m as variables.

4.1. Case A (Variable: ϕ)

Based on the proposed ZVS analysis model in Table 1, the MATLAB simulation results of the ZVS operation region of the TAB converter using ϕ as a variable are shown in Figure 5. The red area is the region where ZVS is performed in the leading leg, the blue area is the lagging leg, and the purple area is the region where ZVS is performed on all legs. The empty area represents the hard-switching region.

Figure 5a shows the ZVS region when the voltage modulation ratio of the three FBs is 1:1:1, and the ZVS operation is assured in all FBs except under extremely light-load conditions. Figure 5b shows the ZVS region under the condition that the voltage modulation ratios of FB 2 and FB 3 are 1.5 and 0.75 based on FB 1. In FB 2, which has a higher equivalent voltage than the other two FBs, ZVS is ensured regardless of the phase shift ratios, but in FB 1 and FB 3, hard switching occurs depending on the phase conditions. In particular, FB 3, which had the lowest voltage, caused hard switching over a wide region. A comparison of Figure 5a,b confirms that soft and hard switching are critically determined by the voltage modulation ratio.

Meanwhile, the effective value of the switching pole voltage can be adjusted through the inner phase shift ratio α as (3). The fundamental value of each FB switching pole voltage can be matched through a simple calculation using arccosine. However, because only the fundamental wave components matched, the actual effective voltage could be reduced. Figure 5c shows the ZVS region when the magnitude of the fundamental value of all FB's switching pole voltages is matched by applying an inner phase shift ratio of 0.46π and 0.67π to FB 1 and FB 2, respectively. By matching the magnitudes of the switching pole voltages, ZVS operation under light-load conditions was ensured in all FBs. However, the ZVS regions between the leading and lagging legs differ because of unstimulated switching. This results from the phase of the inductor current: the leading leg facilitates the ZVS operation when transmitting power, and the lagging leg facilitates the ZVS operation when receiving power.

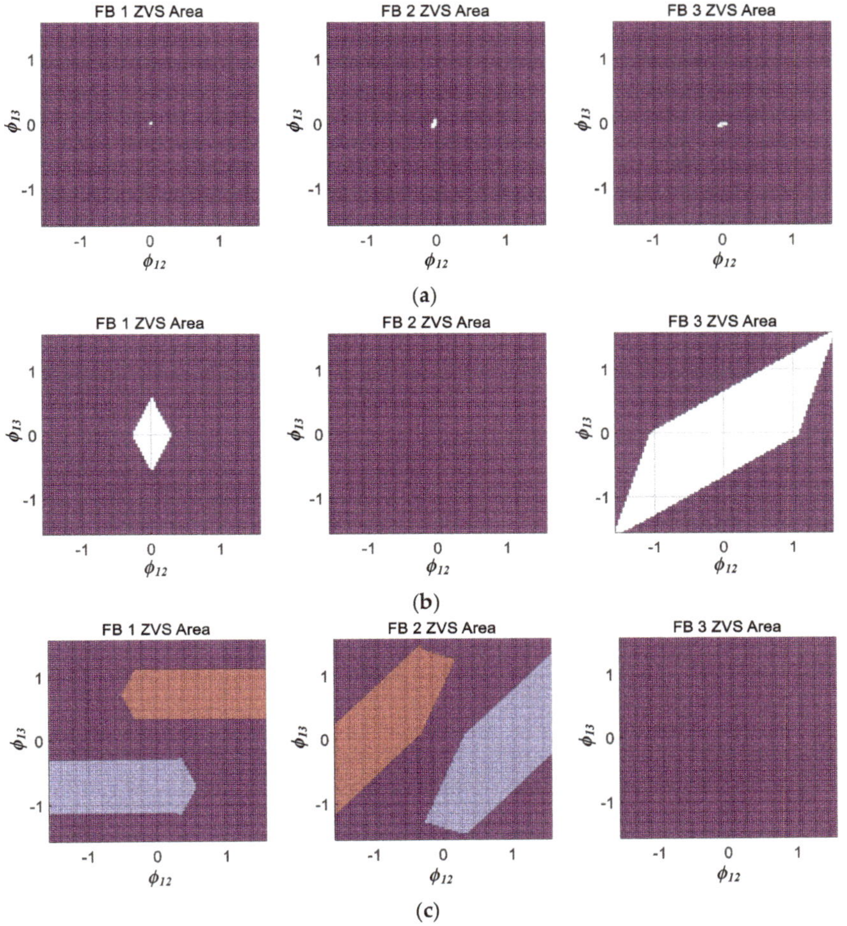

Figure 5. ZVS operation region according to outer phase shift ratio ϕ variation. (**a**) $m_{12} = m_{13} = 1$ and $\alpha_1 = \alpha_2 = \alpha_3 = 0$. (**b**) $m_{12} = 1.5$, $m_{13} = 0.75$ and $\alpha_1 = \alpha_2 = \alpha_3 = 0$. (**c**) $m_{12} = 1.5$, $m_{13} = 0.75$ and $\alpha_1 = 0.46\pi$, $\alpha_2 = 0.67\pi$, $\alpha_3 = 0$.

Figure 6 shows the simulation results of the operation waveform using MATLAB to verify the proposed GHA-based ZVS analysis model shown in Figure 5. In a switching leg, indicating the upper switch turn-on as '1' and the lower switch turn-on as '−1', the corresponding switching pole voltage is listed in Table 2. In Figure 6a–c, ϕ_{12} and ϕ_{13} are set as -0.05π and 0.05π, respectively, to confirm the operation waveform in the light load area where ZVS operation is difficult.

Table 2. Switching pole voltage according to the switching state.

Leading Leg	Lagging Leg	Switching Pole Voltage
1	−1	V_x
−1	−1	0
−1	1	$-V_x$
1	1	0

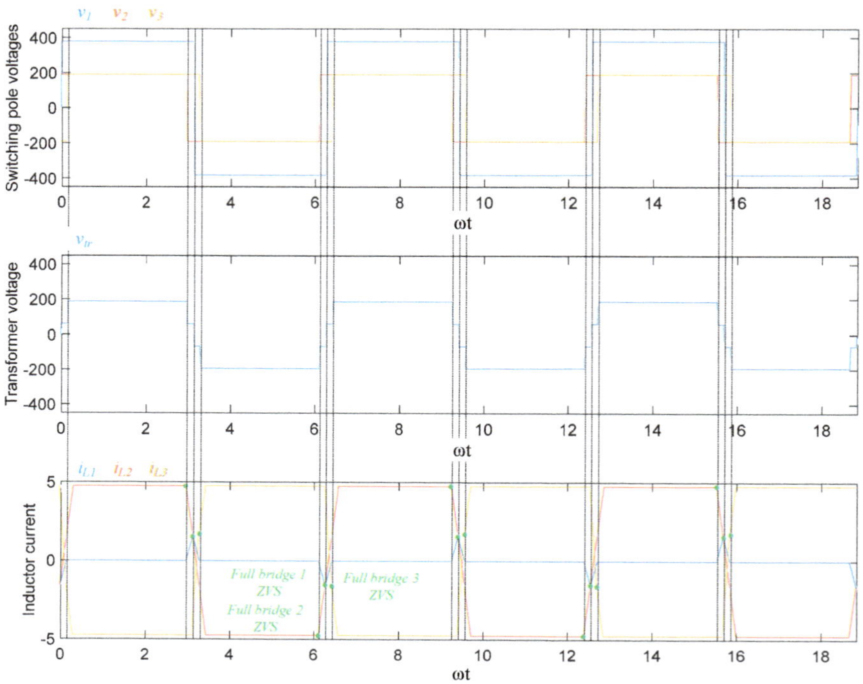

(a)

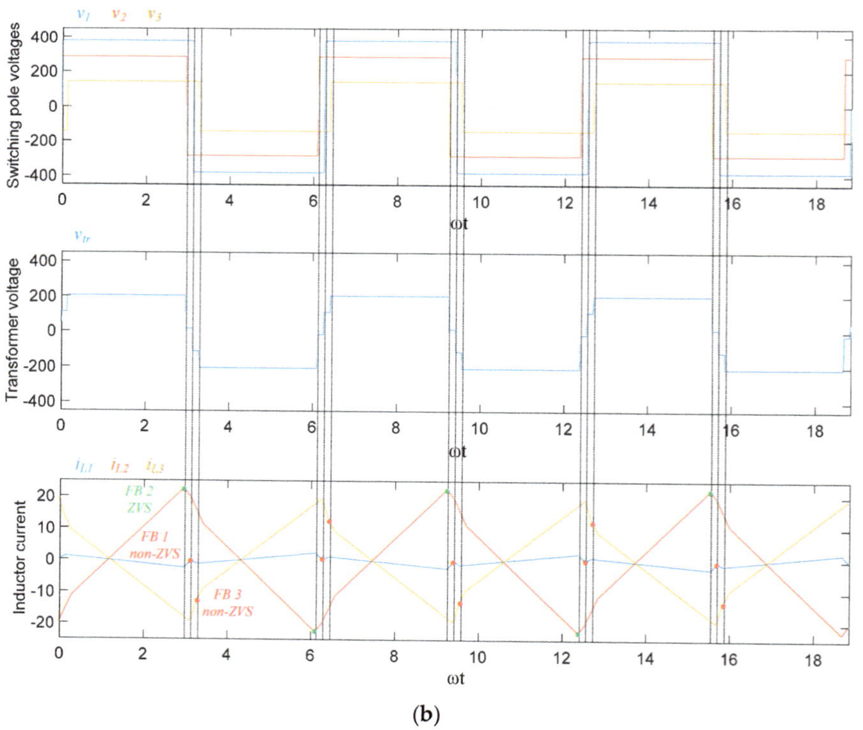

(b)

Figure 6. *Cont.*

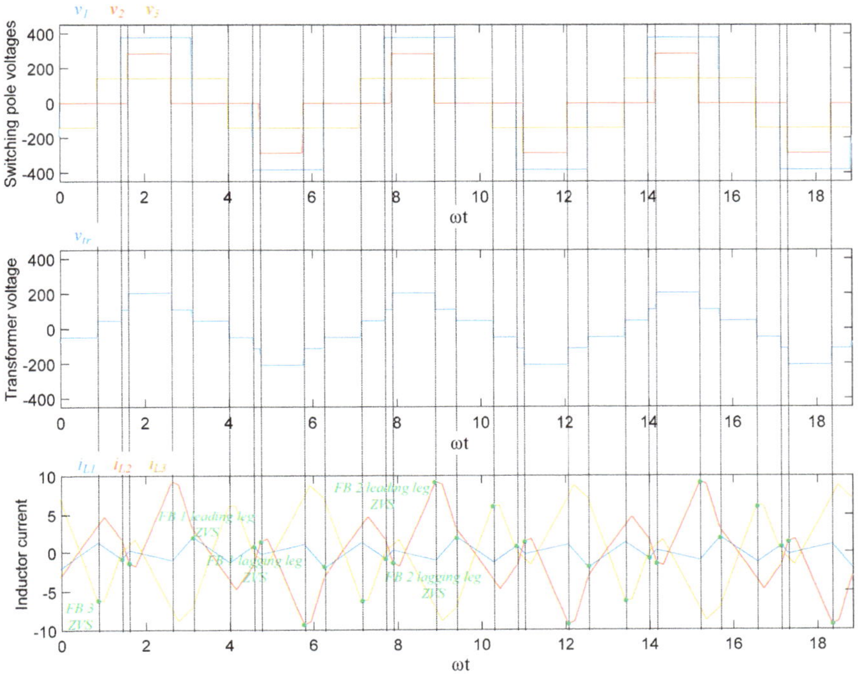

(c)

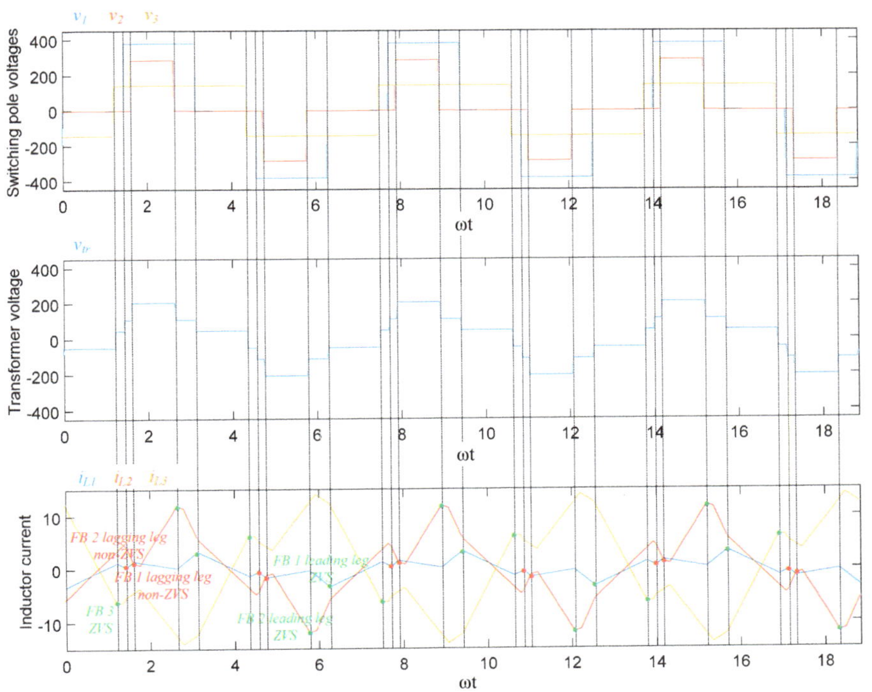

(d)

Figure 6. *Cont.*

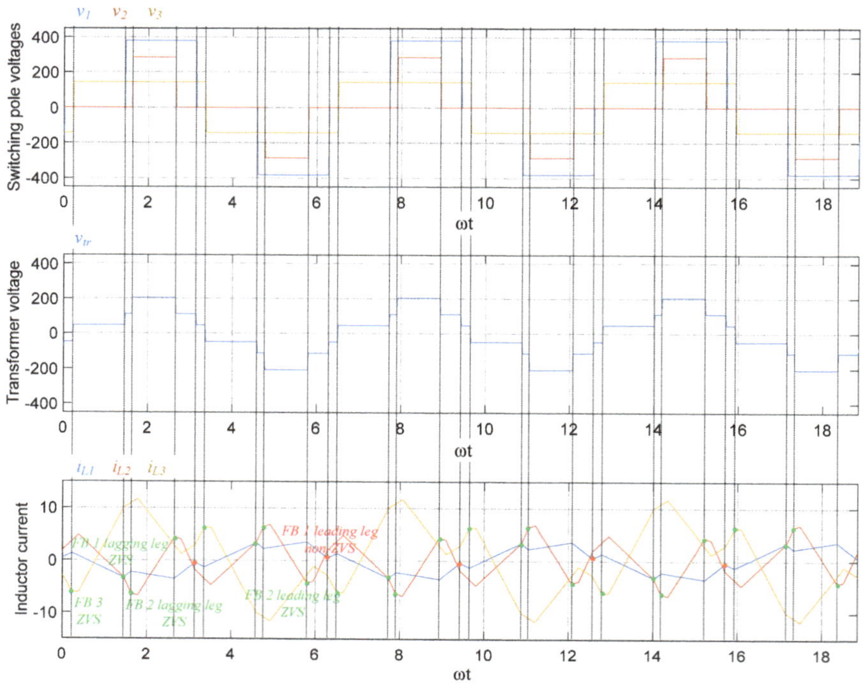

(e)

Figure 6. TAB converter operation waveform. (**a**) $\alpha_1 = \alpha_2 = \alpha_3 = 0$, $m_{12} = m_{13} = 1$, $\phi_{12} = -0.05\pi$, $\phi_{13} = 0.05\pi$. (**b**) $\alpha_1 = \alpha_2 = \alpha_3 = 0$, $m_{12} = 1.5$, $m_{13} = 0.75$, $\phi_{12} = -0.05\pi$, $\phi_{13} = 0.05\pi$. (**c**) $\alpha_1 = 0.46\pi$, $\alpha_2 = 0.67\pi$, $\alpha_3 = 0$, $m_{12} = 1.5$, $m_{13} = 0.75$, $\phi_{12} = -0.05\pi$, $\phi_{13} = 0.05\pi$. (**d**) $\alpha_1 = 0.46\pi$, $\alpha_2 = 0.67\pi$, $\alpha_3 = 0$, $m_{12} = 1.5$, $m_{13} = 0.75$, $\phi_{12} = -0.05\pi$, $\phi_{13} = 0.16\pi$. (**e**) $\alpha_1 = 0.46\pi$, $\alpha_2 = 0.67\pi$, $\alpha_3 = 0$, $m_{12} = 1.5$, $m_{13} = 0.75$, $\phi_{12} = -0.05\pi$, $\phi_{13} = -0.16\pi$.

Figure 6a shows the operating waveform under the conditions shown in Figure 5a; owing to the voltage ratio of 1:1:1, all FBs perform ZVS operation without an inner phase shift. Figure 6b shows the operation waveform under the condition of Figure 5b; except for FB 2, the remaining FBs cannot perform the ZVS operation. In particular, in the case of FB 3, which has the lowest voltage ratio, the inductor current deviates significantly from the condition for performing ZVS. Figure 6c shows the operating waveform when the effective voltage of the switching pole was adjusted by applying an internal phase shift to the FB, as shown in Figure 5c. Because of the asynchronous operation of the leading and lagging legs, the switching-pole voltage became a quasi-square wave. Therefore, the effective values of the voltage are matched, and ZVS operation is ensured for all FBs in the light-load area. In addition, Figure 6d,e show the operation waveforms at the points in Figure 5c, where the ZVS is operated on only one leg of the FB.

4.2. Case B (Variable: m)

The simulation results for Case B, which used m as a variable, are shown in Figure 7. Figure 7a shows the ZVS region under the rated load conditions. All FBs ensured ZVS operations. However, in Figure 7b, which shows the light load conditions, it can be confirmed that the ZVS operation is not achieved according to the voltage modulation ratio. In particular, these results are noticeable for FB with relatively low voltage ratios. Figure 7c shows the ZVS region when an inner phase shift is applied to the FB with a higher voltage ratio. Compared to FB 1, when the voltage modulation ratios of FB 2 and FB 3 are higher than 1, the inner phase shift is applied and varied in proportion to the voltage modulation ratio. In FB 1, ZVS is ensured in the entire area, and in FB 2 and FB 3, ZVS operations are improved in parts where m_{12} and m_{13} are greater than 1.

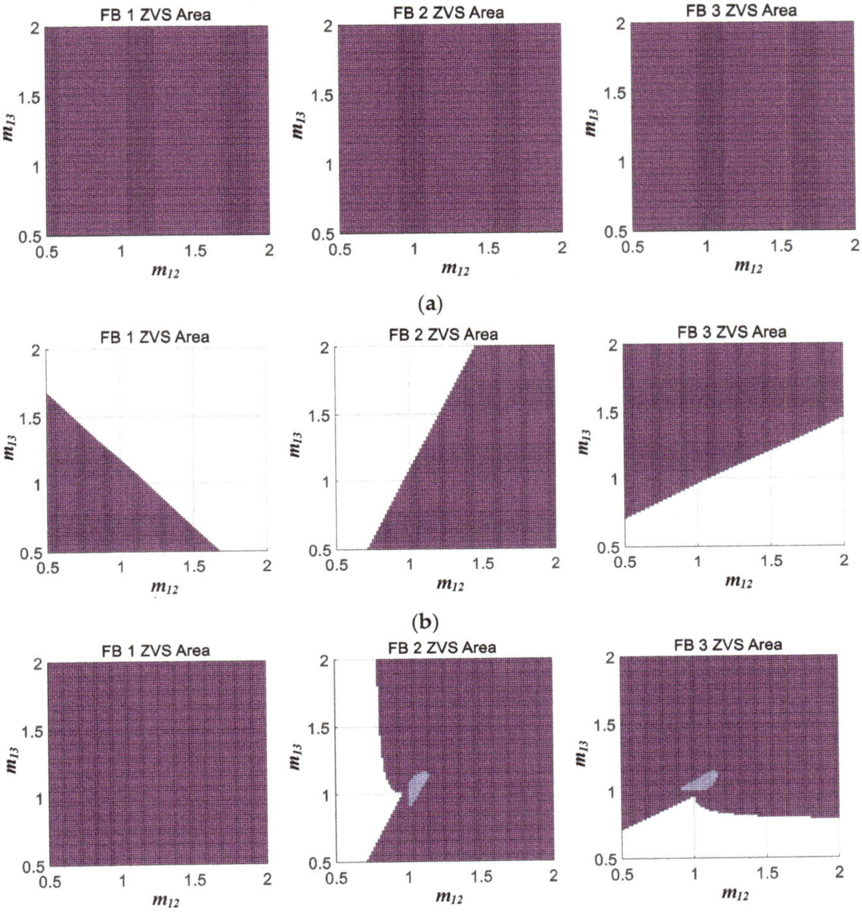

(a)

(b)

(c)

Figure 7. ZVS region according to voltage modulation ratio m variation. (**a**) $\phi_{12} = \phi_{13} = 0.5\pi$ and $\alpha_1 = \alpha_2 = \alpha_3 = 0$. (**b**) $\phi_{12} = \phi_{13} = 0.05\pi$ and $\alpha_1 = \alpha_2 = \alpha_3 = 0$. (**c**) $\phi_{12} = \phi_{13} = 0.05\pi$ and $\alpha_1 = 0$, $\alpha_2 = 2\cos(1/m_{12})$, $\alpha_3 = 2\cos(1/m_{13})$.

Figure 8 shows the operating waveform of the TAB converter used to verify the ZVS analysis model shown in Figure 7. The voltage modulation ratios m_{12} and m_{13} are 0.85 and 2, respectively. The simulation results show the operation waveforms under rated load, light load, and variable inner phase-shift conditions.

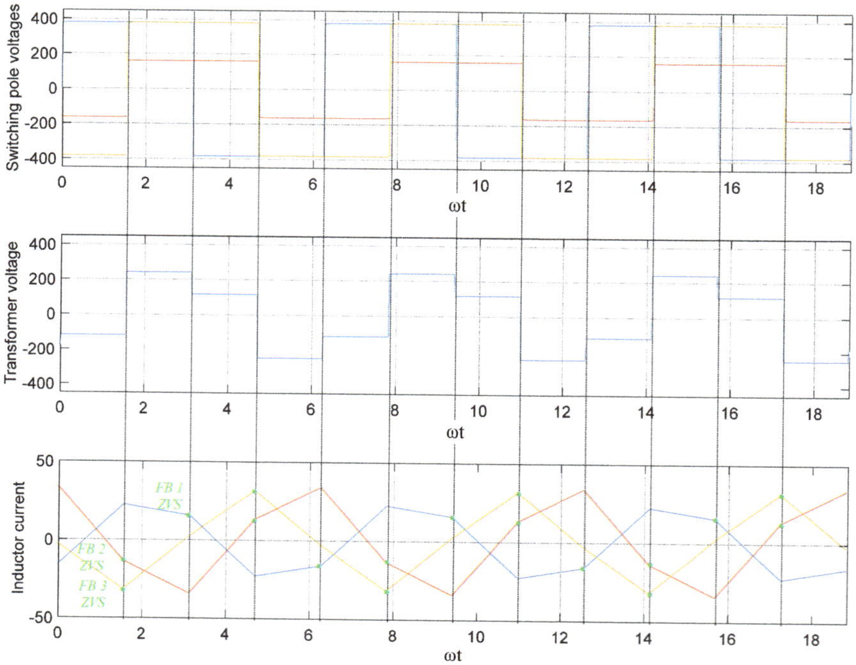

(a)

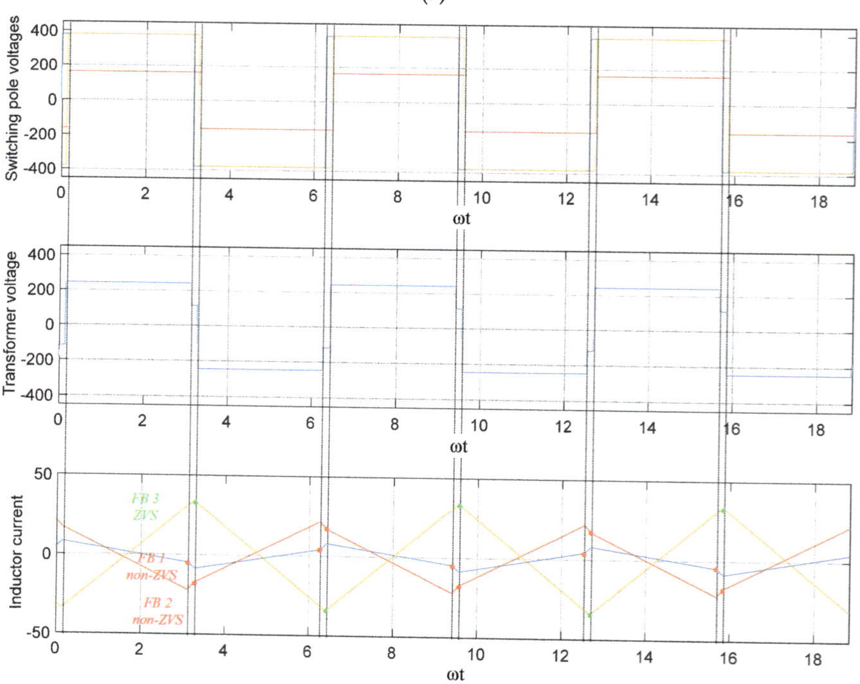

(b)

Figure 8. *Cont.*

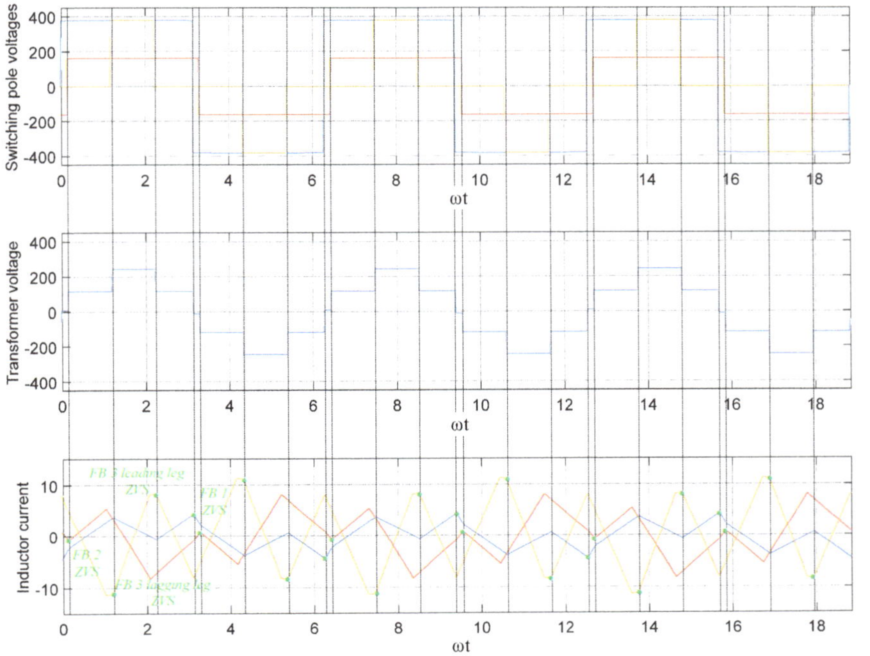

(c)

Figure 8. TAB converter operation waveform. (**a**) $\alpha_1 = \alpha_2 = \alpha_3 = 0$, $m_{12} = 0.85$, $m_{13} = 2$, $\phi_{12} = 0.5\pi$, $\phi_{13} = 0.5\pi$. (**b**) $\alpha_1 = \alpha_2 = \alpha_3 = 0$, $m_{12} = 0.85$, $m_{13} = 2$, $\phi_{12} = 0.05\pi$, $\phi_{13} = 0.05\pi$. (**c**) $\alpha_1 = \alpha_2 = 0$, $\alpha_3 = 0.66\pi$, $m_{12} = 0.85$, $m_{13} = 2$, $\phi_{12} = 0.05\pi$, $\phi_{13} = 0.05\pi$.

Figure 8a shows the simulation results under the rated load conditions, as shown in Figure 7a; all FBs perform the ZVS operation. Figure 8b shows the TAB converter operation waveform under the light-load condition, as shown in Figure 7b. Except for FB 3, which has the highest voltage level, the remaining FBs do not achieve ZVS operation. Figure 8c shows that the operation waveform under the inner phase-shift ratio varies according to the voltage modulation ratio to match the effective voltages of the switching poles, as shown in Figure 7c. An inner phase shift of 0.66π is applied only to FB 3 and owing to the asynchronous operation of the leading and lagging legs, the switching pole voltage becomes a quasi-square wave. Therefore, the effective values of the switching pole voltage are matched in FB 1 and FB 3, and the effective voltage ratios of FB 2 with the other FB are also close to 1, ensuring ZVS operation in all FBs.

4.3. Prototype Experiment Results

An experiment was conducted using a prototype TAB converter to validate the proposed GHA-based ZVS analysis model. The prototype experiment was a 1/10 scale of the simulation, as shown in Figure 6. The switching-pole voltage and inductor current were measured at the operating points of the TAB converter. Moreover, V_{ds} and V_{gs} were measured to observe the ZVS operation. The parameters of the prototype TAB converter used in this study are provided in Table 3.

Table 3. Prototype TAB converter parameters.

MOSFET	$S_{x.1}$~$S_{x.4}$	C3M0060065D
Switching frequency	f_{sw}	100 kHz
Turn ratio	n_1:n_2:n_3	2:1:1
FB 1 series inductance	L_1	33.3 μH
FB 2 and 3 series inductance	L_2 and L_3	8.3 μH
Parasitic capacitance	$C_{oss.x}$	80 pF
FB 1 input voltage	V_1	38 V
Gate-Source voltage	V_{gs}	−5/+15 V

Figure 9 shows the prototype TAB converter and Figures 10–14 show the experimental results for the TAB converter at each operating point in Figure 6. The experimental results were separated according to FB. Under the condition of no inner phase shift, if the ZVS of only one switching leg is assured, the ZVS operation is assured in all MOSFETs; therefore, V_{gs} and V_{ds} were measured only in the leading leg in Figures 10 and 11. Under the condition of an inner phase shift, V_{gs} and V_{ds} were measured for both the leading and lagging legs, as shown in Figures 12–14. V_{ds} is the voltage across the MOSFET and can be used to confirm whether the parasitic capacitor is discharged. Therefore, when the ZVS operation is performed, V_{gs} for the switch turn-on becomes positive after V_{ds} becomes zero. However, if the ZVS operation is not performed, V_{gs} becomes positive before V_{ds} becomes zero.

A slight difference was observed in the experimental results owing to the parasitic resistance of the series inductor and resonance caused by the parasitic inductance and capacitance inside the SiC MOSFET. Overall, the inductor current and switching voltage waveforms are consistent with the GHA modeling-based simulation results shown in Figure 6. It can be seen in Figures 11 and 12 that the voltage modulation ratio between the FBs affects the ZVS operation, and the effective value of the switching pole voltage matching through an internal phase shift can improve the ZVS operation performance. Contrary to the conventional concept that ZVS is assured only when the direction of the inductor current is appropriate during switching, when the inductor current of FB 2 has a negative value during lagging leg switching, the FB 2 lagging leg does not perform ZVS, as shown in Figure 13. This implies that the parasitic capacitor of the MOSFET is not sufficiently discharged. These results are consistent with those of the proposed ZVS analytical model and the simulations shown in Figures 5 and 6, respectively.

Figure 9. Prototype TAB converter.

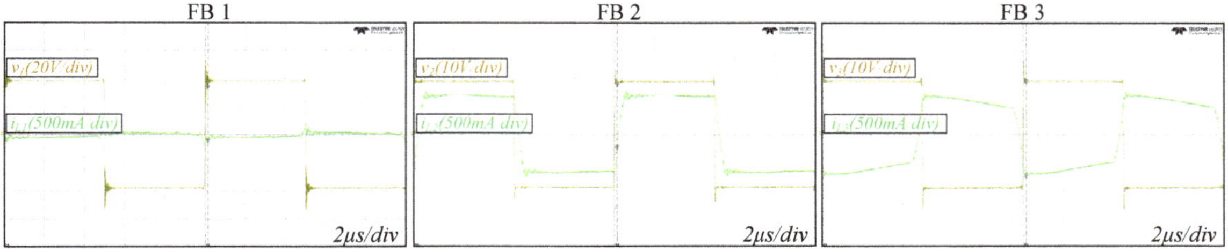

(**a**)

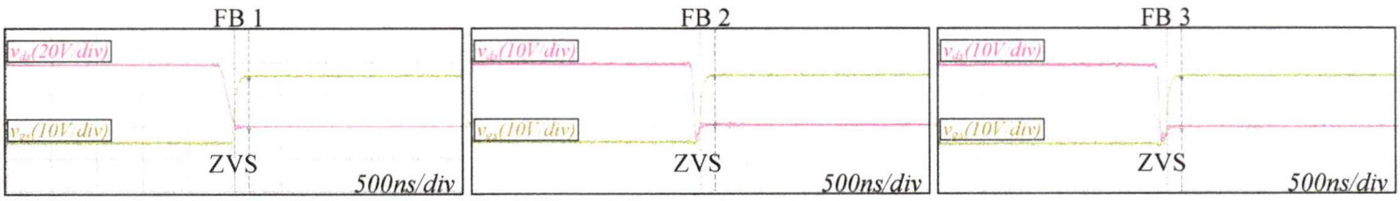

(**b**)

Figure 10. Prototype experiment results of the TAB converter operating point in Figure 6a ($\alpha_1 = \alpha_2 = \alpha_3 = 0$, $m_{12} = m_{13} = 1$, $\phi_{12} = -0.05\pi$, $\phi_{13} = 0.05\pi$). (**a**) Switching pole voltage and inductor current. (**b**) Gate-source and drain-source voltages of MOSFET.

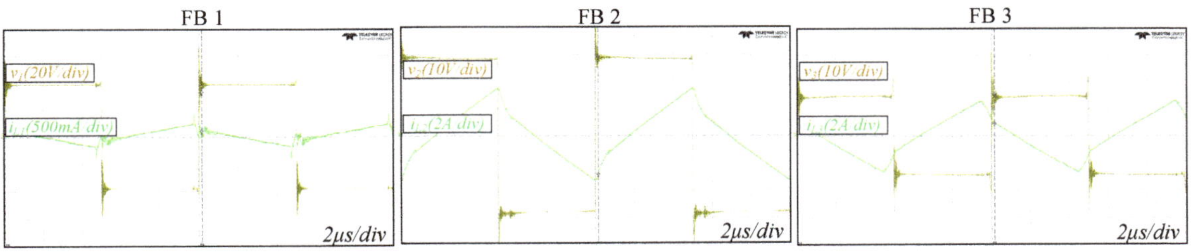

(**a**)

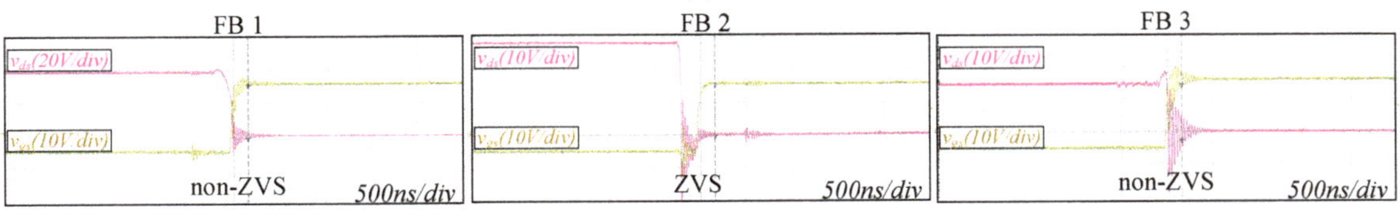

(**b**)

Figure 11. Prototype experiment results of the TAB converter operating point in Figure 6b ($\alpha_1 = \alpha_2 = \alpha_3 = 0$, $m_{12} = 1.5$, $m_{13} = 0.75$, $\phi_{12} = -0.05\pi$, $\phi_{13} = 0.05\pi$). (**a**) Switching pole voltage and inductor current. (**b**) Gate-source and drain-source voltages of MOSFET.

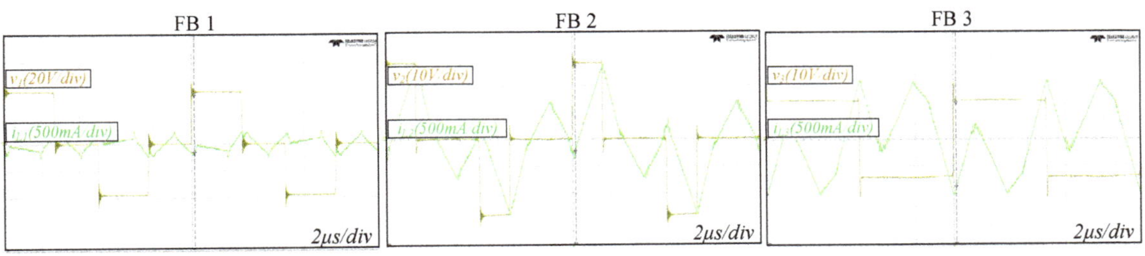

(**a**)

Figure 12. *Cont.*

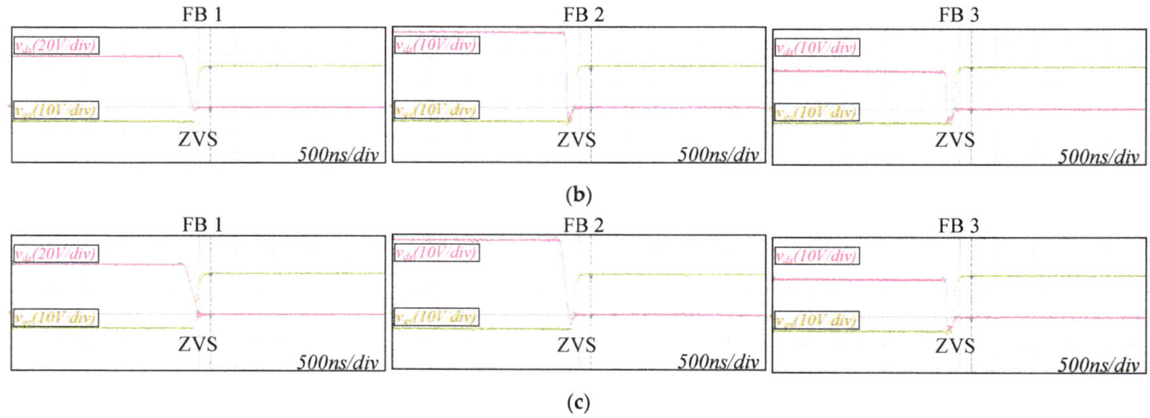

Figure 12. Prototype experiment results of the TAB converter operating point in Figure 6c ($\alpha_1 = 0.46\pi$, $\alpha_2 = 0.67\pi$, $\alpha_3 = 0$, $m_{12} = 1.5$, $m_{13} = 0.75$, $\phi_{12} = -0.05\pi$, $\phi_{13} = 0.05\pi$). (**a**) Switching pole voltage and inductor current. (**b**) Gate-source and drain-source voltages of the leading leg MOSFET. (**c**) Gate-source and drain-source voltages of the lagging leg MOSFET.

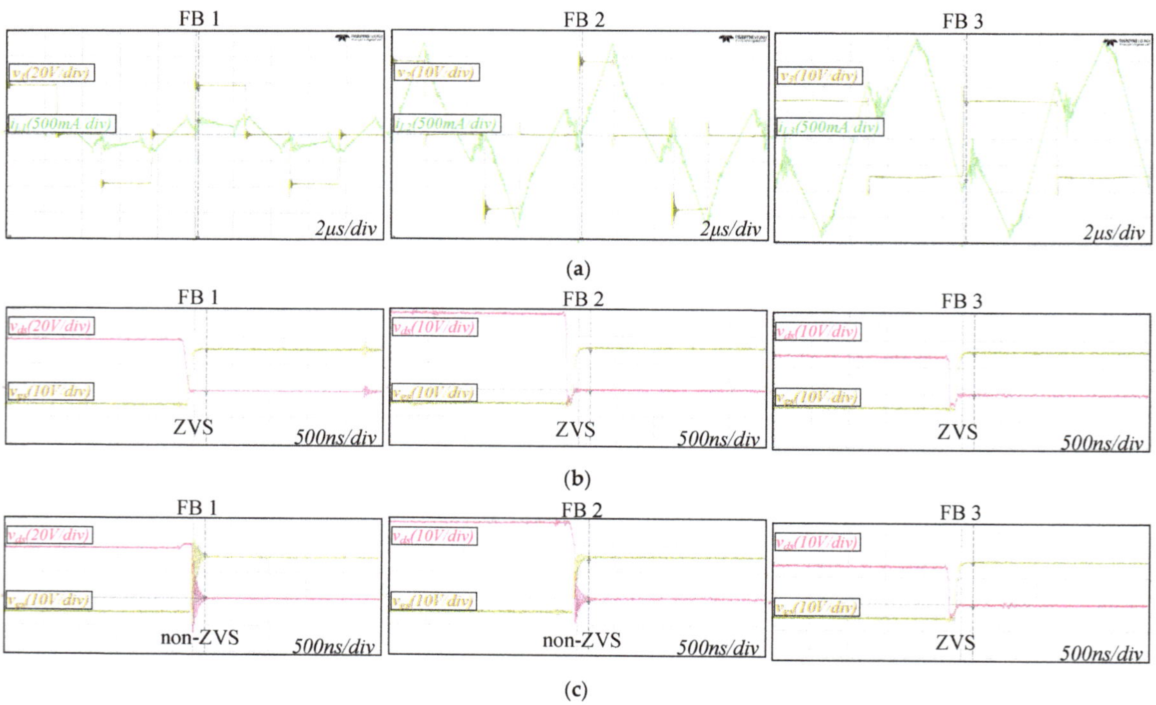

Figure 13. Prototype experiment results of the TAB converter operating point in Figure 6d ($\alpha_1 = 0.46\pi$, $\alpha_2 = 0.67\pi$, $\alpha_3 = 0$, $m_{12} = 1.5$, $m_{13} = 0.75$, $\phi_{12} = -0.05\pi$, $\phi_{13} = 0.16\pi$). (**a**) Switching pole voltage and inductor current. (**b**) Gate-source and drain-source voltages of the leading leg MOSFET. (**c**) Gate-source and drain-source voltages of the lagging leg MOSFET.

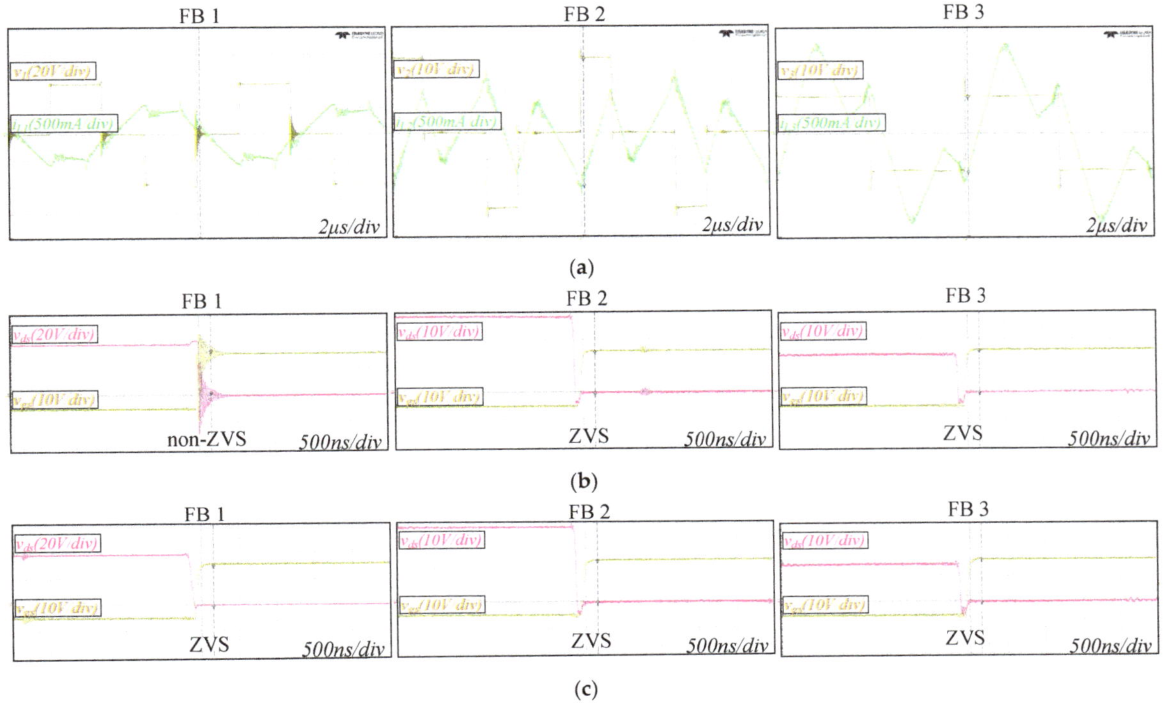

Figure 14. Prototype experiment results of the TAB converter operating point in Figure 6e ($\alpha_1 = 0.46\pi$, $\alpha_2 = 0.67\pi$, $\alpha_3 = 0$, $m_{12} = 1.5$, $m_{13} = 0.75$, $\phi_{12} = -0.05\pi$, $\phi_{13} = -0.16\pi$). (**a**) Switching pole voltage and inductor current. (**b**) Gate-source and drain-source voltages of the leading leg MOSFET. (**c**) Gate-source and drain-source voltages of the lagging leg MOSFET.

5. Conclusions

This study provides a GHA-based ZVS analysis model of a TAB converter. The TAB converter has difficulty modeling and analyzing the ZVS operation in the time domain owing to numerous operation modes using five phase variables. Therefore, in this study, a TAB converter was modeled using a GHA based on the periodicity of the FB switching pole. In addition, a ZVS analysis of the FB was performed using the energy balance equation, and the minimum inductor energy and current conditions were derived. By applying the GHA model to the energy and current conditions for ZVS operation, a ZVS analysis model of the TAB converter was proposed that can be universally applied in all operating modes of the TAB converter. The ZVS operating range was derived using a ZVS analysis model simulation, and the effects of the outer phase shift ratio, inner phase shift ratio, and voltage modulation ratio on the ZVS operation were analyzed. Experiments were performed using a 1/10 scale prototype TAB converter, and the validity of the GHA and ZVS analysis models was verified by matching them with simulation results. As a result of the simulation and 1/10 scale experiment, it can be seen that the lighter the load condition and the lower the effective switching pole voltage ratio, the more difficult it is to perform ZVS operation, and hard switching occurs. The inner phase shift is a method of matching the effective switching pole voltage ratio, lowering the effective voltage size of the FB switching pole with a high voltage ratio, and making it possible to achieve the ZVS of all FBs. Considering that the Y-delta conversion in the TAB converter is a conversion method derived by the superposition of circuits, the DAB converter equivalent circuit can be derived in the multi-active-bridge converter through the principle of superposition. Therefore, the proposed GHA-based ZVS analysis model can be extended and applied to multi-active bridges, such as quadruple-active or penta-active bridges.

Author Contributions: Conceptualization, H.-J.B.; methodology, H.-J.B.; software, H.-J.B., and J.-M.P.; validation, H.-J.B., and J.-M.P.; formal analysis, H.-J.B.; investigation, H.-J.B.; resources, H.-J.B., and J.-M.P.; data curation, H.-J.B., and J.-M.P.; writing—original draft preparation, H.-J.B.; writing—review and editing, H.-J.B., and J.-M.P.; visualization, H.-J.B.; supervision, J.Y., and C.-Y.W.; project administration, J.Y., and C.-Y.W.; funding acquisition, J.-M.P. All authors have read and agreed to the published version of the manuscript.

Funding: This research was supported by grants from the International Energy Joint R&D Program of the Korea Institute of Energy Technology Evaluation and Planning (KETEP), granted financial resources from the Ministry of Trade, Industry, and Energy, Republic of Korea (20208510070060).

Data Availability Statement: Data are contained within the article.

Conflicts of Interest: The authors declare no conflict of interest.

References

1. Dincer, I. Renewable energy and sustainable development: A crucial review. *Renew. Sustain. Energy Rev.* **2000**, *4*, 157–175. [CrossRef]
2. Fuente, D.V.; Rodríguez, C.L.T.; Garcerá, G.; Figueres, E.; Gonzalez, R.O. Photovoltaic Power System with Battery Backup with Grid-Connection and Islanded Operation Capabilities. *IEEE Trans. Ind. Electron.* **2013**, *60*, 1571–1581. [CrossRef]
3. Hadjipaschalis, I.; Poullikkas, A.; Efthimiou, V. Overview of current and future energy storage technologies for electric power applications. *Renew. Sustain. Energy Rev.* **2009**, *13*, 1513–1522. [CrossRef]
4. Bhattacharjee, K.; Kutkut, N.; Batarseh, I. Review of Multiport Converters for Solar and Energy Storage Integration. *IEEE Trans. Power Electron.* **2019**, *34*, 1431–1445. [CrossRef]
5. Litran, S.P.; Duran, E.; Semiao, J.; Diaz-Martin, C. Multiple-Output DC–DC Converters: Applications and Solutions. *Electronics* **2022**, *11*, 1258. [CrossRef]
6. Jiang, W.; Fahimi, B. Multiport Power Electronic Interface—Concept, Modeling, and Design. *IEEE Trans. Power Electron.* **2011**, *26*, 1890–1900. [CrossRef]
7. Farhangi, B.; Toliyat, H.A. Modeling and Analyzing Multiport Isolation Transformer Capacitive Components for Onboard Vehicular Power Conditioners. *IEEE Trans. Ind. Electron.* **2015**, *62*, 3134–3142. [CrossRef]
8. Wheeler, P.; Bozhko, S. The More Electric Aircraft: Technology and challenges. *IEEE Electrif. Mag.* **2014**, *2*, 6–12. [CrossRef]
9. Neubert, M.; Gorodnichev, A.; Gottschlich, J.; De Doncker, R.W. Performance analysis of a triple-active bridge converter for interconnection of future dc-grids. In Proceedings of the 2016 IEEE Energy Conversion Congress and Exposition (ECCE), Milwaukee, WI, USA, 18–22 September 2016.
10. Purgat, P.; Mackay, L.; Schulz, M.; Han, Y.; Qin, Z.; Marz, M.; Bauer, P. Design of a Power Flow Control Converter for Bipolar Meshed LVDC Distribution Grids. In Proceedings of the 2018 IEEE 18th International Power Electronics and Motion Control Conference (PEMC), Budapest, Hungary, 26–30 August 2018.
11. Zhao, C.; Kolar, J.W. A novel three-phase three-port UPS employing a single high-frequency isolation transformer. In Proceedings of the 2004 IEEE 35th Annual Power Electronics Specialists Conference (IEEE Cat. No.04CH37551), Aachen, Germany, 15 November 2004.
12. Koohi, P.; Watson, A.J.; Clare, J.C.; Soerio, T.B.; Wheeler, P.W. A Survey on Multi-Active Bridge DC-DC Converters: Power Flow Decoupling Techniques, Applications, and Challenges. *Energies* **2023**, *16*, 5927. [CrossRef]
13. De Doncker, R.W.A.A.; Divan, D.M.; Kheraluwala, M.H. A three-phase soft-switched high-power-density DC/DC converter for high-power applications. *IEEE Trans. Ind. Appl.* **1991**, *27*, 63–73. [CrossRef]
14. Michon, M.; Duarte, J.L.; Hendrix, M.; Simoes, M.G. A three-port bi-directional converter for hybrid fuel cell systems. In Proceedings of the 2004 IEEE 35th Annual Power Electronics Specialists Conference (IEEE Cat. No.04CH37551), Aachen, Germany, 20–25 June 2004.
15. Bal, S.; Yelaverthi, D.B.; Rathore, A.K.; Srinivasan, D. Improved Modulation Strategy Using Dual Phase Shift Modulation for Active Commutated Current-Fed Dual Active Bridge. *IEEE Trans. Power Electron.* **2018**, *33*, 7359–7375. [CrossRef]
16. Dai, P.; Liu, S.; Fang, S.; Gong, Z. Optimal Asymmetric Duty Modulation for Dual Active Bridge Converters with DC Blocking Capacitors. *Energies* **2023**, *16*, 6674. [CrossRef]
17. Huang, J.; Wang, Y.; Li, Z.; Lei, W. Unified Triple-Phase-Shift Control to Minimize Current Stress and Achieve Full Soft-Switching of Isolated Bidirectional DC–DC Converter. *IEEE Trans. Ind. Electron.* **2016**, *63*, 4169–4179. [CrossRef]
18. Oggier, G.; García, G.O.; Oliva, A.R. Modulation strategy to operate the dual active bridge DC-DC converter under soft switching in the whole operating range. *IEEE Trans. Power Electron.* **2011**, *26*, 1228–1236. [CrossRef]
19. Oggier, G.G.; GarcÍa, G.O.; Oliva, A.R. Switching Control Strategy to Minimize Dual Active Bridge Converter Losses. *IEEE Trans. Power Electron.* **2009**, *24*, 1826–1838. [CrossRef]
20. Krismer, F.; Kolar, J.W. Closed Form Solution for Minimum Conduction Loss Modulation of DAB Converters. *IEEE Trans. Power Electron.* **2012**, *27*, 174–188. [CrossRef]

21. Everts, J. Closed-Form Solution for Efficient ZVS Modulation of DAB Converters. *IEEE Trans. Power Electron.* **2017**, *32*, 7561–7576. [CrossRef]
22. Jiang, L.; Costinett, D. A triple active bridge DC-DC converter capable of achieving full-range ZVS. In Proceedings of the 2016 IEEE Applied Power Electronics Conference and Exposition (APEC), Long Beach, CA, USA, 20–24 March 2016.
23. Yan, Y.; Gui, H.; Bai, H. Complete ZVS Analysis in Dual Active Bridge. *IEEE Trans. Power Electron.* **2021**, *36*, 1247–1252. [CrossRef]
24. Kasper, M.; Burkart, R.M.; Deboy, G.; Kolar, J.W. ZVS of Power MOSFETs Revisited. *IEEE Trans. Power Electron.* **2016**, *31*, 8063–8067. [CrossRef]
25. Purgat, P.; Bandyopadhyay, S.; Qin, Z.; Bauer, P. Zero Voltage Switching Criteria of Triple Active Bridge Converter. *IEEE Trans. Power Electron.* **2021**, *36*, 5425–5439. [CrossRef]
26. Shen, Z.; Burgos, R.; Boroyevich, D.; Wang, F. Soft-switching capability analysis of a dual active bridge dc-dc converter. In Proceedings of the 2009 IEEE Electric Ship Technologies Symposium, Baltimore, MD, USA, 20–22 April 2009.

energies

Article

A Simple and Economical System for Automatic Near-Field Scanning for Power Electronics Converters

Sebastiano Grasso [1], Salvatore Bellinvia [1], Nunzio Salerno [2] and Santi Agatino Rizzo [2,*]

1 STMicroelectronics, Stradale Primosole 50, 95121 Catania, Italy; sebastiano-l1.grasso@st.com (S.G.); bellinvias91@gmail.com (S.B.)
2 Department of Electrical Electronic and Computer Engineering (DIEEI), University of Catania, 95125 Catania, Italy; nunzio.salerno@unict.it
* Correspondence: santi.rizzo@unict.it

Abstract: Electromagnetic compatibility issues must be considered from the early steps in the design of electronic devices. A specific topic is the near-field emission generated by the device due to the traces on a printed circuit board and the specific routing. The analysis of near-field emission is essential to detect potential electromagnetic interference with nearby devices. This problem is crucial in high power density applications. Therefore, especially in these applications, it is necessary to optimize the circuit and the layout to minimize the generated noise. The design and construction of systems able to scan volumes to determine the spatial distribution of electrical E and/or magnetic B fields in the near-field region of a device under test is a very complex process. The realization of equipment that explores a given surface at a given distance from the device is easier. The main purpose of this paper is to show how it is possible to build a cheap two-dimensional scanner, starting from simple hardware not explicitly designed for near-field scan operations. The presented firmware and software solution can map, with good accuracy, the spatial distribution of fields B and E on a fixed plan close to the board. Finally, the developed system has been used in a GaN-based bi-directional DC/DC Converter.

Keywords: electromagnetic compatibility; near-field scan; power electronics

Citation: Grasso, S.; Bellinvia, S.; Salerno, N.; Rizzo, S.A. A Simple and Economical System for Automatic Near-Field Scanning for Power Electronics Converters. *Energies* **2023**, *16*, 7868. https://doi.org/10.3390/en16237868

Academic Editor: Hervé Morel

Received: 9 October 2023
Revised: 20 November 2023
Accepted: 27 November 2023
Published: 1 December 2023

1. Introduction

Electromagnetic compatibility (EMC) today is vital and strategic, considering the growing complexity of any electronic device and the diffusion of electronic modules in every area of human life [1]. Before being put on the market, each electronic device must be tested to be certified as electromagnetically compatible [2]. There are two different aspects to be considered: the first one is the electromagnetic pollution generated in the environment by the equipment under testing (EUT) and the second one is the effects of external electromagnetic fields [3]. Moreover, some are related to electromagnetic pollution through the supply cables or the connections of the EUT and the load or its immunity to conducted noises through those connections. Others are related to electromagnetic pollution spread in the surrounding environment or to the sensibility of the EUT to the external radiated noise [3,4]. There are a lot of characterizations that could be performed to study these effects using electromagnetic simulation flows [5,6] or through mathematical models that are able to predict electromagnetic interference (EMI) phenomena [7,8]. In some cases, an electromagnetic evaluation is performed through a classical measurement approach on a prototype [9].

A specific EMC aspect of the study is the near-field emission generated by EUT [10] due to traces on a printed circuit board (PCB) and the particular routing. The analysis of near-field emission is essential to detect potential electromagnetic interference with nearby devices [11]. This problem is crucial in automotive and, in general, high power density applications. Additionally, it is possible to foresee the far-field emissions, and it could also be possible to optimize the circuit and the layout to minimize the generated noise [12], starting from the near-field estimation.

Near-field emissions generated by EUT can be acquired using commercial equipment that provides accurate measurements and, in some cases, can be used for pre-compliant measurements. However, commercial equipment has a very high cost that ranges from some tens to one hundred thousand dollars. This is due to the complexity of designing and constructing systems that are able to scan volumes to determine the spatial distribution of electrical and/or magnetic fields in the near-field region of the EUT.

When the scanner is used for a rough analysis of electromagnetic emissions produced by a board and/or its use is sporadic, the investment in such expensive equipment cannot be justified. Therefore, in these cases, it is impossible to perform any electromagnetic measurements that could be useful for a first estimation of the board emission. In such circumstances, the realization of equipment that explores a given surface at a given distance from the device or the board under testing is useful.

From this perspective, the main purpose of this paper is to show how it is possible to build a homemade two-dimensional scanner that is able to map the spatial distribution of fields B and E on a fixed plan. For this purpose, the scanner has been obtained starting from simple hardware not specifically designed for near-field scan operations, whose bill of materials is less than two thousand dollars. This strong cost reduction compared to commercial tools is obtained thanks to the customization of the proposed system to the specific application. Hence, the customization has enabled the realization of a simpler and more economical scanner than the commercial ones, whose high complexity and cost are due to the many functionalities and tasks they are designed to perform.

The proposed approach has practical use in near-field scanning in the context of power electronics because it enables everyone to roughly estimate electromagnetic emissions by adopting solutions like the one proposed in this paper. These kinds of simple and economic systems can be used for tuning electromagnetic simulation parameters in order to validate finite element analysis that, in turn, can be used to optimize circuit layouts by estimating the electromagnetic emissions of different layouts before the prototyping stage.

The paper is organized as follows. In Section 2, the near-field scan equipment is described in detail; in Section 3, it is applied to a bi-directional DC/DC converter; and finally, conclusions are provided in Section 4.

2. Near-Field Scan Equipment

An electromagnetic near-field scan is a measurement process to determine the spatial distribution of electrical and/or magnetic fields in the near-field region of a device under test or on a board. Those fields are sensed through specific antennas, which convert the strength of the field into a voltage available at the 50 Ω port of those sensing elements. In the simplest systems, the spatial distribution is on a surface; in the most complicated ones, it is on a specific volume (Figure 1).

Generally, the voltage as a function of time or frequency is measured. It should be highlighted that the DUT may be considered as any object radiating or storing electromagnetic field energy intentionally or unintentionally. The voltage pattern is usually mapped on planar, cylindrical or spherical geometrical surfaces as a collection of a finite number of spatial samples.

These samples are usually converted into a color intensity map that lets it easily understand in which zone, on a volume or surface, the electromagnetic field generated by a specific source (a device or one or more parts of a board) is concentrated.

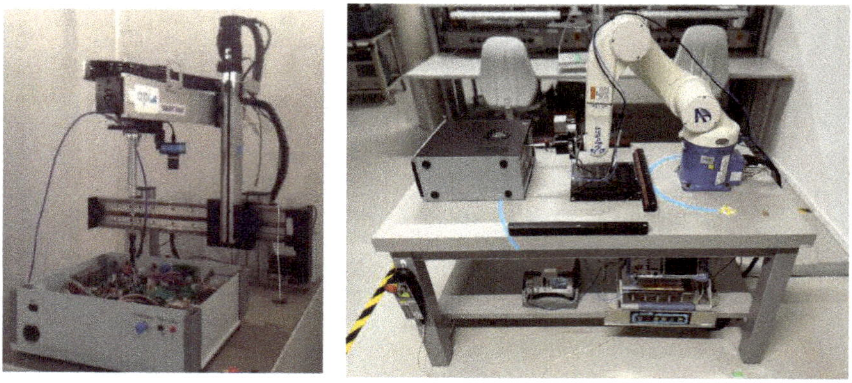

Figure 1. 3D Scanner equipment by two different suppliers.

2.1. General Equipment Overview and Antenna Specifications

The scanner has been developed from existing laser engraver equipment, whose components have been repurposed as follows. The aluminum chassis has been used as the main mechanical support, while the plexiglass parts have been used to complete the aluminum structure, including the legs. The stepper motors are two motors used for movement along the x- and y-axis. The main control board hosts the stepper motors' driver boards and connectors. Toothed belts and wheels have been used for the movement transmission together with motors. Finally, the power adapter has been used to generate 12 V DC from the mains. The laser unit has been not repurposed because it is not useful for the scanner. Similarly, the Arduino Nano board has been not repurposed because its firmware was programmed for laser printing purposes only and was not designed to be controlled by custom software.

A nylon 3D-printed head holding the antennas in the desired positions has substituted the laser (Figure 2). So, the customized firmware has been properly written in the EEPROM of a new Arduino Nano board to maintain the same pinout and the native connectors.

The test equipment used is listed below:

- ESRP7 Rohde & Schwarz Emi test receiver
- GPIB—USB—HS National Instruments GPIB to USB converter
- TDK-Lambda power supply
- Beehive Electronics near-field probes

The adopted Beehive Electronics near-field probes are commercial antennas specifically designed to detect proximity electric and magnetic fields. They are cheap detectors, adapted to an impedance of 50 Ω. A stub antenna (diameter of head 0.08 inches), suitable to measure the electric field, and three different sizes of loop antennas appropriate to measure the magnetic field have been used. The smallest size allows the best spatial resolution and has a diameter of 0.25 inches. In Figure 2, the antennas and the nylon head of the machine with a mounted antenna are shown.

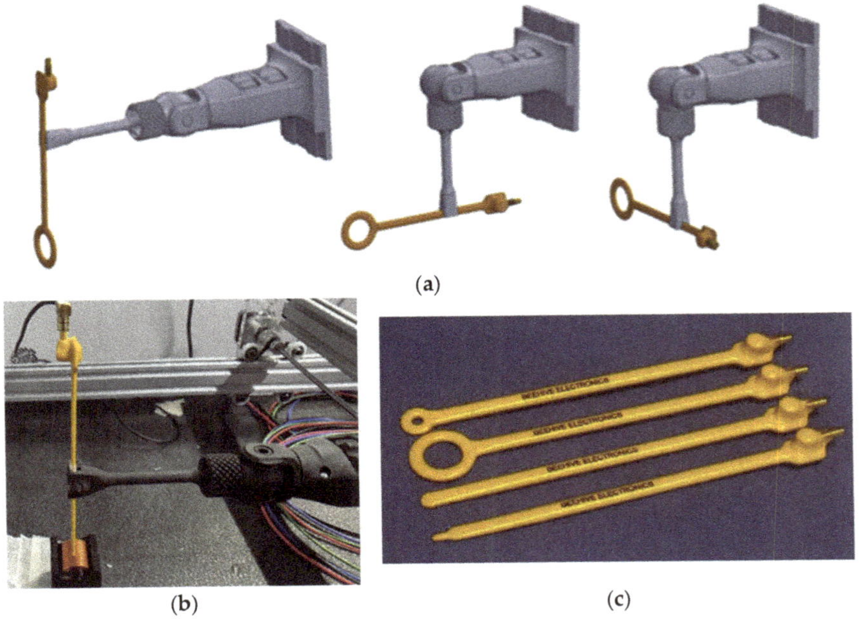

Figure 2. Scanner head (**a**) designed movements; (**b**) 3D-printed tool) and (**c**) antennas.

Figure 3 illustrates the conversion curves to compute the E field in V/m and the B field in T starting from the spectrum acquired in dBm with an EMI test receiver or a spectrum analyzer. In the datasheet [13], besides the curves, the formulas needed to convert the dBm measured by the spectrum analyzer into T or V/m, and taking into account the frequency dependence, are given as:

$$E_{\left(\frac{V}{m}\right)} = 10^{\frac{Pout_{(dBm)}+113.2-20log_{10}\left(F_{(MHz)}\right)}{20}} \tag{1}$$

$$B_{(T)} = 10^{\frac{Pout_{(dBm)}-X-20log_{10}\left(F_{(MHz)}\right)}{20}} \tag{2}$$

where X is a number whose value is reported in Table 1.

The equipment can execute a full scan on an XY plane once the desired distance of Z0 between the board and the antenna in the direction of the z-axis has been fixed. The developed scanner has been customized for the automatic scanning of a rectangular area at a preselected height (Z0) from the board. However, the lack of z-axis handling is not a limitation for the considered application. Finally, the minimum Z0 is bounded by the highest board component because the aim is field measurement in a rectangular area; thus, it is unnecessary for any stepper motor to height change during the measurements.

Table 1. X factor in Equations (1) and (2) for each antenna.

Antenna Type	X	3 dB Frequency (MHz)	First Resonance (MHz)
100C (large loop)	85.1	50	500
100A (medium loop)	65.2	1000	2600
100B (small loop)	42.2	3100	>6000

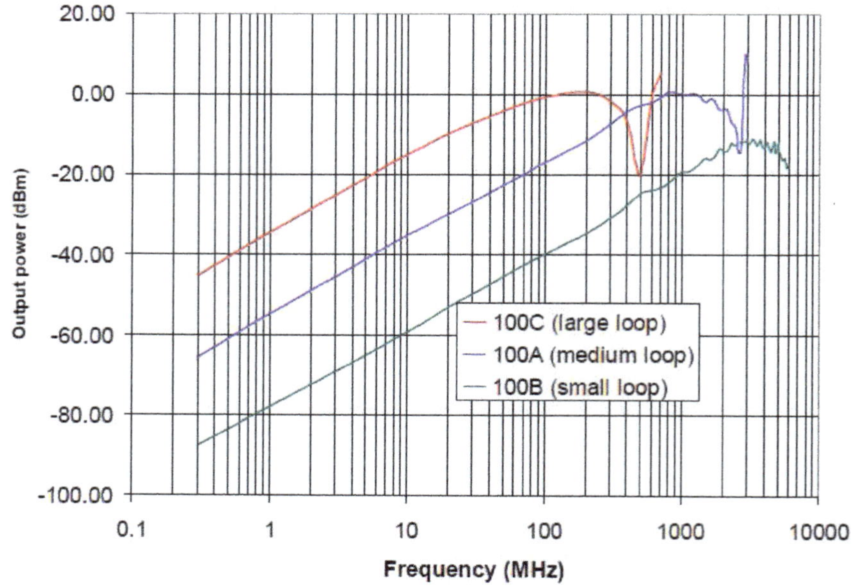

(**a**)

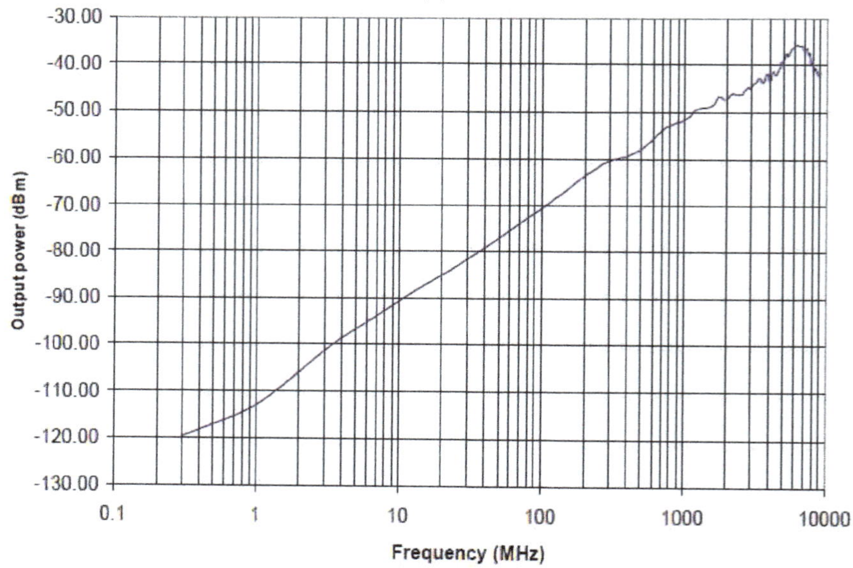

(**b**)

Figure 3. Antenna conversion curves: (**a**) output power vs. frequency at 1 μT field strength; (**b**) output power vs. frequency at 1 V/m field strength.

Distance Z0 is the center of the loop for the magnetic field measurements and the bottom edge for the electric field measurements. Starting from an initial point (X0, Y0, Z0), two stepper motors move the antenna along a preset path parallel to the board surface, as shown in Figure 4a, thus covering an area equal to a rectangle of width X and height Y. Figure 4b shows a picture of the LABVIEW software (2016) interface, which manages the scanning process. This LABVIEW scanning window assigns the dimensions of the scanning area, the scanning step, the starting position and the type of antenna.

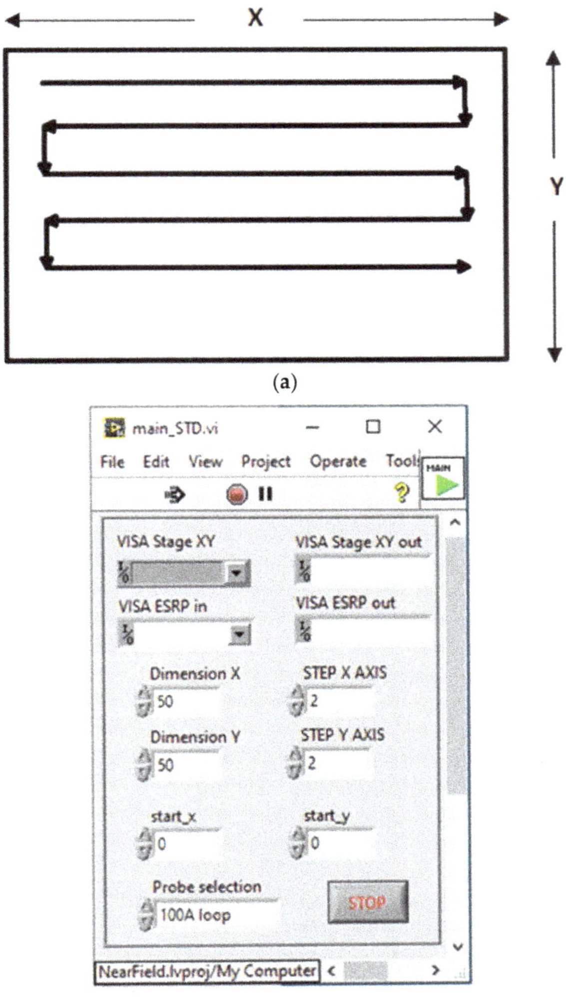

Figure 4. Pattern of the scanning path (**a**) and LABVIEW control panel for the scanning process (**b**).

Then, for each position of the antenna along the path, the LABVIEW software acquires a full spectrum of the signal (Figure 5) in the chosen frequency band by means of an EMI test receiver or a spectrum analyzer.

The acquisition is fully automatic and managed by the software. The LABVIEW software communicates with the machine controller as if it were a VISA (Virtual Instrument Software Architecture) instrument; the Arduino Nano board is already equipped with a USB to UART (serial) converter, which allows the PC to communicate directly with the on-board Atmel microcontroller. A list of simple commands has been defined to request information from the board about the head position, moving the head of a specific step and setting some tuning parameters. For example, some commands are needed to tune the size of the step, both in the X and Y directions. Since the head is not equipped with position encoders, the microcontroller saves the position of the head into its EEPROM (electrically erasable programmable memory) every twenty seconds.

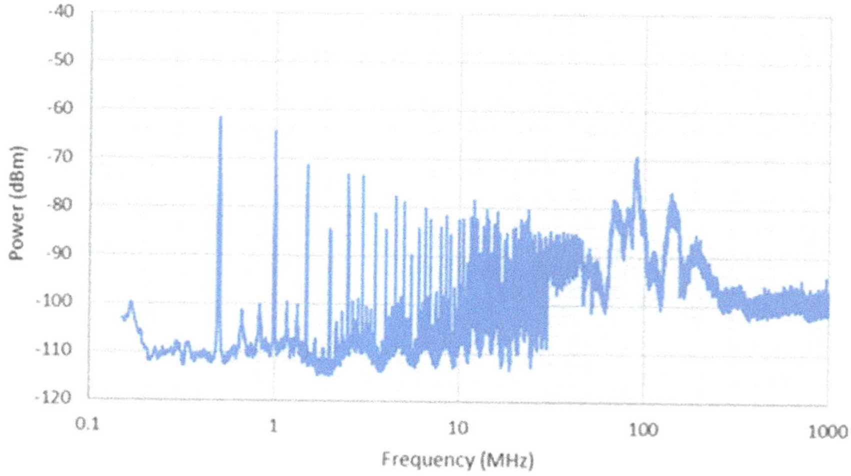

Figure 5. Example of spectrum acquired at a point in the scanned pattern grid.

To avoid damaging the EEPROM (which has a limited number of writing repetitions) in a short time, the firmware uses a smart algorithm that is able to write a location only if the byte to be written is different from the one already stored there, and it is also able to use a pointer in order to write cyclically in the whole memory and not in the same locations all the time. The firmware communicates with the PC when a requested operation is concluded and a new one can be executed; this is useful to synchronize the LABVIEW and the firmware operations. In Figure 6a, a full image of the scanning equipment is shown.

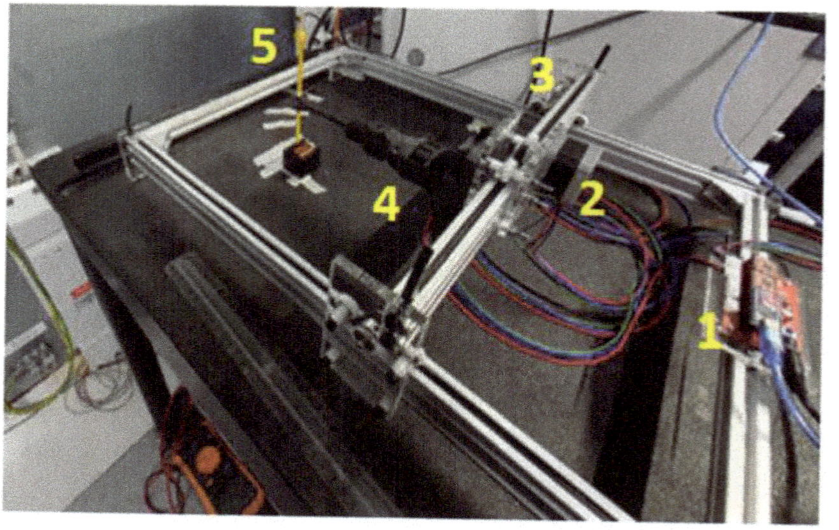

Figure 6. Equipment overview: 1—microcontroller Arduino Nano board and A4988 stepper motor drivers; 2—stepper motor for Y direction; 3—stepper motor for X direction; 4—nylon 3D printed head; 5—magnetic field antenna.

The spectrum and coordinates are saved for post-processing purposes for each point of the scanned surface. For example, using a color map to show the field intensity or one of their components, it is possible to also show the map of a single component of B or E magnitude on an XY plane. In the same post-processing window, it is possible to see the spectrum of the fields or that of its components. In this regard, portable Python 3 software has also been developed; this can be distributed together with the data files generated by the EMI receiver and allows the easy exploration of the fields' maps for each orientation and each frequency (Figure 7).

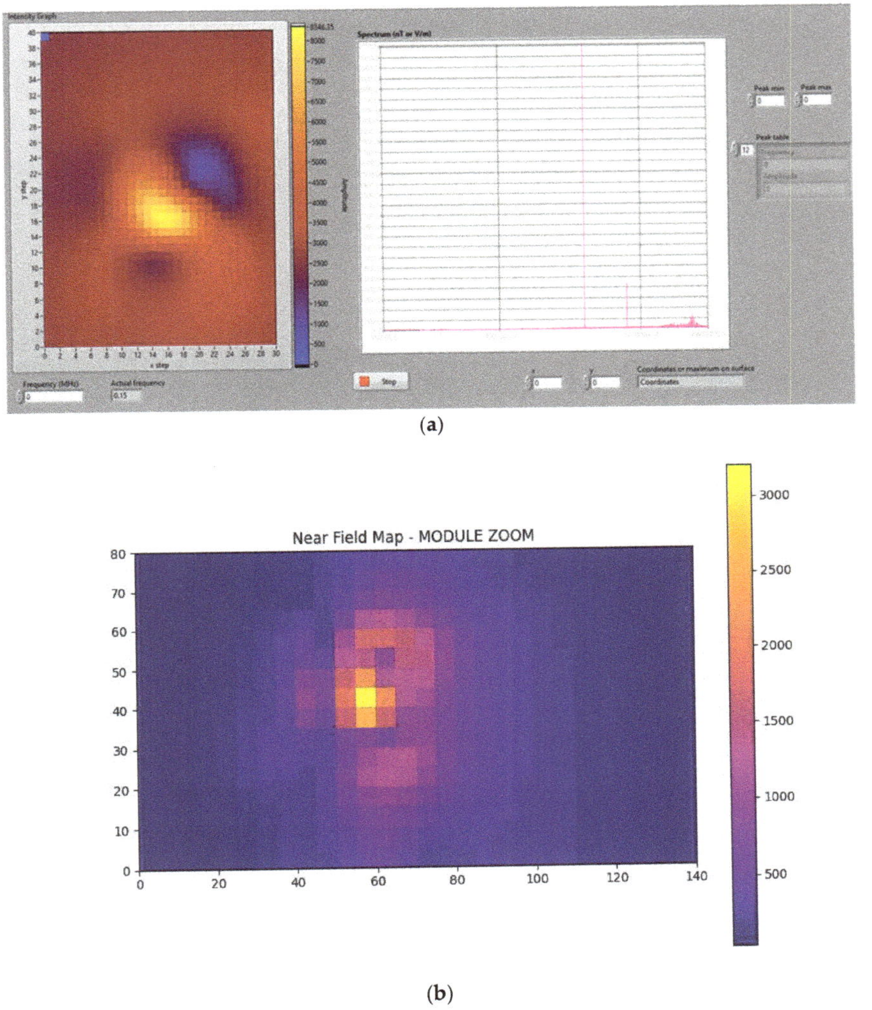

Figure 7. Near-field map. (**a**) LABVIEW post-processing view, (**b**) Python 3 post-processing software results.

Obviously, the proposed scanner is not conceived for applications where more accurate and extensive analyses have to be performed, especially for certification purposes. As mentioned before, the proposed solution enables the performance of a rough measurement of electromagnetic emissions in all the cases where, usually, no measurements are performed due to the lack of an expensive scanner. Therefore, although the accuracy is probably below that of commercial scanners, a rough comparison among different board solutions can be performed in these cases. Moreover, the accuracy depends on different factors, especially the accuracy of the antennas, which is often not declared in the datasheet (as the ones used in the case study of this work).

The spatial resolution depends on the minimum movement possible along the x- and y-axis of the chassis and the antenna dimension. For example, the minimum step is 1 mm with 100 B and 100 D antennas. Finally, a sampling rate of one step per second has been measured when the frequency is in the interval [1 MHz–1 GHz].

2.2. Controller Board Description

The controller board is made with an Arduino Nano microcontroller. The Arduino drives the two A4988 cheap stepper motor controllers through two pins. One sets the direction by pulling the DIR inputs of the A4988 boards up to 3.3 V or pulling them down to 0, and the other sends a square wave to the STEP pins (one period for one step). The microcontroller board also pulls down the ENABLE input in order to switch ON or OFF the modules. In Figure 8a, a simplified schematic of the controller is shown, while in Figure 9, the A4988 driver photo and pinout are shown.

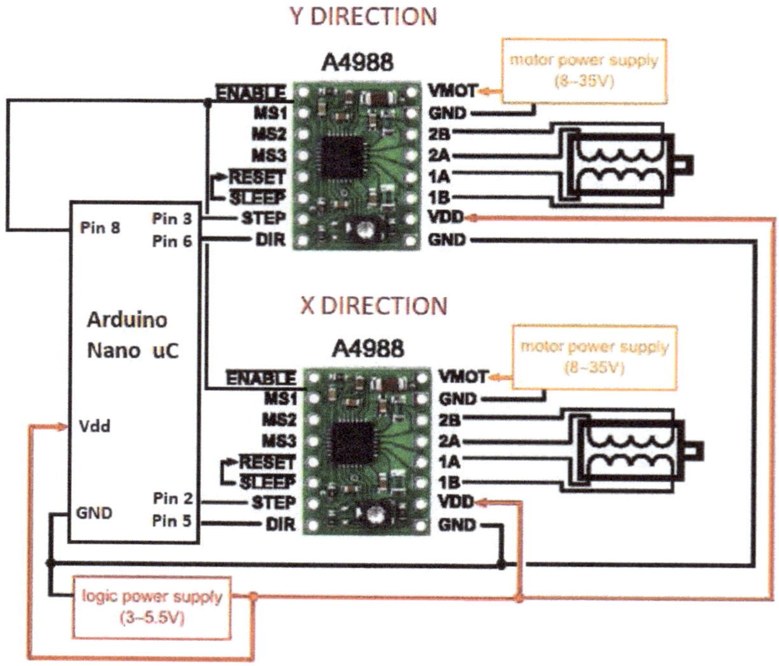

Figure 8. Simplified controller schematic.

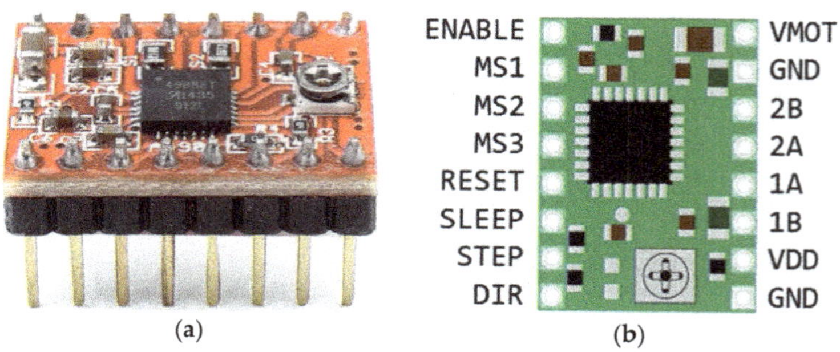

Figure 9. A4988, stepper motor controller: (**a**) driver photo and (**b**) its pinout.

The microcontroller and the logical circuitry of the A4988 boards are supplied by USB. The power circuitry of each A4988 board is instead supplied by an external power supply adapter.

The pins that drive the stepper motor are 1A 1B and 2A 2B. The motor supply is connected to the pin VMOT and its closest GND. The pins SLEEP and RESET are generally connected together. MS1, MS2 and MS3 are digital inputs used to tune the micro-step resolution (Table 2).

Table 2. Micro stepping resolution truth table.

MS1	MS2	MS3	Resolution	Excitation Mode
L	L	L	Full step	2 Phase
H	L	L	Half step	1–2 Phase
L	H	L	Quarter step	W1–2 Phase
H	H	L	Eighth step	2W1–2 Phase
H	H	H	Sixteenth step	4W1–2 Phase

The typical signal that drives the module and the corresponding current on the motors is shown in Figure 10.

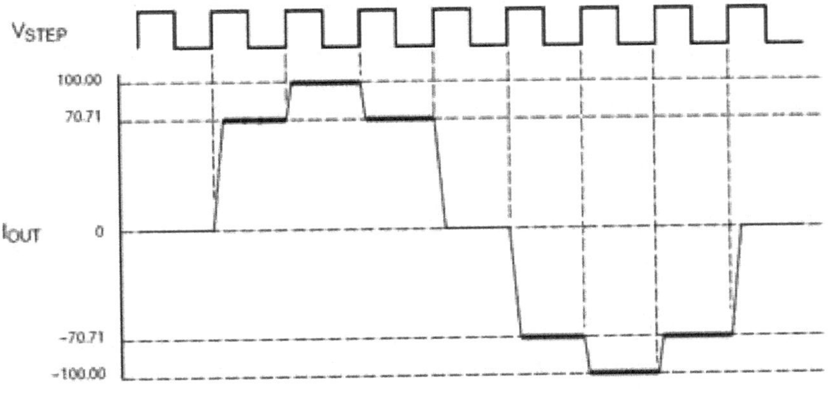

Figure 10. Typical signal on stepper motors.

2.3. *Equipment Firmware Description*

The conceived method for developing the firmware is an original contribution because it enables the equipment control from software that is different from those used in this work. More specifically, the firmware has been structured as a command-based firmware for machine control similar to commercial equipment, but, in contrast, it enables the control of the scanner using software different from the one developed in LABVIEW for this work.

A list of commands to set some tuning parameters and manage the head's movement has been defined. For some commands, the related queries have been defined to obtain the value for that parameter instead of setting it. Table 3 shows a complete list of commands and their descriptions.

Each command returns the information related to the end of its execution, which is useful for synchronizing the firmware with LABVIEW software. Hence, LABVIEW routines wait for an answer from the microcontroller before executing the next scheduled operation.

Some tuning parameters that are not present in the original firmware and are useful to increase accuracy have been introduced in the developed firmware. In detail, they enable the setting of the stepper motor movements before the first use of the equipment and during the calibration. In addition, the steps along the x- and y-axis can be set separately, thus improving the functionality of the equipment.

Table 3. Equipment virtual Com port commands.

Command	{dir}	{Query}	{Value}	Short Description
STEP_SIZE_{dir}<space>{value} STEP_SIZE_{query}	X, Y	X?, Y?	nnn	Size of single step setting/query
MOV_{dir}<space>{value}	X, Y, 0	No query	+nnn, -nnn	Move to set position, "+" forward, "-" back, "0" to the axis origin
DELAY_{dir}<space>{value} DELAY_{query}	X, Y	X?, Y?	nnn	Set the speed of movement
TRIM_XY<space>{value} TRIM_XY{query}	No dir	?	nnn	Set a multiplication factor to adjust stem size
MAX_{dir}<space>{value} MAX_{query}	X, Y	X?, Y?	nnn	Set maximum steps for each direction
POS_{query}	No dir	X?, Y?		Ask for position as number of steps

A serious challenge has been the lack of position encoders and limit switches, which involve potential damage to the motors and transmission belts. The developed firmware pre-sets the maximum x and y positions and stores the current position of the scanning head in the EEPROM to solve this problem. The optimal management of EEPROM writings has been another challenge because the microcontroller's lifetime reduces with increasing writing numbers. A smart save/retrieve algorithm has been developed to face this issue. Considering an EEPROM size of 1024 bytes, the first 21 bytes are reserved for quantities that change infrequently, such as the step size and maximum X and Y. The position of the scanning head, instead, is stored every 20 s, starting from byte 22 of the EEPROM. Each time the position has to be written, a pointer "*" is moved forward of 8 bytes, and then six bytes are saved: three encode the X position and three encode the Y position. In this way, the firmware can know the current position of the pointer by reading the EEPROM interval 22-1024. Once the pointer is found, it knows that the six bytes that follow contain the XY position of the head. To avoid unnecessary operations, the firmware checks the value of a byte in the EEPROM before writing the new one. When it is equal to the old one, the firmware avoids writing it; otherwise, it writes the new value.

2.4. Equipment LABVIEW Software Description

Figure 11 shows the developed LABVIEW main panel, in which some general parameters can be assigned. In detail, they are:

- the VISA address of the ESRP7 EMI test receiver;
- the COM port of the Arduino Nano controller of the position equipment;
- the X and Y dimensions of the scanning area;
- the minimum step along the X and Y directions.

In the same panel, three switches allow, respectively, the scanning of the module, the post-processing of the module measurements, to move the head to a specified position (e.g., the starting point for the scanning process). The panel enables the selection of the type of antenna that, in turn, determines the conversion formula that will be used to transform the field measurements according to Equations (1) and (2).

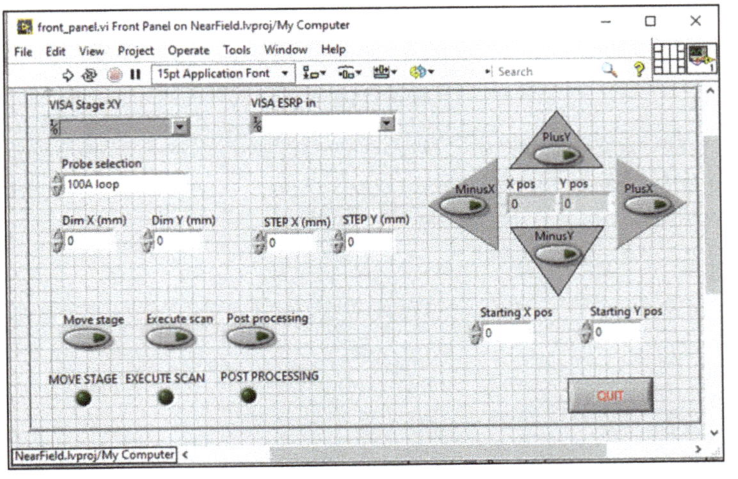

Figure 11. LABVIEW control software main panel.

The panel routine waits in the IDLE state, which is repeated until the STOP command is selected or any other choice is taken. A complex flow is executed if the command button named "Move stage" is selected. More specifically, if a specific position is specified with X, Y or both values are higher than zero, the head moves to these XY coordinates. Otherwise, it moves itself according to the value reached by pressing one of the arrow commands. The process also checks if the X and Y coordinates exceed the maximum values or are negative and, in that case, corrects them. When the scanning routine is started by the front panel, this moves the head to the (0, 0) position and then to the position (X0, Y0) chosen as the starting point for the scanning process. Then, the software also asks the microcontroller to know the max X and Y values stored in its EEPROM. They are the maximum allowed values that can be chosen to avoid the stepper motors forcing the head to move too close to the edges of the system. The total number of steps is also computed before entering the main scanning cycle. First, the head is moved along the X direction of one step, starting from the X0 position, until the maximum value 'max X' is reached. Then, a step along the Y position is performed, and the head starts to move along the X direction again but in the opposite way until the position X0 is reached. A new Y step is performed, and so on, in accordance with Figure 4a. The scanning operation continues cyclically until the max X and max Y positions are reached. At each point, the complete spectrum is measured by the ESRP7. Once the acquisition is performed, the next step is executed. At the end, the spectrum converted into values of field intensity is stored in a matrix, with one column for each pixel in the scanned surface. When the scanning process is finished, the matrix hosting all the measured values of the field and the coordinates of each point on the scanned surface is saved into a file for post-processing purposes.

Figure 12 shows a flowchart with more information about the data acquisition algorithm. Concerning the data processing, a conversion routine has been embedded in the post-processing software used to relate the field magnitude with colors. A post-processing LABVIEW routine containing some automatic analysis capabilities has also been developed. In particular, it shows the maximum magnitude of the fields for each frequency of the analyzed spectrum. Moreover, the map for the frequency where higher fields (B or E) have been found can be requested. The developed Python post-processing software automatically provides the field intensity by positioning the pointer at the desired X–Y map pixel. The possibility to separately show the three components of the B or E fields and the map magnification are additional features.

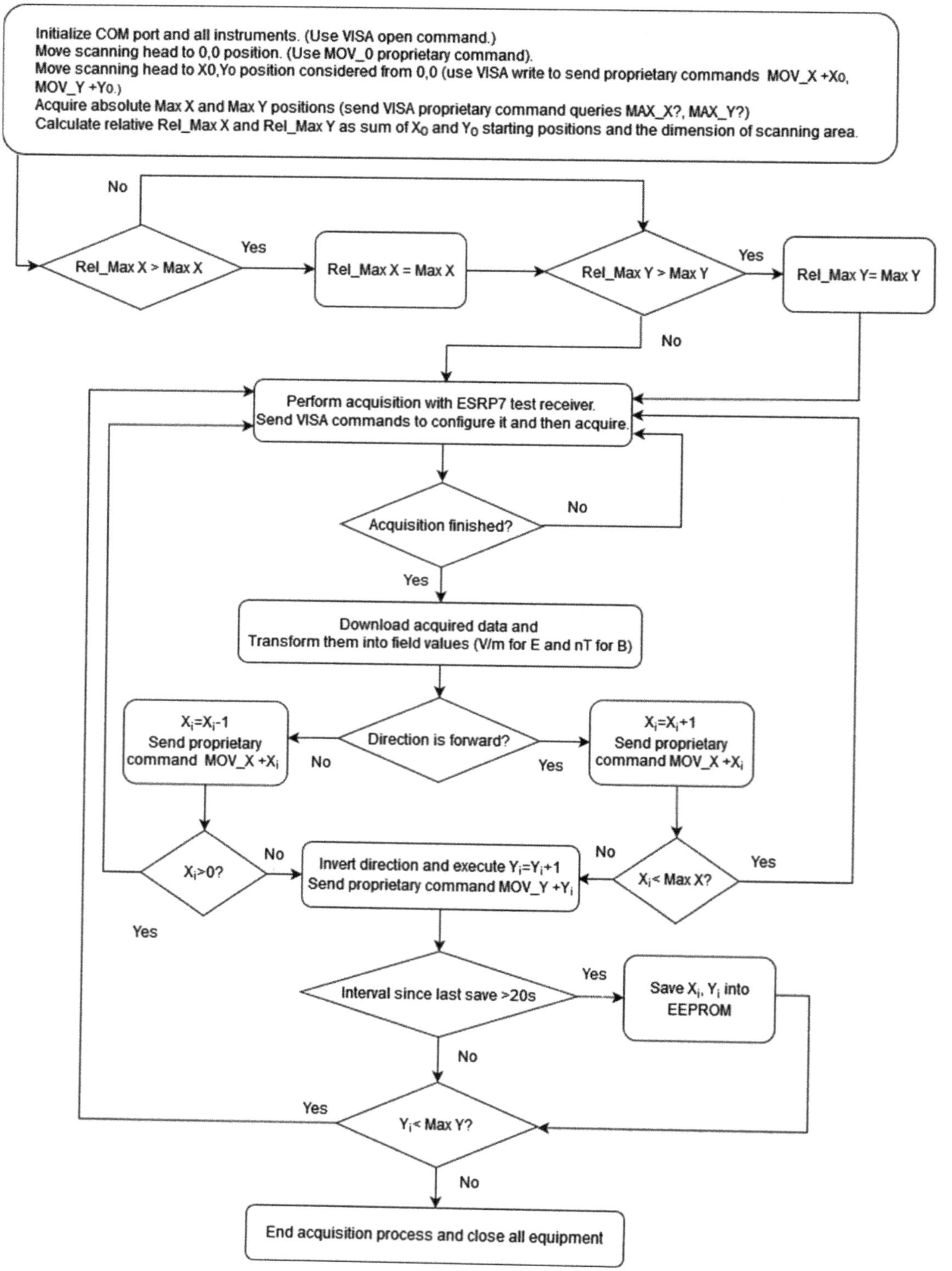

Figure 12. Data acquisition algorithm.

3. A Case Study: Bi-Directional GaN DC/DC Converter

The gallium nitride transistor market is increasing due to the advantages in terms of efficiency and power density [14]. These good features are enabled by their high switching speed (a few nanoseconds) [15,16]. On the other hand, extremely high dv/dt (over 100 V/ns) of the associated voltage waveforms induce severe electromagnetic interference [17–20]. Hence, the proposed inexpensive system for automatic near-field scanning has been used to analyze the electromagnetic noise generated by the output inductor in a GaN-based bi-directional DC/DC switching converter board. The converter has been used in the 48V–12V step-down buck configuration; a simplified schematic is shown in Figure 13.

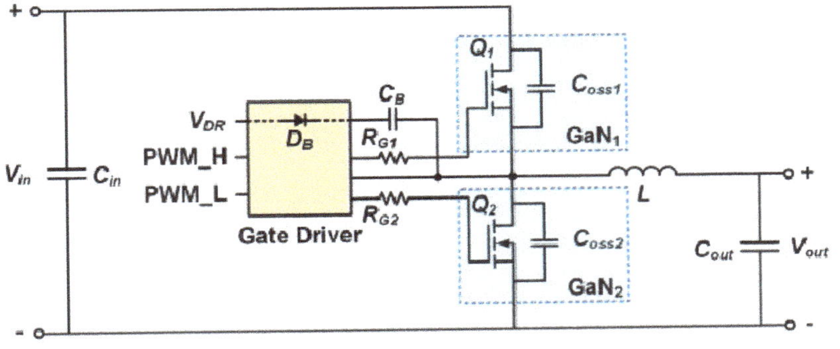

Figure 13. Schematic of a generic GaN buck DC–DC converter.

A picture of a board where the scanning process has been executed is provided in Figure 14. The application is managed by a driver controller that is connected, through resistors, to the GaN half-bridge integrated power stage and the output LC filter.

Figure 14. Board of the GaN buck DC–DC converter: (**a**) front view, (**b**) rear view.

For this device, when working in the synchronous buck mode, oscillations (Figure 15) of the switch node caused by parasitic inductances and fast commutations were observed. The ringing generates an electromagnetic field and, therefore, problems of electromagnetic noise. However, to test the scanning equipment, only magnetic induction B above the big filtering output circuit inductor has been mapped.

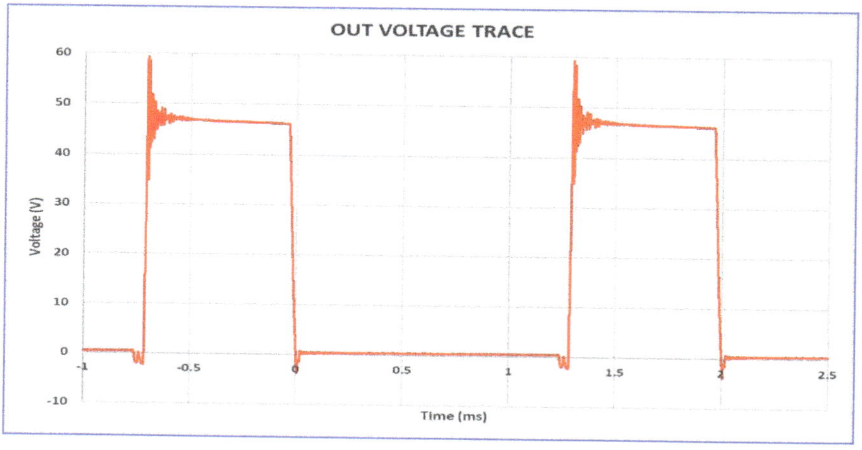

Figure 15. Switching node voltage.

A scan along each axis has been performed in order to obtain the three components of magnetic induction B (Figure 16). The antenna with the smallest loop (type 100B, Table 1) has been used. For each scanned surface unit, the EMI test receiver measured and stored the emissions in a range of frequencies from 150 kHz to 1 GHz.

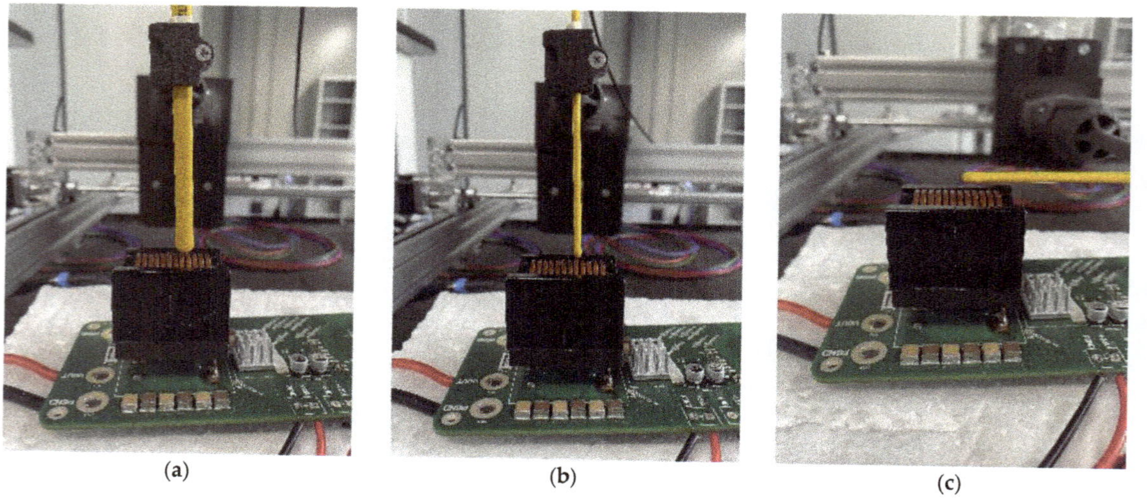

(a) (b) (c)

Figure 16. View of magnetic field acquisition along (**a**) x direction, (**b**) y direction and (**c**) z direction.

By means of the post-processing software, the image of the magnitude of the magnetic field was obtained. Once the colored image of the B magnitude on an XY plane (at a given distance from the scanned element) has been created, it is possible to set an opportune level of transparency and superimpose this image on the real image of the part of the board scanned. The maximum magnitude of the magnetic field at the converter frequency (500 kHz) was 3.83 μT, and it was located almost in the center of the inductor. The minimum, located at the scanning area edge, was about 300 nT. In the inductor area, the average field was about 2.5 μT. At 30~40 MHz, the magnetic field was mainly concentrated on the exposed ferrite parallel parts (left and right sides of the coil). The maximum magnitude of the magnetic field was about 300 μT. In the range 140–150 MHz, a very uniform B distribution was observed. Figure 17 shows a map of the magnetic field magnitude at 500 kHz.

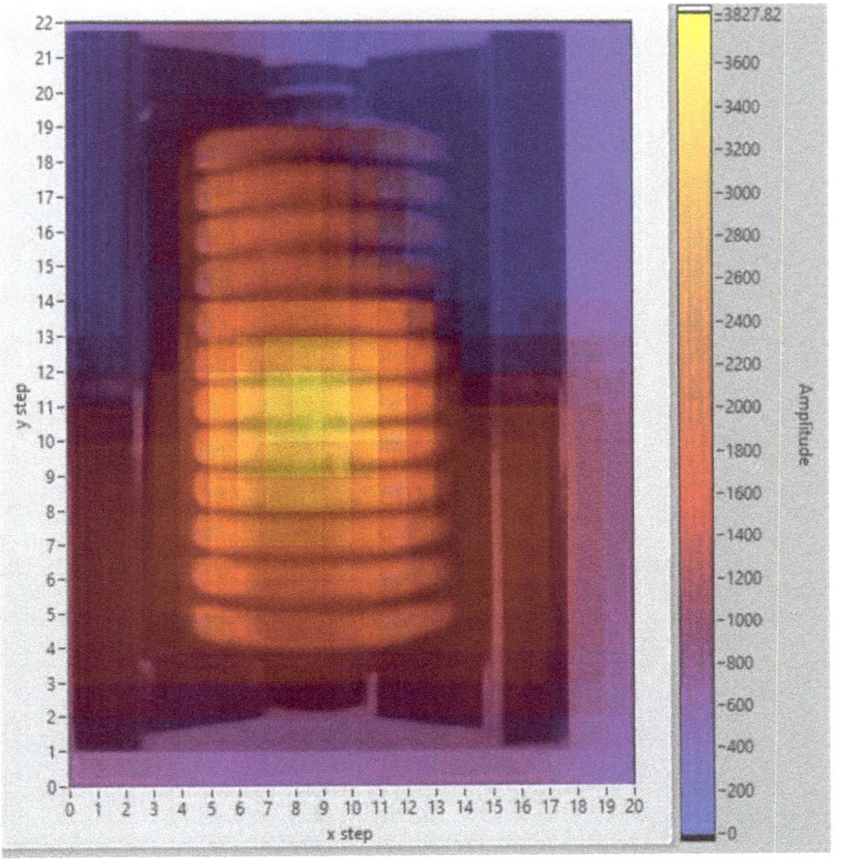

Figure 17. B magnitude detected on the output filter inductor of the bi-directional GaN DC/DC buck converter.

4. Conclusions

The near-field scanner described in this document is an inexpensive system that is able to map the spatial distribution of electric and magnetic fields. Thanks to the head, which is specifically designed for the purpose, it is possible to set distances ranging from a few millimeters up to several centimeters. The resolution along the X and Y directions is accurate enough to guarantee very defined maps of the electromagnetic field over a PCB. The vertical distance of the head from the board is limited only by the higher component in the board itself. The application to different boards and comparing the scanner measurements with the simulation of benchmarks are potential future developments of this activity.

The map acquired allows for an understanding of where the highest emissions on the board surface are located and finding the components and/or traces that generate them, thus allowing the optimization of its layout. The cost of the developed system is less than two thousand dollars, thus making it useful in any application where an investment in expensive commercial equipment is not justified.

Author Contributions: Conceptualization, N.S.; Methodology, S.A.R.; Investigation, S.G. and S.B.; Resources, S.G.; Data curation, S.B.; Writing—original draft, S.G. and S.A.R.; Writing—review & editing, N.S.; Supervision, S.A.R.; Funding acquisition, S.G. All authors have read and agreed to the published version of the manuscript.

Funding: This research received no external funding.

Data Availability Statement: Data are contained within the article.

Conflicts of Interest: The authors declare no conflict of interest.

References

1. Aswin, A.M.; Sundar, S. Electromagnetic Interference (EMI) and Electromagnetic Compatibility (EMC) in Real Time Applications–Survey. *Int. J. Mech. Eng. Technol.* **2018**, *9*, 474–482.
2. Electromagnetic Compatibility (EMC) Directive. Available online: https://single-market-economy.ec.europa.eu/sectors/electrical-and-electronic-engineering-industries-eei/electromagnetic-compatibility-emc-directive_en (accessed on 23 September 2023).
3. Paul, C.R. *Introduction to Electromagnetic Compatibility*, 2nd ed.; Wiley-Interscience: Hoboken, NJ, USA, 2006.
4. Li, C.; Ma, Q.; Tong, Y.; Wang, J.; Xu, P. A survey of conductive and radiated EMI reduction techniques in power electronics converters across wide-bandgap devices. *IET Power Electron.* **2023**, *16*, 2121–2137. [CrossRef]
5. Minardi, G.; Greco, G.; Vinci, G.; Rizzo, S.A.; Salerno, N.; Sorbello, G. Electromagnetic Simulation Flow for Integrated Power Electronics Modules. *Electronics* **2022**, *11*, 2498. [CrossRef]
6. Wunsch, B.; Skibin, S.; Forsström, V.; Stevanovic, I. EMC Component Modeling and System-Level Simulations of Power Converters: AC Motor Drives. *Energies* **2021**, *14*, 1568. [CrossRef]
7. Radchenko, A.; Khilkevich, V.; Bondarenko, N.; Pommerenke, D.; Gonser, M.; Hansen, J.; Keller, C. Transfer function method for predicting the emissions in a -cispr 25 test-setup. *IEEE Trans. Electromagn. Compat.* **2014**, *56*, 894–902. [CrossRef]
8. Mariscotti, A.; Ogunsola, A.; Sandrolini, L. Survey of models and reference data for prediction of e.m. field emissions from electrical machinery. In Proceeding of 8th IEEE Symposium on Diagnostics for Electrical Machines, Power Electronics & Drives (DEMPED), Bologna, Italy, 5–8 September 2011. [CrossRef]
9. Rajagopal, N.; DiMarino, C.; DeBoi, B.; Lemmon, A.; Brovont, A. EMI Evaluation of a SiC MOSFET Module with Organic DBC Substrate. In Proceedings of the IEEE Applied Power Electronics Conference and Exposition (APEC), Phoenix, AZ, USA, 14–17 June 2021; IEEE: Piscataway, NJ, USA, 2021.
10. Chen, H.; Liu, Q.; Li, Y.; Huang, C.; Zhang, H.; Xu, Y. Research on the Method of Near-Field Measurement and Modeling of Powerful Electromagnetic Equipment Radiation Based on Field Distribution Characteristics. *Energies* **2023**, *16*, 2005. [CrossRef]
11. Zhang, B.; Wang, S. A Survey of EMI Research in Power Electronics Systems With Wide-Bandgap Semiconductor Devices. *IEEE J. Emerg. Sel. Top. Power Electron.* **2020**, *8*, 626–643. [CrossRef]
12. Chen, H.; Wang, T.; Feng, L.; Chen, G. Determining Far-Field EMI From Near-Field Coupling of a Power Converter. *IEEE Trans. Power Electron.* **2014**, *29*, 5257–5264. [CrossRef]
13. Beehive ELECTRONICS 100 Series EMC Probes Datasheet. Available online: https://beehive-electronics.com/datasheets/100SeriesDatasheetCurrent.pdf (accessed on 23 November 2023).
14. Scrimizzi, F.; Cammarata, F.; D'Agata, G.; Nicolosi, G.; Musumeci, S.; Rizzo, S.A. The GaN Breakthrough for Sustainable and Cost-Effective Mobility Electrification and Digitalization. *Electronics* **2023**, *12*, 1436. [CrossRef]
15. Letellier, A.; Dubois, M.R.; Trovao, J.P.; Maher, H. Gallium Nitride Semiconductors in Power Electronics for Electric Vehicles: Advantages and Challenges. In Proceedings of the 2015 IEEE Vehicle Power and Propulsion Conference (VPPC), Montreal, QC, Canada, 19–22 October 2015.
16. Musumeci, S.; Mandrile, F.; Barba, V.; Palma, M. Low-Voltage GaN FETs in Motor Control Application; Issues and Advantages: A Review. *Energies* **2021**, *14*, 6378. [CrossRef]
17. Derkacz, P.B.; Schanen, J.-L.; Jeannin, P.-O.; Chrzan, P.J.; Musznicki, P.; Petit, M. EMI Mitigation of GaN Power Inverter Leg by Local Shielding Techniques. *IEEE Trans. Power Electron.* **2022**, *37*, 11996–12004. [CrossRef]
18. Rajeswari, P.; Manikandan, V. Analysis of effects of MOSFET parasitic capacitance on non-synchronous buck converter electromagnetic emission. *Ain Shams Eng. J.* **2023**, *14*. [CrossRef]
19. Raciti, A.; Chimento, F.; Musumeci, S.; Privitera, G. A New Thermal Model for Power Mosfet Devices Accounting for the Behavior in Unclamped Inductive Switching. *Microelectron. Reliab.* **2016**, *58*, 3–11. [CrossRef]
20. Middelstaedt, L.; Strauss, B.; Lindemann, A. Analyzing EMI issues in a DC/DC converter using GaN instead of Si power transistors. In Proceedings of the 2019 10th International Conference on Power Electronics and ECCE Asia (ICPE 2019-ECCE Asia), Busan, Republic of Korea, 27–30 May 2019.

energies

Article

A New Double-Switch SEPIC-Buck Topology for Renewable Energy Applications

Walid Emar [1,*], Haitham Issa [1], Hasan Kanaker [2], Osama Fares [3] and Hani Attar [1]

[1] Energy Engineering Department, Engineering Technology Faculty, Zarqa University, Zarqa 13110, Jordan; hissa@zu.edu.jo (H.I.); hattar@zu.edu.jo (H.A.)
[2] Cyber Security Department, Information Technology Faculty, Isra University, Amman 11622, Jordan; hasan.kanaker@iu.edu.jo
[3] Electrical (Communications and Electronics) Engineering Department, Engineering Faculty, Isra University, Amman 11622, Jordan; osama.fares@iu.edu.jo
* Correspondence: wemar@zu.edu.jo

Abstract: In addition to their conventional use in electric motor drives, DC-DC converters have a variety of other uses, such as energy storage, energy conversion, cyber security systems, uninterruptible power supplies, and renewable energy systems. An innovative DC-DC converter is suggested in this article. Designing a new, high-gain DC-DC converter scheme known as a double-switch SEPIC-buck converter (DSSB) is possible after making some adjustments to the SEPIC converter that is currently known in accordance with accepted techniques. The output voltage magnitude of the proposed converter is either larger than or less than the input voltage magnitude and is the same sign as the input voltage. According to the theoretical and analytical study that has been supported by the real-world application, high voltage gain, low switching stress, and low inductor current ripple are the main characteristics of the proposed DSSB converter. The related small-signal model was also used to build the closed-loop system. The frequency response and output voltage behavior were investigated when the input source voltage abruptly changed as a step function. Based on the comparison study with other DC-DC converters, the DSSB converter outperforms currently known DC-DC converters such as Buck, SEPIC, Boost, Buck-Boost, and other SEPIC converter topologies in terms of voltage gain, harmonic content, normalized current ripple, dynamic performance, and efficiency. Additionally, the frequency response and control of the proposed converter using an alternate current (AC), small-signal, analysis-based, current-mode control technique are both provided. Thus, the DSSB is regarded as safe in overcurrent situations because of the small-signal analysis with the current control strategy. As a result of the verification of the proposed control technique, the resistance to changes in the DSSB parameters, improved dynamic performance, and higher control accuracy are further advantages of current-mode control based on small-signal analysis over other control approaches (PI controllers). Finally, the experimental and simulation results from Simplorer 7 and MATLAB/Simulink are used to validate the findings of the analytical and comparative investigation.

Keywords: double-switch SEPIC-buck converter (DSSB); buck-boost; SEPIC; bode diagrams; PI controllers

Citation: Emar, W.; Issa, H.; Kanaker, H.; Fares, O.; Attar, H. A New Double-Switch SEPIC-Buck Topology for Renewable Energy Applications. *Energies* **2024**, *17*, 238. https://doi.org/10.3390/en17010238

Academic Editor: Salvatore Musumeci

Received: 15 September 2023
Revised: 27 October 2023
Accepted: 10 November 2023
Published: 2 January 2024

1. Introduction

Power electronic converters are essential in many industrial applications, as they produce a DC voltage supply stabilized to a certain value from a DC voltage supply. These converters usually offer an adequate size, weight, and volume for low-, medium-a, and high-power operations ranging from tens of watts to hundreds of kW [1,2].

Power electronic converters are also utilized to get the proper stepping-up/stepping-down voltages to utilize combined renewable energy resources for intelligent microgrid and power system conversions as efficiently as feasible [3,4].

Many industrial applications require the use of high voltage gain DC-DC converters, such as vehicle headlamps, fuel cells, solar cells, and battery backup systems for uninterruptible power sources, autonomous and intelligent residential sector electricity storage systems, telecommunication power supplies, cyber security systems, and voltage regulators for energy conversion systems [5,6].

Traditional DC-DC converters such as bucks, boosts, and buck-boosts are the pre-select when switching to step-up/step-down voltage regulation, respectively, mainly due to their simple structures and their low number of electronic/electrical components. However, these converters' duty cycles (switching ratio) must be at their highest levels to provide a little higher voltage gain, which is insufficient to meet the industrial demands of the above-mentioned applications [7].

Therefore, the wide conversion range DC-DC converter has recently been the subject of much research in power electronics, leading to the visualization of many structures with deep-rooted advantages and disadvantages in terms of switching duty ratios, the number of components, and the associated voltage stress of their semiconductor devices. In addition, most of these studies focus on the high-voltage boost conversion process using different methods such as cascade and parallel stages, coupled inductors (inductors), parallel coupled capacitors, and so on [1–8]. Although there are many research articles on DC/DC converters, there are still many related problems that need to be solved, such as maximum power point tracking, efficiency, voltage gain, current ripple minimization, buck-boost operation regime, quadratic boost, and buck converters [8,9].

For instance, the decrease of current ripple and the voltage stress on the converter switches are studied for an interleaved three-phase boost converter with magnetically coupled inductors that is utilized to power a brushless DC motor fed by solar energy at its highest power point. As the study demonstrates, the suggested converter system has demonstrated advantages over a noncoupled three-phase interleaved boost converter, including low ripple content on both the load and supply sides, improved efficiency, and reduced switch stress Vinay Kumar et al. [10].

Al Attar H. et al. [11], also studied the current ripple reduction, reduced voltage stress, and improved efficiency for bidirectional electric vehicle (EV) chargers. As demonstrated by Al Attar H et. al., 2023, bidirectional DC-DC converters can be effectively employed for applications needing high power density, such as bidirectional electric vehicle (EV) chargers. Other control systems combine different modulation techniques with a small- or large-signal model. In this industrial application, a DC-DC converter couples an AC-DC converter, a high-voltage battery pack, and a DC bus link. The use of DC-DC converters with bidirectional electric vehicle (EV) chargers reduces voltage stress, increases efficiency, and reduces the current ripple, according to Al Attar H et al. [11].

According to Solis-Rodriguez et. al. [12], the Low Energy Storage Quadratic Boost Converter (LES-QBC) with the Particle Swarm Optimization (PSO) technique offers a lower Output Voltage Ripple (OVR) than a standard quadratic boost converter with capacitors of the same characteristics. The Particle Swarm Optimization (PSO) technique was used to investigate the optimization of a recently developed DC-DC converter capacitor selection [12].

Because the suggested converter offers a solution to the problems raised by Attar and many others, including low voltage gain, voltage stress on the converter devices, harmonic content, efficiency, and effectiveness of these converters with particular systems, such as PV solar systems and wind turbines, this motivates the authors of this article to research and develop a novel modified SEPIC-Buck converter with double switches (DSSB).

The features of the proposed converter are as follows:

1. When compared to the conventional SEPIC-Buck converter, the suggested converter has an additional transistor (MOS2).
2. When compared to the duty cycle of the first switch (MOS1), the duty cycle of the second transistor (MOS2), which is used for the converter bucking regime, is shifted by T/2.

3. The proposed converter employs more capacitive/inductive components than the previous DC-DC conventional converters in order to discharge its stored energy into the load while the switches are off and to boost the voltage gain of the converter.
4. The proposed converter's normalized voltage gain is larger than traditional boost, modified Sepic, buck-boost, and other DC-DC converters.
5. Furthermore, the proposed design can lower the harmonic content generated at the output and input, resulting in a larger DC output voltage range and higher efficiency when compared to the current SEPIC, boost, and buck converter circuits.
6. As the reference [12–17] shows, the optimum application of cascaded and shunted capacitors, as well as inductors, precludes any voltage or current surges in the DSSB circuit.

The converter main switches use an appropriate control strategy and a switching optimization technique. Here, small-signal analysis is employed with a current-mode control technique with two dual lead PI controllers for the load current and voltage. Small-signal analysis is used to determine the appropriate switching ratio based on the least amount of error between input and output powers and to select the proper gain value for the PI controllers [10–12].

Furthermore, the verification of the proposed current-mode control technique based on small-signal analysis along with the converter frequency response demonstrates that the additional advantages of the DSSB include its resistance to changes in its parameters, improved dynamic performance, and higher control accuracy.

The converter was thoroughly investigated in terms of mathematical analysis, design, experimental simulation, and theoretical waveforms.

According to the comparative study that has been prepared in this article, the DSSB converter surpasses the currently popular DC-DC converters, including buck, SEPIC, boost, buck-boost, and modified multiple SEPICs in terms of voltage gain, harmonic content, normalized current ripple, dynamic performance, and efficiency.

This research does not, however, take into account the impact of parasitic components on the dynamic performance of the power stages of the suggested converter. The impacts of parasitic components can have a major impact on both steady-state and dynamic performance, as has been thoroughly demonstrated in the literature (see, for instance, [15,16]).

The suggested DSSB converter's primary characteristics are high voltage gain, low switching stress, and low inductor current ripple, according to the theoretical and analytical research validated by practical application. Therefore, the development of a MATLAB/Simulink 7.0, and Simplorer 7.0 simulations were essential steps taken to validate the theoretical analysis. The functioning prototype of the proposed converter, which is intended to enhance the maximum power point tracking of PV solar systems, will be the primary focus of part II of this study.

2. The Essentials of the DSSB Converter Analysis

Figure 1 shows the basic structure of the new, modified, double-switch SEPIC-buck converter (DSSB). The DSSB may comprise a DC source (solar cell system), input smoothing inductor L1, power electronic switches MOS1 and MOS2, capacitive filters (Cs, Cm, C1, and Co), three freewheeling diodes (Dm1, Dm2, and Dm3), three smoothing inductors (L1, L2, and L3), and DC resistive load, R.

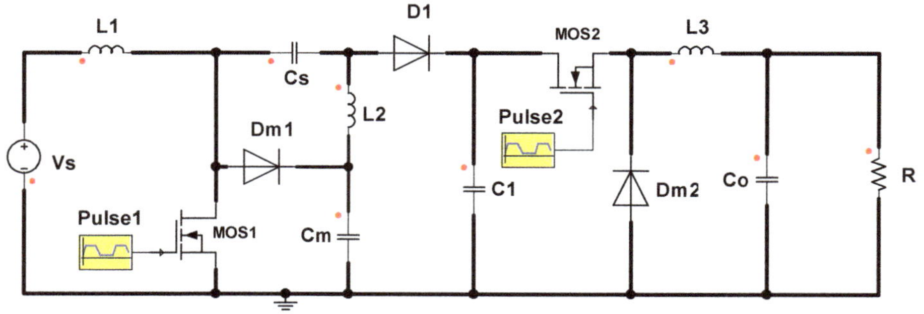

Figure 1. DSSB converter.

2.1. Working Principle of the DSSB

For a better understanding of the dynamic and static behavior of the DSSB, let us assume that a pure DC voltage, V_s, is provided at the DSSB input to supply a resistive load of 10 that represents a heater in an electric drive (tramway). As shown during the mathematical analysis and simulation, the purpose of the additional LC components is to aid in reducing the ripple of the output current and voltage, and correspondingly, the input current and voltage.

2.1.1. Quasi-Stationary State of the DSSB

Based on the volt-second area balance principle, the inductor voltages have zero average values, and the inductor currents of such a DC-DC converter are continuous. Other observations are that the capacitor currents have zero average values under steady-state conditions, and the capacitor voltages have almost smooth DC and continuous waveforms.

The inductor currents in the energy transfer never approach zero values in the continuous conduction mode. The main switch, MOS1, is therefore on for a time $t_{on} = DT$ and off for the remaining time $t_{off} = (1 - D)T$ during the proposed converter's operation in continuous conduction mode [1–3]. Parameter D symbolizes the DC duty ratio of the converter. The time switch MOS2 is configured in a certain way so that this switch works with a time shift $T/2$ as compared to MOS1.

The parameters related to the simulation of the DSSB operating in the continuous conduction mode (CCM) are listed in Table 1 as follows:

Table 1. Technical parameters of the DSSB.

Parameter	Symbol	Real Value
Smoothing Inductors	L	30 mH
Solar cell supply voltage:	V_s	50 V
Load resistance:	R	10 Ω
The capacitance of all capacitors:	C	1 mF
Switching frequency:	f	4 kHz

2.1.2. Buck Regime of the DSSB

The working principle of the DSSB in the buck regime is better explained using the following circuit diagrams for the region of $0 < D < 1/2$:

Based on Figure 2a–c, the DSSB is operating in four stages:

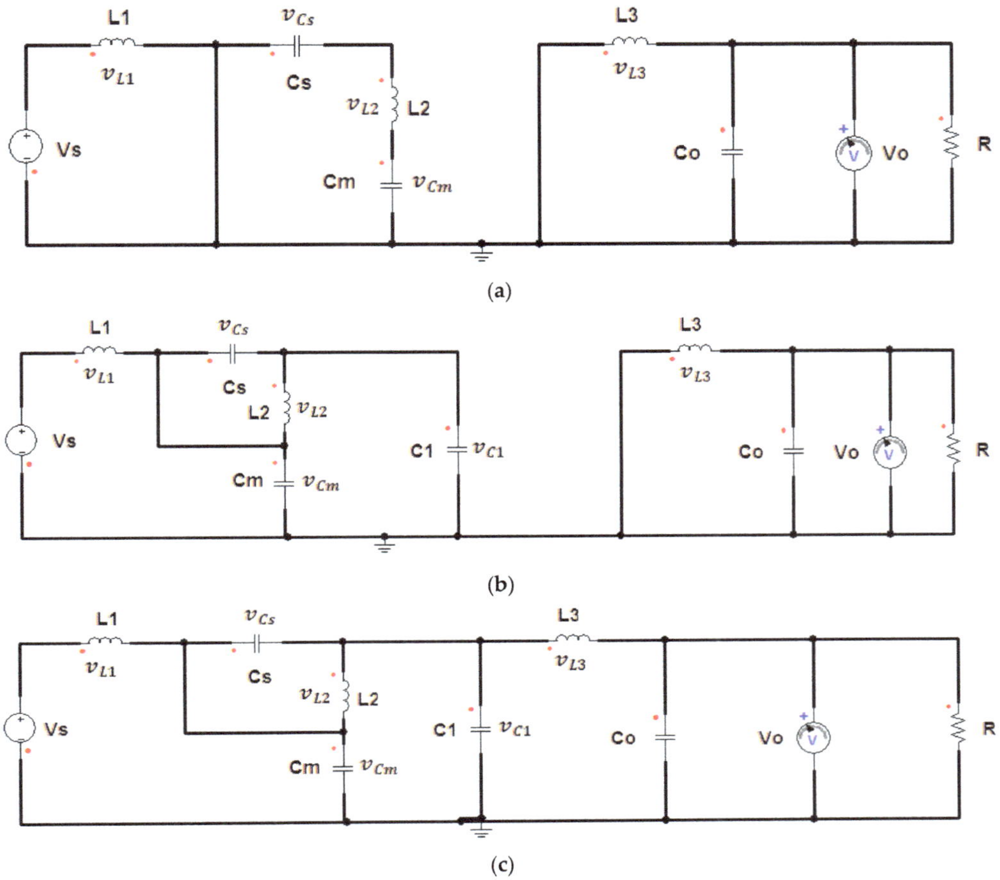

Figure 2. (**a**) On-state of MOS1 and off-state of MOS2 in the operating region of $0 < D < 1/2$, (**b**) Off-state of MOS1 and MOS2 in the operating region of $0 < D < 1/2$, (**c**). Off-state of MOS1 and on-state of MOS2 for the operating region of $0 < D < 1/2$.

The first stage is shown in Figure 2a [0, DT], during which MOS1 is on. Using the inductor currents as state variables in the following equations to get an expression for the voltage conversion ratio, one may find:

$$\begin{array}{c} v_{L1} = L_1 \frac{di_1}{dt} = V_s \\ v_{L2} = L_2 \frac{di_2}{dt} = -v_{Cs} - v_{Cm} \\ v_{L3} = L_3 \frac{di_3}{dt} = -V_o \end{array} \quad (1)$$

The second stage is shown in Figure 2b [DT, T/2], during which MOS1 and MOS2 are off. The resultant equations describing this stage are:

$$\begin{array}{c} v_{L1} = L_1 \frac{di_1}{dt} = V_s - v_{Cm} \\ v_{L2} = L_2 \frac{di_2}{dt} = -v_{Cs} = v_{C1} - v_{Cm} \\ v_{L3} = L_3 \frac{di_3}{dt} = -V_o \end{array} \quad (2)$$

The third stage is shown in Figure 2c [T/2, T/2 + DT], during which MOS1 is off and MOS2 is on. The resultant equations describing this stage are:

$$\begin{aligned} v_{L1} &= L_1 \tfrac{di_1}{dt} = V_s - v_{Cm} \\ v_{L2} &= L_2 \tfrac{di_2}{dt} = -v_{Cs} = v_{C1} - v_{Cm} \\ v_{L3} &= L_3 \tfrac{di_3}{dt} = v_{C1} - V_o \end{aligned} \tag{3}$$

The fourth stage is shown again in Figure 2b [T/2+DT, T], during which MOS1 is off and MOS2 is off. The resultant equations describing this stage are:

$$\begin{aligned} v_{L1} &= L_1 \tfrac{di_1}{dt} = V_s - v_{Cm} \\ v_{L2} &= L_2 \tfrac{di_2}{dt} = -v_{Cs} = v_{C1} - v_{Cm} \\ v_{L3} &= L_3 \tfrac{di_3}{dt} = -V_o \end{aligned} \tag{4}$$

Applying Kirchhoff's voltage law during these operating stages for the average values of the inductor voltages under steady-state conditions, one finds the following set of equations:

$$\begin{aligned} V_{L1} &= V_s - (1-D)\, V_{cm} \\ V_{L2} &= (1-D)V_{C1} - DV_{Cs} - V_{Cm} \\ V_{L3} &= -V_o + D\, V_{C1} \\ V_{Cs} &= V_{Cm} - V_{C1} \end{aligned} \tag{5}$$

Based on the volt-second area balance, the voltages across the inductors have zero average values; therefore, one may find the following relations for the capacitive voltages:

$$\begin{aligned} V_{Cm} &= \tfrac{1}{1-D}\, V_s \\ V_{Cs} &= -DV_{Cm} = \tfrac{-D}{1-D}\, V_s \\ V_{C1} &= \tfrac{V_o}{D} \end{aligned} \tag{6}$$

After rearrangement, one may find for the output voltage conversion, V_o, regarding the supply voltage V_s, in the region of $0 < D < 1/2$, the following expression:

$$V_o = \frac{D\,(1+D)}{(1-D)}\, V_s \tag{7}$$

The voltage gain characteristics in the bucking regime of the DSSB reach the maximum value when $V_o = V_s$. Thus:

$$\frac{V_o}{V_s} = \frac{D\,(1+D)}{(1-D)} = 1 \Longrightarrow D = -1+\sqrt{2} \tag{8}$$

Hence, the output voltage magnitude in the bucking regime is less than that of the supply for $0 < D < -1+\sqrt{2}$ and otherwise, is greater than the supply. Therefore, the DSSB combines the capabilities of the buck and boost converters. The DSSB output and input voltages and currents in the bucking regime are shown in Figure 3a, while the output and input powers with a 98% converter efficiency are shown in Figure 3b. Figure 3c displays the steady-state DSSB voltages and currents waveforms to demonstrate the harmonic content produced in the currents and voltages at low frequency (1 kHz) and supply voltage (25 V).

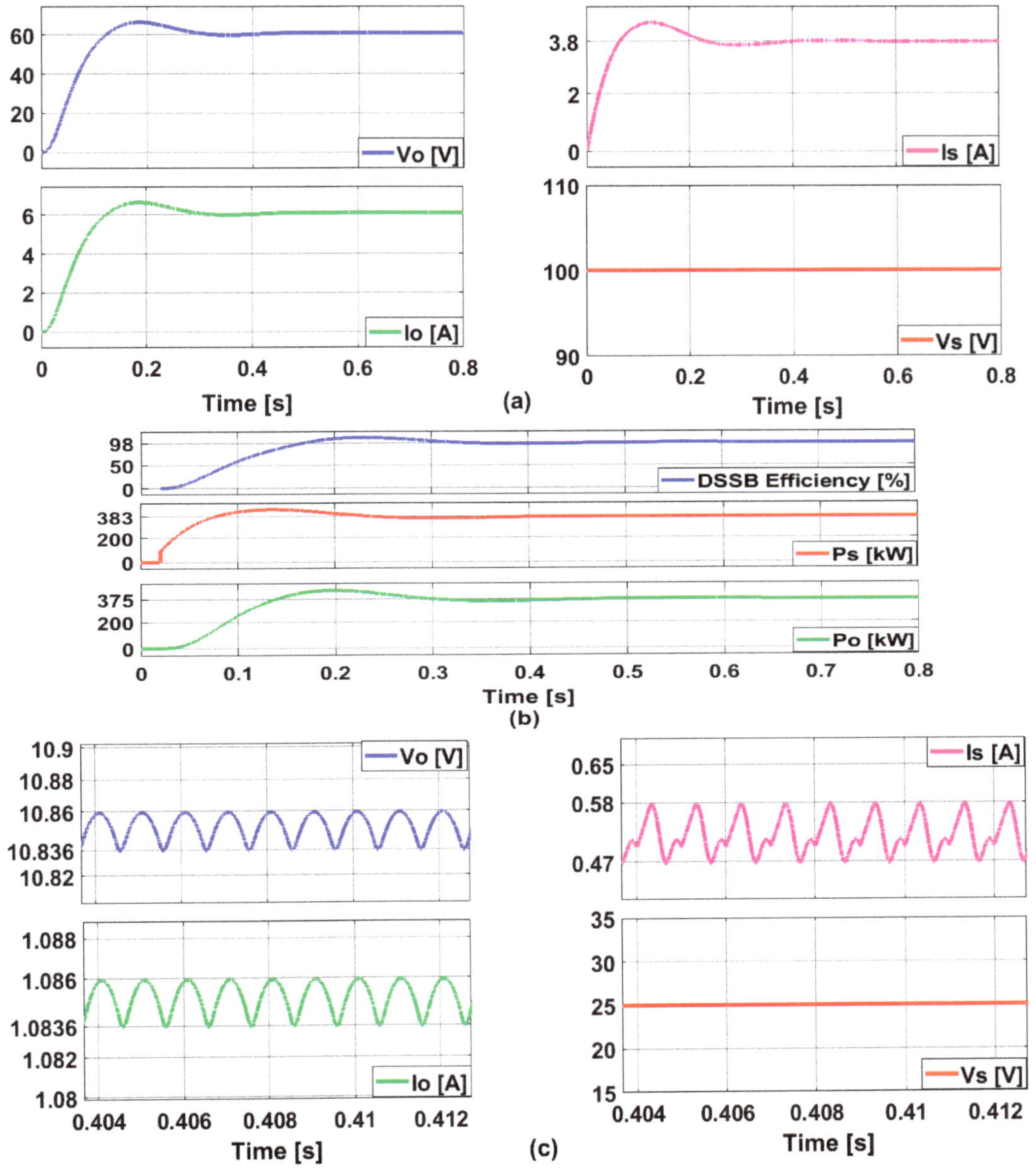

Figure 3. MATLAB/Simulink simulation of the DSSB in the bucking region of $0 < D < \sqrt{2} - 1$: (**a**) Output and input voltages and currents, (**b**) Output and input powers with the converter efficiency of 95%; the switching frequency of the DSSB is $f = 10$ kHz. (**c**) Waveforms of steady-state current and voltage for the DSSB at $f = 1$ kHz and low supply voltage, $V_s = 25$ V .

The voltage and current waveforms for $D = 1/4$ in CCM converter mode are examined and shown in Figure 3a.

2.1.3. Boost-Regime of the DSSB

The second possible operating area of the DSSB is in the region of $\sqrt{2} - 1 < D < 1$. In this region, the converter is operating exclusively in the mode of boosting regime because its voltage gain can only be greater than unity. The switching times of MOS1 and MOS2 within this region may overlap. Therefore, based on Figure 2a,c and Figure 4a, one may obtain again several stages of operation for the DSSB under the premise of optimal steady-state operating conditions of the DSSB:

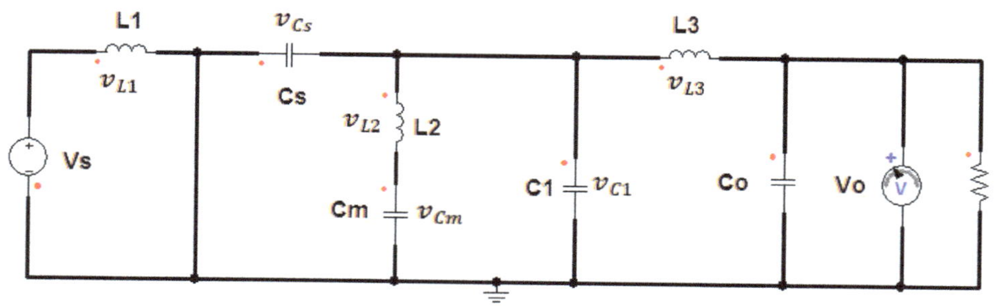

Figure 4. On-state of MOS1 and MOS2 for the region of $\sqrt{2}-1 < D < 1$.

The first stage is shown in Figure 4a [0, (2D-1) T]: MOS1 and MOS2 are on during this time interval. To obtain an expression for the voltage conversion ratio, one may use the inductor currents as state variables in the following equations:

$$\begin{aligned} v_{L1} &= L_1 \tfrac{di_1}{dt} = V_s \\ v_{L2} &= L_2 \tfrac{di_2}{dt} = -v_{Cs} - v_{Cm} \\ v_{L3} &= L_3 \tfrac{di_3}{dt} = v_{C1} - V_o \end{aligned} \tag{9}$$

The second stage is shown in Figure 2a [0, (1-D) T]: MOS1 is on and MOS2 is off during this time interval. The resultant equations describing this stage are:

$$\begin{aligned} v_{L1} &= L_1 \tfrac{di_1}{dt} = V_s \\ v_{L2} &= L_2 \tfrac{di_2}{dt} = -v_{Cs} - v_{Cm} \\ v_{L3} &= L_3 \tfrac{di_3}{dt} = -V_o \end{aligned} \tag{10}$$

The third stage is shown in Figure 2c [0, (1-D) T]: MOS1 is off and MOS2 is on during this time interval. The resultant equations describing this stage are:

$$\begin{aligned} v_{L1} &= L_1 \tfrac{di_1}{dt} = V_s - v_{Cm} \\ v_{L2} &= L_2 \tfrac{di_2}{dt} = -v_{Cs} = v_{C1} - v_{Cm} \\ v_{L3} &= L_3 \tfrac{di_3}{dt} = v_{C1} - V_o \end{aligned} \tag{11}$$

Using the same principle of volt-second area, the average voltages across the inductors have zero values, which means:

$$\begin{aligned} V_{L1} &= V_s - (1-D)\ V_{cm} \\ V_{L2} &= -DV_{Cs} + (1-D)V_{C1} - V_{Cm} \\ V_{L3} &= D\,V_{C1} - V_o \end{aligned} \tag{12}$$

Finally, the voltage conversion ratio can be obtained as follows:

$$\begin{aligned} V_{Cm} &= \tfrac{1}{1-D} V_s \\ V_{Cs} &= -DV_{Cm} = \tfrac{-D}{1-D} V_s \\ V_{C1} &= \tfrac{V_o}{D} \end{aligned} \tag{13}$$

Thus:

$$V_o = \frac{D\,(1+D)}{(1-D)}\,V_s \tag{14}$$

The Matlab/Simulink simulation waveforms for $-1+\sqrt{2} < D < 1$ are shown in Figure 5. The converter efficiency is again 98%, and the switching frequency is $f = 10$ kHz.

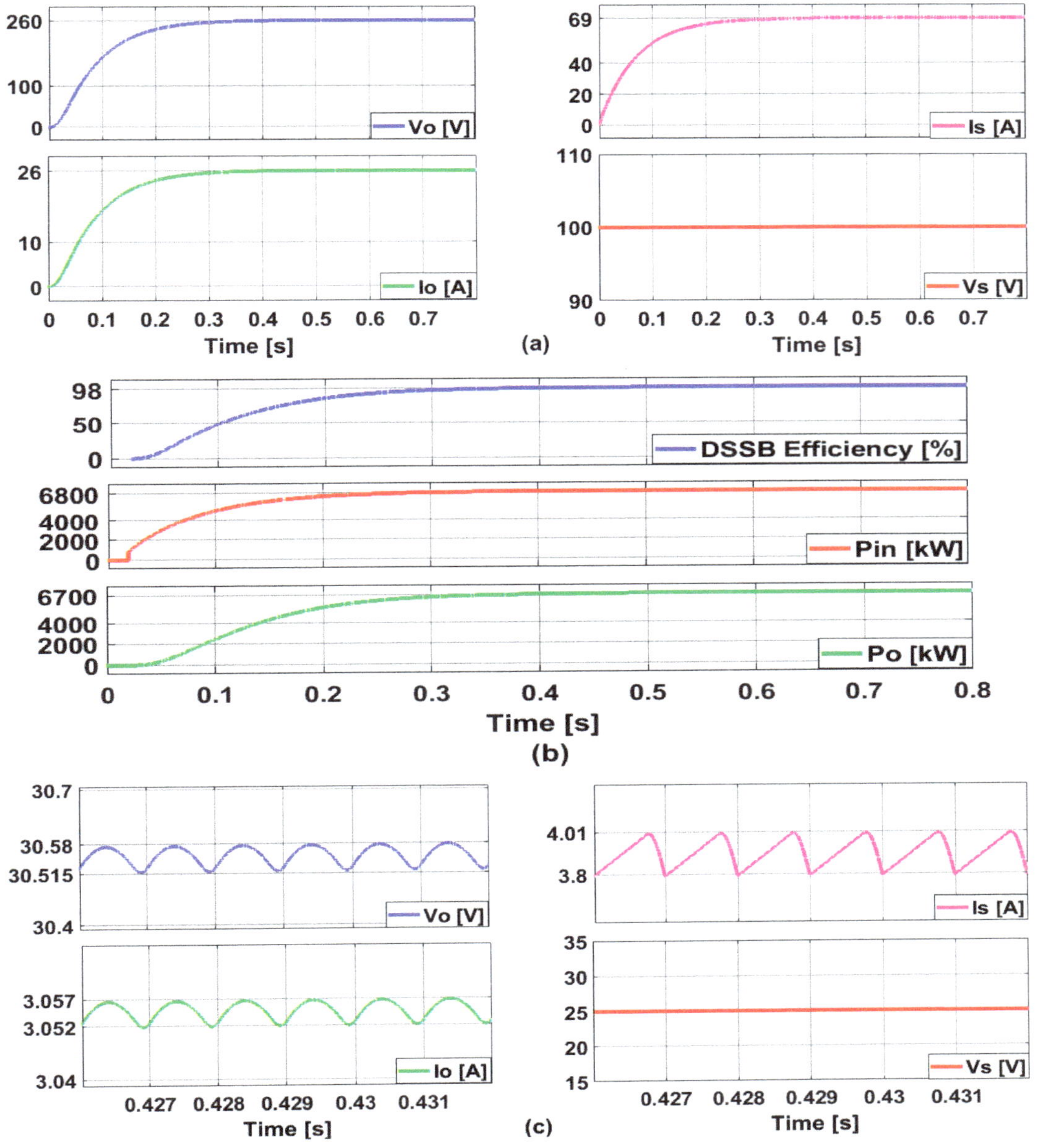

Figure 5. MATLAB/Simulink simulation of the DSSB in the boosting region of $\sqrt{2}-1 < D < 1$: (**a**) Output and input voltages and currents, (**b**) Output and input powers with the converter efficiency of 98%; the switching frequency of the DSSB is $f = 10$ kHz. (**c**) Waveforms of steady-state current and voltage with a 1 kHz target frequency and low supply voltage, $V_s = 25$ V.

According to Equation (14), the DSSB is operated in the range of $0 \leq D \leq -1+\sqrt{2}$ in the bucking regime to step down the load voltage, V_o, in proportion to the supply voltage, V_s. The DSSB is used to step up the load voltage, V_o, in proportion to the supply voltage, V_s, while the converter is operating in the region of $-1+\sqrt{2} \leq D \leq 1$.

As a result of the analysis described above, it can be concluded that the output voltage increases when boosting and only slightly decreases when operating in bucking mode. This is because most of the capacitors and inductors in the circuit are accumulated at the input side before the second switch MOSF2. In order to solve this issue, a third MOSFET with the appropriate switching strategy might be used instead of the diode Dm1, although doing so would complicate the structure design of the converter.

2.1.4. The Inductor Current Ripple

The following problems were found by analyzing the previous analogous circuits in light of the inductor current ripple:

In the region of 0 < D < 1/2:

$$\begin{aligned} \Delta i_1 &= \frac{DV_s}{f\,L} \\ \Delta i_2 &= \frac{D(-V_{Cs}-V_{Cm})}{f\,L} = -\frac{DV_s}{f\,L} \\ \Delta i_3 &= \frac{-D\,V_o}{f\,L} = \frac{-D^2\,(1+D)}{f\,L\,(1-D)}\,V_s \end{aligned} \tag{15}$$

In the region of 1/2 < D < 1:

$$\begin{aligned} \Delta i_1 &= \frac{DV_s}{f\,L} \\ \Delta i_2 &= \frac{D(-V_{Cs}-V_{Cm})}{f\,L} = -\frac{DV_s}{f\,L} \\ \Delta i_3 &= \frac{-D\,V_o}{f\,L} = \frac{-D^2\,(1+D)}{f\,L\,(1-D)}\,V_s \end{aligned} \tag{16}$$

where f is the switching frequency of the DSSB and L is the common inductance value for all inductors.

Figure 6a shows an illustration of the DSSB's normalized inductor current ripple curves for various switching ratio D values. The normalized source current ripple ($i_s = i_1$) of the DSSB is represented by the brown line, and the output current ripple ($i_o = i_3$) is represented by the green line.

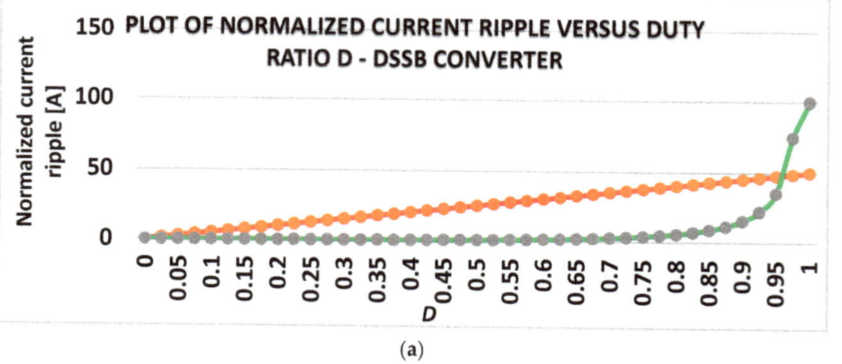

(a)

Figure 6. *Cont.*

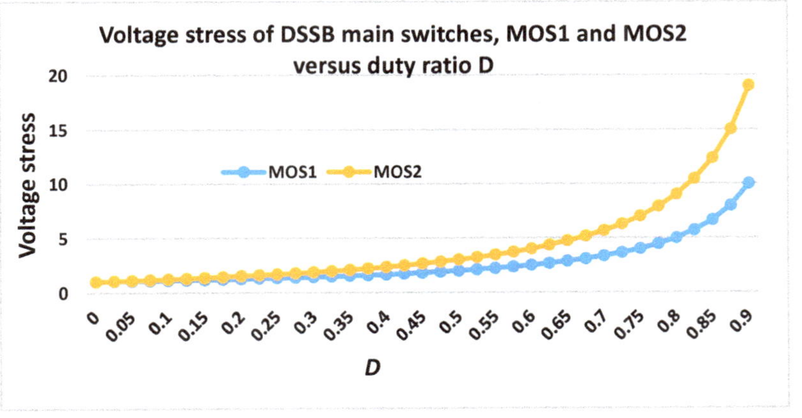

(b)

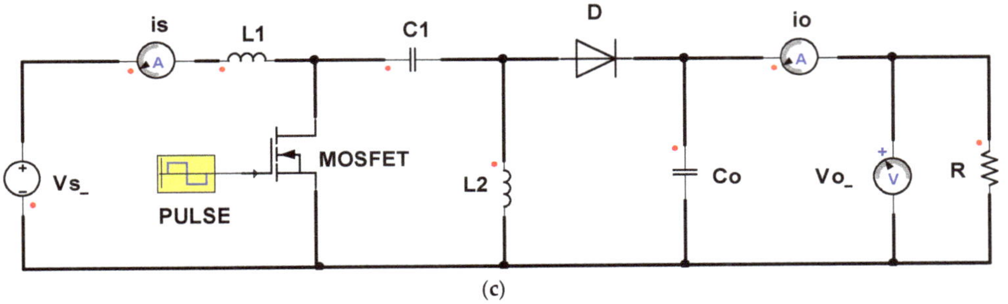

(c)

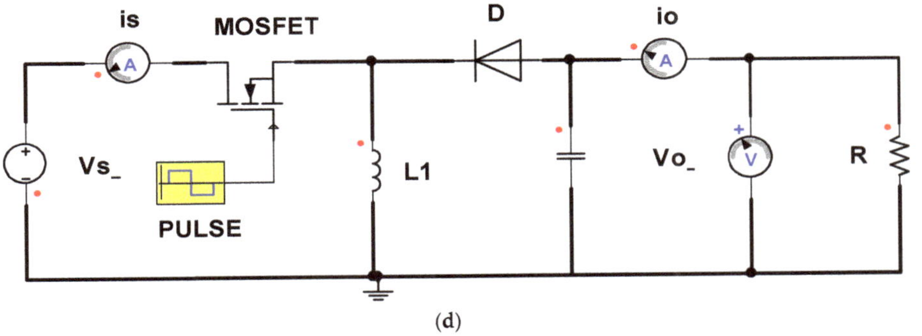

(d)

Figure 6. (**a**) A plot of the normalized inductor current ripples of the DSSB converter, (**b**). A plot of voltage stress of the DSSB's main switches, MOS1 and MOS2, (**c**) A conventional SEPIC converter. (**d**) A switched-mode inverting buck-boost converter.

2.2. Normalized Voltage Stress of the DSSB Main Switches, MOS1 and MOS2

As a ratio between the reverse voltage of MOS1 and the DSSB supply voltage, V_s, the normalized voltage stress of the DSSB converter switch, MOS1 is described as follows:

$$V_{MOS1}/V_s = V_{cm}/V_s = V_{Cm}/V_s = \frac{1}{1-D} \tag{17}$$

However, when MOS2 is in the off-state, it has the following normalized voltage stress:

$$V_{MOS2}/V_s = V_{c1}/V_s = V_{C1}/V_s = \frac{(1+D)}{(1-D)} \tag{18}$$

Figure 6b illustrates the normalized voltage stress of the DSSB's main switches as a function of the increasing switching ratio D using Formulas (17) and (18). When choosing the technical parameters of these switches, it should be remembered that the voltage stress of MOS1 is lower than that of MOS2 based on these figures.

It is wise to compare the voltage stress of the DSSB's input switch, MOS1, with that of other switched-mode step-down/step-up DC-DC converters to confirm that it is the best option for the industrial applications indicated above. The following expressions can be generated for comparison with the standard SEPIC, depicted in Figure 6c.

When the SEPIC converter main switch, MOSFET, is open, the diode is on. Applying Kirchhoff's voltage law around the closed loop of V_s, L_1, C_1, and V_o gives:

$$-V_s + v_{L1} + v_{C1} - V_o = 0 \tag{19}$$

Because the value of C_1 is intentionally very large to keep a voltage across it at the mean of V_s, we get:

$$v_{L1} = -V_o \tag{20}$$

Additionally, for the closed loop V_s, L_1, with the main switch MOSFET, we get:

$$-V_s + v_{L1} + v_{MOSFET} = 0 \Rightarrow v_{MOSFET} = V_s + V_o \tag{21}$$

Substituting from [1] for $V_o = \frac{D}{1-D}$ into Equation (21), one may get the voltage stress ratio of MOSFET:

$$v_{MOSFET}/V_s = \frac{1}{1-D} \tag{22}$$

Similar presumptions to those established for the operation of the DSSB converter are assumed about the voltage stress on the primary switch of the buck-boost converter. The inductor voltage turns on the diode when the main switch is open, allowing current to flow into the load resistor and capacitor. The inductor voltage is at a value of:

$$v_{L1} = V_o = -\frac{D}{1-D}V_s \tag{23}$$

Applying Kirchhoff's voltage law around the closed loop of V_s and L_1 yields:

$$v_{MSOFET} = V_s - v_{L1} = \frac{1}{1-D}V_s \Rightarrow v_{MOSFET}/V_s = \frac{1}{1-D} \tag{24}$$

According to Formulas (17), (22), and (23), the voltage stress ratio of the conventional SEPIC and buck-boost converters' main switches is the same as for the DSSB main switch MOS1.

Similar expressions can be obtained for calculating the diode voltage stresses of the DSSB converter and comparing them with conventional SEPIC and boost/buck converters.

3. Comparison of the DSSB Performance with That of Other DC-DC Converters

The most popular non-isolated DC-DC converters are contrasted with the suggested converter in this study. The voltage gain ratio, voltage switch stress, voltage ripple, efficiency, cost, harmonic content, and ease of implementation are crucial elements that were considered throughout the comparison.

3.1. In Comparison to SEPIC and BB Converters

In this section, PWM modulation techniques are used to compare inverting buck-boost (BB), conventional SEPIC, and DSSB converters, in terms of structure complexity, mean value, rectified mean value, total root-mean-square value (RMS) at the fundamental frequency, RMS value of the AC components only (RMS_{AC}), crest factor, ripple factor, harmonic factor, harmonic content distortion, power efficiency, and power factor. The comparison is performed based on the simulation parameters mentioned in Table 1.

The simulation technique used in this paper can be used to calculate distortion and the RMS magnitude at a specified fundamental frequency using real-time analysis and fast Fourier transform (FFT). The root-mean-square value of the fundamental and other components may be calculated either with a Discrete Fourier Transform or with a Least Squares Parameter Estimate. In either case, after determining RMS1 and RMSAC, the root-mean-square value of any component other than the fundamental component is:

$$RMS_k = \sqrt{(RMS_1)^2 + (RMS_{AC})^2},\ k = 1,\ 2,\ \ldots \tag{25}$$

where RMS_{AC} is the RMS value of all higher-order components added together, and RMS_1 is the RMS value of the fundamental component obtained from the Fourier series analysis.

The total harmonic distortion (THD) is a measure of the harmonic content for any simulation quantity. However, it is necessary to define a period that is used for the computation of the total harmonic distortion. Thus:

$$THD = \sqrt{\frac{RMS_{AC}^2 - RMS_1^2}{RMS_1}} \times 100\% \tag{26}$$

THD indicates the harmonic content in the tested quantity without any indication of the contributing effect of each harmonic quantity. When a filter circuit is used to withdraw higher harmonics, it is important to know the frequency and amplitude of the remaining higher harmonics to reduce unwanted harmonics without specifying the values of the second-order load filter [3]. This is the result of the harmonic factor (HF), which is a measure of the influence of higher-order components on the harmonic content and is given in the following:

$$HF = \frac{\sqrt{\sum_{k=2}^{\infty}(RMS_k)^2}}{RMS_1} \times 100\% \tag{27}$$

The crest factor is a function that returns the crest factor (peak/RMS) for the selected simulated quantity, and it is used to measure the peaks in the waveform of the tested quantity. Thus:

$$KF = \frac{\sqrt{\sum_{k=2}^{\infty} k^2(RMS_k)^2}}{\sqrt{\sum_{k=2}^{\infty}(RMS_k)^2}} \times 100\% \tag{28}$$

In this study, the effectiveness of the PWM modulation technique for the BB, SEPIC, and DSSB converters is evaluated in terms of the performance characteristics mentioned above as well as those shown in reference Tables 2 and 3. In Figure 7a,b, the graphical illustration plots of Tables 2 and 3 provide comparisons of the converter's dynamic performance.

Table 2. Performance characteristics of both converters, BB and DSSB, in terms of load and source currents, $D = 1/2$.

Converter Type	BB		SEPIC		DSSB	
Parameter	**Load Current i_o [A]**	**Source Current i_s [A]**	**Load Current i_o [A]**	**Source Current i_s [A]**	**Load Current i_o [A]**	**Source Current i_s [A]**
Mean value:	4.87	5	4.8	4.8	6.5	10.9
RMS value:	4.9	7.11	4.8	9.4	6.75	11.16
Crest factor:	1.38	2.7	1.54	2.65	1.12	1
Ripple factor:	118 m	1	62.5 m	1.69	286.5 m	221.8 m
Harmonic factor:	1	1.43	1	1.2	1	1
Harmonic content:	117 m	715.24 m	62.4 m	860 m	275.4 m	216.5 m

Table 3. Performance characteristics of BB, SEPIC, and DSSB, in terms of their output voltage dynamic characteristics, where $V_s = 50$ V, $D = 1/2$.

Converter Type	BB	SEPIC	DSSB
Parameter	**Output Voltage [V]**	**Output Voltage [V]**	**Output Voltage [V]**
Mean value:	48.8	48	64.88
RMS value:	49	48	67.5
RMS AC:	4.23	3	18.6
Crest factor:	1.42	1.54	1.12
Ripple factor:	86.74 m	62.48 m	286.5 m
Harmonic factor:	1	1	1
Harmonic content:	86.42 m	62.36 m	275.4 m

Table 4. Performance characteristics of the BB, SEPIC, and DSSB, in terms of RMS and average (mean) DC values of the converters' input and output powers, P_s and P_o, power factor, and efficiency, where $V_s = 50$ V, $D = 1/2$.

Converter Type	BB		SEPIC		DSSB	
Parameter	**Output Power, P_o [W]**	**Input Power, P_s [W]**	**Output Power, P_o [W]**	**Input Power, P_s [W]**	**Output Power, P_o [W]**	**Input Power, P_s [W]**
Mean value (dc):	240	244	232	240	542	594
RMS value (ac):	240	350	234	414	544	595
Power factor = Po(dc + ac)/Ps(dc + ac)%	82%		92%		92%	
Efficiency = Po(dc)/Ps(dc)%	98%		92%		92%	

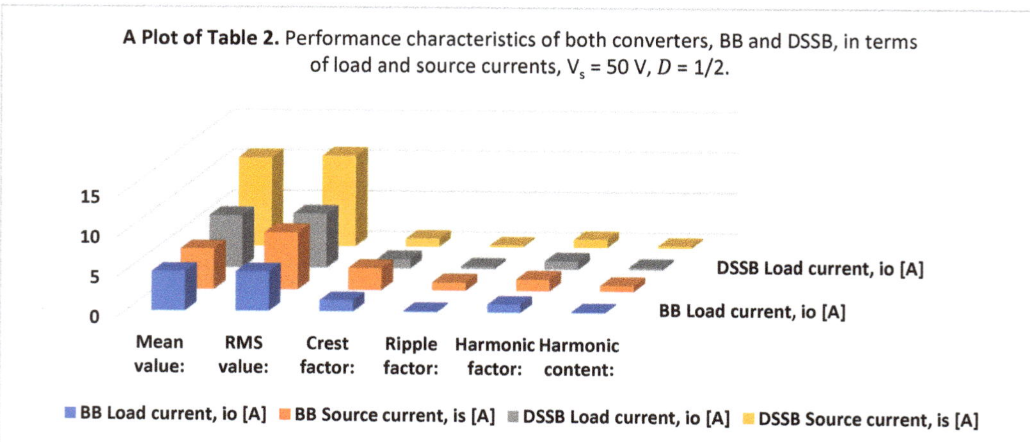

Figure 7. *Cont.*

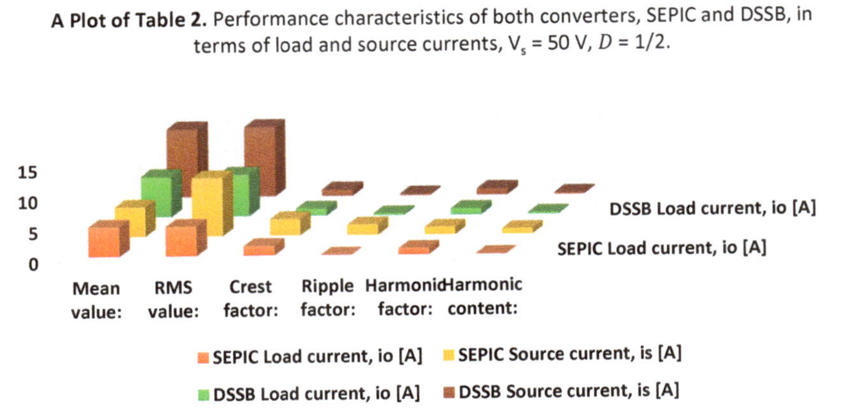

(a)

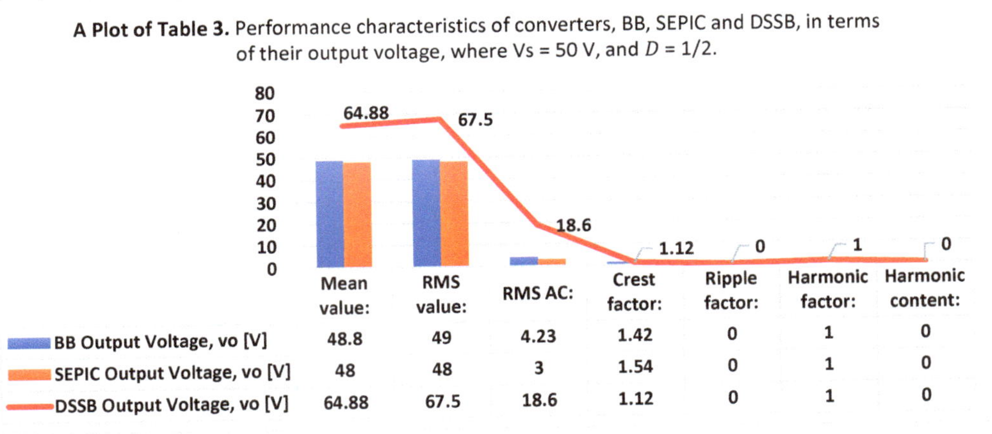

(b)

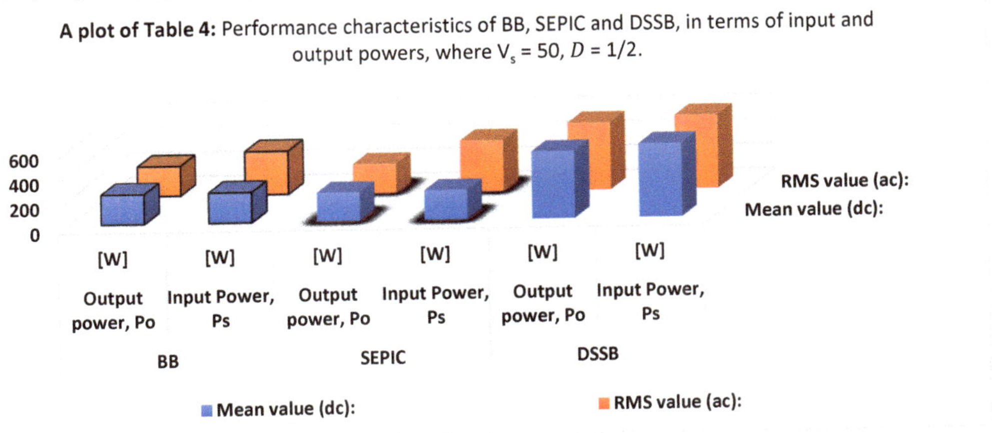

(c)

Figure 7. *Cont.*

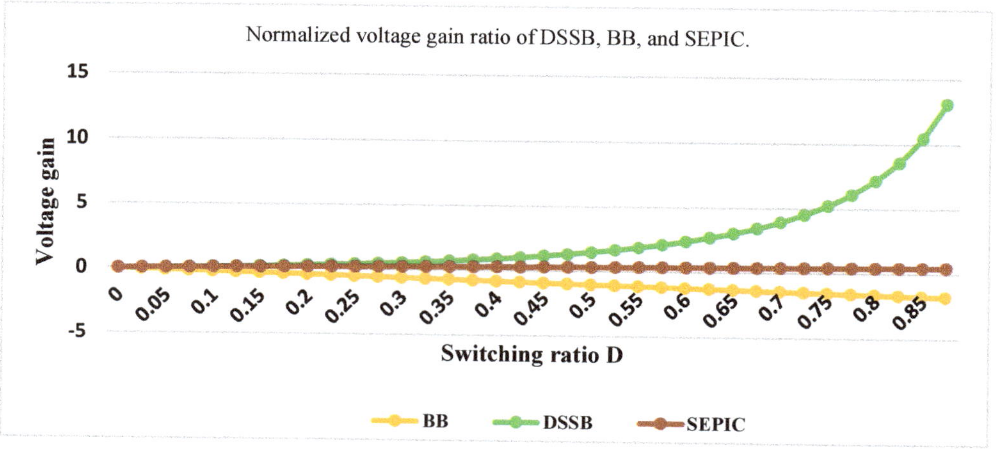

(d)

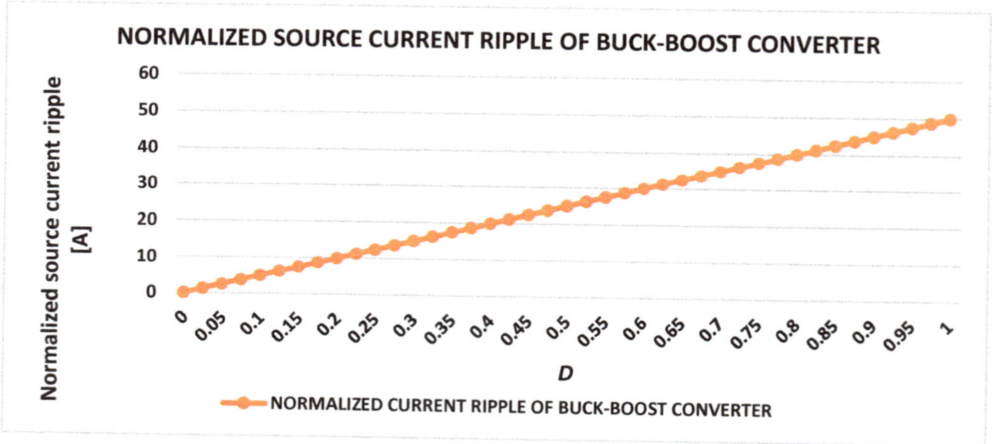

(e)

Figure 7. (**a**) A graphical representation of Table 2, (**b**) A graphical representation of Table 3, (**c**) A graphical representation of Table 4, (**d**) A plot of the normalized voltage gain ratio (V_o/V_s) of DSSB, BB, and SEPIC, (**e**) A plot of the normalized source current ripple of the BB converter.

DC-DC power electronics are frequently used to create pure DC power from pure AC supply systems via rectifier circuits or from DC supply systems that operate periodically. As a result, in addition to the DCs' average components, the power generated at their outputs also contains various AC and RMS components. Because of this, the converter efficiency—which measures the ratio of DC active power created at the output to supply power—is frequently represented as a percentage of the total DC input power generated. The power factor (PF) of the converter is the proportion of the total input power (RMS and DC average power) to the total output power (RMS and DC average power).

The actual RMS and average values of the input and output powers P_s and P_o of the BB, SEPIC, and DSSB converters are calculated and shown in Table 4 and Figure 7c. The average DC power is defined as the amount of power consumed by a DC load, such as a mobile phone battery or DC machinery. The RMS power, also referred to as AC power, is the actual power used by loads such as resistive loads, heaters, lamps, etc. Although having equal and identical DC and RMS power is the primary objective, they are never the same in power electronic systems. In terms of efficiency, harmonic content produced in the converter currents, and voltage gain, the suggested converter DSSB outperforms the BB and conventional SEPIC converters overall.

The normalized voltage gain ratio (V_o/V_s), which is illustrated in Figure 7d, is defined as the ratio of the generated output voltage to the converter input voltage. To further contrast the three converters, the switching ratio, D, which is changed from zero to unity, is employed. The normalized voltage gains of the BB and SEPIC converters are used in accordance with [18–20], whereas the normalized voltage gain of the DSSB is mentioned in Equation (8).

$$\frac{V_o}{V_s} = \frac{-D}{(1-D)} = 1 \Longrightarrow \text{Buck} - \text{boost converter (BB)} \tag{29}$$

$$\frac{V_o}{V_s} = \frac{D}{(1-D)} = 1 \Longrightarrow \text{SEPIC converter} \tag{30}$$

The suggested converter DSSB has a substantially larger voltage gain ratio than the other two converters, BB and SEPIC. The BB converter's implementation and use are limited to specific applications because the output voltage is the opposite of the input voltage. As shown in Table 4 and Figure 7d, the power capacity and current/voltage ratings of the DSSB are much larger than those of the BB and SEPIC converters, and it can be used particularly with PV module arrays with higher power [21–25].

Additionally, it would make sense to compare the normalized current ripple for the DSSB source current to that for the BB and SEPIC converters found in [1]. Equations (15) and (16), visually depicted in Figure 7e, clearly match the normalized current ripple produced by the DSSB with that of the BB and SEPIC.

3.2. In Comparison to Other SEPIC Converter Topologies

The key observations shown in Table 5 are used to highlight the preference for and technical advantages of the DSSB in light of the aforementioned analysis when compared to alternative SEPIC converter topologies, such as standard and modified multiplier SEPICs. Based on the converter principle of operation requirements, the duty ratio is an adjustable parameter that indicates that the degree of inductor current ripple reduction should be also adjustable.

Table 5. DSSB Comparison to traditional and multiplier SEPICs.

Performance	**Type of SEPIC**		
	Traditional Sepic as Explained in [4]	**Modified Multiplier [4]**	**DSSB Figure 1**
Normalized voltage gain:	$V_o = \frac{D}{1-D} V_s$	$V_o = \frac{D^2}{1-D} V_s$	$V_o = \frac{D(1+D)}{1-D} V_s$
Example: $V_s = 50$ V	$D = \frac{1}{4} \Rightarrow V_o = \frac{V_s}{3} = 16.6$ V $D = \frac{1}{2} \Rightarrow V_o = V_s = 50$ V $D = \frac{3}{4} \Rightarrow V_o = 3V_s = 150$ V ... this topology of SEPIC is an up/down converter if $D < 0.5$ or $D > 0.5$, respectively.	$D = \frac{1}{4} \Rightarrow V_o = 4.16$ V $D = \frac{1}{2} \Rightarrow V_o = 25$ V $D = \frac{3}{4} \Rightarrow V_o = 112.5$ V ... this topology of SEPIC is an up/down converter if $D < 0.618$ or $D > 0.618$, respectively.	$D = \frac{1}{4} \Rightarrow V_o = 20.8$ V $D = \frac{1}{2} \Rightarrow V_o = 75$ V $D = \frac{3}{4} \Rightarrow V_o = 262.5$ V ... this topology of SEPIC is an up/down converter if $D < 0.414$ or $D > 0.414$ respectively, with a much greater value of step-up voltage than others.
Normalized inductor source current ripple:	$\Delta i_L = \frac{V_s}{fL} D$ for all values of duty ratio D.	$\Delta i_L = \frac{V_s}{fL} D$ for all values of duty ratio D.	$\Delta i_L = \frac{V_s}{fL} D$ for all values of duty ratio D.
Number of Capacitors and Power Electronic Devices	One transistor, one diode, two capacitors, and two inductors	One transistor, three diodes, three capacitors, and three inductors	Two transistors, three diodes, four capacitors, and three inductors

Figure 8 shows that, when compared to the other SEPIC topologies listed in Table 5, the normalized voltage gain with the updated DSSB conversion processor is much better. The conversion ratio $D = 0$ produced the lowest value of this ripple, whereas $D = 1$ theoretically produced the highest. Additionally, as the duty cycle, D, gets closer to unity, the maximum ripple percentage of the output voltage content increases dramatically.

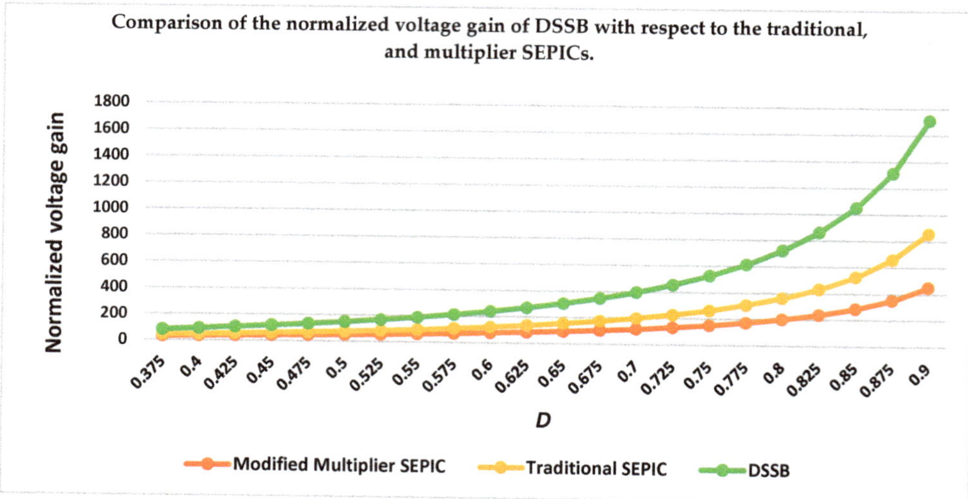

Figure 8. A plot of the normalized voltage gain of the DSSB with some other SEPIC topologies.

4. Control Method of DSSB Converter

This section outlines the proposed DSSB converter control strategy and illustrates a power factor correction strategy based on the regulation of the input-output current and the voltage of the DSSB. As a result, the converter efficiency is improved, and the overall harmonic content of the input-output voltages and current is decreased.

4.1. AC Small-Signal Analysis Technique

The AC small-signal technique is used to approximate the behavior of the converter, such as a linear time-invariant system around a certain switching cycle of the switches as the operating point of interest. This AC small-signal principle enables the DSSB duty cycle representations through open-loop and closed-loop transfer functions [25–28].

Once the frequency response of the DSSB simulation model is determined in Simplorer 7, its open-loop and closed-loop transfer functions are estimated. A suitable compensator is then designed and evaluated based on the linear model. By repeating the small-signal technique for different operating points (for example, different switching (duty) cycle ratios or desired output voltage levels), the gain of the controller was selected for the DSSB operation in the desired operating range. The small-signal model may be created and verified in algorithm form as follows [29,30]:

- Several perturbation input voltages and currents are selected to perform the DSSB small-signal analysis.
- A compensator for the resulting linear model is designed and tuned using techniques such as automatic PI tuning or interactive loop shaping, with the original locus and Bode plots.
- A programmed gain compensator is created to operate the power electronic system under all possible operating conditions.
- The design of the controller is verified and tried by simulating it against the non-linear power electronic system model.

- Instead, a simulation built using the non-linear DSSB model is used to verify and validate the implementation of the controller design.
- A MATLAB and Simplorer 7 models are automatically generated for rapid prototyping and production control system implementation.

Based on the algorithmic procedure explained above, the converter output quantities (source voltages, and currents) are partitioned into the averaged DC and low-frequency AC components as follows:

First, the average values of their present and previous instantaneous values for every switching cycle, $T = \frac{1}{f}$, are calculated as follows:

$$\begin{aligned} X(n) &= \tfrac{1}{T}\int_{nT}^{(n+1)T} xdt \\ X(n-1) &= \tfrac{1}{T}\int_{(n-1)T}^{(n)T} xdt \end{aligned} \tag{31}$$

Applying the circuit averaging method and after neglecting the nonlinear terms, the small-signal voltages of inductors L_1, L_2, and L_3 can be written as follows:

$$\begin{aligned} \tilde{v}_{L_1}(t) &= \left[V_s + \tilde{v}_s(t)\right] + [D-1] \\ &+ \left[V_{CM} + \tilde{v}_{CM}(t)\right] + \tilde{d}(t)V_{CM} \end{aligned} \tag{32}$$

$$\begin{aligned} \tilde{v}_{L_1}(t) &= \left[V_s + \tilde{v}_s(t)\right] + [D-1] \\ &\left[V_{CM} + \tilde{v}_{CM}(t)\right] + \tilde{d}(t)V_{CM} \end{aligned} \tag{33}$$

$$\begin{aligned} \tilde{v}_{L_2}(t) &= -D\left[V_s + \tilde{v}_s(t)\right] - D\left[V_{Cs} + \tilde{v}_{Cs}(t)\right] \\ &+[1-D]\left[V_{C1} + \tilde{v}_{C1}(t)\right][1+D] \\ &\left[V_{CM} + \tilde{v}_{CM}(t)\right] \\ &+\tilde{d}(t)[V_{CM} - V_s - V_{Cs} - V_{C1}] \end{aligned} \tag{34}$$

$$\begin{aligned} \tilde{v}_{L_3}(t) &= -\left[V_o + \tilde{v}_o(t)\right] + \\ &D\left[V_C + \tilde{v}_C(t)\right] + \tilde{d}(t)V_C \end{aligned} \tag{35}$$

where the term $\tilde{v}$ indicates the small-signal component of the respective voltage. Thus, the open-loop voltage gain (transfer function) is given as:

$$\mathrm{G}_v(\mathrm{s}) = V_o/V_{ref} = k_{pv} \times G_a(s) = 12.5 \times \frac{8 \times 10^{-6}}{(1 + 0.02S)^3} \tag{36}$$

The open-loop current gain (transfer function) is given as:

$$\mathrm{G}_i(\mathrm{s}) = I_o/I_{ref} = k_{pi} \times G_b(s) = 25 \times \tfrac{1 \times 10^{-4}}{S\,(1+0.01S)^2} \tag{37}$$

Second, the average present and prior values of the relevant load currents and voltages are multiplied to determine the values of the present and the preceding load powers.

Thus, the DC load voltage, current, and mutual product at each moment, as well as the duty cycle D, may therefore be perturbed at a frequency that is lower than the converter's switching frequency. All other conversion variables will therefore oscillate at this frequency around their respective DC values. As a result, the switching method depicted in Figure 9 is used to determine the averaged load voltages, currents, and duty cycle.

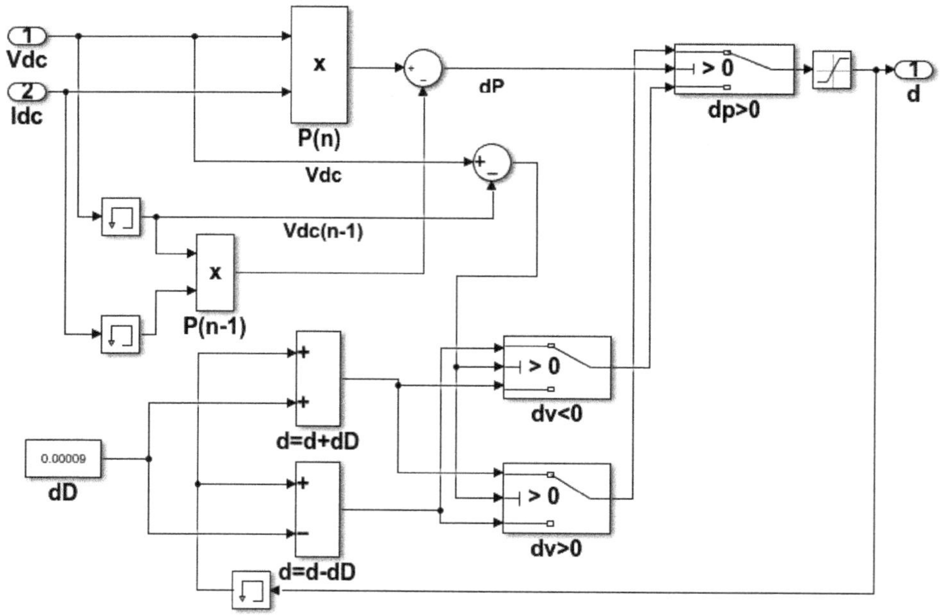

Figure 9. A control algorithm representing a small-signal AC technique for generating the desired duty cycle for the DSSB converter switches.

The DSSB control algorithm constantly monitors how rapidly the power changes in proportion to variations in load voltage and current. The DC-DC DSSB converter is optimized to provide the least amount of load current and voltage ripple and to achieve the highest possible power transfer efficiency. The output of this AC small-signal controller is then delivered to the converter PWM signal generator, to produce a duty ratio that is $D \leq 1$.

4.2. *Current-Mode Technique Based on AC Small-Signal Analysis*

The term "AC small-signal analysis with voltage-current mode technique" or simply "current-mode technique based on AC small-signal" refers to the second method of controlling the power and voltage in the DSSB system. The control structure of such a technique requires two additional loops in addition to the AC small-signal technique, one internal current source loop, and one external voltage source loop. In this approach, a small-signal algorithm is first provided as explained earlier, and then the DSSB load voltage is controlled by a dedicated control scheme using a PI controller. The current-mode controller operates on the difference between the load voltage setpoint and its measured actual value. Then, the load current is controlled by comparing its measured value with its desired value generated from the load voltage control scheme [31,32].

Thus, the DSSB input inductor currents and voltage are sensed and controlled to produce the required output power with a value corresponding to the acceptable and possible normalized input power based on the small-signal technique. Then, to achieve the desired load current and load output voltage with very low ripple, the gate signals for the DSSB switches are generated using a current-mode control technique using proper PWM with two ramp signals phase-shifted by 180° in sequence.

The control block diagram for the DSSB with two PI controllers to regulate the load current and voltage is shown in Figure 10a. The DSSB circuit is depicted in Figure 10b together with the current-mode control and the AC small-signal analysis control blocks.

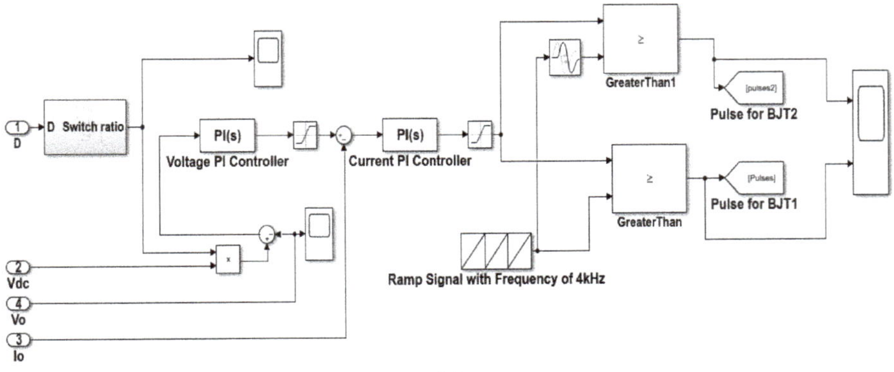

(a)

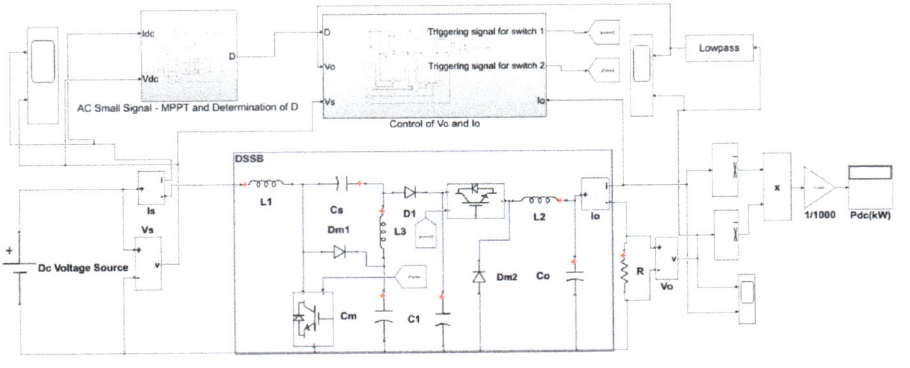

(b)

Figure 10. (**a**) DSSB Current-mode controller for controlling the load voltage and current, (**b**) MATLAB/Simulink implementation circuit block diagram for the DSSB with controllers.

5. Simulation Results

To validate the analytical study and theoretical simulation, a thorough investigation into the DSSB converter used in the CCM using the parameters listed in Table 6 was conducted. The waveforms of the source/load current and voltage shown in Figure 11 are entirely identical to the simulated waveforms shown in Figure 5b. The converter operates in direct current mode (CCM) because its current does not reach zero during the entire conversion period, while the ripple of each current corresponds to the device specifications.

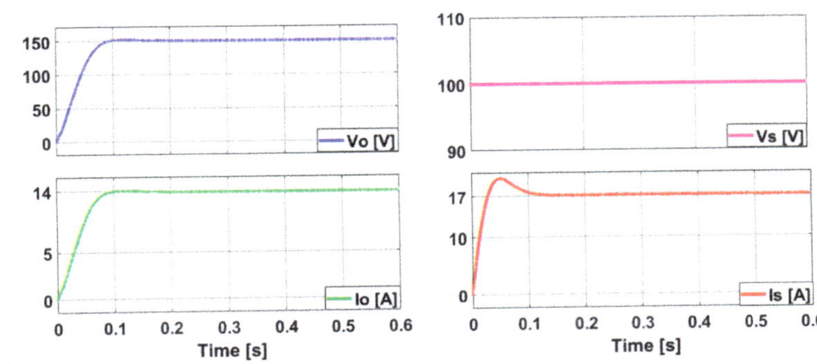

Figure 11. Waveforms prepared in the environment of MATLAB/Simulink for the given designed DSSB specifications.

The important aspect to be investigated is the frequency response of the system that is used to determine the stability of the system. Here, one can use two quantities, gain margin and phase margin, which indicate the amplitude of the system before it becomes stable or unstable. The bode plots of the open-loop current gain transfer function I_o/I_s and voltage gain V_o/V_s for the DSSB is shown in Figures 12–15.

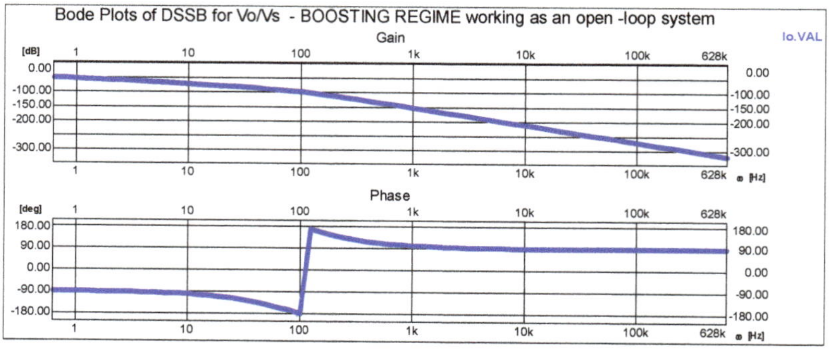

Figure 12. Bode plots of the DSSB representing the current gain *Io/Is* during the bucking regime—open-loop system.

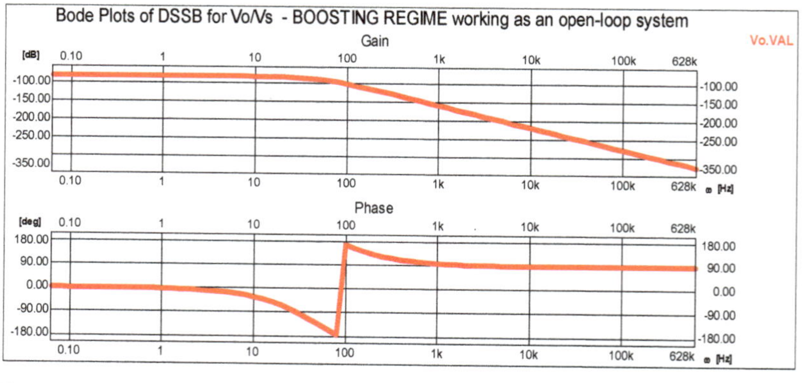

Figure 13. Bode plots of the DSSB representing the voltage gain *Vo/Vs* during the bucking regime—open-loop system.

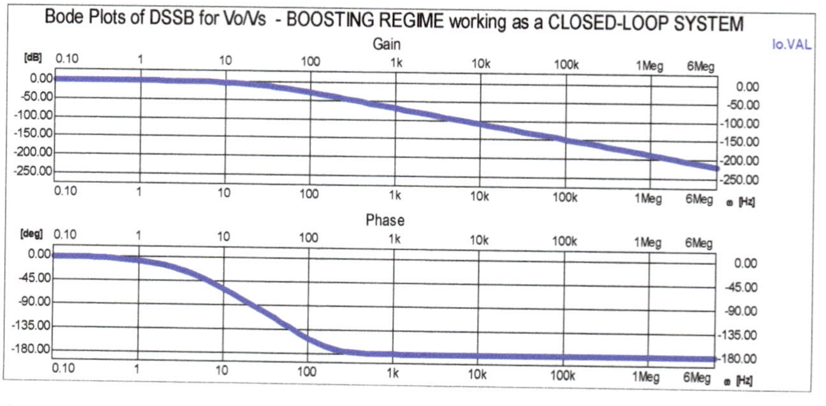

Figure 14. Bode plots of the DSSB representing the current gain *Io/Is* during the boosting regime—closed-loop system.

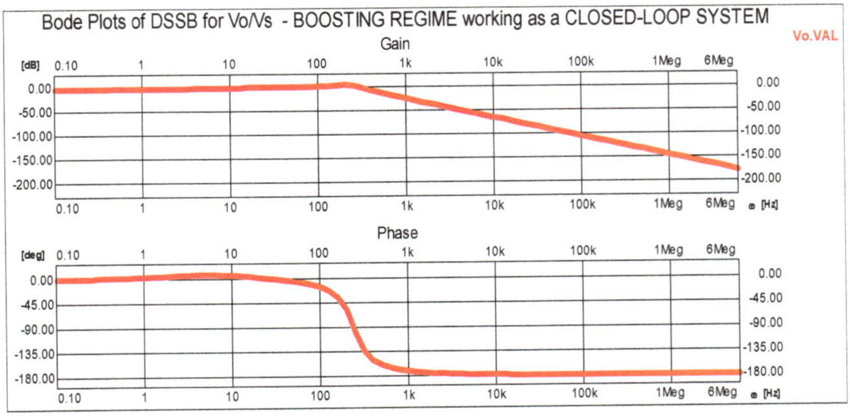

Figure 15. Bode plots of the DSSB representing the voltage gain Vo/V_s during the boosting regime—closed-loop system.

The phase margin for a closed loop system is the additional amount of phase delay required for the phase of the open loop system to reach −180° at a frequency where the magnitude of the open loop system is 0 dB.

The Bode diagrams of the closed-loop system in Figures 11–14 and MATLAB/Simulink together with Simplorer 7 demonstrate that the DSSB system is stable since the zero crossing of 0 dB happened at a phase angle (phase margin) that is less than 180°, or 55°. Additionally, as seen in Figure 14, the gain margin at the −180° crossing is greater than zero and is around 82 dB.

Additionally, the closed-loop system output voltage is constructed so that its bandwidth, as indicated by the closed-loop transfer function, is 355 Hz. The phase crossover frequency for the voltage gain transfer function is around 23 kHz, and for the current gain, it is approximately 700 Hz, as can be seen in Figures 13 and 14. For frequencies below 1 MHz, the gain roll-off rate (slope) is −40 db/decade, which is equal to a rate of −24 db/octave. From the current loop plot, the converter may behave as an underdamped system with two real negative poles and one pole at the origin, which can be observed from the current gain transfer function [33,34].

Table 6. DSSB design specifications [35].

Parameter	Symbol	Real value	Type
Input Voltage	V_s	50 V	DC power source
Input Inductor	L_1	(100–400) mH	Core NEE-65/33 by Thornton, N_{L1} = 180 turns − 4 × AWG 27
Input inductor current ripple	Δi_{L1}	0.12 A	-
Output inductor current ripple	Δi_{L3}	0.08 A	-
Shunt Inductor	L_3	20 mH	Core NEE-65/33 by Thornton, N_{L2} = 180 turns − 4 × AWG 27
Output Inductor	L_2	20 mH	core NEE-65/33 by Thornton, N_{L3} = 45 turns − 5 × AWG 24
Smoothing capacitor	C_s	940 uF	Electrolytic capacitor by Hitano
Shunt capacitor	C_m	940 uF	Electrolytic capacitor by Nichicon
Shunt capacitor	C_1	940 uF	Electrolytic capacitor by Nichicon
Output capacitor	C_o	3.33 mF	Electrolytic capacitor by Hitano
Switching ratio	D	0.4	-
Output voltage:	V_o	25 V	-
Load resistance:	R	10 Ω	Ceramic resistor by LW
Switching frequency:	f	4 kHz	Signal generator

Figure 16 depicts the dynamic response of the DSSB input/output voltage and current waveforms for a variable step response of the input source voltage using the parameters listed in Table 6. Figure 16b displays a compressed timeline of the output voltage and the current transient response to demonstrate the time specifications and requirements of the response, and Figure 16c is used to demonstrate the effectiveness of the current-mode controller based on small-signal analysis.

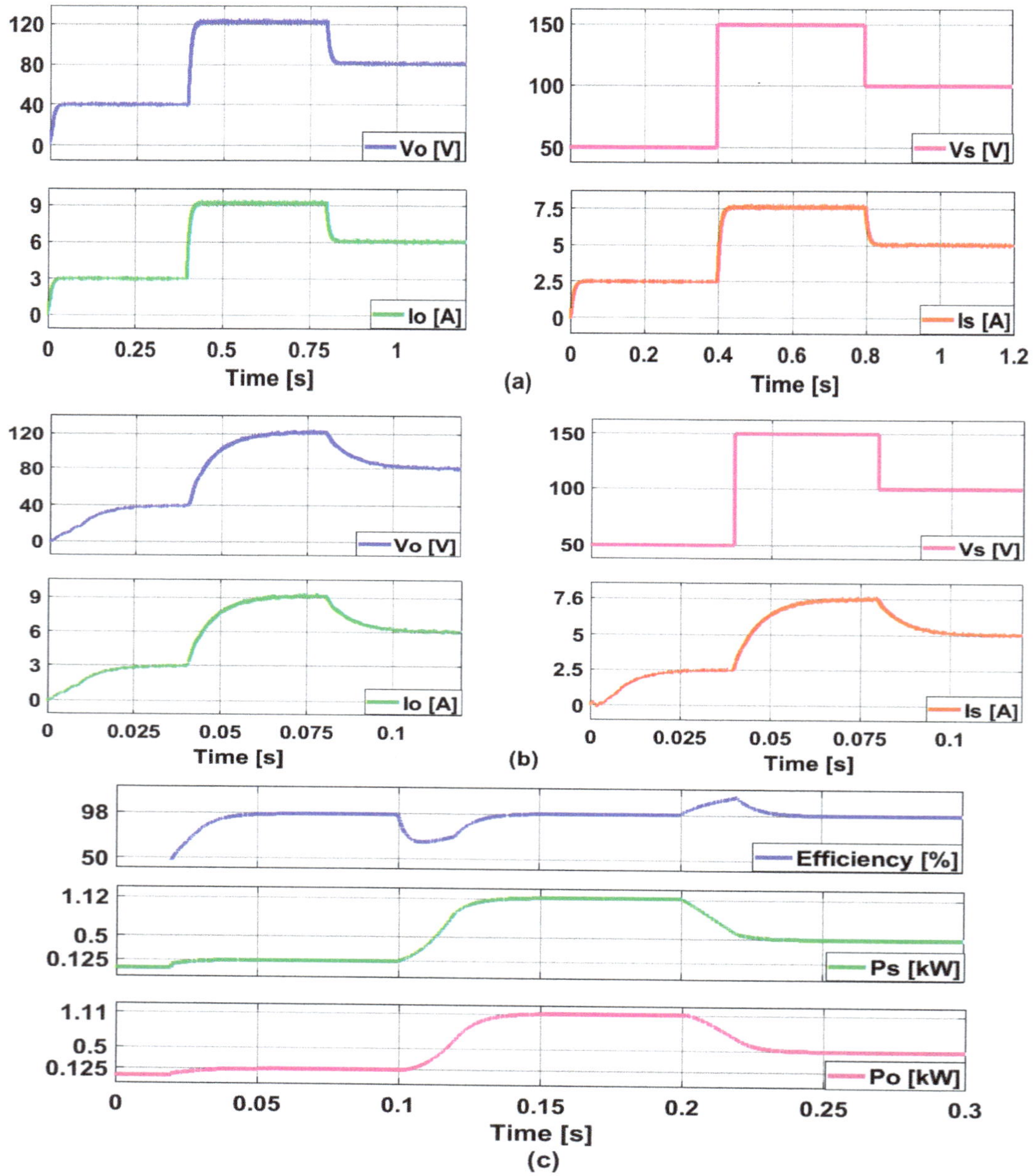

Figure 16. Simulated waveforms prepared in the environment of Simulink/MATLAB for the given designed DSSB specifications: (**a**) Output and input currents and voltages, (**b**) transient behavior of currents and voltages within small time scale, (**c**) Output and input power with efficiency.

The simulation of Figure 16 shows that the DSSB responds very well in terms of output voltage and current simulated waveform because the output voltage and current waveform precisely match the step change in the input voltage and current. Slow overshoot, decreased settling time, and rise time are all characteristics of DSSB-responsive behavior, which should lead to better frequency response. This is evident from the narrower timeframe for a temporary (transitory) response, which is also displayed in Figure 16 for the output voltage and current.

6. Conclusions

This study introduces a double-switch SEPIC-buck converter (DSSB) with a new topology. This new topology can be applied to step-up applications as well as step-down applications of PV solar systems, wind turbine systems, and other similar industrial applications. The converter design employs two active switches, which causes minimal sag and ripple in the input and output inductor currents and voltages. The proposed DSSB converter produces minimal harmonic content with consistent input/output power and efficiency, in contrast to traditional DC-DC converters.

The suggested converter dynamic performance is optimized using a current-mode, control-based, small-signal analysis with two dual-lead PI controllers. A control circuit that incorporates both an open-loop and closed-loop DSSB frequency response is also suggested and shown to work properly. The small-signal technique has been implemented in the DSSB simulation model to represent its dynamic behavior through the frequency response and transfer functions.

There are several advantages to be gained from representing a small-signal analysis with a current-mode control technique through a frequency response:

- Using the current-mode control technique, it is possible to regulate the inductor current's peak-to-peak ripple as well as its mean value, which is an advantage of the current-mode control technique based on small-signal analysis.
- Another advantage is that the higher-order system control used to produce the frequency response of such a converter is reduced to the second-order frequency response.

The proposed current-mode control technique based small-signal analysis is validated by subjecting the proposed converter to unique operating conditions, such as providing it with a variable step input and analyzing its dynamic response. The outcomes demonstrate the DSSB converter's robustness to changes in its parameters, as well as its improved dynamic performance and increased control precision, which are further advantages of this converter.

Additionally, a comparison study between the DSSB converter and the different topologies of the SEPIC converter, including some other DC-DC converters like buck, boost, buck-boost, and others, has also been given. The complexity, voltage stress on the converter active switches, mean and rectified mean values, RMS values, crest and ripple factors, harmonic factors, and harmonic content distortion are all compared. Because of its superior input/output power response, lower ripple factor, lower harmonic content, highest output efficiency, close to unity power factor, reduced current ripple, and overall better dynamic performance, the study concludes that the proposed converter, designated DSSB, might be a competitive option for industrial applications.

Author Contributions: Conceptualization, W.E.; methodology, W.E.; software, W.E.; validation, W.E., H.I., H.K., O.F. and H.A.; formal analysis, W.E. and O.F.; investigation, W.E. and H.K.; resources, W.E.; data curation, W.E.; writing—original draft preparation, W.E.; writing—review and editing, W.E. and H.A.; visualization, W.E. and H.I.; supervision, W.E.; project administration, W.E.; funding acquisition, W.E., H.I., H.K., O.F. and H.A. All authors have read and agreed to the published version of the manuscript.

Funding: This research received no external funding.

Data Availability Statement: All data generated or analyzed during this study are included in this published article.

Acknowledgments: All authors would like to thank Ayman Amer for financial support.

Conflicts of Interest: The authors declare no conflict of interest.

References

1. Hart, D.W. *Power Electronics, International Edition*; McGraw Hill: New York, NY, USA, 2011.
2. Rahbar, K.; Xu, J.; Zhang, R. Real-time energy storage management for renewable integration in microgrid: An off-line optimization approach. *IEEE Trans. Smart Grid* **2014**, *6*, 124–134. [CrossRef]
3. Rahbar, K.; Chai, C.C.; Zhang, R. Energy cooperation optimization in microgrids with renewable energy integration. *IEEE Trans. Smart Grid* **2016**, *9*, 1482–1493. [CrossRef]
4. Vadivel, S.; Ragupathy, U.S. Modeling and Design of High-Performance Converters for Optimal Utilization of Interconnected Renewable Energy Resources to Micro Grid with GOLRS Controller. *Int. J. Control Autom. Syst.* **2021**, *19*, 63–75. [CrossRef]
5. Li, H.; Liu, C.; Zhang, X.; Guo, Z.; Zheng, T.Q. Stability Analysis for Two-Stage Cascaded DC-DC Converters System Based on Describing Function Method. In Proceedings of the 2018 IEEE Energy Conversion Congress and Exposition (ECCE), Portland, OR, USA, 23–27 September 2018; pp. 4141–4147. [CrossRef]
6. Azer, P.; Emadi, A. Generalized State Space Average Model for Multi-Phase Interleaved Buck, Boost and Buck-Boost DC-DC Converters: Transient, Steady-State and Switching Dynamics. *IEEE Access* **2020**, *8*, 77735–77745. [CrossRef]
7. Yalçin, F.; Yazici, İ. Robust Discrete-Time Hybrid Controller for Non-Inverting Buck-Boost DC-DC Converter. *Sak. Univ. J. Sci.* **2023**, *27*, 68–82. [CrossRef]
8. Wu, J.; Mo, M.; Li, J.; Zhang, Y.; Zhang, X.; Fan, X. Inverse decoupling internal model control for multilevel buck converter with constant power load. *Energy Rep.* **2023**, *9* (Suppl. 7), 1181–1189. [CrossRef]
9. Hao, Y.; Guo, K.; Liu, L.; Cai, W.; Liu, F. improved cubic boost converter based on voltage closed-loop control. *Energy Rep.* **2022**, *8* (Suppl. 6), 87–95. [CrossRef]
10. Vinay Kumar, C.; Rao, G.M.; Raghu Ram, A.; Prasanna Kumar, Y. Designing of Neuro-Fuzzy Controllers for Brushless DC Motor Drives Operating with Multiswitch Three-Phase Topology. *J. Electr. Comput. Eng.* **2022**, *2022*, 7001448. [CrossRef]
11. Al Attar, H.; Hamida, M.A.; Ghanes, M.; Taleb, M. Review on Modeling and Control Strategies of DC–DC LLC Converters for Bidirectional Electric Vehicle Charger Applications. *Energies* **2023**, *16*, 3946. [CrossRef]
12. Solis-Rodriguez, J.; Rosas-Caro, J.C.; Alejo-Reyes, A.; Valdez-Resendiz, J.E. Optimal Selection of Capacitors for a Low Energy Storage Quadratic Boost Converter (LES-QBC). *Energies* **2023**, *16*, 2510. [CrossRef]
13. Kalimuthu; Ramkumar, M.S.; Amudha, A.; Balachander, K.; Krishnan, M.S. A High Gain Input-Parallel Output-Series Dc/Dc Converter with Dual Coupled-Inductors. *Math. Comput. For. Nat. Resour. Sci.* **2019**, *11*, 242–246.
14. Dagal, I.; Akın, B.; Akboy, E. MPPT mechanism based on novel hybrid particle swarm optimization and salp swarm optimization algorithm for battery charging through simulink. *Sci. Rep.* **2022**, *12*, 2664. [CrossRef] [PubMed]
15. Seyezhai, R.; Mathur, B.L. A Comparison of Three-Phase Uncoupled and Directly Coupled Interleaved Boost Converter for Fuel Cell Applications. *Int. J. Electr. Eng. Inform.* **2011**, *3*, 394–407. [CrossRef]
16. Reatti, A.; Corti, F.; Tesi, A.; Torlai, A.; Kazimierczuk, M.K. Effect of Parasitic Components on Dynamic Performance of Power Stages of DC-DC PWM Buck and Boost Converters in CCM. In Proceedings of the 2019 IEEE International Symposium on Circuits and Systems (ISCAS), Sapporo, Japan, 26–29 May 2019; pp. 1–5. [CrossRef]
17. Zhang, W.; Liu, Y.; Li, Z.; Zhang, X. The dynamic Power Loss analysis in Buck Converter. In Proceedings of the 2009 IEEE 6th International Power Electronics and Motion Control Conference, Wuhan, China, 17–20 May 2009; pp. 362–367. [CrossRef]
18. Trakuldit, S.; Tattiwong, K.; Bunlaksananusorn, C. Design and evaluation of a quadratic Buck Converter. *Energy Rep.* **2022**, *8* (Suppl. 1), 536–543. [CrossRef]
19. Emar, W.; Huneiti, Z.; Hayajneh, S. Analysis, Synthesis and Simulation of Compact Two-channel Boost Converter for Portable Equipments Operating with a Battery or Solar Cell. *Procedia Comput. Sci.* **2015**, *65*, 241–248. [CrossRef]
20. Emar, W. Analysis, modeling and simulation of step-up converter using Matlab-Simulink and simplorer. *Int. J. Model. Simul. Sci. Comput.* **2016**, *7*, 1650004:1–1650004:13. [CrossRef]
21. de Dieu Nguimfack-Ndongmo, J.; Ngoussandou, B.P.; Goron, D.; Asoh, D.A.; Kidmo, D.K.; Nfah, E.M.; Kenné, G. Nonlinear neuro-adaptive MPPT controller and voltage stabilization of PV Systems under real environmental conditions. *Energy Rep.* **2022**, *8* (Suppl. 9), 1037–1052. [CrossRef]
22. Zaid, M.; Khan, S.; Mahmood, A.; Ali, M.; Sarwar, A.; Khalid, M. A New High Gain Boost Converter with Common Ground for Solar-PV Application and Low Ripple Input Current. *Arab. J. Sci. Eng.* **2023**, *48*, 14655–14669. [CrossRef]
23. Raj, T.A.B.; Ramesh, R.; Maglin, J.R.; Vaigundamoorthi, M.; Christopher, I.W.; Gopinath, C.; Yaashuwanth, C. Grid Connected Solar PV System with SEPIC Converter Compared with Parallel Boost Converter Based MPPT. *Int. J. Photoenergy* **2014**, *2014*, 385720. [CrossRef]
24. Attar, H.; Kamarposhti, M.A.; Solyman, A.A.A. Impacts of integration of wind farms on voltage stability margin. *Int. J. Electr. Comput. Eng. (IJECE)* **2022**, *12*, 4623–4631. [CrossRef]
25. Iqteit, N.A.; Yahya, K.; Makahleh, F.M.; Attar, H.; Amer, A.; Solyman, A.A.A.; Qudaimat, A.; Tamizi, K. Simple Mathematical and Simulink Model of Stepper Motor. *Energies* **2022**, *15*, 6159. [CrossRef]

26. Bairabathina, S.; S, B. Design and Validation of a SEPIC-Based Novel Multi-Input DC-DC Converter for Grid-Independent Hybrid Electric Vehicles. *Energies* **2022**, *15*, 5663. [CrossRef]
27. Iqbal, D.; Siddique, M.; Chaudhary, A.; Liaqat Bhatti, K.; Abrar, M.; Hussain, M. Novel Concept of Reducing OVR at the Output of SEPIC Converter using Programmable Capacitors. *Int. J. Electr. Eng. Inform.* **2021**, *13*, 477–494. [CrossRef]
28. Komathi, C.; Umamaheswari, M.G. Analysis, and design of genetic algorithm-based cascade control strategy for improving the dynamic performance of interleaved DC–DC SEPIC PFC converter. *Neural Comput. Appl.* **2020**, *32*, 5033–5047. [CrossRef]
29. Senthilkumar, R.; Justin Sunil Dhas, G. Fractional order controller design for SEPIC converter using metaheuristic algorithm. *J. Intell. Fuzzy Syst.* **2018**, *35*, 6269–6276. [CrossRef]
30. Kwon, J.-M.; Choi, W.-Y.; Lee, J.-J.; Kim, E.-H.; Kwon, B.-H. Continuous-conduction-mode SEPIC converter with low reverse-recovery loss for power factor correction. *IEE Proc. Electr. Power Appl.* **2006**, *153*, 673–681. [CrossRef]
31. Dalimunthe, A.M.A.; Sara, I.D.; Tarmizi. Adaptive Control for SEPIC Converter. In Proceedings of the 2020 4rd International Conference on Electrical, Telecommunication and Computer Engineering (ELTICOM), Medan, Indonesia, 3–4 September 2020; pp. 92–96. [CrossRef]
32. Hu, J.; Sagneri, A.D.; Rivas, J.M.; Han, Y.; Davis, S.M.; Perreault, D.J. High-Frequency Resonant SEPIC Converter with Wide Input and Output Voltage Ranges. *IEEE Trans. Power Electron.* **2012**, *27*, 189–200. [CrossRef]
33. Weitz, N.; Utzelmann, S.; Ditze, S.; März, M. A Resonant Push–Pull DC–DC Converter With an Intrinsic Current Source Behavior for Radio Frequency Power Conversion. *IEEE Trans. Power Electron.* **2022**, *37*, 7001–7012. [CrossRef]
34. Yang, Y.; Liu, Y.; Zhao, Y.; Liu, J.; Zhu, B. High frequency and high-power density bipolar DC–DC converter with GaN HEMT. *Energy Rep.* **2023**, *9* (Suppl. 7), 617–624. [CrossRef]
35. Amer, A.; Attar, H.; As'ad, S.; Alsaqoor, S.; Colak, I.; Alahmer, A.; Alali, M.; Borowski, G.; Hmada, M.; Solyman, A. Floating Photovoltaics: Assessing the Potential, Advantages, and Challenges of Harnessing Solar Energy on Water Bodies. *J. Ecol. Eng.* **2023**, *24*, 324–339. [CrossRef]

MDPI
St. Alban-Anlage 66
4052 Basel
Switzerland
www.mdpi.com

Energies Editorial Office
E-mail: energies@mdpi.com
www.mdpi.com/journal/energies